T0336886

HOLOGRAPHIC DUALITY IN CONDENSED MATTER PHYSICS

This pioneering treatise presents how the new mathematical techniques of holographic duality unify seemingly unrelated fields of physics. Morphing quantum field theory, general relativity and the renormalisation group into a single computational framework, this book is the first to bring together a wide range of research in this rapidly developing field. Set within the context of condensed matter physics and using boxes highlighting the specific techniques required, it examines the holographic description of thermal properties of matter, Fermi liquids and superconductors, and hitherto unknown forms of macroscopically entangled quantum matter in terms of general relativity, stars and black holes.

Showing that holographic duality can succeed where classic mathematical approaches fail, this text provides a thorough overview of this major breakthrough at the heart of modern physics. The inclusion of extensive introductory material using non-technical language and online Mathematica notebooks ensures the appeal to students and researchers alike.

JAN ZAANEN is Professor of Theoretical Physics at the Instituut-Lorentz for Theoretical Physics, Leiden University, the Netherlands where he specialises in the physics of strongly interacting electrons. He is a recipient of the Dutch Spinoza Award and fellow of the Dutch Royal Academy of Sciences and the American Physical Society.

YA-WEN SUN is a Postdoctoral Researcher at the Institute for Theoretical Physics at the Universidad Autónoma de Madrid where she works on applications of AdS/CFT to condensed matter theory, QCD and hydrodynamics as well as other aspects of quantum gravity.

YAN LIU is a Postdoctoral Researcher at the Institute for Theoretical Physics at the Universidad Autónoma de Madrid where he specialises in high-energy theoretical physics including gauge/gravity duality and AdS/CMT.

KOENRAAD SCHALM is Professor of Theoretical Physics at the Instituut-Lorentz for Theoretical Physics, Leiden University, the Netherlands. His research focusses on how string theory may be detected in laboratory experiments or cosmological observations. He is the recipient of Innovative Research Incentives Awards of the Netherlands Organisation for Scientific Research.

HOLOGRAPHIC DUALITY IN CONDENSED MATTER PHYSICS

JAN ZAANEN

Universiteit Leiden, the Netherlands

YA-WEN SUN

Universidad Autónoma de Madrid, Spain

YAN LIU

Universidad Autónoma de Madrid, Spain

KOENRAAD SCHALM

Universiteit Leiden, the Netherlands

CAMBRIDGE
UNIVERSITY PRESS

CAMBRIDGE
UNIVERSITY PRESS

University Printing House, Cambridge CB2 8BS, United Kingdom

One Liberty Plaza, 20th Floor, New York, NY 10006, USA

477 Williamstown Road, Port Melbourne, VIC 3207, Australia

4843/24, 2nd Floor, Ansari Road, Daryaganj, Delhi - 110002, India

79 Anson Road, #06-04/06, Singapore 079906

Cambridge University Press is part of the University of Cambridge.

It furthers the University's mission by disseminating knowledge in the pursuit of education, learning and research at the highest international levels of excellence.

www.cambridge.org
Information on this title: www.cambridge.org/9781107080089

© Cambridge University Press 2015

First published 2015

A catalogue record for this publication is available from the British Library

Library of Congress Cataloging in Publication data
Zaanen, Jan, 1957– author.
Holographic duality in condensed matter physics / Jan Zaanen (Universiteit Leiden, the Netherlands), Yan Liu (Universidad Autónoma de Madrid, Spain), Ya-Wen Sun (Universidad Autónoma de Madrid, Spain), Koenraad Schalm (Universiteit Leiden, the Netherlands).
pages cm
Includes bibliographical references and index.
ISBN 978-1-107-08008-9 (hbk.)
1. Condensed matter. 2. Holography. I. Liu, Yan (Postdoctoral researcher), author. II. Sun, Ya-Wen, author. III. Schalm, Koenraad, 1971– author. IV. Title.
QC173.457.H65Z36 2015
530.4′12–dc23
2015021137

ISBN 978-1-107-08008-9 Hardback
ISBN 978-1-107-43947-4 Paperback

Additional resources for this publication at www.cambridge.org/9781107080089

Contents

Contents

Preface

Not so long ago, two large and quite old fields in physics, string theory and condensed matter physics, were more or less at the opposite ends of the physics building. During the 40 or so years of its history, string theory has developed into a high art of "mathematical machine building", propelled forwards by the internal powers of mathematics as inspired by physics. Yet, it has suffered greatly for the shortcoming that its theoretical answers are always beyond the reach of experimental machinery. Modern condensed matter physics is in the opposite corner. It has been propelled forwards by continuously improving experiments, which have delivered one serendipitous discovery after another during the last few decades. However, its interpretational framework rests by and large on equations developed 40 years or so ago. There has been an increasing sense that it is these that fall short in trying to explain the strongly interacting quantum many-body systems as realised by electrons in high-T_c superconductors and other unconventional materials.

All this changed dramatically in 2007 when physicists started to feed condensed matter questions to the most powerful mathematical machine of string theory: the holographic duality in the title of the book, also known as the "anti-de Sitter/conformal field theory" (AdS/CFT) correspondence. This book introduces the explosion of answers that has followed since then.

The first (Jan) and last (Koenraad) of this book's authors are from such opposite corners. As soon as the seminal work of Herzog, Kovtun, Sachdev and Son in 2007 showed that these two subjects have dealings with each other, Jan and Koenraad recognised the potential and met up, almost literally half-way up the stairs. As has been characteristic for the development at large, it took us remarkably little effort to get on speaking terms, despite our superficially very different backgrounds. Shrouded by differences in language, string theory and condensed matter had already been on a collision course for a while, meeting each other on the common ground of quantum criticality/conformal field theory. In the years that followed this dialogue only intensified and the upbeat tone of this book is a testimony of the great time we had together. The largest part of that time was shared with the two middle authors (Ya-Wen and Yan), who came to Leiden as postdocs in 2010, freshly graduated from the Chinese Academy of Science in Beijing. The seeds for this book were planted when Jan received the invitation to become the 2012 Solvay Professor in Brussels, with the request to organise an AdS/condensed-matter-theory lecture course. What you see before you grew from the notes of this course, joined together with lecture notes by Koenraad at the 2012 Cargèse and 2013 Crete schools.

This is an incredibly fast-moving field, and many pages had to be added to describe the developments that happened since the summer of 2012. In January 2014 we stopped playing catch up, and we decided to get it out as quickly as possible, given the high demand for such a text at this moment. The first nine months of 2014 turned into a writing frenzy for all of us, and the result is lying in front of you. We are well aware that in certain regards the book therefore has its limitations and that the text will already be obsolete as soon as it appears. Examples of significant developments that occurred since our cut-off are a holographically inspired theory of incoherent metals[1] and a holographic solution of the anomalous temperature scaling of the Hall angle as observed in high-T_c superconductors.[2] Nor do we claim this to be a comprehensive review that does justice to all of the papers which have been published on the subject. What we have done is to provide an introduction to serve a non-expert readership that wishes to be informed about the main developments. Our aim has been to catch the mainstream, those developments where one discerns a consensus in the expert community that these are the most significant accomplishments. As authors we found it quite obvious how to make this selection and we sincerely believe that our choices will be approved by the AdS/CMT experts. We felt that we just had the role of humble narrators working on the chronicles of a monumental physics odyssey. We wish to take you on board and we hope you will enjoy it as much as we do!

We are in the first instance indebted to numerous holographists who contributed to our understanding of the correspondence. We are particularly grateful to Andrea Amoretti, Steffen Klug, Richard Davison, Andrey Bagrov, Petter Sæterskog and Balázs Meszéna for their thorough proofreading of the manuscripts and their many helpful suggestions, and to Mihael Petač for his help on the figures. Both the Leiden and Madrid physics departments gave us all the room to concentrate on the writing of this book. We acknowledge financial support of various funding organisations, in particular the Solvay Foundation and the Dutch Foundation of Fundamental

---

[1] S. Hartnoll, *Nature Phys.* **11**, 54 (2015), arXiv:1405.3651.
[2] M. Blake, A. Donos, *Phys. Rev. Lett.* **114**, 021601 (2015), arXiv:1406.1659.

Research on Matter (FOM) in the initial stages of this project, as well as the Spanish MINECO's "Centros de Excelencia Severo Ochoa" Programme under grant SEV-2012-0249, the Netherlands Organisation for Scientific Research/Ministry of Science and Education (NWO/OCW), and a grant from the Templeton Foundation: the opinions expressed in this publication are those of the authors and do not necessarily reflect the views of the John Templeton Foundation. Jan Zaanen and Koenraad Schalm acknowledge the hospitality of various institutions during the writing process: the Aspen Center of Physics, supported by the National Science Foundation under Grant No. PHYS-1066293, the Kavli Institute for Theoretical Physics, supported by the National Science Foundation under Grant No. NSF PHY11-25915, and the physics department of Harvard University in particular.

Jan Zaanen, Ya-Wen Sun, Yan Liu and Koenraad Schalm
Leiden and Madrid

1

Introduction

1.1 A tour guide of holographic matter

A quake is rumbling through the core of physics. Suddenly apparently unrelated areas appear to have a common ground, showing an eerie capacity to fertilise each other. In physics such occasions are invariably propelled by novel mathematical machinery and the present case is no exception. This new mathematical contraption is "holographic duality" (or "anti-de Sitter/conformal field theory correspondence"), which was originally discovered in string theory in the 1990s. Until recently its use was limited to the historic scope of string theory – particle physics and quantum gravity. At a breathtaking pace it has since rolled out over many of the subject areas of modern fundamental physics, even yielding new insights into old subjects such as the nineteenth-century theory of hydrodynamics.

Several books of this kind could be written, and are being written, highlighting how anti-de Sitter/conformal field theory (AdS/CFT) impacts on various fields in physics. This book will focus on a prominent area where the developments have been particularly stunning. This is the application to equilibrium condensed matter physics. This started in 2007, and in a matter of a few years condensed matter theory was rewritten in a different mathematical language. This language is the one that one would perhaps least expect: general relativity. On its own a rewriting of condensed matter might not sound like a great advance, no matter how unconventional the language. However, the correspondence makes it possible to explore regimes of quantum many-body physics that are completely inaccessible with conventional techniques. In particular we refer to non-Fermi-liquid states of matter formed in finite-density systems of strongly interacting fermion systems. The holographic mathematics here becomes particularly expressive, suggesting that a general principle of a new kind is at work. It appears that this principle relates to the physics of compressible quantum matter: the notion that the nature of this state of matter is governed by a macroscopic quantum entanglement involving all of its constituents. This discovery is not just remarkable on its own. "Holographic strange matter" also has tantalisingly suggestive resemblances to the mysterious phenomena observed experimentally in strongly interacting electron systems that have been realised in special materials such as the high-T_c superconductors. First seen around thirty years ago, these have defied any reasonable explanation despite countless attempts resting on the available mathematical techniques. Could it be that holography supplies the mathematical equations that will shine light on this number-one mystery of condensed matter physics?

The correspondence also has its limitations. Though much evidence has been collected in favour, the jury is still out both with regard to the quantum information aspects and regarding the empirical relevancy. At present holography is an exciting research programme with a potential to alter the fundaments of the theory of physics. This should be of interest to the physics community at large, but there is an intrinsic difficulty. String theory famously has to be a unified theory of physics, but by the discovery of AdS/CFT this was changed into the "unification of the *theories* of physics". The implication is that the physicist has to pay the price that he/she has to be expert in all fanciful, modern areas of physics at the same time! A head-on encounter with the application of holography often entails a nearly seamless switch between string theory and high-energy-style quantum field theory, sophisticated general relativity including the latest in black-hole physics, via modern condensed matter theory, all the way to the tedious data sorting that one encounters in dealing with the real world studied by experimental physics.

Our aim has been to lower the barrier of entry to this field and make it readily accessible. At present the required knowledge is still predominantly scattered throughout a hard-to-penetrate research literature. We have done our best to write a comprehensive overview of the main current of anti-de Sitter/condensed matter theory (AdS/CMT). It is not exhaustive: we had to make choices, and we hope that expert readers will agree with our choice of the most substantive contributions. At the same time, we also had the ambition that the text can be employed as a textbook, for students who want to master the skill of doing the actual computations.

Above all we wanted to provide a text that is an optimally user-friendly access point to the subject, also for those readers who are not in a position to make the big investments of effort required in order to acquire the full holographist's skill set, but who are eager to get a well-informed impression of the big picture.

To accommodate these conflicting requirements, we have chosen to structure the text in a layered fashion, employing the time-tested use of *boxes*. The main text is rather descriptive, explaining what goes on conceptually and how the computations work. Equations are used only insofar as they are instrumental in getting the big picture across. We then supplement the main text by employing boxes found at the end of sections, where we present the actual calculations in some detail. For those readers who find even the main text too heavy, we have added in addition dual "rule boxes" summarising the punchlines of the neighbouring main text.

The body of the book is formed by chapters 6–13 focussed on the application of holography to condensed matter physics. We introduce these with several initial chapters of background material. Chapters 2 and 3 are intended as a collection of points from condensed matter physics that are of particular relevance to holography. Although these chapters are aimed in the first instance at string theorists and physicists of other fields, we strongly encourage also condensed matter experts to read them, since our presentation deviates significantly from the standard treatment in textbooks. Chapters 4 and 5 are intended to constitute a tutorial in the AdS/CFT apparatus, geared towards its use in the condensed matter context. We have avoided as much as possible the heavy string-theoretical machinery, and instead focussed on the pragmatic use of the correspondence. A reader equipped with a background acquired from entry-level graduate courses in general relativity, quantum field theory and condensed matter physics should be able to comprehend this book in some depth. One issue we had to consider is that detailed calculations in holography often resort to numerics. The level of these is fortunately not very challenging, and to assist the reader we will make the Mathematica codes for several basic calculations available via the website accompanying this book: **www.cambridge.org/9781107080089**.

To offer the reader a first grip on the storyline of this book, let us use the remainder of this introductory chapter to present a grand vista on the AdS/CMT landscape.

1.2 The AdS/CFT correspondence: unifying the theories of physics

The story that we will tell started in the mid 1990s. These years found the string-theory community in a euphoric state since it had become clear that there was much more to string theory than had previously been realised. The culmination of this second string revolution was the discovery in 1997 by the young theorist Juan Maldacena of what has become known as the "AdS/CFT correspondence" [1]. All along, string theory had been propelled by the insight that somehow general relativity is part of the quantum theory of relativistic strings, and that it therefore carried the inherent promise that it would eventually reveal the theory of quantum gravity. AdS/CFT was a great leap forward in this regard.

Quantum field theory (QFT) and general relativity (GR) are the two grand theories of physics, but their mutual relationship is complicated, if not even antagonistic. Maldacena's discovery, however, connected these two pillars in a way that nobody had foreseen. He showed that in a special limit these theories can be two sides of the same coin! The two sides of this unification of quantum physics with general relativity are in a mathematical sense as opposite as can be. This refers to the meaning of "duality" in the title of this book. GR and QFT are in a *dual* relation with each other in the sense of the particle–wave duality of quantum mechanics. Particles and waves are opposites in the sense that they are related by Fourier transformation. But at the same time the particle and wave representations form a wholeness revealing what quantum mechanics is. Depending on the question one asks, either the particle or the wave description is the better viewpoint. In the same sense GR and QFT "merge in their oppositeness", albeit the resulting "wholeness" is much richer than quantum mechanics: in a sense it seems to encapsulate all theories of physics. Soon after Maldacena's discovery Gubser, Klebanov, Polyakov [2] and independently Witten [3] (GKPW) came up with a set of tight and general mathematical rules demonstrating precisely how one can *quantitatively* relate results in one description to the other side. The unveiling of this dictionary launched an enormous research effort: thousands of papers revolving around checks and double checks of the correspondence, applying the new viewpoint to pernicious open problems and diversifying it to a variety of physics subjects were published. This book is dedicated to perhaps the least anticipated success of this headlong dash of exploration: the application to condensed matter physics.

Despite all the progress the correspondence is still shrouded in mystery. Metaphorically it is like the oracle of Greek mythology: upon throwing questions at the correspondence, it delivers answers that make sense, but it is far from clear why it works so well. This is the prevalent moral of the developments we will describe in this book – particularly in the condensed matter context there is a real potential to challenge and mobilise experimental physics to check whether some of the most enigmatic answers of this oracle are actually correct. The mystery of AdS/CFT is rooted precisely in the *quantum* gravity side. In full generality, AdS/CFT relates stringy *quantum* gravity to certain quantum field theories. Stringy quantum gravity remains very poorly understood. However, in a special limit the stringy quantum gravity side reduces to the solid ground of classical general relativity. This limit has a corresponding incarnation in the dual quantum field theory. It requires that the field theory contains matrix-valued fields of rank N and the limit means that one considers the system both in the regime of the "large $N \to \infty$" limit and at "strong coupling". A typical example is a very strongly coupled $SU(N)$ Yang–Mills theory in the limit of a large number of colours N. This sounds rather remote from the real world of condensed matter physics. To make matters even worse, to have full mathematical control in its string theory origin one also has to take a supersymmetric theory. Fortunately, in trying to apply the lessons of AdS/CFT more broadly, supersymmetry appears to be less crucial. The large-N limit is the serious mathematical obstacle to consider. Our main goal will be to study field-theoretical problems that cannot be handled with existing field-theoretical techniques. Within the large-N limit one can now use this AdS/CFT dictionary to call upon the great mathematical quality of Einstein's theory of general relativity to arrive at solutions that with the help of the dictionary can be translated back into field-theoretical answers. But the large-N limit implies that this is trustworthy only on departing from an extremely *symmetric* physics at high energy, which is very different from the mundane "chemistry" of e.g. electrons in solids at the Ångström scale. One would therefore like to lower the N-fold symmetry all the way to the puny symmetries governing the interacting electrons. In principle this is also possible in AdS/CFT. In practice, however, upon doing so the quantum-gravity hell breaks loose in the gravitational dual and, despite a very intense effort, we are still pretty much in the dark.

How can AdS/CFT still deliver? Remarkably, in spite of the large-N obstacle, the oracle seems to deliver answers to questions that do not depend sensitively on these matters. The string theorists refer to the relevant context by invoking "UV-independence". This is coincident with the notion of "strong emergence" of the condensed matter physicists. Both are about the theme that the "whole is more than the sum of the parts", which is familiar to all physicists, where we are interested in the circumstances under which the wholeness has such a strong logic of its own that the detailed nature of the parts is no longer of relevance. It is the independence of macroscopic phenomena from the details of the microscopic physics. This started seriously with Boltzmann's formulation of statistical physics in the nineteenth century for the description of classical matter. It triumphs in the form of an understanding of the solid, fluid and gas phases of simple thermal matter, including the microscopic origins of the phenomenological elasticity and Navier–Stokes theories. This was generalised to the zero-temperature "quantum" realms in condensed matter physics in the form of Landau-style order-parameter theories describing superfluids and superconductors, as well as the Fermi-liquid theory. The next great step was taken by the Wilsonian renormalisation-group revolution in the 1970s, merging together the methodological underpinning of the description of the critical state realised at continuous phase transitions and the fundamental quantum field theories of high-energy physics.

The correspondence builds further on this. Its "magical" power lies in its capacity to express the mathematical structure of the "strongly emergent" theories of matter through the very different geometrical structure of GR. We will unfold this story step by step in chapters 6–12, following closely the historical development, up to the state of the art involving predictions for states of matter that are ruled by macroscopic quantum entanglement.

1.3 AdS/CFT, the geometrisation of the renormalisation group
and the quantum critical state

What is actually the meaning of the abbreviations AdS and CFT, and especially the adjective "holographic" in the title of the book? This is all tied together with a mathematical relation that is seen by many as the most beautiful and stunning of all "emergence physics–general relativity" relations. It can be written as an "equation"

$$RG = GR. \tag{1.1}$$

Here RG is short for the renormalisation group and GR is of course general relativity. The renormalisation group refers to the property of field theory that by integrating out short-distance degrees of freedom one induces a flow describing how the theory changes as one lowers the scale to longer and longer wavelength. This is enumerated in terms of differential equations expressing the running of coupling constants. What AdS/CFT miraculously tells us is that this scaling "direction" turns into an extra *geometrical dimension* in the gravitational dual. The scaling flow in the field theory is now encoded in the purely geometrical properties of this higher-dimensional gravitational space-time, which in turn is governed by solutions of the Einstein equations. When the field theory lives in a $(d + 1)$-dimensional space-time, the corresponding bulk has $d + 2$ dimensions; this extra dimension is often called the "radial direction". This is metaphorically like a hologram: one has a two-dimensional photographic plate with interference fringes (the field theory) and by shining through a laser beam (the dictionary of the AdS/CFT correspondence) a three-dimensional (3D) image (the bulk) is constructed. The miracle that makes this work is the quantum-gravitational "holographic principle" originating in black-hole physics (see section 4.1). It insists that the counting of degrees of freedom in a gravitational theory behaves similarly to a quantum field theory in one dimension fewer. This is why the members of the large family of AdS/CFT-type correspondences are called "holographic dualities": in the present context this metaphor acquires an extra appeal since we will see that the physics in the gravitational "bulk" is quite recognisable (the 3D images) while the field-theory side is quite abstract and counterintuitive (the interference patterns on the plate).

How precisely do we stitch these different worlds together? It is now the moment to explain the abbreviation. It stands for the particular configuration in which the duality between QFT and GR was first realised. The first part AdS stands for anti-de Sitter space. This is the space-time on the general-relativity side of the correspondence. This is a solution to Einstein's theory characterised by a negative cosmological constant. Geometrically it is the Lorentzian higher-dimensional generalisation of a hyperboloid. The properties of hyperbolic space are famously represented by Escher in a series of drawings, tessellating this space with fishes, devils and reptiles. Formally it appears that such hyperbolic spaces are infinitely large, but here one of the marvels of relativity comes into play. This is not true for light-like propagation. They reach the edge of space-time in a finite time. This means that one has to supplement the gravity theory with specific boundary conditions and boundary information. Since the boundary is naturally of one dimension fewer than the "bulk", one can now imagine that the boundary is correlated with the space-time where the field theory lives, while its RG flow is associated with the extra radial direction moving from this boundary towards the centre of AdS: the "deep interior". This intuitive viewpoint is correct and will serve as a great support in understanding the quantitative dictionary between the two sides of the correspondence. The second abbreviation, CFT, stands for "conformal field theory". In the first explicit realisations of the correspondence the quantum theories were always of a very special type. They were conformally invariant. This means that "everything stays the same under arbitrary scale transformations that preserve angles". This is the category of the theories which are of great interest to contemporary condensed matter physics. Conformally invariant theories explain the universal behaviour at second-order phase transitions. Here, however, these conformal field theories arise naturally as zero-temperature relativistic field theories. They are "quantum critical theories" describing the universal physics near a zero-temperature "quantum phase transition" controlled by another external parameter. Now precisely the notion of quantum criticality appears to play a central role in all the big puzzles of strongly correlated matter revealed by experiment, most notably the strange metals found in high-T_c superconductors and other exotic materials. All along, quantum criticality will be a central motive linking the holographic-duality activities to the laboratory floor.

These are still only glimpses of the beautiful relationship between quantum field theories and general relativity. But, using the pieces just revealed, anti-de Sitter spaces, conformal field theories and the quantum critical state, and the extra direction as somehow encoding the RG flow, we can already give a remarkably intuitive description of the dictionary relating the two sides.

Let us start from the field-theory side of the correspondence, and take a relativistic conformal field theory as our starting point. The natural description of such a CFT is that it lives in a flat, $(d + 1)$-dimensional non-dynamical Minkowski space-time, i.e. quantities that respect special relativity are expressed in terms of four-vectors/four-tensors with inner products defined by the metric

$$ds^2 = \eta_{\mu\nu}\, dx^\mu\, dx^\nu = -dt^2 + d\mathbf{x}^2. \qquad (1.2)$$

Here $\mu, \nu = 0$ is the time direction, and $\mu, \nu = 1, \ldots, d$ are spatial directions. In addition to manifest invariance under global Lorentz transformations, Minkowski

space-time is invariant under time and spatial translations. Correspondingly one
has conservation of energy and momentum.

A generic field theory is also subject to renormalisation. Following Wilson, we
can integrate out short-distance degrees of freedom consistently, provided that
we change the coupling constants g_i of the theory under scale transformations
according to differential equations that are local in the RG scale u,

$$u \frac{\partial g_i(u)}{\partial u} = \beta(g_j(u)). \tag{1.3}$$

Right at a critical point, however, the beta functions vanish by definition, $\beta = 0$,
and the physics becomes scale-invariant. The combined space and time scale transformation $x^\mu \to \lambda x^\mu$ is now also a symmetry. For a relativistic Lorentz-invariant
theory, scale invariance together with unitarity is conjectured to imply invariance
under the full set of conformal transformations (see e.g. [4] for a review), i.e.
all transformations that preserve angles but not necessarily lengths. These include
so-called special conformal transformations that combine with scale and Lorentz
transformations to form the group $SO(d + 1, 2)$.

Picture now a generic field theory evaluated at different values of the
renormalisation-group scale r, and put them in a sequence (Fig. 1.1). If we
consider the label r of each theory as a continuous variable, we get a new $(d + 2)$-
dimensional "space-time" that has one extra dimension that parametrises the RG
flow. On the field-theory side, i.e. for each field-theory slice, the RG scale r is of
course a non-geometrical entity that lets us know how the field theory behaves as
we change the scale. The essence of AdS/CFT is that the *full family* of theories has
an alternative dual geometric description in terms of a *real* $(d + 2)$-dimensional
space-time. To reflect the scaling properties of the underlying field theory, this
space-time cannot be flat but has to have a curved shape, exactly as is familiar
from Einstein's theory of general relativity. In such curved space-times, distances
are measured with the help of the local metric $ds^2 = g_{\mu\nu}(x)dx^\mu dx^\nu$. Remarkably,
if the field theory we started from was indeed a very special field theory with conformal and scale invariance, we can deduce what the form of the metric should be,
solely from the symmetries of a conformal field theory. Generically any specific
metric tensor $g_{\mu\nu}(x)$ also corresponds to a particular "gauge choice" associated
with a preferred coordinate frame. However, if the space-time has a true physical
symmetry, then the metric should respect this symmetry irrespective of the coordinate choice. Such a symmetry is called an isometry, to distinguish it from the
choice-of-coordinate-frame symmetries (general covariance, or diffeomorphism)
that are the basis of GR. Since the field theory we wish to describe is conformal
and invariant under scale transformations, we must insist that its holographic gravitational dual is so too, if the two sides of the coin are to match. We must therefore

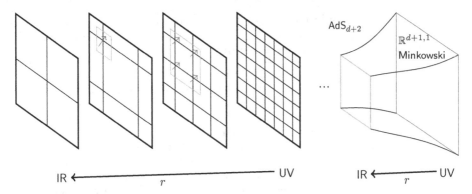

Figure 1.1 Consider the set of copies of a field theory generated by successive coarse-graining steps. Geometrically this set naturally groups together to form a space with one extra spatial direction. The essence of AdS/CFT is that for a conformal field theory (CFT) this grouping can be made mathematically exact. The built-up space-time is an anti-de Sitter space (AdS) in which the extra dimension r in the bulk is naturally interpreted as the renormalisation-group scale in the boundary field theory. Figure source [5].

find a metric in such a way that the total $(d + 2)$-dimensional space-time has the scaling symmetry $x^\mu \rightarrow \lambda x^\mu$ as an isometry. The "stacking" picture of our field theory, which is Lorentz- and translation-invariant, already tells us that the metric must be of the form

$$ds^2 = f(r)\eta_{\mu\nu}\, dx^\mu\, dx^\nu + g(r)dr^2. \tag{1.4}$$

If we then use the coordinate freedom to set r equal to the energy scale of the field theory, i.e. under scale transformations r should scale as $r \rightarrow r/\lambda$, the unique invariant $(d + 2)$-dimensional metric is

$$ds^2 = \frac{r^2}{L^2}\eta_{\mu\nu}\, dx^\mu\, dx^\nu + \frac{L^2}{r^2}\, dr^2. \tag{1.5}$$

This is precisely the metric of $(d + 2)$-dimensional *anti-de Sitter space*. We now see the explicit connection between AdS and CFT. A more detailed study of the AdS metric reveals not only that its *full* set of isometries is Lorentz plus scaling but also that it actually forms the group $SO(d + 1, 2)$. This matches exactly to the conformal angle-preserving symmetry of a $(d + 1)$-dimensional CFT, including the special conformal transformations.

Let's inspect this geometry more closely. The anti-de Sitter AdS_{d+2} geometry is thus a family of copies of Minkowski spaces parametrised by the "radial coordinate" r, whose size is seen to shrink when r decreases from the UV of the field theory at $r \rightarrow \infty$ to the IR at $r \rightarrow 0$. The "UV" region where $r \rightarrow \infty$ is the earlier-mentioned boundary of the space-time, which light-like objects can reach

Table 1.1 *Notations used in this book*

Label	Physical meaning
d	the space dimension of the field theory
$d+1$	the space-time dimension of the field theory
$d+2$	the space-time dimension of the gravity theory
L	AdS radius
r	AdS "energy" radial coordinate, $r = 0$ $(r_{\rm h})$ is the interior/horizon and $r = \infty$ is the boundary
z	AdS "length" radial coordinate, $z = \infty$ $(z_{\rm h})$ is the interior/horizon and $z = 0$ is the boundary
$x_i, \; i = 0, \ldots, d$	time/spatial coordinate of the field theory, and coordinates of AdS transverse to the radial direction

in a finite time. One therefore needs to supply boundary conditions in addition to the dynamical laws, and this will play a very important role throughout. The "IR" region of the field theory is the deep interior of AdS. The free parameter L with the dimension of length is called the "AdS radius", and its meaning for the field theory will become clear later. One can also make the extra holographic dimension an RG "length" by changing coordinates to $z = L^2/r$, for which the metric reads

$$ds^2 = \frac{L^2}{z^2} \left(\eta_{\mu\nu}\, dx^\mu\, dx^\nu + dz^2 \right). \tag{1.6}$$

We shall use the notations introduced here throughout. For convenience, we summarise them in Table 1.1.

This simple geometrical picture (see Fig. 1.1) already teaches us that the field theory has an intricate relationship with a non-trivial Einstein space-time. The fact that we took a critical scale-invariant theory as the starting point imposed strong restrictions on the emergent geometry of the $(d + 2)$-dimensional space-time. A critical theory is very special, however. If we now deform the conformal field theory such that we create a non-trivial flow to a new IR, it is clear that the geometrical structure must change, i.e. the space-time must be able to respond to the physics. The simplest dynamical gravitational theory that fulfils the Landau criteria of satisfying the symmetry requirements (in this case general covariance) with a minimal number of derivatives, and that also has the AdS space-time as a solution, is the Einstein–Hilbert action with an added negative cosmological constant,

$$S = \frac{1}{16\pi G} \int d^{d+2}x \sqrt{-g} \left[R - 2\Lambda + \cdots \right], \tag{1.7}$$

where $g = \det g_{\mu\nu}$, R is the Ricci scalar obtained from the metric, and G is the gravitational coupling constant, while $-2\Lambda = d(d + 1)/L^2$ is a negative cosmological constant, parametrised in terms of the AdS radius L. The ellipsis means that one can add matter (scalar, vector, fermion fields) and/or higher-derivative gravity terms, if one so desires. The cosmological constant is formally the same as that which accounts for the expansion of the Universe discovered in 1998, except for its sign. With a positive as opposed to a negative cosmological constant the resulting solution to Einstein's equations is the de Sitter space-time which describes an exponentially expanding universe. Let us remark that, even though the precise notion of holography which we will explain in chapters 4 and 5 is thought to work for all gravitational space-times, currently we know how to apply it only for (asymptotically) anti-de Sitter ones. The Einstein–Hilbert action Eq. (1.7) that should govern the gravitational response for a deformation of the CFT has introduced a new parameter: Newton's constant. In the natural units ($\hbar = c = 1$) we are implicitly using above G has dimensions $G = \ell_{\text{pl}}^d$, where ℓ_{pl} is known as the Planck length. Explicitly $\ell_{\text{pl}}^d = G\hbar/c^3$. Since Newton's constant is the coupling constant of general relativity, controlling the fluctuations of space-time, we learn immediately that the AdS radius of curvature needs to be very faint, $L \gg \ell_{\text{pl}}$, in order to avoid large quantum-gravity effects. If the curvature becomes larger, our limitation to a minimal number of derivatives becomes invalid, and indeed the explicit constructions of AdS/CFT dual pairs from string theory actually tell us to add such corrections to the Einstein–Hilbert action when the curvature increases (section 1.4).

The meaning for the field theory of this inherent limit that the curvature must be small in Planck units is the large-N matrix limit mentioned earlier. This follows directly from explicit string-theoretical "top-down" constructions of AdS/CFT dual pairs. Generally, the AdS gravity theory corresponds to full string theory including all higher-derivative corrections. The simplification to the classical and weakly coupled Einstein gravity of Eq. (1.7) occurs only when the number of degrees of freedom in the CFT is appropriately large. For instance, for a $(3 + 1)$-dimensional CFT dual to a $(4 + 1)$-dimensional AdS space, one finds that the AdS curvature L in units of the Planck length ℓ_{pl} is related to the effective number of degrees of freedom in the CFT N,

$$\frac{L^4}{\ell_{\text{pl}}^4} \sim N. \tag{1.8}$$

Thus we need $N \gg 1$ in order to have a reliable dual classical gravitational description.

There is an additional further requirement if we are to trust Eq. (1.7). For the perturbative string-theory starting point to be reliable, the intrinsic string coupling constant must be small. This coupling g_s is proportional to the ratio of the gravitational Planck length ℓ_{pl} to the intrinsic string length ℓ_s. It also controls the coupling in the CFT. In the example of a $(3 + 1)$-dimensional CFT dual to a $(4 + 1)$-dimensional AdS space this translates into the condition

$$\frac{\ell_{\text{pl}}^4}{\ell_s^4} \sim g_s \simeq g_{\text{CFT}}^2 \ll 1. \tag{1.9}$$

A weak string coupling therefore suppresses the true scale at which quantum-gravity effects become important. Microscopically, it is the string length ℓ_s that sets the scale at which deviations from Einstein's relativity Eq. (1.7) become important, rather than the Planck length. The more precise version of Eq. (1.8) is therefore

$$\frac{L^4}{\ell_s^4} \sim g_{\text{CFT}}^2 N \gg 1. \tag{1.10}$$

One thus finds that to rely on classical gravity in the bulk one has to make the number of degrees of freedom (N) in the field theory very large, so large that the combination $g_{\text{CFT}}^2 N$ with $g_{\text{CFT}}^2 \ll 1$ is also large. It is a well-known fact in field theory that such large-N limits generically rearrange the perturbation expansion. Not coincidentally, in the top-down models derived from string theory the natural variables are the rearranged large-N variables. An example is $(3 + 1)$-dimensional maximally supersymmetric $SU(N)$ Yang–Mills theory. In the large-N limit the new effective 't Hooft coupling in the field theory is precisely the combination $g_{\text{CFT}}^2 N$ in the earlier equation (1.9). Because the large-N limit rearranges the perturbation expansion, this means that the microscopic coupling g_{CFT}^2 is no longer an accurate gauge of the theoretical control of the theory, even though it is small. What controls the theory is this new effective coupling, and it *has* to be very large for the correspondence to work. This *duality* that strongly coupled field physics in the field theory corresponds to weakly coupled general relativity is what makes AdS/CFT so powerful. It gives us insight into regimes of the field theory that were previously inaccessible. The first dictionary rule of the correspondence is

Rule 1   AdS/CFT can map strongly coupled non-gravitational physics to a weakly coupled perturbative gravitational problem.

The converse is also true: strongly coupled gravitational physics maps to a weakly coupled field theory, but this aspect is not the focus of this book.

At this point one can still wonder whether the observation that one can draw a "geometrised" picture of renormalisation flow is no more than an interesting metaphor, even with the tantalising strong/weak-duality backbone. To render

AdS/CFT a real computational framework one needs real equations. Gubser, Klebanov and Polyakov [2] and independently Witten [3] made the crucial step by realising how the physics on each side of the duality precisely translates into that on the other. Here we will try to introduce the essence from the bottom up; we will discuss the full mathematical "Gubser–Klebanov–Polyakov–Witten" (GKPW) rule in chapter 5.

Let us consider the following puzzle: take as given that the renormalisation group can be "geometrised" such that a quantum critical scale-invariant state can be encoded in terms of classical gravity in an anti-de Sitter geometry. How does one, for instance, "string together" the two-point correlation functions in the field theory from this starting point? Two-point correlation functions in a conformal field theory obey a very simple rule. For a Lorentz-invariant conformal field theory at zero temperature, their form is fixed by conformal and Lorentz invariance with the scaling dimension of the operator as the only free parameter. Consider two points separated by a distance x in Euclidean space-time. Given a "bare" (UV) scalar operator $\mathcal{O}(x)$ of the conformal field theory with conformal dimension Δ, i.e. under a scale transformation it transforms as $\mathcal{O}(\lambda x) = \lambda^{-\Delta}\mathcal{O}(x)$, the two-point propagator will be

$$\langle \mathcal{O}(x)\mathcal{O}(0)\rangle = \frac{1}{|x|^{2\Delta}}. \tag{1.11}$$

Here we have fixed the arbitrary normalisation to one. This is probably familiar to the reader. The correlation function can only depend on the invariant distance $|x| = \sqrt{c^2\tau^2 + x^2 + y^2 + \cdots}$ with τ the Euclidean time. More importantly, given scale invariance it has to be the case that any response becomes an algebraic function of distance.

To interpret experiments we are usually interested in the response as a function of energy and momentum. We can Fourier transform this expression, and after a Wick rotation from imaginary to real time with the appropriate prescription we obtain the retarded correlation function (chapter 5)

$$\langle \mathcal{O}(\mathbf{k}, \omega)\mathcal{O}(0)\rangle_{\mathrm{R}} \sim k^{2\Delta-d-1}$$
$$\sim (-\omega^2/c^2 + \mathbf{k}^2)^{\Delta-d/2-1/2}. \tag{1.12}$$

These are the "branch-cut" propagators of the quantum critical state, as we know them in condensed matter physics (section 2.1). Focussing on the spectral function $A(\mathbf{k}, \omega)$ – the imaginary part of Eq. (1.12) – one finds at zero momentum a pure power law, $A \sim \omega^{2\Delta-d-1}$, whereas at finite momentum there is no spectral weight until $\omega = ck$, after which the function tends asymptotically to the pure power law at very high frequency.

The task is now to reconstruct such two-point correlators from objects that we can identify in the AdS space-time. The operators in the CFT are associated with the "bare" high-energy degrees of freedom. According to the geometrisation roadmap of the RG these should be associated with going as far out as possible in the AdS space. Then the following statement appears to be obvious:

Rule 2   The $(d + 1)$-dimensional space-time of the (bare UV) field theory is coincident with the boundary of the gravitational $(d + 2)$-dimensional AdS bulk space-time.

At its deepest level the map between the AdS gravity and the CFT is more abstract and in principle one should not identify a location in AdS with the CFT. Nevertheless, it will be convenient and almost always perfectly safe to picture it in this simple fashion.

Which object in the AdS gravity theory can correspond to the scalar operator $\mathcal{O}(x)$ of the CFT? The most straightforward guess is just to consider a scalar field ϕ in AdS, and find out what can be done with it. The simplest action one can write down for the dynamics of this scalar in AdS is a quadratic action minimally coupled to the space-time,

$$S = \int d^{d+2}x \sqrt{-g} \left(-\frac{1}{2} g^{\mu\nu} \partial_\mu \phi \, \partial_\nu \phi - \frac{1}{2} m^2 \phi^2 + \cdots \right), \qquad (1.13)$$

where g and $g^{\mu\nu}$ are the determinant and the inverse of the metric $g_{\mu\nu}$, respectively. To a first approximation we can consider this action as a classical field theory in the curved AdS space-time, with the metric Eq. (1.6) as the fixed background. Thus the equation of motion for ϕ is

$$\frac{1}{\sqrt{-g}} \partial_\mu \left(\sqrt{-g} g^{\mu\nu} \partial_\nu \phi \right) - m^2 \phi = 0. \qquad (1.14)$$

We now Fourier transform to frequency–momentum space, but only with regard to the flat space-time shared by the boundary and the bulk, and Wick rotate to a Euclidean signature ($t \to -i\tau$, $\omega \to i\omega_E$). For two-point correlators at space-like momentum we can do this straightforwardly, since it is safe to Wick rotate back to real time at the end of the computation:

$$\begin{aligned}
\phi(x^\mu, r) &= \int \frac{d\omega \, d^d\mathbf{k}}{(2\pi)^{d+1}} f_k(r) e^{ik_\mu x^\mu} \\
&= \int \frac{d\omega \, d^d\mathbf{k}}{(2\pi)^{d+1}} f_k(r) e^{-i\omega t + i\mathbf{k}\cdot\mathbf{x}} \\
&= i \int \frac{d\omega_E \, d^d\mathbf{k}}{(2\pi)^{d+1}} f_k(r) e^{-i\omega_E \tau_E + i\mathbf{k}\cdot\mathbf{x}}.
\end{aligned} \qquad (1.15)$$

On choosing to use the "RG-energy" coordinates for AdS,

$$ds^2 = \frac{L^2}{r^2} dr^2 + \frac{r^2}{L^2} \eta_{\mu\nu} \, dx^\mu \, dx^\nu, \tag{1.16}$$

one obtains an equation of motion for the dependence on the radial direction of the Fourier components f_k,

$$\left[\frac{\partial^2}{\partial r^2} + \frac{d+2}{r} \frac{\partial}{\partial r} - \left(\frac{k^2 L^4}{r^4} + \frac{m^2 L^2}{r^2} \right) \right] f_k(r) = 0, \tag{1.17}$$

where $k^2 = \omega_E^2/c^2 + \mathbf{k}^2$ (Euclidean signature). The equation is an ODE, but to avoid confusion with the d denoting the number of spatial dimensions, we have kept the partial-derivative notation. An important aspect is that, on redefining $r/k = \zeta$, the solution is a function of ζ alone. We can immediately infer the momentum dependence of the solution from the dependence on the radial coordinate r. This is the notion that r parametrises the energy scale in the field theory in action.

Now we use rule 2: the field theory can be thought to live at the boundary. With the Ansatz that $f_k(r/k) = f_k^{(0)}(r/k)^\alpha + \cdots$ behaves as a power near the boundary $(r \to \infty)$, we find the two independent solutions to the second-order equation of motion

$$f_k(r/k) = (r)^{-d-1+\Delta} \left(A(k) + \mathcal{O}\left(\frac{k}{r} \right) \right)$$

$$+ (r)^{-\Delta} \left(B(k) + \mathcal{O}\left(\frac{k}{r} \right) \right) \quad (r \to \infty), \tag{1.18}$$

where

$$\Delta = \frac{d+1}{2} + \sqrt{\frac{(d+1)^2}{4} + m^2 L^2}$$

and the coefficients $A(k) = ak^{d+1-\Delta}$ and $B = bk^\Delta$ have a fixed power-law dependence on k. This directly follows from the fact that the right-hand side must be a function of $\zeta = r/k$ alone. Since our field is real, the exponents had better be real, which implies what is known as the Breitenlohner–Freedman (BF) bound [6],

$$m^2 L^2 \geq -\frac{(d+1)^2}{4}. \tag{1.19}$$

Note that the mass squared can be negative. This is a special property of the AdS space-time. As long as this BF bound is satisfied, the AdS space itself is stable even in the presence of the somewhat negative mass-squared field ϕ. For large negative mass squared $m^2L^2 < -(d+1)^2/4$, the complex exponents do signal a conventional linear instability ("tachyon") of this space-time; the field will acquire a finite amplitude, which will in turn backreact on the geometry of the space-time. To give a brief preview of chapter 10, such BF-bound violations will essentially be the way to gravitationally encode spontaneous symmetry breaking in the boundary field theory.

Returning to our puzzle regarding how to determine the two-point correlation function in the AdS dual of a conformal field theory, we note that the information available near the boundary is contained in the universal asymptotes of the solutions to the equation of motion Eq. (1.18). The boundary field theory should not contain information regarding the "unphysical" dimension r, but we note that the leading coefficients $A(k)$ and $B(k)$ of the two independent solutions each have a simple algebraic (monomial k^n) dependence on the frequency and momentum of the field theory. Using the bulk solutions to $\phi(r)$, we can try to simply engineer a way to reconstruct the known answer in the CFT, in the form of the correlator Eq. (1.12). Given that the integration constants a and b are undetermined, and that the overall normalisation should not play a role, one arrives at

$$\frac{B(k)}{A(k)} = \frac{b}{a}k^{2\Delta-d-1}. \tag{1.20}$$

Note that the mass that we put in by hand turns out to determine the scaling dimension of the boundary operator. The *mass–scaling dimension* relation is one of the essentials of the correspondence. This simple engineering result, which is coincident with the two-point correlation function of a Euclidean conformal scalar field $\mathcal{O}$ with conformal dimension $\Delta = (d+1)/2 + \sqrt{(d+1)^2/4 + m^2L^2}$ in momentum space Eq. (1.12), is in fact the right dictionary rule. A similar mass–scaling relation exists for operators with spin, e.g. for massive vectors one finds $\Delta_{\text{vector}} = (d+1)/2 \pm \sqrt{(d-1)^2/4 + m^2L^2}$ [7].

Rule 3 The CFT two-point correlation function for scalar operators with scaling dimension Δ can be expressed in terms of the ratio of the coefficients $A(k)$ and $B(k)$ of the leading and sub-leading asymptotes of the corresponding AdS massive scalar waves $\phi(r, k) = A(k)r^{-d-1+\Delta} + B(k)r^{-\Delta} + \cdots$ in the near-boundary region: $\langle \mathcal{O}(-k)\mathcal{O}(k)\rangle = B(k)/A(k)$. The scaling dimension $\Delta = (d+1)/2 + \sqrt{(d+1)^2/4 + m^2L^2}$ is set by the mass of the AdS scalar wave.

Let us continue with our engineering. The most basic object of the boundary theory is the relativistic energy–momentum tensor $T_{\mu\nu}$. This operator is always there as the generator of time and space translations. In contrast with the scalar operator, one has to address the tensor indices [8]. Fortunately, due to the fact that $T_{\mu\nu}$ as a symmetry generator is conserved, $\partial_\mu T^{\mu\nu} = 0$, the result for its two-point correlation function is strongly constrained. One finds

$$\langle T_{\mu\nu}(x) T_{\rho\sigma}(y) \rangle = \frac{c_T}{s^{2d+2}} J_{\mu\alpha}(s) J_{\nu\beta}(s) P_{\alpha\beta,\rho\sigma}, \tag{1.21}$$

where

$$s = x - y,$$

$$J_{\mu\nu}(x) = \delta_{\mu\nu} - 2\frac{x_\mu x_\nu}{|x|^2},$$

$$P_{\mu\nu,\rho\sigma} = \frac{1}{2}(\delta_{\mu\rho}\delta_{\nu\sigma} + \delta_{\mu\sigma}\delta_{\nu\rho}) - \frac{1}{d+1}\delta_{\mu\nu}\delta_{\rho\sigma}$$

and c_T is a constant. One notices that the conformal dimension of the energy–momentum tensor is fixed to be $\Delta_{T_{\mu\nu}} = d + 1$. This is just the requirement that the energy is subjected to "engineering scaling", because it is set by the volume. What object on the gravitational side encodes the dynamics of this ever-present operator? Given that $T_{\mu\nu}$ is a spin-2 tensor, there is only one candidate. It must be the graviton encoding the dynamical nature of space-time. It is the unique consistent interacting spin-2 field one can write down. The symmetries and tensor properties of $T_{\mu\nu}$ ensure that after repeating the above procedure for the gravitons one recovers precisely Eq. (1.21). An essential difference is that we know the mass of the gravitons a priori: by the same symmetry principle they are massless.

We have uncovered here a further fundamental aspect of the holographic correspondence. The energy–momentum tensor $T_{\mu\nu}$ is the "Noether current" associated with *global* space and time translations. In the conformal field theory on the boundary it also generates the additional *global* scale and special conformal transformations. But in the previous section we showed that these global conformal transformations translate into isometries of the bulk geometry: they are part of the *local* coordinate transformations ("diffeomorphisms", "general covariance") which are at the heart of the structure of Einstein theory. Now we see that on the gravity side it is indeed the *gauge* field of the local symmetries, the graviton, that encodes the dynamics of the conserved current of the global symmetries in the boundary field theory. This is a very deep feature of the correspondence:

Rule 4 AdS/CFT is a local–global duality. Global symmetry in the boundary field theory dualises in a local gauge symmetry in the bulk. The gauge bosons are the gravitational objects describing the dynamics of the boundary conserved currents.

One is naturally led to infer that the same should happen for the conserved currents associated with global internal symmetries in the boundary. These should be encoded in Maxwell or Yang–Mills gauge fields in the bulk.

1.4 Holographic duality and the nature of matter

These four rules should already allow the reader to understand a great deal of many of the holographic results in principle. Here we arrived at them with intuitive engineering, but they can be placed on a much firmer footing. We will do so in chapter 4 and chapter 5 in particular. These chapters will give a far more thorough background for the correspondence, including the precise quantitative GKPW rule which prescribes how to compute quantities in the field theory from the AdS gravity theory. This GKPW rule, an equality between the partition functions of the two sides, is the beating heart of the correspondence: it is what gives it its computational power. If one wants to understand how to practice holography, it is essential that one understands these chapters describing the dictionary thoroughly. Even then it is useful to have a translational aid readily available, and to this end we have included a literal dictionary table at the end of chapter 5.

The obvious question, however, is how to apply this dictionary to condensed matter puzzles. At first sight AdS/CFT seems far removed from this subject, since the relevance of the conformal boundary field theory to the real world is far from obvious. How can we connect this to the material world tested in the laboratory? It is hard-wired into the present formulation of holographic duality that conformal invariance controls the physics at short distances (the UV). The geometry in the bulk near the boundary has to be AdS in order for us to formulate the GKPW rule. However, there is no such restriction in the deep interior. The correspondence leaves us room to alter the geometry in this interior, as long as this satisfies the Einstein equations. Equivalently, one has complete freedom to break the conformal invariance in physically meaningful ways in the long-wavelength "infrared" (IR) regime of the boundary system. In this way one can address the question regarding the kind of collective "emergence" physics that governs the matter formed from the microscopic strongly interacting CFT degrees of freedom. To what extent do the phenomenological theories describing the collective physics in the IR remain dependent on the detailed nature of the UV degrees of freedom, and to what extent are these just reflecting ubiquitous principles governing the emergent physics? In a surprising development that nobody expected, the AdS/CFT "oracle" turns out to have much to report regarding these matters. This is the story we wish to tell.

Before we delve into the specifics, let us first spell out the restrictions we imposed on ourselves in the writing of this book. There has been so much progress made with holography in recent years that a number of largely complementary books could be written. Indeed, while this book was being written, the first textbooks dedicated to the gauge–gravity duality on its own [9, 10] and to its application to QCD in the context of the quark–gluon plasma studied in heavy-ion collisions appeared [11, 12]. There are in addition numerous excellent reviews on AdS/CFT or gauge/gravity duality more generally (e.g. [7, 9, 13, 14, 15, 16]), and even on the topic of this book, AdS/CMT (e.g. [17, 18, 19, 20, 21, 22]).

Our goal has been to provide a very pedagogical approach to holography that makes it easily accessible to a wide audience and at the same time places the recent developments into a coherent framework of modern condensed matter physics. We shall focus on *equilibrium* physics in the first place. Holography is actually unique in its capacity to deal with the non-equilibrium time-dependent physics of strongly interacting quantum systems as well. Technically, real-time evolutions of the boundary field theory amount to considering non-stationary gravitational physics in the dual bulk gravity description. At present the non-equilibrium agenda is rapidly developing, helped by exciting technical developments in numerical general relativity. These make it possible to reliably compute the non-stationary space-time evolution in the bulk AdS gravity, including dynamical black-hole for-mation. Both the physics and the methodology are complementary to what we cover here, and deserve their own treatises [23]. The equilibrium physics in the boundary of interest to us, on the other hand, is dual to stationary gravity in the bulk, and this is intrinsically easier to handle.

Secondly, this book has a "bottom-up" orientation. This should be distinguished from "top-down" approaches, where the full firepower of string theory is mobilised to derive "high-precision" holographic correspondences. The first historical exam-ple of a top-down holographic duality is the correspondence found by Maldacena [1]: from a specific string theory one can deduce precisely the nature of both the boundary field theory (the maximally supersymmetric large-N Yang–Mills theory) and its corresponding AdS bulk gravity (10-dimensional supergravity compactified on $AdS_5 \times S^5$). In principle one even knows all the numbers here. This has been extended to a portfolio of top-down constructions that do play an important role in the background of the developments we will discuss. They serve both as sanity checks and as sources of inspiration.

The "bottom-up" approach is the technically much simpler "phenomenologi-cal" approach. We will present the dictionary of the AdS/CFT correspondence in chapter 5. In combination with symmetry principles and the intrinsic consistency of general relativity, this makes it possible to "engineer" unique and simple constructions for the mathematical structure of particular holographic correspondences. These have a similar quality to the phenomenological theories of physics such as the Landau–Ginzburg theory of phase transitions in that the structure of the equations is universal but the numbers occur as free parameters. Nevertheless, the bottom-up constructions cannot be seen as independent of the top-down portfolio. The latter provides an invaluable validation of physics encountered in bottom-up phenomenology and, when called for, we will refer to the top-down constructions in this book. To describe top-down constructions in detail, however, one needs much of the heavy formalism of string theory. This is beyond our scope, but we have included an extra chapter at the very end of the book (chapter 13) as a first entry into this subject for the reader who wants to know more about such top-down constructions.

Finally, with regard to the potential applications to empirical physics, we focus on the typical questions arising in the context of strongly interacting electron systems in solids, and in closely related areas such as systems of cold atoms. The dream that has been motivating this particular development is that holographic duality can shed mathematical light on the mysterious behaviours discovered by condensed matter experimentalists during the last thirty years in strongly interacting electron systems as realised in high-T_c superconductors, the heavy-fermion systems and so forth. This context is extensively discussed in chapters 2 and 3. The implication is that there will be a strong focus on what holography has to tell regarding these *finite*-density systems. This is in contrast with the other area where holography has proven to be of relevance, namely the study of the physics of QCD, where the specifics of the zero-density Yang–Mills theory are in the foreground. There is some quite useful overlap between these different areas, and we will present a very short summary of AdS/QCD in section 6.3. A full exposition of this area of holography can be found in the recent companion book [12].

Within the context of *equilibrium, bottom-up, finite-density* holography AdS/CFT has given us some radical new perspectives on condensed matter physics. To give the reader an inkling of the excitement of these holographic perspectives and to aid him/her on his/her road through this book, let us now present a brief tour through the core of this book. In this way, a first-time visitor to the holographic universe will acquire an idea of what our destination looks like.

1.4.1 Zero-density conformal matter at a finite temperature
(chapters 6 and 7)

The first question any condensed matter physicist will raise is that of how to address systems at finite temperature. The AdS/CFT answer is to introduce *a black hole* in the gravitational description. Startling as this may sound, it is easy to understand. As Hawking famously discovered, quantum mechanically black holes are not black but radiate with a purely thermal spectrum. In the Euclidean time prescription it is then straightforward to demonstrate that an AdS black hole corresponds to a boundary field theory placed in a heat bath at the temperature of this black-hole radiation. It has long been known that one can ascribe not just a temperature but rather a full set of thermodynamic quantities to a black hole. Each of these now directly translates to the boundary field theory. In particular, a black hole possesses a Bekenstein–Hawking entropy that scales not with the volume enclosed but with the area of the black-hole horizon (section 6.1). Since the field theory lives on the boundary, it therefore scales precisely in the right way to explain its entropy. This area law is in fact the origin of the quantum-gravitational holographic principle.

This black-hole-gravity description of thermal physics in the field theory vividly illustrates the notion that it is natural to alter the deep interior of AdS to address the physics in the IR. The temperature involves a scale that is typically small compared with the UV cut-off. This scale breaks the conformal invariance in the boundary, which means that the AdS geometry has to be altered in the deep interior. Through the identification of the extra radial dimension with the RG scale, it therefore has to be encoded in the deep interior, and this is where the black hole resides.

The remarkable consequence of the holographic encoding of thermal physics in a black-hole background is that from the perspective of a "boundary physicist" (e.g. the condensed matter specialist) one can get away with computing a simple black-hole problem in the bulk, instead of the hard work of enumerating Boltzmann partition sums for the thermal problem. Moreover, using the quantitative AdS/CFT dictionary one can compute the full free energy of the boundary field theory, and this in turn makes it possible to compute thermal phase diagrams. In AdS it turns out that many of the black-hole uniqueness theorems do not hold. Later in the book we will show numerous examples where black holes "un-collapse" to a state of lower free energy (section 6.2, chapters 10 and 11). This is the gravitational dual of a thermal phase transition. The message to take home is that classical statistical physics has an impeccable gravitational dual. The caveat is that in the classical gravity approximation almost all thermal transitions are mean field. This is associated with the large number (N^2) of degrees of freedom of the large-N limit. In principle $1/N$ corrections will restore the order-parameter fluctuations in the usual way, and this can be computed in specific cases.

Having understood the thermodynamics, one can subsequently ask how to compute the hydrodynamical behaviour. This is one of the most striking successes of holography. It has to be the case that the long-time physics of any system at finite temperature is described by the universal theory governing classical fluids, namely the Navier–Stokes theory of hydrodynamics – as long as translational symmetry is unbroken. The difficulty is that this is a dissipative affair that yields to the second law of thermodynamics. This hydrodynamic emergence is hard to capture in conventional methods, but holography does so effortlessly. In chapter 7 we show the stunning result by Bhattacharyya, Hubeny, Minwalla and Rangamani that non-stationary gravity in AdS, as controlled by the gradient expansion, precisely dualises to the Navier–Stokes theory in the boundary (section 7.2). More than anything else, this effectively illustrates the claim that holography is capable of generating the universal theories of strong emergence physics.

The fluid/gravity correspondence fully includes all the dissipative physics, with the transport coefficients fixed by the specific holographic model. The values of these holographic parameters are non-standard, but the discovery thereof is another success story on its own. In 2002 Policastro, Son and Starinets computed the viscosity of a strongly coupled conformal fluid holographically. They arrived at a specific ratio of viscosity to entropy density $\eta/s = (1/(4\pi))\hbar/k_B$ (section 7.1). This by-now-famous ratio is extremely small compared with any empirically determined viscosity. The striking aspect is that the dimension of this dissipative quantity is set by Planck's constant. This is nowadays well understood as a manifestation of the notion of "Planckian dissipation", the fact that in the finite-temperature quantum critical-state quantities relax extremely rapidly (see section 2.1). The nearly perfect fluid behaviour that follows from such minimal viscosity has been observed in the quark–gluon plasma created at the heavy-ion colliders, as well as in the unitary fermionic cold-atom gas, while it is an important motive in the quest for understanding the transport properties of the strange metals formed in electron systems.

1.4.2 *The finite-density holographic strange metals (chapters 8 and 9)*

It is an amazing fact that the physics of thermal matter can be captured in the language of general relativity, even though in principle the laws governing the finite-temperature behaviour of the boundary field theory are precisely known – the success story of statistical physics resting on the Boltzmannian principle. Starting around 2008 the powers of the correspondence were unleashed to explore the uncharted territory of the boundary where the conventional methods of field theory are impotent. This refers to the physics of *strongly interacting matter at a finite charge density*. This is one of the most fundamental problems of condensed matter physics: conventional approaches to finite-matter quantum field theory fail due to the infamous "fermion signs".

In chapter 8 we will present firm evidence that *holographic duality effortlessly processes the fermion signs in a fully controlled, general manner*, at least within the limitations of its large-N UV. Holography is the first and only mathematical machinery we know that is able to do so. This might be regarded as the most significant accomplishment of the correspondence in the condensed matter context.

Using holography, one obtains finite-density strongly coupled boundary duals that represent an extensive family of zero-temperature *metallic* states of matter. These metals are radically different from conventional Fermi-liquid metals, but do have suggestive traits in common with the "non-Fermi-liquid" metallic states observed experimentally in strongly correlated condensed matter systems such as the high-T_c superconductors. Inspired by the terminology of the experimental community, these are called holographic *strange metals*. The gravitational description of the simplest such strange metal follows again from the uniqueness of general relativity. There is only one solution in Einstein–Maxwell gravity: the *charged Reissner–Nordström black hole* (section 8.1). Generically such a black hole describes a finite-temperature state in the boundary but it can be tuned to the limit where all its energy is carried by the electromagnetic field. This *extremal* Reissner–Nordström (RN) black hole corresponds to the zero-temperature state of the boundary. Following the RG = GR rule, one can zoom in close to this black hole to understand the IR of the field theory. The surprising result is that the new geometry realised in the deep interior becomes a *scaling geometry*. Through the dictionary this implies that the field theory exhibits an *emergent quantum critical phase* in the finite-density system. Moreover, in the RN setting this scaling is only in the time direction: spatial directions stay put. The RN metal is therefore a state that is best called "local quantum critical", characterised by a dynamical critical exponent relating the scaling of space and time that tends to infinity $z \to \infty$ (section 8.2). Remarkably this is quite suggestive of the laboratory strange metals: the expression "local quantum criticality" was introduced in the 1990s to describe precisely this kind of scaling behaviour that appears to be present in various physical strange metals. The simplest RN metal does have a property to be viewed with suspicion: the extremal RN black hole still has a finite horizon area and this implies that the field theory is characterised by a finite *ground-state entropy*. There are, however, straightforward extensions of these systems that include the flow of the most relevant scalar operator in the theory (section 8.4). In this context the RN metal is just a limiting case of a large family of strange metals. All of these are characterised by an emergent quantum critical phase with a dynamical critical exponent $1 \leq z \leq \infty$, and many naturally exhibit *hyperscaling violation*. This is the failure of the free energy to scale near criticality with the relevant length scale raised to the power of the space-time dimension. The meaning of hyperscaling violation of this kind in the boundary is as unclear as it is tantalising; the only known example of a theory with a finite hyperscaling-violating exponent is the Fermi liquid, where it becomes equal to the dimension $d - 1$ of the Fermi surface. It again suggests that strongly coupled fermion physics plays a crucial role.

Using the results from chapter 7, one can now study the physical responses of such metals (section 8.3). Of foremost interest is the most basic property of any metal: its optical conductivity. The dictionary translates this directly into the computation of the propagation of photons in the bulk gravitational background. The result looks not that remarkable: it could be easily confused with the optical conductivity of a simple free-electron band-structure system. The reason is that the gross features of the conductivity are governed by general "hydrodynamic" principles. The holographic examples effectively convey the message that collective properties governed by conservation laws are just not very sensitive to the gross effects of interactions. This theme is taken up again in chapter 12, which focusses on the effects of the breaking of the translational invariance; in this recent development the relevancy of holography to transport in real solids becomes truly alive.

Photoemission studies of strange metals do reveal remarkable properties, however. What these "fermion sign-full" probes see is the subject of chapter 9. Generically, the fermion spectra deep in the strange-metal phase bear no resemblance of any kind to the usual Fermi-liquid results dominated by dispersing quasiparticle poles. There are no Fermi surfaces, and in the extreme case of the local quantum critical states the spectral functions become power-law functions of energy, while the exponents are dependent on momentum. By changing the parameters of the system, it is possible to tune the system into a regime where a Fermi surface emerges (section 9.2). Furthermore, the damping of the quasiparticles can also be tuned. Depending on the parameter choice, a quite literal "marginal Fermi liquid" can be realised, where the quasiparticles are on the verge of becoming over-damped, much like the marginal Fermi-liquid phenomenology introduced in the late 1980s to describe the normal state of high-T_c superconductors. By tuning further, one can realise states that are still characterised by a Fermi surface, while the quasiparticles are completely over-damped.

When it was first observed in 2009, this controlled emergence of (non-)Fermi liquids caused a splash, accelerating the development of AdS/CMT. In the developments that followed it became clear, however, that these early signals are to a degree misleading. In the probe limit one ignores the fact that the bulk fermions should backreact on the geometry. It turns out that the backreaction becomes large precisely in the regime where the holographic Fermi surfaces are seen in the probe limit (section 9.4). For these parameters the strange-metal state is thus a false vacuum and the large backreaction will drastically change the state of the field theory. As we will discuss next, the reorganisation of the bulk includes the formation of fermionic matter in the bulk, resulting in a genuine Fermi-liquid state in the boundary dual.

1.4.3 Cohesive holographic matter: superconductors and Fermi liquids (chapters 10 and 11)

This instability of holographic strange metals is most manifest when probed with a scalar order parameter. Chapter 10 describes how this encodes spontaneous symmetry breaking in holography. We have already seen in rule 4 that the global symmetry on the boundary dualises to gauge symmetry in the bulk. The "breaking" of this gauge symmetry on the gravity side occurs according to the textbook: it is accomplished via the Higgs mechanism. The required Higgs field in turn dualises back to the order parameter in the boundary. It turns out that this order sets in at a classic second-order thermal phase transition. Upon lowering temperature in the RN metal, at some point the mass of the scalar field violates the BF bound of the deep IR geometry, Eq. (1.19), the AdS equivalent of negative mass squared. Physically this triggers a spontaneous discharge of the vacuum due to the strong curvature near the horizon. The effect is that at lower temperatures the RN black hole acquires an "atmosphere" of finite-amplitude Higgs field: the "black hole acquires scalar hair". The dictionary now delivers like an impeccable clockwork mechanism: the asymptotic contributions of the scalar field as discussed in the second section have only a sub-leading component. This means that one finds that there is a VEV without a source and this is precisely the definition of spontaneous symmetry breaking in the boundary field theory (section 10.1).

Following its discovery in 2008 by Gubser, Hartnoll, Herzog and Horowitz, holographic superconductivity turned into a major research area, and at present it is quite thoroughly understood. Empirically the low-temperature and -energy properties of unconventional high-T_c superconductors behave much like those of conventional superconductors. The holographic superconductor is similarly strikingly successful at reproducing this "BCS fixed-point physics". It is characterised by a gap, it supports long-lived Bogoliubov fermions in a large-parameter regime, it reproduces the Ginzburg–Landau phenomenology and so forth (section 10.2). The differences are found at higher energies and temperatures. In contrast to the Eliashberg/BCS version, the holographic superconductor emerges from a non-Fermi-liquid metal. In most experiments, the differences are not sharp enough to distinguish this strange-metal superconductivity from the BCS variety. Starting from some elementary considerations regarding the mechanism of superconductivity, we will argue that these matters can be decided in principle by a new type of experiment (section 10.3).

A similar spontaneous discharge instability can be recognised for holographic fermions in an RN metal background. However, though the mechanism is qualitatively similar to holographic superconductivity, we discuss in chapter 11 why there are actually vast differences between the holographic description of bosonic and fermionic instabilities. The technical challenge is the inherent-quantum-mechanical nature of the bulk fermions; they cannot be treated classically as the scalar field of holographic superconductivity. In order for such fermions to contribute to the stress-energy in the bulk, as a prerequisite for computable gravitational backreaction these have to form finite-density charged fermion matter. Self-gravitating fermion matter was famously solved in the 1930s by the work of Tolman, Oppenheimer and Volkoff addressing neutron stars. This involves the crucial simplification of a high-density fluid limit, where the fermion matter is described in terms of a simple Thomas–Fermi equation of state. With this Fermi-fluid approach one does indeed find that in the regime where the strange metal exhibits Fermi surfaces in the probe limit, the preferred ground state is rather a charged-fermion ("electron") star, to which the RN black hole must have "un-collapsed" (section 11.2). The fluid limit comes with a price. Each mode in the extra radial direction corresponds to a Fermi surface in the boundary field theory (chapter 9). In the Thomas–Fermi limit the level spacing and hence the number of modes have been taken to infinity. For condensed matter purposes, one is instead interested in the regime where one has at most a few isolated Fermi surfaces. One must now solve the technically challenging problem of a fully quantised Fermi gas in the bulk in a curved geometry that is self-consistently determined. Instead we shall discuss two modified configurations that are easier to follow: a "hard-wall" model where a mass gap is imposed by hand (section 11.1) and one where the flow of the most relevant neutral operator is also taken into account (section 11.3). Each in its own way can be argued to holographically capture the physics of Fermi liquids characterised by isolated Fermi surfaces.

1.4.4 Condensed matter realism: the breaking of translational symmetry (chapter 12)

The absence or presence of translational symmetry is at the heart of realistic transport behaviour. If momentum is strictly conserved, all metals turn into perfect metals. In the Galilean continuum the presence of a lattice or small-scale disorder may appear irrelevant, but its breaking of translational invariance has qualitative consequences for transport. In chapter 12 we discuss the effects of translational symmetry breaking in holographic set-ups and its consequences for transport properties.

For *weak* translational symmetry breaking through either lattices or impurities, the DC and low-frequency transport properties are special in the sense that they can be addressed in terms of simple universal principles. The key observation is that in strongly coupled systems local equilibrium is established long before the momentum will relax (section 12.1). This contrasts sharply with conventional quasiparticle systems, where momentum conservation is destroyed already at microscopic times. To address DC transport one can now rely on the hydrodynamical *memory-matrix formalism* which addresses this regime. This results in Drude behaviour at low frequencies, characterised by momentum relaxation rates that can be addressed quantitatively in principle.

Momentum relaxation in the boundary has a very specific interpretation in the gravitational bulk. The corresponding gauge field – the graviton – should be "Higgsed": one has a theory of massive gravity. Vice versa, any (consistent) theory of massive gravity in the bulk encodes the translational symmetry breaking (section 12.5). This has been used to demonstrate that in a local quantum critical phase with no other scale than momentum relaxation, the DC resistivity has to be proportional to the entropy. This could offer an appealingly simple explanation for the famous linear-in-temperature resistivity found experimentally in the normal state of the high-T_c superconductors: the electronic entropy in these systems, which are believed to be locally quantum critical, is linear in temperature.

For strong potentials or high-frequency transport, holography provides a unique perspective, since there is no alternative method to describe such physics. This is an active area of research, with so far mostly "black box" numerical results; we summarise this development briefly. A particularly striking aspect of strong potentials is that they can drive the strange metals into new phases. One finds unconventional metal–insulator transitions that occur only in the direction of the periodic potential breaking the translational symmetry (section 12.4).

This all deals with the explicit breaking of translational symmetry. However, the most common everyday form of spontaneous symmetry breaking is associated with the solid state of matter. As we will discuss in chapter 12, it is actually known from top-down constructions of how to encode the spontaneous breaking of translations in the holographic bulk. Although it is firmly established that at some point the holographic liquid will undergo a phase transition to a crystalline state, no fully backreacted description of a holographic crystal has as yet been brought totally under control. It is a bit of an irony that the first subject discussed in an introductory solid-state physics course (the crystal) ends up being treated last in this holography book.

1.5 Holography, condensed matter physics and quantum information

The core of this book is a special exhibit dedicated to the physics of forms of condensed matter, where all its show pieces have been fabricated by the holographic duality. But can this exhibit be related to the condensed matter realised under physical circumstances and observed in condensed matter laboratories? During the last thirty years the experimentalists have produced one serendipitous discovery after another, made possible by the spectacular progress in measurement techniques used to study the electron systems realised in high-T_c superconductors, heavy-fermion systems and so forth. Time after time the theorists were taken by surprise by the new facts. To understand what is going on in these quantum many-body worlds one needs the power of mathematics, and up to now the available equations just do not seem to work. Could it be that holography will eventually yield the system of equations that will crack puzzles like the empirical strange-metal behaviour, the origin of superconductivity at a high temperature, the principles behind the heaviness of the quasiparticles in the heavy-fermion systems and other such mysteries?

Answering this question is the grand challenge faced by AdS/CMT. The excitement in this regard is driven by the suggestive resemblances between the novel physical properties exhibited by holographic matter and the enigmatic experimental results. Intriguingly, in most cases these resemblances were recognised *after* the holographic computations had been completed. But the resemblance is still much less precise than one would like to see: it cannot be excluded at all that it is merely coincidental.

1.5.1 Is holography relevant for empirical condensed matter?

To help readers to form their own thoughts on these matters we have included in chapter 3 some background material pertaining to the physical reality of the exotic electron systems. The connections with the main subject of the book are loose at best, if at all existent. However, there is a bare minimum of facts and ideas one has to know when one wants to apply holography to this part of the real world. One has to be familiar with the special "UV physics" in these electron systems that is widely believed to establish the conditions for their anomalous IR. This so-called "Mottness" is entirely different from the strongly coupled conformal fields determining the UV in holography, and we are at present completely in the dark regarding whether this matters or not.

An important theme is the origin of superconductivity at a "high" temperature. One has to be aware that a consensus has emerged that this is no longer a problem of principle. It is now understood that large-energy-scale bare electronic repulsive interactions can be responsible for the pairing. To quite a high degree this consensus is rooted in a long-lasting effort that culminated in ingenious numerical methods. These seem to overcome the sign problem at least to such an extent that their often surprising results are increasingly trusted. They offer information complementary to experiment, and this can be of quite some interest to the holographist: a case in point is the locality of the quantum dynamics revealed by the dynamical mean-field-theory methods, which is strikingly similar to the locality at work in the holographic strange metals as outlined in section 1.4.2.

We end the discussion in chapter 3 with a summary of what matters most in this context. This is what experiment tells us about strongly interacting quantum systems. This research culminated in quite a large number of "anomalies", which seem to defy conventional understanding. The holographic community has been mainly preoccupied with explanations concerning the strange metal realised in the best high-T_c superconductors. The strange metal is perhaps the "mother of all anomalies", but there are many other strange behaviours that need to be explained. We will present a list of the most obvious anomalies which have been collected over the course of time in the best-documented systems with unconventional superconductivity and strange-metal behaviour, namely the heavy-fermion intermetallics and the cuprate high-T_c superconductors. Any such list is tentative to a degree and likely to be incomplete, given that one is biased by the lack of a real understanding in sorting the facts. But it will probably suffice as a benchmark list: when holography has managed to deliver a profound explanation for all these items, it will be legitimate to claim that this product of string theory is a theory that actually describes physical phenomena.

1.5.2 Quantum matter: entanglement on the macroscopic scale

There is yet another reason to be interested in what AdS/CFT has to tell us about the nature of matter. In a way this supersedes the question of whether its results can be directly applied to explain condensed matter experiments. The very fundaments of physics are at stake. Matter takes a central place in nature, of course, but in recent years awareness has been growing that we have only seen the tip of the iceberg when it comes to understanding what matter actually is. This awareness is to a large degree inspired by the advent of the mathematics of quantum information. In this language it becomes manifest that the forms of matter discussed in textbooks are almost without exception of a special kind: these are "short-range entangled tensor-product states". This is just a precise way of saying that this stuff behaves entirely according to the principles of classical physics. But this need not be the case: there should be a much broader class of forms of matter where the effects of entanglement are manifest on the macroscopic scale, and with entirely different physical properties. This is called "*quantum matter*".

This theme of quantum matter has come increasingly to the foreground in modern condensed matter physics. Even in the context of conventional condensed matter physics, the language of quantum information offers a generalised view that is particularly helpful when dealing with holographic matter. Chapter 2 is intended as a reference frame for this viewpoint on condensed matter physics, with the various subjects selected to aid the comparison with holographic results later in the book. It is written with emphasis on this quantum-matter theme, and the discussion therefore deviates markedly from those in the classic condensed matter textbooks. We first discuss the relatively well-understood quantum critical states realised in bosonic systems, which are closely related to the holographic results at zero density. We then turn to the generalities of quantum matter, highlighting the incompressible "topologically ordered" states that are at present eliciting great interest in condensed matter physics.

Of particular significance is the realisation that the Fermi liquid is a genuine long-range entangled compressible state of matter, albeit a very simple one. We recommend that the reader should absorb this material in detail since it is quite helpful in appreciating the holographic strange metals. Although these are non-Fermi liquids, the Fermi liquid acts in a number of regards as a helpful metaphor to get a sharper focus on the strange metals. This is even more true for the classic problem of a $(2 + 1)$-dimensional Fermi gas coupled to a quantum critical boson, where recently much progress has been made, as we discuss at the end of chapter 2.

Similarly, the BCS superconductor has also some surprising traits when viewed from a quantum information perspective. The key element of the BCS theory is the Cooper mechanism insisting that exclusively electron pairs form, which is a deeply fermionic affair. A generalised form of the Cooper mechanism appears to be at work in holographic superconductivity, and in chapter 2 we will lay the groundwork in this context, as a preliminary for chapter 10. However, there is yet another side to the BCS superconductor that is not found in the standard textbooks. The BCS superconductor ought to be seen as a seminal, paradigm-setting affair in the context of long-range entangled states of matter that can be formed in bosonic systems. We will sketch the arguments in chapter 3, since our discussion departs from the resonating-valence-bond ideas of Philip Anderson and evolved in the realisation that the BCS state can be viewed as a spin liquid with a topological order enumerated by Ising gauge theory. The message is that the BCS state is not at all uniquely tied to the Fermi liquid normal state.

From the perspective of quantum information, all these considerations point to the existence of genuine long-range entangled states as a class. The true puzzle is the nature of such finite-density quantum matter in the absence of the simplifying condition of a topological gap, or beyond the adiabatically free Fermi liquid. The strange metals as predicted by holography might just be of this kind. We will conjecture that the *holographic states of matter characterised by an infrared scaling in the physics represent densely entangled compressible states of quantum matter.* The idea that these emergent quantum critical states predicted by holography represent a general class of long-range entangled quantum matter is speculative, and far from generally accepted. Nevertheless, there are tantalising indications that

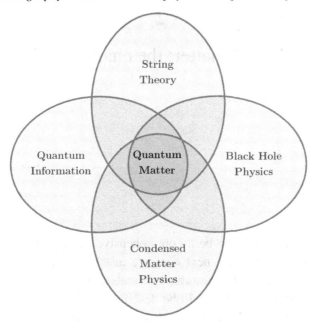

Figure 1.2 Four of the fundamental areas of physics all overlap on the question of long-range entangled quantum matter.

holography is able to do this. The most striking revelation of holography is that its strange metals constitute quantum critical *phases*. The formation of such quantum critical phases is just impossible in bosonic systems, since the principles of statistical physics insist that scale invariance can only emerge by fine tuning to isolated points in coupling-constant space in higher-dimensional systems. These holographic quantum critical phases do require a finite density, and at finite density the fermion signs sever the connection with statistical physics, while the simple Fermi liquid already demonstrates that the fermion signs are a powerful guardian of long-range entanglement. This conjecture has profound implications for both sides of the correspondence. Most prominently it exhibits how the question of quantum matter lies at the heart of not one, but four different fundamental areas of physics (Fig. 1.2). It therefore deserves a place in this book, and in chapter 14 we will discuss these ideas in more detail.

2

Condensed matter: the charted territory

This book does not pretend to be a comprehensive treatise of condensed matter physics. This chapter (and the next one) are just intended to lay down a common ground for physicists of various backgrounds relating to the condensed matter aspects of holography. The style will be descriptive, if not sketchy: there will be no boxes with detailed computations and so forth and instead we intend to guide the reader into the condensed matter literature. This chapter is on solid ground, dealing with those aspects of the physics that are at present quite well understood, while being at the same time of relevance to the holographic constructions in the later chapters. In the next chapter we will delve into the unknown where the challenges for holography reside. We do urge also the condensed matter expert to have a look, since the relations with the AdS/CMT results of chapters 6–13 often involve a rather modern understanding of the condensed matter canon, which goes beyond the standard textbooks.

We start with a subject that is regarded as advanced in condensed matter physics, but where the connections with holography are most obvious and unproblematic: the zero-density "bosonic" quantum field theories describing the physics of quantised order-parameter fields (section 2.1). In this long section we will first introduce the repertoire of such theories, emphasising the physical context where they arise. Mainly for synchronisation purposes we will then briefly discuss the popular (in condensed matter) Abelian–Higgs or "(particle–)vortex" duality in $2 + 1$ dimensions. We perceive this as an instructive metaphor for the condensed matter physicist to appreciate the weak–strong duality property of the holographic duality. It should also be of interest for the high-energy physicist to learn about the context where such dualities are in the foreground in condensed matter. We then turn to a theme that is at the core of AdS/CMT: the strongly interacting conformal quantum field theories that arise as the effective description of the physics at continuous zero-temperature quantum phase transitions. The quantum critical state realised right at the phase transition is the theatre of the "un-particle physics" which is ruled by the principle of scale (or even conformal) invariance of the quantum dynamics which is also the grand motive underlying much of the holographic description of matter. This imprints itself on the classical fluid realised in the finite-temperature quantum critical state, characterised by an extreme ("Planckian") dissipative power. This underlies, among other things, the "minimal viscosity" property, as first predicted by AdS/CFT and triumphantly confirmed by measurements on the quark–gluon plasma (QGP) created in heavy-ion collisions. The first contact between condensed matter and string theory was established in this context [24], and it is a prominent theme throughout this book (especially chapters 6 and 12).

We then turn to a very modern subject: "quantum matter", referring to the notion that there exist forms of matter that are in the grip of highly collective quantum entanglements that imprint on the nature of this stuff on the macroscopic scale (section 2.2). Although we will leave it rather implicit in the body of this text, we do believe that the strange matter predicted by holography is of this kind. In recent years a quite deep understanding of how this works in *incompressible* systems (those characterised by an energy gap) has been achieved in the form of the "topological order" of fractional quantum Hall states and so forth. Although there is not much direct contact between this theme of "topological insulators" and holography, we find these notions quite inspirational when dealing with the mysteries of the finite-density *compressible* matter as predicted by holography. In the outlook chapter at the very end of this book (chapter 14) we will return to this theme, with some speculations regarding the role which might be played by this "material spooky action at a distance".

The remainder of this chapter deals with the conventional understanding of "matter at a finite fermion density". Its central pillar is the Fermi liquid. In section 2.3 we will discuss this from a perspective that is quite different from what is generally found in textbooks. For a simple reason (anti-symmetrisation) the Fermi gas can be regarded as a genuine long-range entangled compressible form of quantum matter. Although there is nothing wrong with the textbook version of Fermi-liquid theory, in this perspective it becomes evident why the Fermi liquid is so different from the genuinely classical states of matter. Although the focus in the later chapters will be on the "non-Fermi-liquid" strange metals predicted by holography, on a number of occasions the Fermi liquid will turn out to be a source of informative metaphors for the interpretation of this strange matter (e.g. the zero sound and the hyperscaling violation of chapter 8).

The other pillar of the condensed matter canon is the BCS theory of superconductivity, which in turn rests on the "Hartree–Fock" mean-field theory for fermions. Having arrived at holographic superconductivity (chapter 10), it will turn out that there is a quite useful commonality between the BCS mechanism of superconductivity, which departs from the Fermi liquid, and the holographic mechanism, which in turn arises as an instability of the strange metals. This similarity becomes manifest in the language of time-dependent field theory ("RPA") that we discuss in this guise in section 2.4.

We finish this chapter with the "merger" of the bosonic quantum critical state of section 2.1 and the fermion physics of sections 2.3 and 2.4: the celebrated "Hertz–Millis" theory of the quantum critical metal (section 2.5). This is quite a popular affair in condensed matter physics, while it is of direct relevance to some of the real-world quantum critical metals. It departs from the assertion that the electrons in the first instance form a tranquil Fermi liquid. However, this is coupled to a bosonic order-parameter that turns quantum critical. By itself, this is quite a rich problem, which has been studied intensely over the years. Very recently the understanding has leaped forward: it now appears that, when certain conditions are satisfied ($2 + 1$ dimensions, particular order-parameter symmetries), a metallic state is realised right at the critical point which is still characterised by a Fermi surface, while the Fermi-liquid quasiparticles have turned into truly critical fermionic fluctuations. Perhaps the only serious caveat is that there is now good evidence for pairing and superconductivity taking over near the critical point.

2.1 Quantum field theory in condensed matter: the fluctuating order of bosons

The traditional point of contact between condensed matter physics and the quantum field theory developed in the high-energy tradition is rooted in the quantum dynamics of order parameters. This is often called the "Ginzburg–Landau–Wilson paradigm". This departs from the great insight by Landau that the collective behaviour of a system of an infinite number of interacting microscopic degrees of freedom can be captured by the order parameter: quite a simple object, exerting absolute power in the macroscopic realm. At finite temperature its behaviour is governed by the probabilistic rules of Boltzmann's statistical physics and it describes the phases of classical matter, including the high-temperature disordered and low-temperature ordered phases of matter. "Ginzburg–Landau" refers to the stable phases of matter which are well described in terms of a mean field, and the adjective "Wilson" is added, referring to the special nature of the critical state realised at continuous phase transitions where the beauty of the Wilsonian renormalisation group comes alive. This can be promoted to the description of the zero-temperature quantum realms using the Euclidean path-integral mapping: the quantum problem turns into an equivalent statistical-physics problem in a space-time with Euclidean signature. The temperature of the classical problem turns into the coupling constant of the quantum problem controlling the severity of the quantum fluctuations of the order parameter. Temperature in the quantum problem is in turn governed by the radius of the imaginary-time "circle", which is shrinking inversely proportionally to the physical temperature. This thermal field theory is limited to equilibrium conditions, but one can still compute the linear-response observables by Wick rotation of the Euclidean correlation functions into the real-time propagators of the quantum theory.

This grand scheme underlies much of the understanding of quantum field theory, both in high-energy physics and in condensed matter physics: statistical physics is a powerful tool in the hands of the theorist, and this branch of physics is fundamentally understood. There are still loose ends, but they are mainly of a quantitative nature. An example that is of relevance in the context of this book is the difficulty of addressing the hydrodynamical regime, as defined by the measurement frequency being small compared with the temperature. The Wick rotation becomes here very hazardous, where even state-of-the-art quantum Monte Carlo computations on simple systems fall short in terms of delivering results. However, it is relatively straightforward to deduce the qualitative nature of the physics in this regime on the basis of simple scaling arguments. We will give here a sketch, while the authoritative book by Sachdev [25] is mandatory material for anybody who wants to dig deeper into the way in which this comes alive at the quantum phase transitions of condensed matter physics (see also [26]).

2.1.1 The repertoire of bosonic field theories

The central pillar of the Landau paradigm is the notion of *spontaneous symmetry breaking*. The system as a whole acquires a lower symmetry than the one governing the parts, as a conspiracy that works only in the thermodynamical limit. The profundity lies in its consequence that the system as a whole automatically acquires a rigidity, a "capacity to push back", which is non-existent on the microscopic scale. A vivid daily-life example is the breaking of translations and rotations of Galilean space, which goes hand in hand with the capacity of such "crystalline" matter to exhibit a reactive response towards an external shear stress.

Having the general principle in hand, one can easily extrapolate to corners of reality that are way beyond direct human observation. The case in point is the Higgs mechanism, with the Higgs condensate being responsible for giving mass to the elementary particles. The story is well known: consider the Ginzburg–Landau equations describing the free-energy change caused by the breaking of the global $U(1)$ symmetry associated with the conservation of the number of electrons in a superconductor, as encoded in the complex scalar order parameter Φ,

$$\Delta F = \int d^d x \left[|(\vec{\nabla} - ie\vec{A})\Phi|^2 + r \left(\frac{T - T_c}{T_c} \right) |\Phi|^2 + u|\Phi|^4 + (\vec{\nabla} \times \vec{A})^2 \right],$$

$$(2.1)$$

coupled to the (spatial) $U(1)$ gauge field $\vec{A}$ describing electromagnetism.

This is easily lifted to the status of a relativistic quantum field theory, by embedding this in the $(d + 1)$-dimensional Euclidean space-time of thermal field theory. The extra dimension corresponds to the imaginary time, forming a circle with radius $R_\tau = \hbar/(k_B T)$. The quantum partition sum is

$$\mathcal{Z} = \text{Tr}\left[e^{-\beta H}\right] = \int \mathcal{D}\Phi \ \mathcal{D}\Phi^* \ \mathcal{D}A_\mu \ e^{-S/\hbar},$$

$$(2.2)$$

$$S = \int_0^{R_\tau} d\tau \int d^d x \left[|(\partial_\mu - ieA_\mu)\Phi|^2 + m^2|\Phi|^2 + u|\Phi|^4 + F_{\mu\nu}F^{\mu\nu} \right]$$

in an obvious relativistic notation, expressing that the dynamics is isotropic in Euclidean space-time. This is the Abelian–Higgs field theory written down by Peter Higgs in 1964.

On the one hand, this vividly illustrates an outlook that is common to string theorists and condensed matter physicists: the Standard Model of particle physics is an effective field theory associated with strong emergence. Turning it around, it also means that the principles governing the emergence have a life of their own, rather independently of the detailed nature of the microscopic constituents. Higgs and Ginzburg–Landau are basically the same theory, although electrons in solids are very different from Planck-scale physics. The string theorists call this appropriately "UV-independence" and this notion lies at the heart of the AdS/CMT pursuit. The microscopic theories that are processed by AdS/CFT are very different from the mundane "chemistry" of electrons in oxides and so forth, but the AdS/CFT correspondence acts as a "generating functional" of phenomenological theories governed by (one hopes) general emergence laws.

Under which circumstances is the quantum Abelian–Higgs system realised in condensed matter systems? The canonical example is the Bose–Hubbard system. One imagines a system of bosons living on a periodic lattice formed from potential wells ($b_i^\dagger$ creates a boson at site i), which can delocalise by tunnelling between nearest-neighbour sites with a "hopping" t. In addition, these bosons are subjected to a tunable local repulsive interaction U (for $U > 0$) and a chemical potential μ,

$$H = -t \sum_{\langle ij \rangle} \left(b_i^\dagger b_j + b_j^\dagger b_i \right) + U \sum_i n_i(n_i - 1) + \mu \sum_i n_i \qquad (2.3)$$

with $n_i = b_i^\dagger b_i$.

This system can be quite literally engineered by the cold-atom experimentalists, by loading bosonic atoms in an optical lattice and using the Feshbach resonance to tune the repulsion U [27]. Imposing a chemical potential such that there is on average an integer number of bosons per site, one immediately infers that a zero-temperature *quantum* phase transition has to occur at a critical value of U/t. For small U the system will form a superfluid at $T = 0$, but when U becomes too large a Bose Mott insulator will form: just a traffic jam of bosons since any hop departing from a background with an integer number of bosons per site will cost an additional energy $\sim U$. Close to this continuous (second-order) phase transition the correlation time and length scales become long compared with the UV scales, and here the relativistic continuum field-theory description becomes appropriate in the IR. Technically, one rewrites the lattice system Eq. (2.3) in phase-number representation (using coherent-state path-integral techniques), and after naive coarse graining one obtains the neutral version ($e = 0$, no gauge fields) of the Abelian–Higgs theory Eq. (2.2), where Φ encodes for the quantum fluctuating superfluid order parameter, while the role of the velocity of light is taken by the zero sound velocity of the superfluid [25].

This simple system is in particular very interesting in $2 + 1$ dimensions, where it has the status of the "fruit fly of quantum criticality". In the first place, close to the quantum phase transition where the field-theoretical description becomes literal, the theory is characterised in this special dimension by a famous weak–strong (Kramers–Wannier) "Abelian–Higgs duality". This is at the same time also a global–local duality: the Mott insulator turns out to be equally well described by the gauged Higgs theory, although this "dual superconductor" is formed from the topological excitations (vortices) of the superfluid (see the next subsection). The field theory becomes truly alive right at the quantum phase transition itself, where it is instrumental to describe the quantum critical state. Again the dimensionality is crucial: in order to share general properties with the conformal field theories of the correspondence (see section 7.3), it has to be "strongly interacting": the dimensionality has to be below the upper critical space-time dimension [25], which is $3 + 1$ for the simple $U(1)$ symmetry.

One can now further generalise such field theories. A first step is to enlarge the complex scalar order parameter to an N-component vector, with components Φ^a. For the neutral case

$$S = \int d^d x \ d\tau \Big[\text{Tr}[(\partial_\mu \Phi^a)^2] + m^2 \ \text{Tr}[(\Phi^a)^2] + u \ \text{Tr}[(\Phi^a)^2]\text{Tr}[(\Phi^a)^2] \Big]. \quad (2.4)$$

For $N = 3$ this is the "soft spin" version of the $O(3)$ non-linear sigma model which describes the quantum phase transition of (unfrustrated) Heisenberg antiferromagnets. One can control the critical theory by considering the $N \to \infty$ limit, which is complementary to the large-d limit, in order subsequently to study $1/N$ corrections. These "large-vector-N" theories have quite some history in condensed matter physics [25], but the drawback is that in the large-N limit one finds a free (Gaussian) theory, and it is hard work to derive from this the properties of the strongly interacting critical state.

In holography "large-N" is of crucial importance, but this refers to a quite different type of field theory, which is quite unfamiliar in the condensed matter context: it is a *matrix* large-N theory. The "order parameter" is now an $N \times N$ matrix with entries Φ^{ab} and the (neutral) theory is typically written as

$$S = \int d^d x \; d\tau \left[\text{Tr}[(\partial_\mu \Phi^{ab})^2] + m^2 \; \text{Tr}[(\Phi^{ab})^2] + u \; \text{Tr}[(\Phi^{ab})^2]\text{Tr}[(\Phi^{ab})^2] \right].$$
(2.5)

AdS/CFT is rooted in the physics of such quantum field theories, in the case that they are very strongly coupled while N is large, as discussed at length in chapter 4. Typical examples of physical relevance are Yang–Mills theories, where the entries of the matrices code for gluons that are exchanging colours between the quarks. Differently from the vector variety, in the *matrix* large-N limit the theories can continue to be strongly interacting, while due to the "non-renormalisation theorems" some maximally supersymmetric versions have an appetite to form quantum critical phases automatically, without fine tuning to isolated points in coupling-constant space. This is the unfamiliar microscopic stuff (for the condensed matter physicist) found at the short-distance cut-off of the field-theory physics which is described literally by the correspondence.

To complete this overview of "simple" field theories of relevance to condensed matter, one can also identify incarnations of field theories containing fermions. Well-behaved cases depart from Dirac fermions at zero density, since this evades the fermion-sign troubles. An appropriate condensed matter metaphor is to consider a theorist's version of graphene where the electron–electron interactions can be made larger than in the real system. At some critical coupling this electron system will become unstable with respect to a symmetry-breaking transition that will gap the Dirac spectrum. Near the continuous quantum phase transition an effective field theory of the form

$$S = \int d^d x \; d\tau \left[\bar{\psi}(\gamma^\mu \; \partial_\mu)\psi + \frac{1}{2} \; \partial^\mu \phi \; \partial_\mu \phi + V(\phi) - g\phi\bar{\psi}\psi \right]$$
(2.6)

will be realised, where ψ is a massless Dirac fermion and ϕ is a bosonic field that goes critical, coupled by the Yukawa coupling g. It can now be demonstrated that the Yukawa coupling between the fermions and the critical boson is finite at the quantum phase transition [28]. This has the ramification that the fermions are also "pulled on the critical surface", turning them into "un-particle" fermionic critical excitations of the kind found in zero-density AdS/CFT constructions, as discussed in chapter 9. Notice that the anomalous dimensions have been computed only to one-loop order in the ϵ expansion (i.e. for $d = 3 - \epsilon$ spatial dimensions, because $d + 1 = 4$ is the upper critical space-time dimension) [28].

This is the main repertoire of basic quantum field theories of relevance to condensed matter physics that are "sign-free", mapping onto a Boltzmannian statistical-physics problem in Euclidean space-time. Away from their quantum critical points, these theories are very well understood. Right at the quantum phase transition and below the upper critical dimension one is dealing with the strongly interacting critical state governed by emergent conformal theory. Although much is known regarding "the critical state", it is far from having been completely enumerated mathematically. The exceptions are the $(1 + 1)$-dimensional CFTs which are famously integrable, but even in these cases there are still big gaps [29, 30, 31]. For instance, theories with a central charge $c > 1$ are not brought under control, and this is in the holography context quite a bottleneck, since the relevant $(1 + 1)$-dimensional CFTs have $c \propto N^2$ with N large. Lacking a truly mathematical enumeration, there are still many deep questions of principle in higher-dimensional CFTs. A typical example is the generalisation of Zamolodchikov's c-theorem [32] for the $(1 + 1)$-dimensional theories, proving the "one-way renormalization flow" from the UV to the IR, to higher dimensions, which is at present the subject of an intensive effort [33, 34, 35]. But there are also still problems of a more pragmatic kind related to the properties of the fluid formed in the quantum critical regime at finite temperature, a problem that we will discuss in section 2.1.3.

2.1.2 Abelian–Higgs duality in $2 + 1$ dimensions

Holographic duality is, among other things, also a weak–strong duality that relates gauge theory in the bulk to a theory controlled by global symmetry in the boundary. Although a lot more is going on, it shares these features with a duality that is often used for various purposes in modern condensed matter physics: the "particle–vortex" (or "vortex") duality in $2 + 1$ dimensions which is known in the high-energy literature as the Abelian–Higgs duality (for a pedestrian introduction see Ref. [36]). We find this just a useful metaphor, offering some common language bridging the culture gap between the string-theory and condensed matter communities. The reader familiar with the theme can safely skip this subsection.

Dualities are quite ubiquitous in quantum field theories and statistical physics. They are very powerful, being at the heart of non-perturbative field-theoretic physics. The simplest of all dualities, as known by all physicists, is the particle–wave duality of quantum mechanics. This already captures the essence of the word "dual": the particle (position) and wave (momentum) descriptions are in a way exact opposites (the Fourier transformation), but they form a "harmonious whole". A full knowledge of quantum mechanics requires an understanding of both sides, as well as the nature of their "oppositeness". This quantum-mechanics wisdom was generalised to field theory *"avant la lettre"* by Kramers and Wannier's demonstration in the 1940s of the self-duality property of the two-dimensional Ising system (in quantum incarnation: the (1+1)-dimensional transverse-field Ising model) [25]. Depart from low temperature and (nearly) all the Ising spins point in the same direction, forming an ordered ferromagnet. The unique agents of disorder in this state are its topological excitations: the domain walls. Upon increasing the temperature these occur in closed loops, which grow gradually in size until at the transition these loops "unbind" into free domain walls. Since one free domain wall suffices to destroy the long-range order everywhere, this signals the transition to the high-temperature disordered state. But this disordered state is actually a condensate of the domain walls! This is in turn precisely described by an Ising model in terms of dual Ising variables where just the coupling constant $J/(k_B T)$ is inverted: the two-dimensional Ising model is characterised by the rare property of self-duality.

Increasing the complexity by one step, weak–strong duality becomes less powerful but more interesting. Consider the Bose–Hubbard system at "zero chemical potential" in $2+1$ dimensions (Eq. (2.3)), which is equivalent to classical *XY* spins in 3D. Although rigorous mathematical proof is still lacking, there is abundant evidence for the correctness of the conceptual picture which we will now sketch. Let us start deep in the superfluid phase, realised at a small coupling constant. Writing the order-parameter field in amplitude/phase representation as $\Phi = |\Phi| \exp(-i\phi)$, the amplitude is frozen out and the effective action acquires the Josephson form; in shorthand,

$$S \sim g \int d^d x \ d\tau \ (\partial_\mu \phi)^2, \quad \text{mod}(2\pi). \tag{2.7}$$

The superfluid sound velocity $c_s = 1$ plays here the role of the velocity of light. Importantly, the phase field ϕ is compact with periodicity 2π ("mod(2π)"). The intriguing property of this simple problem is that its dual is coincident with the Coulomb phase of non-compact quantum electrodynamics in $2+1$ dimensions – this is special to this particular dimension (e.g. see [37]). This is easy to demonstrate. Arbitrary field configurations can be separated into smooth ("Goldstone") and multivalued ("vorticity") contributions as $\phi(\vec{r}) = \phi_{sm}(\vec{r}) + \phi_{MV}(\vec{r})$ ($\vec{r}$ refers to Euclidean space-time). One now Legendre transforms Eq. (2.7), which can be done conveniently by using Hubbard–Stratonovich auxiliary fields $J_\mu(\vec{r})$. The result is

$$S_{\text{dual}} \sim \int d^d x \ d\tau \left(\frac{1}{g} J_\mu^2 + i J^\mu \ \partial_\mu (\phi_{sm} + \phi_{MV}) \right). \tag{2.8}$$

One infers immediately that after integrating the *J*s one recovers Eq. (2.7). Notice that in this step already the coupling constant *g* is inverted, exchanging the sense of weak and strong coupling.

The smooth part of ϕ is integrable and one can "pull it through" the derivative to obtain $J^\mu \, \partial_\mu \phi_{sm} \rightarrow \phi_{sm} \, \partial_\mu J^\mu$. The smooth phase configurations can now be integrated out, acting as a Lagrange multiplier imposing the conservation law $\partial_\mu J^\mu = 0$. The phase variables have been transformed into the "momentum-like" super-currents J_μ: this is just the continuity equation governing the conserved superflow, with an action $\sim J_\mu^2$. The conservation law can be imposed in $2 + 1$ dimensions by parametrising the current in terms of a non-compact $U(1)$ gauge field

$$J_\mu = \varepsilon_{\mu\nu\lambda} \, \partial^\nu A^\lambda, \tag{2.9}$$

with the effect that $J_\mu^2 = F_{\mu\nu} F^{\mu\nu}$, the Maxwell action: in $2 + 1$ dimensions the capacity to propagate forces of the superfluid vacuum is identical to that of the (non-compact) QED vacuum! But what are the sources? We still have the multivalued parts of the phase field, and upon inserting Eq. (2.9) $i J^\mu \, \partial_\mu \phi_{MV} \rightarrow i A_\mu J_\mu^V$, where $J_\lambda^V = \varepsilon_{\mu\nu\lambda} \, \partial_\mu \partial_\nu \phi_{MV}$. Using Stokes' theorem, one immediately recognises that J_μ^V represents the vortex currents. On collecting the various pieces one obtains the dual action

$$S_{dual} \sim \int d^2x \, d\tau \left(\frac{1}{g} F_{\mu\nu} F^{\mu\nu} + i A^\mu J_\mu^V \right). \tag{2.10}$$

The vortices in the superfluid have a long-range interaction in $2 + 1$ dimensions, being *identical* to those between electrically charged particles.

This is all very precise deep in the superfluid. But what happens when the coupling constant (the U/t ratio of the lattice model) increases such that the system approaches the 3D *XY* universality-class quantum phase transition to the Bose Mott insulator? This cannot be handled with rigour, but the following qualitative considerations are surely accurate. Initially closed loops in space-time of vortex–antivortex pairs will "seed in the vacuum", but since these are bound in pairs they will not destroy the superfluid order. These loops will increase in size getting closer to the transition, to "blow out" right at the quantum phase transition where the (anti)vortices unbind, becoming as large as the system itself. Moving into the Mott-insulator regime, the proliferated (anti)vortices form a system of delocalised, relativistic bosons subjected to the long-range gauge interaction. This is literally described by the Abelian–Higgs action Eq. (2.2), with the caveat that the order parameter $\Phi \rightarrow \Phi_V$ describes now the relativistic vortex condensate. Henceforth, the quantum disordered superfluid (Mott insulator) is at the same time an *ordered* relativistic superconductor: the literal Abelian–Higgs phase. The Mott gap corresponds to the Higgs mass of the dual and it is easy to demonstrate [37] that the "massive photons" of the dual superconductor are coincident with the "holons" and "doublons" of the lattice Bose–Hubbard model Eq. (2.3).

This Mott insulator has as its defining property that charge is locally quantised. Upon going to the large-U limit, $b_i^\dagger b_i |\Psi_0\rangle = n_i |\Psi_0\rangle$, where $|\Psi_0\rangle$ is the (trivial) vacuum of the strongly coupled Bose Mott insulator and n_i is the precisely quantised number of bosons per site. This emergent local charge quantisation can be imposed by the famous "stay at home" compact $U(1)$ gauge fields ϕ_i which are at the heart of the slave theories which will be briefly discussed in section 3.2: $b_i^\dagger \to e^{i\phi_i} b_i^\dagger$ leaving $b_i^\dagger b_i$ invariant. As explained in Ref. [37], this is just the "remnant" of the global $U(1)$ phase which is broken in the superfluid ($b_i^\dagger \to \langle |b_i^\dagger| \rangle e^{i\phi}$), which acquires an emergent gauge symmetry because it is "stirred" by the vortex condensate, forming a coherent quantum superposition of "disordering" events in position space. The message is that this universal aspect of Mottness does not need a lattice. What matters is that number (charge) gets locally quantised, and this is just precisely encoded in the continuum field theory where this quantisation becomes precise at distances larger than the correlation length associated with the dual-superconductor fixed point. This is an observation to keep in mind when one is dealing with the potential capacity of the far more complicated continuum field theories associated with the AdS/CFT correspondence to encode also some form of Mottness, as argued by Phillips and co-workers [38].

The conclusion is that the strongly coupled (large U/t, Mott insulator), severely fluctuating state associated with the weakly coupled superfluid is just the same as the tranquil, ordered Higgs phase of the gauged superconductor. Of course, it also works the other way around: the strongly coupled state associated with the superconductor is the deconfining Coulomb phase of the Higgs model. As we will discuss at length in chapters 4 and 5, the same pattern, albeit in a grossly generalised form, lies at the foundations of AdS/CFT. The bulk is like the tranquil physics of the orderly dual superconductor, which is, however, no longer described by classical gauge and phase fields, but instead by general relativity in combination with classical "material" fields (Maxwell, Higgs, ...) all governed by gauge symmetries. The dual lives at the boundary and is in a sense as strongly coupled as can be imagined, while the quantities that can be computed are in the final analysis controlled by global symmetry.

2.1.3 The bosonic quantum critical state

Much of the interest in using the AdS/CFT correspondence in condensed matter physics is rooted in the popular and modern subject of quantum criticality. This started around 25 years ago, inspired by experimental discoveries. Here we will entirely focus on the relatively well-chartered territory of quantum phase transitions in bosonic systems. This theme was launched in earnest in the early days of high-T_c superconductivity through the seminal work by Chakravarty, Nelson and

Halperin [39]. Motivated by the relatively quantal physics of the antiferromagnet realised in insulating cuprates, they studied the physics of the $O(3)$ quantum non-linear sigma model, the $O(3)$ incarnation of Eq. (2.4). In Sachdev's hands this has grown into a vast subject that deserves a book by itself [25, 26], and in this section we will just outline some of the main ideas which are of particular relevance in the AdS/CMT context.

The reader is assumed to be familiar with thermal (or "Euclidean") field theory. For the purpose of studying the quantum field-theoretical problem in equilibrium, both at zero temperature and at finite temperature, one analytically continues real time (Minkowski signature) to imaginary time (Euclidean signature). The imaginary-time direction forms a circle with a compactification radius set by the inverse temperature,

$$R_\tau = \frac{\hbar}{k_B T}. \tag{2.11}$$

We emphasise this simple equation since it will play a remarkable role in the "magic" of the strongly interacting quantum critical state.

The "bosonic" quantum critical state is operationally defined by the prescription that the field theory is just coincident with some Boltzmann probabilistic statistical-physics problem in this Euclidean space-time. This implies that one can mobilise the powerful machinery describing the thermal critical state as developed in the 1970s revolving around the Wilsonian renormalisation group. The "dictionary" translating this to the quantum case is simple and elegant. The zero-temperature coupling constant of the quantum problem is equivalent to temperature in the classical problem, while temperature enters the quantum realms through Eq. (2.11). The Euclidean quantum propagators are just like the correlation functions of the classical problem, but for them to become meaningful as linear response functions one has to Wick rotate back to real time. We shall see that this operation is largely responsible for the gross shift in physical interpretation compared with the classical problem, while it is also the origin of technical difficulties.

One has to pay special attention to identifying the effective statistical-physics problem in Euclidean space-time that codes for the quantum field theory. The simplest fluctuating order problems just turn into simple relativistic field theories like the vector theories Eq. (2.4). These are characterised by an emergent Lorentz invariance (isotropy of Euclidean space-time) that is quite natural when one is dealing with non-conserved order parameters. However, since the Lorentz invariance is emergent, it needs protection, and this fails generically in the presence of other "heat-bath" degrees of freedom. This is typically captured at the quantum critical point by the dynamical critical exponent $\mathbf{z}$ relating space and time, or momentum and frequency through $\omega \sim k^{\mathbf{z}}$. Lorentz invariance implies $\mathbf{z} = 1$; for a non-conserved order parameter one expects simple diffusion behaviour, $z = 2$ (also applying to non-relativistic particles at a low density); and $z = 3$ for a conserved order parameter, as in a ferromagnet. The important ramification for the quantum theory is that the effective dimensionality of Euclidean space-time of relevance to the nature of the critical regime becomes $d + z$ (d is the number of space dimensions). We will find out that, for the quantum critical metallic *phases* as predicted by holography, z is a central quantity determining their nature, with the puzzling feature that it can become arbitrarily large (chapter 8). At the next level of sophistication one has to be aware of the Berry-phase terms which might survive the coarse graining to the semi-classical theory using the coherent-state path integrals [40]. An elementary example is the Berry phase responsible for the encoding of the conserved status of the ferromagnetic order parameter, which causes, among other things, the quadratic dispersion of the "ferro-magnon". A spectacular example is the effect of the Berry phase at the quantum phase transition from an antiferromagnet to a valence-bond solid phase found in $(2+1)$-dimensional "$J_1 - J_2$" frustrated Heisenberg spin problems. These Berry phases encode for the breaking of translations in the valence-bond solid but, since these are dangerously irrelevant, they disappear at the critical point. It turns out that this is the effect of fractionalising the critical fluctuations: they carry $S = 1/2$ spin quantum numbers instead of the UV $S = 1$ quantum numbers [41] (see also section 2.4).

After these preliminaries, let us now turn to the gross features of this "statistical-physics-style" quantum criticality. We will assume that the reader is at home in the celebrated theory of the classical critical state [26]. Given that we will encounter some gross generalisations of it later in this book, let us consider the very basics. The essence of the critical state is that the system becomes scale-invariant, and often also conformally invariant, the extra symmetry being associated with the property that angles are invariant under scale transformations. This is a very powerful symmetry principle, to the degree that the gross features of the physics of the critical state follow directly from the symmetry itself. It is not at all a symmetry of the UV of the condensed matter systems: it is in a way quite remarkable that it is "dynamically generated" in the IR. But in dimensions higher than $1+1$ dimensions (or two dimensions), this is, within the limitations of the rules of Boltzmann statistical physics, possible only at the isolated unstable fixed points associated with continuous phase transitions. A main motive in this book will be that holography is somehow insisting that in non-Boltzmannian finite-density systems scale invariance emerges in the deep IR without invoking fine tuning, describing quantum critical *phases* (see especially chapter 8).

The idea of renormalisation flow is believed to be a universal meta-principle, and the Wilsonian renormalisation group deals with how this works near the isolated critical points. This is an overly familiar story: on departing from the scale of the lattice constant right at the critical coupling, one first encounters a complicated flow where a large variety of operators encoding for the scale-full lattice physics will fade away upon descending in energy – these are the irrelevant operators. At some scale one enters the basin of attraction of the IR unstable fixed point and all remaining operators are subjected to a marginal flow: one enters the scale-invariant, critical regime. A crucial question is now whether interactions remain finite in the critical regime. In the statistical-physics systems this is governed by dimensionality: above the upper critical dimension the order-parameter self-interaction (like $u|\Phi|^4$) is irrelevant and one just ends up with massless-free-field (Gaussian) theory right at the fixed point. Upon ascending from this fixed point the physics of its critical regime is governed by the perturbative corrections coming from the irrelevant operators and is thereby non-universal.

Below the upper critical dimension the interactions continue to be finite, and this defines the "strongly interacting critical state". This is coincident with the notion of the "strongly coupled" nature of the conformal field theories treated by holographic duality, and this should not be confused with the conventional notion of "strongly interacting particles". The very notion of a "particle" does not make sense in the strongly interacting quantum critical state. The meaning of a "particle" is that some quantum numbers can be localised in some small space-time volume while subsequently the worldlines associated with this "quantum number lump" are to be summed up. In the strongly interacting critical state the scale/conformal invariance takes over in an absolute sense, destroying this sense of locality: by scale transformation the localised quantum numbers of particle physics can be spread over a macroscopic volume and the scale invariance insists that it is all the same. This is genuine "un-particle physics": a "quantum soup" is formed where any sense of "individuality" has vanished and where the physics is governed by a quantum collectivity of platonic perfection.

Exact solutions of such strongly interacting critical states are typically not available. However, much can be learned by just exploiting the powers of scale invariance: the art of scaling analysis, which is very powerful as a phenomenological framework to process experimental information when a real theoretical understanding is lacking. A first highlight in this regard is the "quantum critical wedge" found in the coupling-constant–temperature diagram (Fig. 2.1). This just rests on the assumption that one is dealing with a strongly interacting quantum critical state controlled by an isolated zero-temperature critical point, while the critical state is behaving like a classical critical state in Euclidean space-time. On deviating from the critical coupling g_c by a small amount $\delta g = |(g - g_c)/g_c|$, a relevant flow to the stable fixed points on either side of the quantum critical point will set in. A stronghold of the classical theory is the existence of a correlation length ξ governed by the correlation length exponent ν such that $\xi \sim 1/(\delta g)^{\nu}$. The meaning of

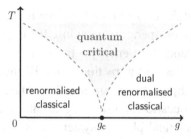

Figure 2.1 The "quantum critical wedge" in the coupling-constant (g) – temper-ature (T) plane associated with a continuous zero-temperature quantum phase transition at $g = g_{\rm c}$. In this quantum critical regime the physics is that of a finite-temperature fluid that is governed by the quantum critical "ultraviolet" even at values away from the critical coupling $g_{\rm c}$. At a temperature $k_{\rm B}T \simeq |(g - g_{\rm c})/g_{\rm c}|^{z\nu}$, with z and ν the dynamical critical and correlation-length expo-nents, respectively, there occur crossovers (dashed lines) to conventional regimes where the long-time finite-temperature physics is controlled by the physics of the stable, classical states of matter on either side of the quantum phase transition.

the correlation length is that at shorter distances one is still dealing with the criti-cal state, while at longer distances the relevant flow to the stable states takes over. This translates into an (imaginary) time scale $\tau_\xi \sim 1/(\delta g)^{z\nu}$, turning into an energy scale $\Delta_\xi = \hbar/\tau_\xi$ after Wick rotation.

At zero temperature the system will respond as if it is still right at the criti-cal coupling, when it is measured at energies larger than Δ_ξ, while below this scale it will be ruled by the physics of the stable state. The coupling constant acts as the temperature of the equivalent statistical-physics problem, but how does physical temperature enter the quantum problem? It does so through the funda-mental postulate of thermal field theory: the inverse temperature sets the radius of the Euclidean time circle and thereby enters the equivalent classical problem as a finite-size scaling. It follows immediately that at finite δg a crossover will occur at a temperature $k_{\rm B}T_\xi \simeq \Delta_\xi$, from a low-temperature regime that is governed by the finite-temperature physics of the stable state to a high-temperature regime that is indistinguishable from the finite-temperature quantum critical liquid realised pre-cisely at the quantum critical point. This explains the "quantum critical wedge" in the coupling-constant–temperature diagram illustrated in Fig. 2.1. This behaviour is often seen in experiments in the laboratory (section 3.6), and it is in practice an important diagnostic revealing the presence of a quantum phase transition.

This odd rule that temperature takes the role of finite-size scaling has quite a number of other profound implications. The thermodynamics is a basic affair that will be addressed repeatedly in the later chapters. As a reference let us sketch how the thermodynamic scaling works dealing with the conventional quantum critical points. Assuming that the system has an effective dimension $d + z < d_{\rm uc}$ (upper

critical dimension), hyperscaling will apply and the singular part of the free-energy
density ($F_s = F - F_{\text{reg}}$) will behave under a spatial scale transformation $x \to bx$ as

$$F_s(\delta g, T) = b^{-(d+z)} F_s(b^{y_{\delta g}} \delta g, b^z T), \tag{2.12}$$

where $y_{\delta g} = 1/\nu$ is the scaling dimension of the coupling constant. This is
equivalent to the following scaling forms for the free-energy density [42, 43],

$$F_s(\delta g, T) = -\rho_0 (\delta g)^{(d+z)/y_{\delta g}} \bar{f} \left(\frac{T}{T_0 (\delta g)^{z/y_{\delta g}}} \right) \tag{2.13}$$

$$= -\rho_0 \left(\frac{T}{T_0} \right)^{(d+z)/z} f \left(\frac{\delta g}{(T/T_0)^{y_{\delta g}/z}} \right), \tag{2.14}$$

where ρ_0 and T_0 are non-universal constants while $f(x)$ and $\bar{f}(x)$ are universal
scaling functions. The first form is useful in the low-temperature regime away
from the critical point: $\bar{f}(x) = \bar{f}(0) + g(x)$, where $g(x)$ describes the low-
temperature thermodynamics of the stable phases to the left and right of the
quantum critical point. The second form is useful right at the critical point: since
there is no singularity at $\delta g = 0, T > 0$, the function f can be expanded as
$f(x \to 0) = f(0) + xf'(0) + (1/2)x^2 f''(0) + \cdots$.

The equivalent classical problem in Euclidean space-time is characterised by the
thermodynamical exponents $\alpha, \beta, \gamma, \delta$. Where should one look for these dimensions in the quantum critical incarnation of the same universality class? For
example, let us focus on the specific-heat exponent α. The quantity of interest is
the susceptibility associated with the coupling constant $\chi_{\delta g}(\delta g, T) = \partial^2 F/\partial(\delta g)^2$.
Using Eq. (2.14), it is easy to show it will contain a singular contribution
$\chi_{\delta g}(\delta g, T \to 0) \sim 1/|\delta g|^{\alpha_{\delta g}}$ upon taking δg through zero, where $\alpha_{\delta g}$ is the
specific-heat exponent of the equivalent classical problem in $d + z$ dimensions.
Alternatively, one can tune δg to zero and measure the temperature dependence
of this susceptibility in the finite-temperature quantum critical state, and from Eq.
(2.14) it follows that $\chi_{\delta g}(\delta g, T \to 0) \sim T^{((d+z)\alpha_{\delta g})/(z(2-\alpha_{\delta g}))}$. This exponent shows
up in unexpected corners!

The entropy and/or specific heat is, on the other hand, governed by the response
of the equivalent classical problem to the finite-size scaling associated with the time
circle. The most revealing information is that which one obtains by measuring the
specific heat $C = -T(\partial^2 F/\partial T^2)$ right at the critical coupling. It follows immediately from Eq. (2.14) that $C \sim T^{d/z}$. For instance, for a relativistic ($z = 1$) system
this boils down to the familiar "Debye" behaviour $C \sim T^d$ which holds for any relativistic, scale-invariant theory. This simple scaling behaviour arises simply from
the fact that the free energy is subject to the engineering scaling requirement that it
scales with the volume. This requirement is actually not written in stone. Although
violations are rare in classical systems, the Fermi gas is actually already "violating

hyperscaling": it is characterised by the Sommerfeld specific heat $C \sim T$, regardless of space dimensions. The reason is that its massless excitations occur around the Fermi surface, which itself has a dimensionality equal to $\theta_{FL} = d - 1$, with the effect that the free energy scales like T^2 in all dimensions (section 2.3). A quantum critical state can therefore in principle be characterised by an additional "hyperscaling violation exponent" θ, such that $S \sim T^{(d-\theta)/z}$. We will see in chapter 8 that such an extra "hyperscaling violation" is actually ubiquitous in the conformal strange-metal states described by holographic duality.

To finish the discussion of thermodynamics, a highlight of phenomenological scaling analysis has been the confirmation in several of the heavy-fermion-type quantum critical systems [44] of the prediction that the ratio of the thermal expansion and specific heat (the Grüneisen parameter) should acquire a universal value right at the quantum critical point when the pressure takes the role of tuning the coupling constant [42]. This is a direct ramification of the type of analysis presented above, while it is a remarkably powerful diagnostic demonstrating the existence of quantum criticality and hyperscaling, dealing with these, in other regards, rather mysterious fermionic critical systems (section 3.6.1).

The power of this scaling analysis for thermodynamical purposes is tied to the existence of isolated unstable fixed points and is therefore not of relevance to the conformal *phases* predicted by the correspondence. This is quite different for the dynamical properties in the quantum critical regime which are strongly constrained by the scale invariance, both at zero temperature and at finite temperature. We are in the first instance interested in the linear response, and the quantities of interest are the dynamical susceptibilities associated with the various physical observables. Thermal field theory prescribes how to obtain those: compute the two-point correlators in the equivalent statistical-physics problem in Euclidean space-time and Wick rotate to real time. The Wick rotation is the culprit, being responsible for a dramatic shift in the physical interpretation. It is also the cause of technical difficulties, because of the infamous "loss of information" problem: smoothly varying Euclidean correlators turn into "peaky" real-time propagators, and to reconstruct the latter from the former one needs in practice closed solutions. The small amount of noise unavoidable in even the highest-quality numerical quantum Monte Carlo (QMC) simulations obscures the real-time linear response, especially at long times at finite temperature. However, one can isolate the gross features that matter most, just by relying on scaling considerations [25].

To illustrate this let us consider here the simple emergent Lorentz-invariant ($z = 1$) strongly interacting quantum critical states. Right at the critical coupling and at zero temperature the combination of Lorentz invariance and conformal invariance completely fixes the form of the dynamical susceptibilities. According to thermal field theory, these are computed by first determining the two-point correlation functions in Euclidean space-time $C_\phi(|r|) = \langle \phi(\vec{r})\phi(\vec{0})\rangle$, where ϕ is the field of interest. Euclideanised Lorentz invariance dictates that the correlation function can only depend on the invariant distance $|r| = \sqrt{c^2\tau^2 + x^2 + y^2 + \cdots}$. In turn, scale invariance now imposes that this correlation function should be a power law characterised by a conformal (or anomalous) dimension Δ_ϕ of the operator ϕ,

$$C_\phi(|r|) \sim \frac{1}{|r|^{2\Delta_\phi}}. \tag{2.15}$$

Upon Fourier transformation to Euclidean momentum, namely $|p| = \sqrt{\omega_n^2 + p_x^2 + \ldots}$, where ω_n are the Matsubara frequencies, this turns into $C_\phi(|p|) \sim |p|^{2\nu_\phi}$, where $\nu_\phi = \Delta_\phi - (d+1)/2$. All that remains to be done to obtain the real frequency-retarded propagator is to Wick rotate the Euclidean propagator to real frequency, $\omega_n \to i\omega$. The result for the dynamical susceptibility becomes

$$\chi_\phi(\vec{k}, \omega) \equiv \langle \phi(\vec{k}, \omega)\phi(0, 0)\rangle \propto \left(\sqrt{c^2k^2 - \omega^2}\right)^{2\nu_\phi}. \tag{2.16}$$

These are the "branch cut" response functions which are the fingerprints of the quantum critical state. The imaginary part of χ tells us where the excitations are (the spectral function), and one sees immediately that this χ_ϕ'' is zero for $\omega < ck$, whereas for timelike momenta $\omega > ck$ it is finite, asymptoting to a pure power law $\chi'' \propto \omega^{2\nu_\phi}$ for $\omega \gg k$. For conformal fermions this looks the same except that one has now positive- and negative-energy states bounded by the Dirac cone. The result Eq. (2.16) will be encountered many times in later chapters.

The conformal dimension ν_ϕ is bounded from below by -1 (the unitarity limit), where one recovers the free-particle spectrum $\chi \propto 1/(c^2k^2 - \omega^2)$ characterised by poles at $\omega = \pm ck$. Inspired by standard "perturbative particle physics", one can certainly rewrite Eq. (2.16) as $\chi \propto 1/(c^2k^2 - \omega^2 + \Pi_\phi(\omega, k))$ involving the self-energy Π_ϕ. However, for $\nu > -1$ this is a rather meaningless operation since the "un-particle physics" revealed by the branch-cut propagators is utterly non-perturbative in terms of the "UV" free fields ϕ: as we have already argued, the very notion of "particles" (free fields) that "scatter" (perturbative correction) has become completely meaningless in the strongly interacting quantum critical state.

By tuning away from the critical state, either by δg or by temperature, one introduces scales, and the Euclidean correlators will have the form $C(\tau) = \Phi_\phi(\tau/R_\tau, R_\tau/\xi, \ldots)$, where Φ_ϕ is a universal scaling function with arguments involving the radius of the finite-temperature time circle $R_\tau = \hbar/(k_B T)$ and the correlation length ξ. To obtain these scaling functions in closed form one needs exact solutions, which are usually not available. However, by simple reasoning one can extract the qualitative physics near the quantum critical state. What do the excitations look like, after tuning away from the critical point at zero temperature? The important energy scale is Δ_ξ associated with the correlation length at finite δg. In the regime $\omega > \Delta_\xi$ the probe field cannot discern that the system is away from the critical point and the spectrum is still of the branch-cut form Eq. (2.16). At $\omega \simeq \Delta_\xi$ this comes to an end in a smooth crossover, while at lower energies one finds the massless (e.g. phase mode of the superfluid) or massive (e.g. holons/doublons of the Mott insulator) particle excitations associated with the stable phases. These will appear as if they are bound states "pulled out" of the critical continuum, having close to the quantum critical point a small pole strength, which diminishes when $\delta g \to 0$. But in other regards they will behave like the particle excitations deep inside the stable phase, since at sufficiently long times and distances their properties are governed by adiabatic continuity.

What happens when temperature is switched on? Surely, when $k_B T \ll \Delta_\xi$ the system exhibits at low frequencies the response associated with the thermal physics of the stable states. This story, however, acquires a most interesting twist, on considering the finite-temperature physics of the quantum critical state itself ($k_B T > \Delta_\xi$). Unique to the quantum critical state, the only scale in the problem is temperature itself through the finite length of the imaginary-time axis. When hyperscaling applies, this implies directly that all dynamical quantities should be universal functions of the ratio of energy and temperature: this is the so-called "E/T scaling" which has been used with much effect in several instances dealing with experimental data, in the form of collapses of scaling exhibited by dynamical responses (see section 3.6).

However, also the absolute magnitude of a particular number matters. In Euclidean signature one has an imaginary time scale $R_\tau \simeq \hbar/(k_B T)$, but what does this mean in real time? By application of general principles it has to be that in the "hydrodynamical" regime where the measurement energy is small compared with temperature the liquid that is realised should be a classical, dissipative fluid subject to the second law of thermodynamics. Thermal field theory gets this right, and the magic is in the Wick rotation: as can be demonstrated explicitly in some special cases in the field theory [25], the real-time response functions associated with non-conserved quantities in the hydrodynamical regime acquire a *relaxational* classical behaviour controlled by a relaxation time

$$\tau_\hbar = A \frac{\hbar}{k_B T}, \tag{2.17}$$

where A is an amplitude of order unity set by the universality class of the quantum critical state. This includes energy relaxation, where $\tau_\hbar$ has the meaning of the time it takes to convert work into entropy under linear-response conditions.

This is quite remarkable, since this "Planckian dissipation" [45] implies that in the strongly interacting quantum critical state Planck's constant determines, together with temperature, the rate of the production of heat. This Planckian dissipation time is actually very short, much shorter than one finds in stable phases of matter, where this rate is strongly reduced by the stability scale Δ_ξ. For instance, in a Fermi liquid one finds that this relaxation time is of order $(E_F/(k_B T))\tau_\hbar$, where E_F is the Fermi energy.

In holography the Wick rotation is painless – one computes with the same ease in Lorentzian and Euclidean signature. The principle of Planckian dissipation was rediscovered by the string theorists in 2001 in the form of the "minimal viscosity" [46] as discussed in chapter 7. This result can be understood just in terms of the general wisdoms pertaining to the long-time limit of the finite-temperature quantum critical liquid, as we just discussed. This liquid has to behave like a classical fluid, governed by Navier–Stokes hydrodynamics, while the influence of the quantum critical "ultraviolet" enters entirely through the values of hydrodynamical parameters. The viscosity embodies the dissipative nature of such a liquid. In the presence of a single momentum relaxation rate τ the viscosity of a relativistic liquid is given by $\eta = (\varepsilon + p)\tau$, where ε and p are the energy density and pressure, respectively. The entropy density is $s = (\varepsilon + p)/T$ and therefore the ratio of viscosity and entropy is $\eta/s = T\tau$. In a Planckian dissipator $\eta/s = T\tau_\hbar = A(\hbar/k_B)$, meaning that the gross magnitude of this ratio is set by Planck's constant! In chapter 7 we will see that according to holography $A = 1/(4\pi)$, which is related to the very strongly coupled theory in the boundary.

Finally, notice that the viscosity is a quantity yielding a very sharp contrast between the quantum critical state and the Fermi liquid. This quantity can be directly studied in ^{3}He and it is well known that the viscosity in its Fermi-liquid regime is of order $\eta \simeq nE_F\tau_{FL}$, where n is the density, E_F is the Fermi energy and $\tau_{FL} \simeq E_F\hbar/(k_B T)^2$, is the collision time. The entropy density is Sommerfeld, $s \simeq nk_B(k_B T/E_F)$, and it follows that the viscosity–entropy ratio of the Fermi liquid is $\eta/s \simeq (E_F/(k_B T))^3(\hbar/k_B)$. This ratio diverges like the cube of $E_F/(k_B T)$, the reason being that the microscopic time associated with particle collisions becomes very long in the low-temperature Fermi gas.

2.2 Quantum matter: when entanglement becomes macroscopic

Are all forms of matter described by the Ginzburg–Landau–Wilson paradigm, as we just discussed? There is an increasing awareness that this is not necessarily the case. We have already emphasised that the path integrals describing systems of finite density of fermions, or even bosons in the presence of a magnetic field breaking time reversal, have a non-probabilistic structure. These systems are a priori not like Boltzmann-type statistical-physics problems in "Lorentzian" disguise. Does this mean that the renormalised stuff one finds on the macroscopic scale can be of a completely different nature from the well-charted territory of symmetry broken states described in condensed matter textbooks?

There is yet another way to at least conceptualise this matter, using the modern language of quantum information theory. This revolves around the concept of *entanglement*. This came to the fore by virtue of the demonstration that it can be used to process information in a way that is superior to any classical system: the idea of the quantum computer. However, it is increasingly becoming realised that it also encapsulates the essence of the question regarding the nature of matter. The "classical" forms of matter we are used to are characterised by "short-range entanglement", which is present on microscopic scales. However, there are also forms of matter where the entanglement survives on macroscopic scales, giving rise to new forms of truly "quantum collective" behaviour. "Quantum matter" refers to those macroscopic substances which are characterised by irreducible long-range entanglements.

The study of quantum matter is a real frontier of physics. At present very little is known, and the forms of quantum matter which have been identified are most probably only the tip of the iceberg. We will present in this section a short summary of the currently flourishing development of the topological insulators. These are incompressible states of matter, characterised by a ground state separated from all excitations by an absolute energy gap, which are now understood to carry such a macroscopic entanglement (or "topological order", "quantum order"), giving rise to physical properties described by topological field theory.

2.2.1 The nature of matter versus the "spooky action at a distance"

It appears that in the present era a major reconsideration of quantum physics inspired by the rapidly developing subject of quantum information theory is taking place. As soon as we ask the question of how quantum systems handle information a sharp contrast with classical reality arises. This started a long time ago, when Einstein and Schrödinger protested against the postulates of quantum-mechanical measurement theory. Schrödinger constructed his famous cat, highlighting the "weirdness" of the coherent superposition of a single quantum-mechanical degree of freedom, in terms of the quantum-bit notation

$$|\phi\rangle = \alpha|0\rangle + \beta|1\rangle. \tag{2.18}$$

Subsequently, Einstein took this a step further, illustrating the "spooky action at a distance" by the famous EPR paradox. The story should be familiar to the reader: departing from a Bell pair like

$$|\text{Bell}\rangle = \frac{1}{2}(|0\rangle|1\rangle + |1\rangle|0\rangle),$$ (2.19)

the two bits share some form of non-local information that cannot be used to signal, i.e. the pairs are entangled. In the 1990s it was realised that one could design a quantum computer employing sequential 1- and 2-bit unitary operations on a system of quantum bits that outperforms any classical computer dramatically in certain cases [47]. A famous example is Shor's algorithm, where the prime factors of any integer can be computed in polynomial time, while classically this problem is NP-hard.

What has this to do with the nature of macroscopic matter? Condensed matter physics and quantum field theory share the same perspective with regard to quantum information. Instead of the single bits and Bell pairs of quantum computational engineering, one is now dealing with a thermodynamically large system composed of an infinite number of bits with a Hilbert space that is spanned by a gigantic number of different field configurations. The first question one can ask is how should one classify the entanglements in this Hilbert space of the physical vacua (ground states) of such a system, and what does this mean for the physical properties of such a state? Since temperature is detrimental for entanglement at large (macroscopic) scales, it appears that this question is truly meaningful only at precisely zero temperature. An even harder question is the following one: how should one understand excited states in this way, especially when dealing with quantum non-equilibrium physics? Progress has also been made on this frontier, but we will here ignore these developments. One category is very well understood: the stuff that is called appropriately classical matter. At zero temperature symmetries are broken, either in the original or in the dual, while the physics of these excited states is described in terms of the Ginzburg–Landau effective actions. From the quantum-information perspective these are classified as *short-range entangled product states*. Simple crystals are quite representative. On zooming in all the way to the subatomic scale, one finds a hugely entangled quantum soup involving quarks and gluons confining into baryons, which in turn form nuclei and finally atoms. Having arrived at the atomic scale, and assuming for simplicity that the crystal is a trivial insulator, at all larger lengths the vacuum is simply

$$|\text{Crystal}\rangle = \prod_i X_i^\dagger(R_i)|\text{vac}\rangle,$$ (2.20)

where $X_i^\dagger$ creates the atom in a real-space wave packet $\Psi \sim e^{(-(R_i - R_i^0)^2/\sigma^2)}$ at a site of the periodic lattice R_i^0, while the width of the wave packet σ is small compared with the lattice constant. Similarly, departing from hard-core bosons on a lattice, the superfluid or superconductor can be written as a product state like $\prod_i (u_i + v_i b_i^\dagger)|\text{vac}\rangle$.

However, is it a natural law that all states of matter have to be classical according to the product-state definition? This question can be sharply posed in the following way. The vacuum of a thermodynamically large system with a Hilbert space spanned by all possible field configurations $|\text{config}, i\rangle$ can be written in full generality as

$$|\Psi_0\rangle = \sum_i A_i^0 |\text{config}, i\rangle. \qquad (2.21)$$

Are there physical Hamiltonians having $|\Psi_0\rangle$ as a ground state while $|\Psi_0\rangle$ *cannot be reduced to a product-state form in any representation*. If so, one is clearly dealing with a new form of matter where the "spookiness" of quantum physics imprints on the very nature of matter itself: "quantum matter".

The "representation independence" of the long-range entanglement is a subtle issue and one has to keep one's eyes wide open in order to avoid being lured into believing that some state that looks as if it is long-range entangled is actually a classical state in disguise. In the quantum-information community the dictum is that the "vacuum should not be reducible to a product state by local unitary transformations". This wisdom can be traced back to the quantum mechanics of Bell pairs. A state like $(|0\rangle|0\rangle + |0\rangle|1\rangle + |1\rangle|0\rangle + |1\rangle|1\rangle)/2$ might appear to be quite entangled. However, it is easily shown that this just corresponds to the product state $|+\rangle|+\rangle$ in terms of the single-bit states $|+\rangle = (|0\rangle + |1\rangle)/\sqrt{2}$.

How useful is this notion dealing with quantum field theory? Consider a straightforward Bose condensate formed from interacting bosons, like ^{4}He. Viewed in real-space–number representation, one could call the superfluid a "quantum liquid" since all the bosons are truly delocalised while they are subjected to the infinite-range bosonic exchanges underlying the Bose condensation. However, entanglement has to be representation-independent, and in the dual-phase representation it is just a short-range entangled product state, while the macroscopic physics is enumerated by the classical Ginzburg–Landau theory. But dealing with a complicated state, how can one be sure that one knows all the "duals"? This reflects the immaturity of "field-theoretical quantum information". For the two-bit system the von Neumann entanglement entropy is a mercilessly precise measure of entanglement. However, no such mathematical device is known for field-theoretical systems – as we will discuss in chapter 14, the extension of the bipartite von Neumann entropy to field theory fails painfully in this regard.

2.2.2 Long-range entanglement and the incompressible quantum liquids

A first example of genuine quantum matter was discovered in the early 1980s, although it was realised only much later that this is its secret [48]. We refer here to the fractional quantum Hall effects [49] explained by Laughlin in terms of his divine guess for the wave function of their vacuum,

$$|\Psi_L, m\rangle \propto \prod_{i<j}(z_i - z_j)^m \prod_k e^{-|z_k|^2}. \tag{2.22}$$

Here $z_i = (x_i + iy_i)/(2l_B)$, where $l_B = \sqrt{\hbar c/(eB)}$, parametrises the position of the spin-polarised electron in the two-dimensional electron gas in a large magnetic field in the complex plane. The integer quantum Hall state constructed by filling up the quantum-mechanical Landau orbits corresponds to $m = 1$. As a first example of the explicit use of fermion-sign structure beyond the Fermi gas (see chapter 14), Laughlin realised that the nodes of the wave function are attached to the particle positions in a Landau quantised Fermi gas ($|\Psi\rangle \to 0$ when $z_i \to z_j$). The effects of strong repulsive interactions can be wired in by increasing the power of m, with the effect that the correlation hole becomes deeper.

The Laughlin state is a truly long-ranged entangled state [48], and as a greatly simplifying circumstance it is, on the fractional quantum Hall plateau, *incompressible*: it is isolated from all excitations by an energy gap, and this protects the entanglement. Although there is no order parameter of any kind, every Laughlin state has a definite thermodynamical identity, bearing its own properties. The different Laughlin states are separated by genuine quantum phase transitions ("plateau transitions") that are still far from being understood. Empirical scaling analysis suggests that an emergent modular invariance is realised (see Ref. [50] and references therein).

The $(2 + 1)$-dimensional Chern–Simons topological field theory takes the role of Ginzburg–Landau theory in enumerating the physical properties of these states. It acts as a counting device enumerating the consequences of the irreducible long-range entanglement wired into the vacuum [48, 49]. (a) When the target space is a compact two-dimensional manifold characterised by a genus g, the ground state will be m^g-fold degenerate. (b) This "bulk" carries isolated particles like "quasi-hole" excitations characterised by a fractionalised charge $e^* = e/m$, exhibiting a fractional exchange statistics. For the Abelian Laughlin states these are anyons with statistical angle $\theta = \pi/m$. (c) A most stunning consequence of the structure of the long-range entanglement in the bulk is the *bulk–edge* correspondence. When the topological bulk has an open boundary, gauge invariance of the Chern–Simons fields imposes that at this boundary propagating modes described by a chiral Wess–Zumino–Witten $(1 + 1)$-dimensional field theory are present [48, 49]. There is a precise one-to-one correspondence between the bulk and the edge in the sense that the bulk can be reconstructed when all the data of the edge theory are available and vice versa. For instance, the physical currents run at the edge, and the sharp quantisation of their Hall conductance is in this way inherited from the bulk, while the noise spectra of the edge currents can be used to measure the fractionalisation of the charge. Since this is a book dealing with holography, the reader will notice an eerie resemblance to AdS/CFT: this concerns the $(2 + 1)$-dimensional bulk which is in one-to-one correspondence with the physics on the "holographic screen" in $1 + 1$ dimensions, imposed by gauge invariance that is to some extent similar to the procedures which will be discussed in chapters 4 and 5. This resemblance becomes in fact quite literal on dealing with $(2+1)$-dimensional gravity. Einstein gravity is in this dimension non-dynamical (or "incompressible"), and it was discovered in the 1980s that the $(2 + 1)$-dimensional pure gravity in an anti-de Sitter background can be described in terms of a (Poincaré) $SO(2, 2)$ Chern–Simons field theory [51, 52]. This is at present under active investigation in the context of higher-spin gravity [53, 54, 55]. A major complication is that Chern–Simons is still rather poorly understood when dealing with infinite, non-Abelian gauge groups.

Fractional quantum Hall has established a new paradigm that is at present being rolled out to other systems. A first boost of interest comes from the realisation that the braiding properties of the quasiholes in non-Abelian quantum Hall states (believed to be realised at the $5/2$ plateau) can be exploited for quantum computation, profiting from the protection of the topological vacuum against decoherence [56]. Subsequently, it was discovered that topological states can also be realised in band insulators in the absence of a magnetic field: these are the "topological band insulators" and the "topological superconductors" [57, 58]. These states require protection by time-reversal symmetry, while the topology of the electrons also communicates with the space group of the crystals [59]. They are also realised in three space dimensions, exhibiting the bulk–edge correspondence, while the insulators are expected to accomplish axion electrodynamics [60], and the superconductors can be tailored to form Majorana zero modes that can be used for topological quantum-computational purposes [61]. These systems are much easier to deal with in the laboratory than the fractional quantum Hall states, and they are being subjected to intensive experimental research. On the theoretical side, a concerted effort attempting to classify all possible forms of topological insulators is unfolding. This includes those which cannot be described in terms of the band structure of non-interacting fermions: see Ref. [62], which relies on the vortex duality of section 2.1.2, as well as the spin-liquid-type entanglement as discussed in section 3.2.

Although it is interesting to find out whether fractional quantum Hall states and so forth can be encoded in a gravitational dual, it is less obvious whether holography can be of much help in elucidating the topological order in incompressible systems. The issue is that there is already quite powerful mathematics available in the form of topological field theory and its extensions: it is likely that any holographic construction would need to have recourse to the same overriding principles. However, the situation is quite different for *compressible* forms of quantum matter. Very little is known, while the insights obtained for incompressible quantum matter are suggestive that there are still great surprises to be discovered.

2.3 The remarkable Fermi liquid

Continuing with the quantum matter theme, hardly anything is known when dealing with *compressible* states of matter. There is actually one form of genuinely long-range-entangled compressible quantum matter that has been understood since a very long time ago: the Fermi gas! Fermion statistics is a quite un-classical affair. Because of the requirement that the ground-state wave function is anti-symmetric under the exchange of *any* pair of particles, it is genuinely long-range-entangled. The entanglement associated with the symmetry requirement for the Fock space of bosons can be evaded by choosing a dual-phase representation where it becomes product, but this is impossible for fermions.

We do have the suspicion that the mysterious properties of types of holographic matter are just reflecting the fact that these are new forms of compressible quantum matter. As we will see later, although these are not Fermi liquids, they seem at least in some regards to be more closely related to Fermi liquids than they are to anything else. Having this at the back of our minds, we will review the standard Fermi-liquid lore, albeit it from a perhaps somewhat unusual viewpoint, purposely amplifying its quantum-matter side.

In undergraduate courses one is taught how to deal with the physics of the Fermi gas. However, to be reminded of how "un-product" the state is, let us view it through its first-quantised Euclidean path integral,

$$
\begin{aligned}
\mathcal{Z} &= \mathrm{Tr}\left[e^{-\beta H}\right] \\
&= \int d\mathbf{R}\, \frac{1}{N!} \sum_{\mathcal{P}} (-1)^{\mathcal{P}} \int_{\mathbf{R} \to \mathcal{P}\mathbf{R}} \mathcal{D}\mathbf{R}(\tau) \\
&\quad \times \exp\left\{ -\frac{1}{\hbar} \int_0^{\hbar\beta} d\tau \left(\frac{m}{2} \frac{\partial^2 \mathbf{R}(\tau)}{\partial \tau^2} + V(\mathbf{R}(\tau)) \right) \right\},
\end{aligned}
\tag{2.23}
$$

where $\mathbf{R}$ refers to configuration space (positions of all particles) while $\mathbf{R}(\tau)$ refers to the world histories of the system of wordlines as functions of imaginary time τ. These fermions have a mass m and an interaction potential V. The complication lies in the requirement that one has to sum over all permutations $\mathcal{P}$ and, when the parity of the permutation is uneven, this gives rise to negative "probabilities" (the factor $(-1)^{\mathcal{P}}$): the minus signs that render the Euclidean quantum partition sum unrelated to a Boltzmannian partition sum.

For the non-interacting Fermi gas the path integral can of course be evaluated [63]. Just as for the bosons, one rewrites this as a sum over windings around the imaginary-time circle, finding that at the Fermi temperature the winding numbers become macroscopically large. Instead of signalling Bose condensation, the contributions of configurations with winding numbers n and $n + 1$ with n very large occur with opposite sign, nearly cancelling each other out. One can view this as destructive interferences occurring in real space [64], having the effect of pushing up the zero-point motion energy of the system which translates into the Fermi energy. For free fermions this alternating sum can be solved in closed form, recovering the usual Fermi–Dirac story. However, how should one handle such alternating sums when the interaction potential V becomes so large that one can no longer get away with perturbation theory, while simultaneously instabilities of the BCS/Hartree–Fock type do not occur? The answer is that this sum is mathematically very ill-behaved: it is even claimed to be NP-hard [65]. The fact is that there is no systematic and general mathematical machinery that can handle it. This is the infamous sign problem. Sometimes it is misunderstood as a technical problem in the development of numerical quantum Monte Carlo algorithms. It is not: lacking any working mathematical machinery, humanity is just blind to the physics of strongly interacting fermions at finite density. However, from the modern perspective of quantum matter it turns into an *opportunity*. The sign "problem" means that product states are just wrong, and neither can one rely on the Fermi gas. Even the "crossroads" of the two in the form of electronic mean-field theory (BCS and so forth) falls short: the fermion signs might be the key ingredient causing new forms of long-range entanglement. The message to keep at the back of one's mind when studying the chapters on finite-density holography is that the unfamiliar properties of this holographic matter are just telling a story about the physics behind the fermion-sign brick wall.

But let us first continue highlighting the "quantum-matter weirdness" of the Fermi gas. We saw in the previous section that the effective theory enumerating the collective properties of incompressible entangled matter takes the form of an effectively classical theory, albeit in the form of topological field theory. What is this theory for the Fermi gas? The answer is known of course, but it is again informative to use the thermal field-theory language [64]. Put a system of spinless free fermions in a finite-volume space, and after Fourier transformation one ends up with a lattice of discrete single-particle momenta $\mathbf{k}_i$. It is now easy to show that in momentum configuration space $\mathbf{K} = (\mathbf{k}_1, \dots, \mathbf{k}_N)$ the full density matrix as a function of imaginary time τ can be written in a form where the signs are eliminated, giving a probabilistic Bolzmannian form,

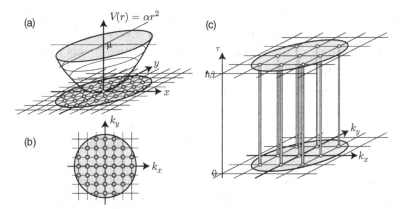

Figure 2.2 (a) A system of classical atoms forming a Mott insulating state in the presence of a commensurate optical lattice of infinite strength, living in a harmonic potential trap $V(r) = \alpha r^2$ of finite strength. (b) The trap in momentum space k_x, k_y instead of real space; the Fermi surface is just the boundary between the occupied optical lattice sites and the empty ones. (c) A grid of allowed momentum states $k = (2\pi/L)(k_x, k_y, k_z, \ldots)$, where the k_i are the usual integers and any worldline just closes on itself along the imaginary-time τ direction $0 \to \beta$: single-particle momentum conservation prohibits anything but the one cycles. Figure source [64].

$$\rho_F(\mathbf{K}, \mathbf{K}'; \tau) = \prod_{\mathbf{k}_1 \neq \mathbf{k}_2 \neq \cdots \neq \mathbf{k}_N}^{N} 2\pi \delta(\mathbf{k}_i - \mathbf{k}_i') e^{-|\mathbf{k}_i|^2 \tau^2/(2\hbar m)}. \tag{2.24}$$

What kind of classical system is this? It is actually a Bose Mott insulator in the classical ("infinite U") limit, living in a harmonic well in momentum space! The lattice of momentum points corresponds to a periodic lattice of infinitely deep potential wells, which is in turn modulated by a harmonic potential (the dispersion relation $\omega \sim k^2$). Every site can be occupied by no more than one particle at a time: this is the Mott constraint which arises on summing over the permutations in momentum space, "eating" the fermion signs. One now has to fill up this Mott insulator in the harmonic potential, and at zero temperature one finds a sharp rim dividing the occupied from the unoccupied states: the Fermi surface (Fig. 2.2). Upon switching temperature on, the sharp rim will become blurred, since one can excite the point particles over the rim: this is the Lindhardt continuum of particle–hole excitations.

One immediately recognises the Fermi–Dirac statistics governing free fermions, but in this thermal field-theory setting it is a bit disorienting. The deep IR of the Fermi gas, as well as the Fermi liquid, is in fact governed by a classical theory, but this theory is in a way quite the opposite of the theories governing classical or Bose gasses. The fermion signs have actually turned it into a traffic jam, albeit in momentum space. A further confusing property of the Fermi gas is that it is a very stable, "cohesive" form of matter. It is protected by a very sturdy scale: the Fermi energy. The simplicity of the Fermi gas is a bit deceptive in this regard, and this issue really comes to life when interactions are switched on, turning the Fermi gas into the Fermi liquid. The modern way to express Landau's principle of adiabatic continuity is by using the functional renormalisation-group language [66, 67]. This amounts to the observation that the free-fermion system with a renormalised kinetic energy is a stable fixed point, since nearly all interaction operators are doomed to be irrelevant upon approaching the deep infrared, with the pair channel being the only potential exception. But there are actually some quite odd things happening when viewed from a genuine "non-entangled" reference frame. In an unusual sense, *the Fermi liquid is characterised by an "order parameter" that "breaks symmetry".* The quotation marks are there for a purpose: this "order" is quite different from the order of classical matter.

The arguments are elementary. The defining property of the Fermi liquid is that its deep IR is controlled by the Fermi surface. The Fermi surface is a $(d-1)$-dimensional manifold with a precise locus in d-dimensional single-particle momentum space. But as soon as interactions are switched on, single-particle momentum is no longer conserved! This is obvious: a typical interaction term has the form $H_1 = \sum_{k,k',q} V(k, k', q) c_{k+q}^{\dagger} c_k c_{k'-q}^{\dagger} c_{k'}$ and this surely does not commute with $n_k = c_k^{\dagger} c_k$. It is a precise statement that in the presence of interactions single-particle momentum is no longer sharply quantised. But how can it be that the system as a whole is still characterised by a sharp surface living in this space which is not supposed to exist? Such sharp quantum numbers that spring into existence only in the thermodynamic limit of an interacting system are of course familiar from the classical context. For instance, single atoms in the Galilean continuum have no right to have a sharply quantised position. They can, however, collectively break translational symmetry, with the effect that the average position of an individual atom becomes very sharp in the crystal.

It is the modern view to consider the Fermi surface of the Fermi liquid as the generalised "order parameter" of the state. One can now interpret the outcomes of diagrammatics (including the Landau phenomenology) from this perspective. Dealing with a classical symmetry breaking, the Goldstone bosons arise automatically as the excitations signalling that there is symmetry to restore. Anything that happens in the deep infrared of the Fermi liquid can be understood in a similar way: everything is about fluctuations of the Fermi-surface "order parameter".

The bare-fermion propagator is unusually powerful in revealing information regarding the collective Fermi-liquid state in the deep infrared. As every student has noticed, it is the central wheel of the perturbative diagrammatics,

$$G_f(k, \omega) = \frac{1}{\omega - \varepsilon_k - \Sigma(k, \omega)}, \qquad (2.25)$$

where all the effects of the interactions are collected in the self-energy $\Sigma(k, \omega)$. As usual, the imaginary part of G_f is the electron spectral function, which is measured more or less directly by (inverse) photoemission. Upon approaching the Fermi surface, this propagator acquires a universal form. Counting momenta relative to k_F ($\bar{k} = k - k_F$) and energy from the chemical potential,

$$G_f(k, \omega) = \frac{Z_k}{\omega - \bar{v}_F\bar{k} - iC_k\omega^2/\bar{E}_F} + G_{incoh}(k, \omega). \qquad (2.26)$$

The spectral weight associated with G_{incoh} is vanishing when $\omega \to 0$ and all the spectral weight is in the quasiparticle piece. This has a number of interesting consequences.

(i) By interrogating the system with bare fermions (as in photoemission) one can directly read off the quasiparticle properties, thereby obtaining nearly all the information one needs regarding the collective Fermi-liquid state, such as the renormalised Fermi energy ($\bar{E}_F$, or Fermi velocity $\bar{v}_F$) and the quasi-particle pole strength Z_k. Given that the Fermi liquid can be very strongly interacting, this is quite exceptional: in any other strongly interacting liquid the low-energy excitations are detached from the UV degrees of freedom by strong emergence and one learns nothing directly from the UV propagators regarding the collective physics.

(ii) Upon moving away from the Fermi surface the quasiparticles acquire a life-time $\sim \omega^2$. This has the effect that single-particle momentum blurs, as it should, and only when $k \to k_F$ does it become sharp. Single-particle-momentum space is restored only right at the Fermi surface. Having this in mind, the Luttinger theorem is particularly remarkable. In the non-interacting limit the momentum-space volume enclosed by the Fermi surface is determined by the fermion density, and the theorem demonstrates that this is still the case in the interacting system as long as a Fermi liquid is formed.

(iii) The quasiparticle pole strength Z_k plays the role of the order parameter. It sets the magnitude of the discontinuity at k_F in the momentum distribution. It becomes unity in the non-interacting limit. The interactions cause the locus of the Fermi surface in momentum space to fluctuate, but it retains a mean position with a magnitude that is gradually decreasing, like a Debye–Waller factor "eating away" the order parameter in a crystal. The analytical structure of the real part of the self-energy $\Sigma(k, \omega)$ ties the pole strength to the renormalisation of the quasiparticle mass, determining in turn the renormalised

Fermi energy and Fermi velocity. Assuming a reasonably smooth momentum dependence, $Z_k = (1 - \partial_\omega \Sigma')^{-1}$ and $m_*/m = 1/Z_k$.

This is a quite deep and general property of the Fermi-liquid "order": when the Fermi surface is on the verge of disappearing because of strong quantum fluctuations ($Z_k \to 0$), the scale that protects the stable state is also disappearing (the renormalised $\bar{E}_F \to 0$). Differently from the bosonic order dynamics, there is the additional manifold of fermionic quasiparticle excitations, governed by a velocity scale that is vanishing on approaching this critical point. When these would survive right at the critical point, the quasiparticles would acquire an infinite mass, meaning that the state would be characterised by a zero-temperature entropy. Actually, Fermi liquids characterised by extreme quantum fluctuations are quite abundant: ^{3}He is an early example ($m_*/m \simeq 10$), and a large class of heavy-fermion systems with mass enhancements as large as by a factor of 1000 has been discovered (see section 3.6.1).

When one is dealing with bosonic and zero-density matter, massless excitations occur usually only at isolated *points* in momentum space. Accordingly, the specific heat will vary as $C \sim T^{d/z}$ at finite temperature. The Fermi liquid is highly exceptional because masslessness occurs in a d-dimensional system on a manifold with a dimensionality of $d - 1$ in momentum space: the Fermi surface. The effect is that the Fermi liquid behaves effectively like a $(1 + 1)$-dimensional system: every point on the Fermi surface supports a particle-like excitation and the (Sommerfeld) specific heat is therefore like that of a ($z = 1$) $(1 + 1)$-dimensional system in any space dimension. In addition, at any finite temperature the fugacity of these particle–hole excitations is finite, with the consequence that "Fermi-liquid order" is limited to strictly zero temperature, like in one space dimension. This is an example of the holographic sense of "hyperscaling violation" which we will encounter when we deal with holographic strange metals, in chapter 8, and the von Neumann entropy, in chapter 14. This is a quite unnatural phenomenon in statistical physics, and it is an example where only the Fermi liquid appears to show some similarity to holographic matter.

The Fermi liquid might also be useful as a metaphor for dealing with the "quantum hydrodynamics" of the holographic strange metals. The quotation marks indicate now a potential semantic confusion: hydrodynamics per se refers to the universal behaviours of finite-temperature fluids. These hydrodynamical principles will in general not fully control the behaviour of zero-temperature *quantum* fluids. One can, however, still ask what happens when this quantum liquid is put in a system of pipes that is cooled to strictly zero degrees Kelvin, and pushed softly with a piston. This physics is called "quantum hydrodynamics" in the condensed matter community.

Besides the "Goldstone-protected" superfluid and superconductor, the only conventional zero-temperature fluid is the Fermi liquid. Its quantum hydrodynamics was deciphered by Landau and co-workers in the 1950s and subsequently rather thoroughly tested, especially in the Fermi-liquid regime realised in ^{3}He. This is again textbook material [68]. The Fermi gas is in fact a singular, unphysical limit. As soon as one adds any finite repulsive interaction the Fermi liquid will develop a massless, propagating longitudinal "zero" (temperature) sound mode exhausting the excitations at long wavelength. Accordingly, there will be a conserved, hydrodynamical flow as in a classical fluid. The Landau Fermi liquid theory is (implicitly) asserting that this is correctly described by time-dependent Hartree–Fock: it is just the re-summation of the particle–hole loops, also known as the random-phase approximation (RPA) in the condensed matter community. In parallel with the pair channel of BCS (section 2.4), one writes the RPA expression for the density susceptibility in terms of the Fermi-gas density susceptibility (particle–hole loop) $\chi_\rho^0(q, \omega)$ and the q-dependent interaction V_q as

$$\chi_\rho(q, \omega) = \frac{\chi_\rho^0}{1 - V_q \chi_\rho^0} . \tag{2.27}$$

The absorptive (imaginary) part χ_ρ'' will reveal at larger momenta the "Lindhardt continuum" of incoherent particle–hole excitations. It is easy to show that for any finite interaction strength a bound state will form in a range of small momenta at an energy above the Lindhardt continuum, associated with the condition $1 - V_q \chi_\rho^0(q, \omega) = 0$ for a pole. This represents the zero longitudinal sound mode, emerging at $q = 0$ as a relativistic mode in the case of short-range interactions. In the charged system, it will turn into the massive-plasmon excitation, just because of the long-range nature of the Coulomb interaction. The sound will eventually "die" by Landau damping when it enters the continuum. Physically, it corresponds to a coherent "breathing" vibration of the Fermi surface, and by application of Luttinger's theorem this implies that it is also a longitudinal density (sound) wave.

This "zero" longitudinal sound is quite distinct from the sound carried by a thermal fluid. The qualitative difference becomes manifest on considering the attenuation (or damping) of the sound at finite momenta and temperature. In the finite-temperature Fermi liquid one finds a crossover from a low-frequency "collision-full" regime associated with the thermal quasiparticle fluid to a high-frequency "collisionless" regime associated with the coherent vibrations of the Fermi surface. In the latter regime one finds the zero sound, which also gets damped by the presence of the thermally excited particles, but its damping is proportional to the quasiparticle collision *rate*, $1/\tau_{\text{col}} \simeq (k_B T)^2/(\hbar E_F)$. The damping of the thermal sound $\sim Dq^2$, where q is the momentum and the diffusivity $D \simeq \eta/(nm)$, where nm is the mass density and η the viscosity. The viscosity is in turn set by the quasiparticle collision time $\eta \simeq n E_f \tau_{col}$, with the result that the attenuation is proportional to $D \simeq (\hbar/m)(E_F/k_B T)^2$. The outcome is the "attenuation maximum" found at the crossover from the quantum regime to the thermal regime, famously confirmed in ^{3}He [69]. Although they are in detail quite different, the holographic strange metals described by holographic probe branes (chapter 13) are also characterised by similar "quantum" and "classical" longitudinal sounds, separated as well by a similar attenuation maximum, while the Reissner–Nordström strange metal of chapter 8 is in this regard different.

There is yet another quantum-hydrodynamical property of the *strongly* interacting Fermi liquid that is very strange when viewed from a classical perspective. Although it was discovered by Landau himself, this appears to be no longer widely realised. Transverse sound propagates in an elastic medium in the form of transverse phonons. In a thermal liquid such shear motions are subject to viscous behaviour, translating the effect into a purely relaxational response. However, Landau discovered that this is no longer true in a strongly interacting Fermi liquid characterised by a substantially enhanced quasiparticle mass. The mass enhancement is controlled by the Landau parameter F_1 and, when this exceeds a critical value of 6, transverse sound starts to propagate! This was experimentally confirmed in ^{3}He [70]: more than anything else, this highlights the strange rigidity accompanying the "Fermi surface order", forcing the liquid which is not breaking translations to behave like an elastic medium!

2.4 The mean-field instabilities of the fermion system: BCS and beyond

The condensed matter textbook would now proceed with the weak coupling instabilities of the Fermi liquid. When interactions are switched on, at some point the Fermi liquid will become unstable towards an order breaking a symmetry. This can be associated with the particle–hole channel, leading to charge-density waves breaking translations (crystallisation of the electrons) or spin-density waves breaking translations and internal spin symmetry (antiferromagnets). Except for $1 + 1$ dimensions, the system has to exceed a critical interaction strength for this to happen, given the absence of a precise "nesting" property. The situation is, however, different for attractive interactions, where, for any finite strength of the coupling, an instability will occur in the pair channel, leading to a BCS superconducting ground state.

The mathematical machinery which is behind this paradigm is the electronic "Hartree–Fock" mean-field theory. This is insisting that, modulo the quantum fluctuations of the order, the order parameter just translates into a potential felt by the fermions. When this potential generates a gap in the fermion spectrum at the Fermi energy, the resulting energy gain will stabilise in turn the ordered state. The speciality of the pair channel is that this gap will span the whole Fermi surface, with the possible exception of point or line nodes dealing with unconventional (non-s-wave) superconductors, even for arbitrarily small interactions. This leads to the famous BCS expression for the gap, where ω_B is the characteristic frequency of the mediator of the attractive interactions (phonons in conventional superconductors)

$$\Delta = \hbar\omega_B e^{-1/\lambda}, \tag{2.28}$$

revealing that the gap is exponentially small in the small coupling $\lambda \simeq N_0 v$, where $N_0 \simeq 1/E_F$ and v are the electronic density of states at E_F and the strength of the attractive interaction, respectively. This "BCS logarithm" is yet again deeply rooted in fermionic physics: only fermions very near the Fermi surface feel the gap, while the vast majority of them are locked in the unperturbed Fermi sea. A large coherence length $\xi = v_F \hbar/\Delta$ is associated with the small energy scale (the "size of the Cooper pair"), having the effect that the bosonic order-parameter fluctuations are frozen out at distances shorter than ξ since the system has returned to the "entangled" Fermi gas. This leads to a dramatic suppression of the order-parameter fluctuations, to the degree that the thermal phase transition in e.g. aluminium behaves as a perfect mean-field transition since the temperature resolution in the laboratory does not suffice to access the true critical regime associated with a complex scalar in three dimensions.

The other deep aspect of the BCS theory is that the pair channel is *uniquely* singled out. Subjecting classical or bosonic matter to attractive interactions does not lead to a system of two-particle bound states but instead to clumping: the formation of an N-particle bound state like a cohesive fluid or solid. The underlying "Cooper mechanism" was historically the key to BCS: because of the *sharpness of the Fermi surface* the electrons will form exclusively *bound pairs*. But how should one deal with this when the normal state is a fermionic system that is a non-Fermi liquid, lacking a real Fermi surface and so forth? This question is of great relevance to the empirical high-T_c superconductors with their non-Fermi-liquid normal states (section 3.6). But it also arises as a theoretical question in the context of holographic superconductivity.

In dealing with holographic matter we will learn that similar transitions occur, involving strange metals at high temperature that turn into ordered states at low temperature, like the holographic superconductors (chapter 10), the holographic crystals (chapter 12) and even the holographic Fermi liquids, albeit in an odd sense (chapter 11). In fact, it appears that holographic superconductivity signals the existence of a *generalised form of the Cooper mechanism*, which is generically at work in the strange metals. As we will highlight in chapter 10, holographic superconductors behave in a number of ways strikingly similarly to the standard BCS variety, but yet the underlying physics is very different, especially the pairing mechanism. The 64-thousand-dollar question is how can one sharply distinguish strange-metal from Fermi-liquid pairing?

At about the same time as holographic superconductivity was discovered, a very simple phenomenological theory that rather precisely addresses this issue was constructed: the "quantum critical BCS" phenomenology [71]. This departs from the following assumptions: (a) for whatever reason, the dominant instability is in the pair channel; (b) the normal state is a strictly conformal metallic phase, such that all propagators are of the "un-particle" branch-cut form as explained in section 2.1.3, including the propagator of the fermion pairs; (c) this scale invariance is exclusively broken by the relevance of the pair channel; and (d) last but not least, the superconducting order parameter is governed by mean-field dynamics.

This catches the essence of holographic superconductivity [72]. The mean-field behaviour of the order parameter is crucial since it makes it possible to compare apples with apples. One should keep in mind, however, that the origin of this mean-field behaviour of holographic order parameters is elsewhere: the matrix large-N limit, which has no obvious meaning in condensed matter systems. One should now focus on how the *dynamical pair susceptibility* evolves, approaching the superconducting phase transition in the normal state. The claim is that this quantity contains the crucial information distinguishing sharply the strange-metal pairing mechanism from anything that eventually relies on the conventional Cooper mechanism. In principle this dynamical pair susceptibility can be measured, using a tunnelling device that was tested already in the 1970s for conventional superconductors [72]. However, for practical (material-related) reasons it turns out to be quite difficult to build such a device for the superconductors of interest.

The dynamical pair susceptibility is familiar as the quantity which is at the centre of the BCS formalism,

$$\chi_p(\mathbf{q}, \omega) = -i \int_0^\infty dt \, e^{i\omega t - 0^+ t} \langle [b^\dagger(\mathbf{q}, 0), b(\mathbf{q}, t)] \rangle, \qquad (2.29)$$

where $b^\dagger(\mathbf{q}, t) = \sum_{\mathbf{k}} c^\dagger_{\mathbf{k}+\mathbf{q}/2,\uparrow}(t) c^\dagger_{-\mathbf{k}+\mathbf{q}/2,\downarrow}(t)$, with $c^{(\dagger)}_{\mathbf{k},\sigma}$ the usual annihilation (creation) operators for electrons with momentum $\mathbf{k}$ and spin σ, while this can be decomposed into the various s, p, d, ... pairing channels. Its imaginary part as a function of frequency is most revealing, and this is also the quantity which is accessible in principle by experiment [72].

Let us find out how this works for a conventional BCS superconductor. One computes first the "bare" pair susceptibility of the Fermi liquid, which is just the particle–particle loop. In Migdal–Eliashberg theory one dresses the single-fermion propagators with self-energy corrections ignoring all vertex corrections, exploiting the smallness of the Migdal parameter [73]. However, the essence is caught by the non-interacting Fermi gas of the weak-coupling "minimal BCS" theory. Let us focus on $\mathbf{q} = 0$, where the instability will develop:

$$\text{Im } \chi_{\text{pair}}^0(\omega, T) = \frac{1}{2E_F} \tanh\left(\frac{\hbar\omega}{4k_B T}\right). \tag{2.30}$$

One now injects an attractive interaction V between the fermions – in the full Migdal–Eliashberg theory this would be computed from the electron–phonon interaction, again ignoring vertex corrections. According to the conventional formalism one "re-sums the bubbles", resulting in the RPA expression for the full dynamical susceptibility of the normal state,

$$\chi_{\text{pair}}(\omega, T) = \frac{\chi_{\text{pair}}^0(\omega, T)}{1 - V\chi_{\text{pair}}^0(\omega, T)}. \tag{2.31}$$

Coming from high temperature, the superconducting instability is signalled by the condition that a pole develops at zero frequency in the full susceptibility. This requires

$$1 - V \text{ Re } \chi_{\text{pair}}^0(\omega = 0, T) = 0. \tag{2.32}$$

According to the Kramers–Kronig transformation,

$$\chi'(\omega = 0) = \int d\tilde{\omega} \, \frac{\chi''(\tilde{\omega}, T)}{\tilde{\omega}}$$

$$\simeq \int_{k_B T}^{\hbar\omega_B} 1/(2E_F) = N_0 \ln((\hbar\omega_B)/(k_B T)), \tag{2.33}$$

where N_0 is the density state, while we have here smuggled in a retardation scale/phonon frequency ω_B. On substituting this into Eq. (2.32), one obtains immediately the famous BCS expression

$$k_B T_c \simeq \hbar\omega_B \exp\left(-1/(N_0 V)\right). \tag{2.34}$$

It is now interesting to follow the evolution of the absorptive part of χ_{pair} in the ω, T plane in the normal state, approaching the transition (see Fig. 10.9(a)). One infers at high temperature the bare Fermi-gas response Eq. (2.30), but upon approaching T_c the denominator switches on, with the result that a peak that moves down in energy develops and sharpens when the temperature decreases, turning into the pole at $\omega = 0$ at T_c. This is the "relaxational peak", which can equally well be understood in terms of the effective order-parameter theory,

$$\mathcal{L} = \frac{1}{\tau_r}\Psi \, \partial_t\Psi + |\nabla\Psi|^2 + i\frac{1}{\tau_\mu}\Psi \, \partial_t\Psi + \alpha_0(T - T_c)|\Psi|^2 + u|\Psi|^4 + \cdots . \tag{2.35}$$

This describes the "Ornstein–Zernike" mean-field relaxational order-parameter dynamics, with the ramification that, at low frequencies close to T_c,

$$\chi_{\text{pair}}(\omega, T) = \frac{\chi'_{\text{pair}}(\omega = 0, T)}{1 - i\omega\tau_r - \omega\tau_\mu}. \tag{2.36}$$

This just describes the over-damped, relaxational response of the order parameter, persisting during a time τ_r in the normal state before it relaxes away. The time τ_μ measures the breaking of the charge-conjugation symmetry at the transition: this is perhaps less familiar since it plays a role only in strongly coupled superconductors where one cannot get away with using a constant density of states on the scale of the gap. The bottom line is that the RPA form Eq. (2.31) for the dynamical pair susceptibility is controlled by the mean field nature of the order parameter: the RPA is well understood as representing time-dependent mean-field theory.

Where does the information regarding the nature of the normal metal enter? This is actually encapsulated in the form of the bare susceptibility Eq. (2.30). At zero temperature the imaginary part is just frequency-independent, and this goes hand in hand with the "BCS logarithm" in the real part Eq. (2.33). Let us now compare this quantity with the general form of a two-point correlator of a conformal system, Eq. (2.16): the bare susceptibility of the Fermi gas is just a conformal propagator characterised by a "marginal" scaling dimension $2\nu_b = 0$. That the pair susceptibility of the Fermi gas behaves as if the system were conformal is an accident. After all, the Fermi gas is scale-full (the Fermi energy), and as soon as one switches on perturbative corrections the conformality (including the energy–temperature scaling) is destroyed since the perturbative corrections pick up the Fermi energy.

We now consider the case that the metal is not a Fermi liquid but instead a truly quantum critical metal, controlled by the conformal invariance highlighted in section 2.1.3. At temperatures high compared with the transition temperature all two-point correlators, including the pair propagator, should be conformal. At zero temperature and high frequency the pair propagator should behave as $1/(i\omega)^{\nu_b}$, turning into scaling functions of ω/T at finite temperature with a precise form that is not known in general. The precise form is, however, not crucial, and in the computations of Ref. [71] and Ref. [72] these are modelled using explicit results from $(1 + 1)$-dimensional conformal-field theory. In the same critical guise one can add a "relevant operator" taking the role of the BCS V. Asserting that the phase transition is a mean-field one, one can just get away with using the RPA expression except that one has to use a bare pair propagator with a scaling dimension $\nu_b \neq 0$. This defines the very simple "quantum critical BCS" phenomenology.

An important consequence is that the expression for the gap and/or T_c differs drastically from the standard BCS equation [71]. Upon incorporating a UV cut-off ω_c and a retardation scale ω_B one finds instead an expression for the gap

$$\Delta = 2\omega_B \left(1 - \frac{1}{\lambda} \left(\frac{2\omega_B}{\omega_c} \right)^{-\nu_b} \right)^{1/\nu_b}, \tag{2.37}$$

where $\lambda = 2(V/\omega_c)(1 + \nu_b)/\nu_b$. Instead of the BCS logarithm one finds, not surprisingly, an algebraic form. This becomes interesting when the pair operator of the conformal metal is *relevant*, i.e. $\nu_b < 0$. Regardless of the precise value of this scaling dimension, one finds that even quite moderate (e.g. electron–phonon) attractive interactions can easily explain T_c of hundreds of kelvins.

This is surely quite an ad-hoc construction. However, as we will see in chapter 10, this describes quite literally how holographic superconductivity works, representing a "microscopic" realisation of this phenomenology. One might now want to have a quick look at the figures showing what the imaginary part of the pair susceptibility looks like in the normal state as a function of energy and temperature (Fig. 10.9 and Fig. 10.10). For those interested in raising T_c, this is the quantity which contains the most direct information on the nature of the pairing mechanism. A "pair-susceptibility spectrometer" should therefore be put at the top of the wish list as the number-one machine to be constructed in this branch of experimental physics.

2.5 The Hertz–Millis model and the critical Fermi surface

Until the advent of holography, all that could be imagined dealing with fermions at a finite density is that they have to be "stored" in one way or another in a Fermi gas. Accordingly, it became a folklore to automatically assume that when fermions are present there will be a Fermi surface as well. Perhaps the central result of AdS/CMT is just the demonstration that there exist states of finite-density fermions that are completely different, lacking any resemblance to the Fermi gas. One should be aware of this conditioning when communicating with the condensed matter mainstream.

This has also shaped the perception of the meaning of a "quantum critical metal": it is automatically assumed that one refers to the Hertz–Millis model, or some variation thereof. This is actually quite an interesting affair in its own right. It could well be more relevant to the phenomena in the laboratory than the much more radical holographic strange metals. Moreover, it has been the subject of a very intensive theoretical effort. Especially during the last few years, new depths have been discovered, rendering the problem theoretically very appealing.

What is "Hertz–Millis"? It just asserts that one can depart from a UV characterised by a Fermi gas, encapsulating the metallic side. In addition, one postulates a bosonic order-parameter field undergoing a quantum phase transition of the kind discussed in section 2.1.3. This critical boson is then in turn coupled to the fermionic excitations by a Yukawa-type coupling. The model is of the form

$$S = \int d^d x \, d\tau \left[(\partial_\mu \vec{\phi})^2 + r|\vec{\phi}|^2 + u|\vec{\phi}|^4 + g\vec{\phi} \cdot \bar{\psi}\vec{\sigma}\psi \right.$$
$$\left. + \bar{\psi}\left(\gamma^0(\partial_\tau - i\mu) + \gamma^i \partial_i + m\right)\psi \right]. \tag{2.38}$$

This is quite like the "graphene" model Eq. (2.6), but now applied to a fermion system at a finite density at a chemical potential μ so that it forms a Fermi gas in the absence of the coupling to the critical bosonic vector field $\vec{\phi}$.

Actually, the problem posed by Eq. (2.38) is the birthplace of the idea of quantum criticality in condensed matter systems. Well ahead of its time, John Hertz introduced the very idea in 1976 [74] (reinterpreting earlier ideas [75]), to a degree already addressing some of the key notions discussed in section 2.1.3, but focussing on the metallic situation. Much later Millis corrected some technical errors in the original formulation [76]. Hertz departed from the weakly coupled, itinerant situation where an instability builds up in the Fermi liquid, but now focussing on magnetic transitions (spin-density waves, "Stoner" ferromagnetism). Insofar as the mechanism of the transition is concerned, it is, at finite temperature, governed by the same time-dependent mean-field/RPA situation that we just discussed for the superconductor, yielding the "loop sum" expression $\chi \simeq \chi^0/(1 - V\chi^0)$. Also for the magnets this describes the development of a relaxational ("paramagnon") pole in the metallic state, signalling the approach to the instability which sets in when the pole hits the real axis. As we emphasised, for thermal transitions this procedure becomes asymptotically correct for dealing with a mean-field critical regime.

For the thermal transition one finds an effective action of the form Eq. (2.35). Departing from the "UV" action Eq. (2.38), one obtains a similar effective action governing the zero-temperature transition, just by "naively" integrating out the fermions. On the metallic side, the massive order-parameter mode is at long wavelength Landau-damped by the decay in electron–hole pairs giving rise to a "$z = 2$" effective action in the case of a non-conserved order parameter (like the antiferromagnet), which becomes, in momentum–frequency space,

$$\mathcal{L} = \left(|\omega| + q^2 + r\right)|\vec{\phi}(q, \omega)|^2 + w|\vec{\phi}|^2 + \cdots, \tag{2.39}$$

ignoring irrelevant terms. The message is that due to the Landau damping the order-parameter dynamics has become relaxational with a dynamical critical exponent

$\mathbf{z} = 2$. For a conserved order parameter like for the ferromagnet, one finds instead $\mathcal{L} = (|\omega|/q + q^2 + r)|\vec{\phi}(q, \omega)|^2 + \cdots$: in this case $\mathbf{z} = 3$.

The difference from the bosonic transitions of section 2.1.3 is therefore that the dynamical critical exponent is increased from the relativistic $\mathbf{z} = 1$ to $\mathbf{z} = 2$ or 3. This has an interesting consequence in the quantum setting: the quantum critical dynamics is governed by an effective dimensionality $d_{\text{eff}} = d + \mathbf{z}$. When $d_{\text{eff}} \geq 4$, one is at or above the upper critical dimension and mean-field theory becomes correct. Given that $\mathbf{z} \geq 2$, one is, in space dimensions $d = 2$ and larger, always dealing with a simple Gaussian quantum critical state, characterised by mean-field exponents and susceptible to perturbative corrections.

This would be the end of the story, were it not that there is an implicit adiabaticity assumption involved, which was ignored in this early work. The fermions can be safely integrated out only when they can be regarded as high-energy degrees of freedom. However, for any finite mass of the order parameter there are always fermionic excitations living at a lower energy. The "backreaction" of the bosonic field on the fermion system should therefore be scrutinised, to make sure that it is devoid of perturbative singularities changing the nature of the fixed point. This has turned into a theoretical subject in its own right in the last twenty years, with a large literature dealing with perturbative diagrammatics (e.g. [77]).

The verdict is that in three or more space dimensions (as is of relevance to e.g. the heavy-fermion systems) one can get away with perturbation theory since the coupling between the massless boson at the critical point and the fermion excitations turns out to be irrelevant. One finds logarithmic corrections to the fixed point, manifesting themselves for instance as the logarithmic divergence of the specific heat for decreasing temperature at the critical point, as is routinely observed in the "good actor" heavy-fermion metals (see section 3.6.1). However, another aspect is that the electrons interact with the critical boson in a way that is similar to how electrons couple to phonons: the critical boson can mediate effective attractive interactions between the electrons. Given that the critical bosons lack an energy scale, while the Fermi liquid with its large Fermi energy is assumed to be still effectively intact, one can identify a "small" Migdal parameter. One can now employ the Migdal–Eliashberg formalism (see Ref. [78] and references therein, and [79]), fully dressing the fermion and boson propagators but ignoring the vertex corrections. This translates in an algebraic "glue function" $\lambda(\omega) \sim 1/\omega^\gamma$, which is as singular as can be in an Eliashberg setting. From such a treatment one finds a superconducting "dome" surrounding the quantum phase transition, with a maximum T_c at the quantum critical point. The perturbative dressings are very strong in the normal state, but they are not of such a kind that the normal state turns into a conformal metal itself. In chapter 10 we will show the result of a *tour de force* computation for the real-frequency dynamical pair susceptibility demonstrating a strong violation of conformality (compare Fig. 10.9 and Fig. 10.10). Especially in the heavy-fermion systems [77], but also in the cuprates, this idea of pairing driven by quantum critical fluctuations is a popular view regarding the origin of the superconducting domes "surrounding" the quantum critical points (section 3.6). This is not yet the end of the story, because recently it was demonstrated that in $2 + 1$ dimensions, in the case of antiferromagnetic and nematic quantum phase transitions, as well as mass-less transverse gauge bosons, there is a subtle but mortal perturbative singularity [80, 81]. The verdict is that the theory flows to strong coupling and a new type of fermionic fixed point is realised. This departs from a small Yukawa coupling, and the theory is by construction still organised "around" the Fermi surface, but the electron–hole excitations moving "on" the Fermi surface can be shown to prolifer-ate in a way that looks like the diagrams of large-N strongly coupled Yang–Mills theory (see chapter 5). The fixed-point physics can be enumerated in terms of the so-called "patch" models which are more amenable to strong-coupling treatments. Using dimensional regularisation in terms of $d = 3 - \epsilon$ [82, 83], it appears that one can demonstrate the existence of a stable non-Fermi-liquid fixed point where the fermion propagators are still organised "around" the Fermi surface, but now acquiring a branch-cut critical form characterised by anomalous dimensions that can be computed in the ϵ expansion. Viewing the Fermi surface as an order param-eter, this might be looked at as if this surface as a whole had turned itself into a truly critical quantum fluctuating entity [84].

Perhaps the most conclusive result associated with this category of models is the one due to Berg *et al.* [85]. This departs from the notion that accidental sym-metries can give rise to sign cancellations. A canonical example is the critical Dirac–Yukawa system at zero density Eq. (2.6), where charge conjugation ensures that the fermion determinants always occur in perfectly matched pairs. Berg *et al.* show that the universal "patch" Hamiltonian can be regularised with a two-orbital lattice-fermion model. This is characterised by an accidental symmetry in "orbital space" having the effect that again the signs cancel out. The effective "bosonic" problem can be handled by quantum Monte Carlo, and it is found that this system is characterised both by Fermi-surface reconstruction and by an unconventional spin-singlet superconducting ground state.

3

Condensed matter: the challenges

If this book ever makes it to a second edition, this is most likely to be the chapter that will have to be most thoroughly rewritten. Is there a need in condensed matter physics for a theory that goes beyond the paradigm which we sketched in rough outline in the previous chapter? If so, would the lessons of AdS/CMT which are found in the later chapters be of any relevance for this purpose?

At present the fog of war is still obscuring the battlefield. This war started some thirty years ago with the discovery of high-T_c superconductivity. Before this event, there was a sense that insofar as metals and superconductors are involved the fundamentals could be understood in terms of the "fifties paradigm" of the previous chapter. In the frenzy that followed the high-T_c discovery, experiments showed that strange things were happening. The reaction of the mainstream was to try to tamper with the established paradigm, to accommodate the anomalies. However, Philip W. Anderson, who was very influential back then, took the lead in insisting that new physics is at work in the copper oxide electron systems [86]. This in turn had a great impact on the research agenda. During the subsequent thirty years the field diversified to other materials, while the repertoire of experimental methods employed to study the electrons in solids greatly expanded. Literally millions of papers were written on the subject. But some of the most basic questions formulated in the late 1980s are still awaiting a definitive answer. It is just impossible to do justice to this large and confusing literature in the present context (see e.g. [87]). We will therefore present here a small selection of subjects, which is intended to form a minimal background for the holographist to communicate with the condensed matter community.

Back in the late 1980s the great puzzle was why the superconducting transition temperature could be as high as 150 K, given that the conventional phonon mechanism runs out of steam at 40 K or so. It was also realised early on that the electron systems in cuprates are characterised by unusually strong inter-electron repulsions. An aspect that is well understood in these systems is the microscopic physics.

Unusually deep ionic potentials have the effect that the electron system behaves to a degree like a system of isolated atoms. This situation is typically encountered in transition-metal salts and metals characterised by partly filled f-shells. The dominant role of the Coulomb repulsions in atomic physics survives to a degree in these solids, having the ultimate consequence that the electron system can turn into an insulator because of the strong atomic-like repulsions – resulting in the Mott insulators. High-T_c superconductivity can be brought about by doping such a Mott insulator formed in the copper oxides. Invariably, one finds anomalous behaviours in systems characterised by this kind of "chemistry", and in section 3.1 we will introduce the canonical Hubbard, t–J and Anderson lattice models forming the point of departure of this theatre of "correlated electron systems".

At present, the sheer magnitude of the superconducting T_c is no longer considered to be a problem of principle. An enormous body of work dealing with the question of whether strong local repulsions can actually be responsible for the superconductivity appeared. Although none of it is truly conclusive, even on departing from very different limits and approximations one invariably finds an appetite for superconductivity. The key is that such repulsion-driven superconductivity is characterised by an unconventional pairing symmetry, like the $d_{x^2-y^2}$ order parameter found in the copper oxides. Although none of the existing methods is good enough to address matters quantitatively, just the fact that the pairing is associated with the large-energy electronic Coulomb interactions implies that T_c can be "high". The holographist should be aware of the condensed matter methods which are believed to render reliable information in this regard. In section 3.3 we discuss the state of the art in the weak-coupling regime, where it becomes particularly transparent why electrons do bind in unconventional Cooper pairs despite the repulsions. When the interactions become strong, only heavy numerical machinery is available as a source of somewhat reliable information. In section 3.4 we discuss the "density-matrix renormalisation group" (DMRG) and the related "tensor-product-state" methods which correspond to variational Ansätze that are based on modern quantum information insights. Of particular interest for the holographist are the "dynamical mean-field theory" (DMFT) methods discussed in section 3.5 which use the limit of large numbers of dimensions to acquire control. This approach shares with holography the intriguing property of a quantum dynamics turning local in space in systems with a finite density of fermions.

The most important question posed by Anderson [86] is after all these years still awaiting a conclusive answer: is the normal state realised above T_c a "non-Fermi liquid", a new metallic state of fermions at a finite density that is qualitatively different from the Fermi liquid? If this is the case, this implies that the pairing mechanism is also distinct from the BCS mechanism that departs from a Fermi-liquid normal state. This is surely the most pressing matter in the present context of holography. When we turn to finite-density holographic matter starting in chapter 8, we will find that such strange metals are ubiquitous. These are quite suggestive with regard to the experimental observations in the cuprates [88]. The cuprates might form quantum critical (or conformal) metallic *phases*, which do not require fine tuning to the quantum critical points of the previous chapter. These phases are in turn very unstable forms of matter that tend to spawn a variety of ordered phases upon reducing temperature, where especially superconducting ground states are quite natural.

It remains to the present day controversial whether the non-superconducting states are non-Fermi-liquid systems. On the theory side, there are just no working conventional methods for describing metallic states lacking even a Fermi surface – the best one can do is invoke the "critical Fermi-surface" state arising in the Hertz–Millis context (section 2.5). The DMFT is, on the computational side, again the holographist's best friend: it is formulated at finite temperature, and the computations show that typically spectral functions rapidly become an incoherent affair when the temperature is raised.

Given the lack of truly powerful mathematical methods, one has to eventually fall back on experimental information to address the non-Fermi-liquid puzzle. The interpretation of any experimental result is in the eye of the beholder: after all these years, the controversy is still raging about whether the experiments signal that the metallic states need a new physics principle or whether everything can eventually be explained by sufficiently complicated perturbative physics around the Fermi liquid. We will end this chapter with the most risky affair found in this book: a discussion of the experimental state of the art both in the cuprates and in the heavy-fermion systems. Given the volatile nature of this subject, we will just do this in the form of a list of the most mysterious features revealed by experiment. This is our list of homework assignments for the student of holography: when a profound explanation for all these items has been found, one can declare victory.

Assuming that non-Fermi liquids exist, the next hurdle faced by the holographist is that the ultraviolet conditions responsible for this behaviour in AdS/CFT appear to be very different from what should be the cause in the laboratory systems. In the holographic version one departs from a UV defined by an extremely strongly interacting large-N Yang–Mills conformal field theory. How does this relate to the mundane "chemistry" of the electrons in copper oxides and so forth at the Ångström scale? It is actually quite hard to avoid Fermi-liquid physics in simple electron systems. The standard answer is that one needs the influence of the strong local repulsions which survive in the strongly correlated metals (the "Mottness") to stand a chance of being able to destroy the Fermi liquid.

There is surely no mathematical theorem available proving that Mottness will obliterate the Fermi liquid. In section 3.2 we will briefly discuss a line of arguments, originally introduced by Anderson [86], suggesting that for quantum statistical reasons it should be impossible to form Fermi liquids with large Fermi surfaces when the Mottness is fully developed. We will then turn to Anderson's resonating valence bond (RVB) construction. This refers to an Ansatz wave function for a superconducting state. Although evidence is still lacking for its literal applicability, it has been greatly influential as a powerful metaphor teaching us how to think differently about superconductivity. A main lesson is that a BCS-like superconducting state can survive despite the Mottness UV that might form an insurmountable barrier for a Fermi-liquid normal state. We will discuss how this evolved into the contemporary understanding of the BCS state as an entangled, topologically ordered state described by BF and Ising gauge theory. The RVB theory has also been the birthplace of the related development which eventually led to the discovery of the various $(Z_2, U(1))$ spin liquids which are in a sense like insulating versions of the BCS superconductor, vividly illustrating how simple $(Z_2, \text{compact } U(1))$ gauge-theory structure emerges in the deep infrared, enumerating the entanglement of the vacuum.

3.1 The repertoire of strongly correlated electron models

Departing from the fermions of condensed matter physics, the Fermi liquid is amazingly resilient when these systems live in the (effective) Galilean continuum. This situation is closely approached by the low-density two-dimensional electron gasses realised in semiconductors. On the microscopic scale there is just the long-range Coulomb interaction and the kinetic energy, with associated energy scales $E_c \sim 1/r_s$ and the bare Fermi energy $E_F \sim 1/r_s^2$, respectively, where r_s is the distance between the electrons. According to experiments, it appears that the Fermi liquid survives up to the very low densities at which $E_c \simeq 40 E_F$ [89]. Going back much further in history, the same theme is behind the great impact of Landau's Fermi-liquid idea. This was first proposed in the context of ^{3}He physics. Around 3 K ^{3}He turns from a gas into a liquid, which is initially a dense classical van der Waals fluid. This classical fluid is a highly correlated affair that can be viewed as a near traffic jam of impenetrable balls. Upon lowering the temperature to the mK range, this miraculously turns into a near-perfect incarnation of a Fermi gas where the classical balls have turned into non-interacting fermions that communicate only via the Pauli principle, at the expense that these effective helium atoms have a mass ten times as high as the real ones. It was Landau's genius that allowed him to recognise this miracle from the available experimental information. In the years since there has been very little progress in the understanding of how this works. The verdict is that the stability of the Fermi liquid as an IR fixed point is thoroughly understood, but how it emerges from an interaction-dominated UV is a problem regarding which we are still completely in the dark.

It appears that invariably Mottness is at work in the laboratory systems where one observes strange-metal behaviour, namely the heavy-fermion systems and the cuprate high-T_c superconductors (sections 3.6.1 and 3.6.2). Mottness refers to Nevil Mott, the theorist who identified the Mott insulator. Distinct from the standard band insulator, the Mott version insulates because of the *electron repulsions* becoming dominant due to the presence of a strong lattice potential. The physics is very simple. When one is dealing with strong ionic potentials, causing the density of the electrons in space to become very inhomogeneous, it is just a better idea to depart from the limit of isolated atoms. Turning two neutral atoms into a negatively and positively charged ion requires quite a large amount of energy (10–20 eV depending on the atom), just because of the Coulomb repulsion associated with pushing the electrons together in the negatively charged ion.

When one is dealing with valence electron states in solids that are still "atomic-like" because of the effectively strong ionic potentials, this atomic-limit repulsion survives in the solid. It is to a degree screened, but not all the way as in the very itinerant metals and insulators discussed in the solid-state physics textbooks. This "remnant" of the atomic Coulomb interaction is particularly strong when one is dealing with 3d transition-metal salts and anything containing 4f (lanthanide) and 5f (actinide) partly filled shells, while it is a factor to keep an eye on in organic materials (2p electrons), 4d and 5d salts and 3d metals. This screened atomic Coulomb interaction is called the "Hubbard U". That it is a particularly significant factor just follows from experiment: to the present day the theoretical understanding of this microscopic physics is far from complete. Its consequences for the microscopic electronic structure in the form of the observation of "Hubbard bands" were nailed down in the early 1980s, using photoemission spectroscopy, which was back then still quite crude [90, 91, 92, 93]. This added legitimacy to a tradition of theoretical toy models that already had a long history in condensed matter physics [87], with as prime example the minimal Hubbard model. This describes a single species of spin-full ($\sigma = \uparrow, \downarrow$) tight-binding fermions hopping with a rate t on a lattice, subjected to a local repulsion, U,

$$H = - \sum_{\langle ij \rangle \sigma} t c_{i\sigma}^{\dagger} c_{j\sigma} + \mu \sum_{i\sigma} n_{i\sigma} + U \sum_{i} n_{i\uparrow} n_{i\downarrow} . \qquad (3.1)$$

This is just the minimal fermionic version of the Bose Hubbard model Eq. (2.3). When one is dealing with the half-filled system (one fermion per site) one finds, as for the Bose version, that the system turns into a "traffic jam" insulator when U exceeds the band width W. The difference is now that a spin system is left behind, and by virtual hopping fluctuations these spins interact via the super-exchange interaction $J = 2|t|^2/U$, forming an antiferromagnetic Heisenberg spin system (for "realistic" systems see Ref. [94]). One can subsequently dope the system away from half-filling, for instance by removing electrons. For large U the low-energy physics can now be captured in a model that is seemingly even simpler than the Hubbard model. Besides the spins forming the Heisenberg spin system, one has the "holes" (missing spins), which can delocalise as long as they do not cause doubly occupied configurations. This is the famous t–J model,

$$H_{tJ} = J \sum_{\langle ij \rangle} \vec{S}_i \cdot \vec{S}_j - t \sum_{\langle ij \rangle \sigma} (1 - n_{i-\sigma}) c_{i\sigma}^\dagger c_{j\sigma} (1 - n_{j-\sigma}). \tag{3.2}$$

Immediately after the discovery of high-T_c superconductivity it became clear that the "parent" materials are antiferromagnetic Mott insulators characterised by a rather large U, while the superconductivity arises when holes are introduced by chemically doping the material. This triggered a gigantic theoretical effort to solve the doped-Mott-insulator problem: according to Google scholar nearly 600,000 papers on the Hubbard model and more than 3,000,000 on the t–J model have been published (see Ref. [87] for a historical perspective). Very differently from the boson version, despite this effort very little has been established in terms of rigorous and general theory. As we will discuss in a moment, the difficulties are deeply rooted in the statistical side: the fermion-sign problem.

To complete this exhibit of "standard models" of correlated electron physics, there is one more important class: the Anderson and Kondo lattice models. These are of relevance to the heavy-fermion systems which are found in metallic systems containing 4f or 5f elements. In the case of the Hubbard model it is assumed that only one species of electrons near the chemical potential is subject to the Hubbard repulsion. In reality there will of course be quite a number of electron "species" in the real band structure. This is also a theme in principle in, for instance, the oxides – next to the transition-metal 3d shell, there are oxygen 2p states and transition-metal 4s states. However, although the reasons may be subtle, these can be ignored in the first instance, since they effectively behave like a single band system [87]. However, this fails badly when one is dealing with real metallic materials. In the heavy-fermion intermetallics there are next to the f-bands also other bands crossing the Fermi energy that would behave as good Fermi liquids were it not that the f states are around. The simplest case is that one is dealing with one species of such normal electrons "c" with a dispersion relation ε_k, hybridising with f states localised at sites i via a hopping-matrix element V_{ik}. These f states are subject to a Hubbard interaction U

$$H_{AL} = \sum_{i\sigma} \varepsilon_f f_{i\sigma}^\dagger f_{i\sigma} + \sum_{k\sigma} \varepsilon_k c_{k\sigma}^\dagger c_{k\sigma} + \sum_{ik\sigma} V_{ik} \left(c_{k\sigma}^\dagger f_{i\sigma} + \text{h.c.} \right) + U \sum_i n_{i\uparrow}^f n_{i\downarrow}^f. \tag{3.3}$$

When one tunes the f-electron system towards half-filling while U is large, this system turns again into a spin system. The difference from the Hubbard Hamiltonian is now that these spins ($\vec{S}_i$) do not directly interact with each other but instead undergo a "Kondo" exchange interaction with the conduction electrons of the form $\sum_{k,k',i} (V_{ki} V_{ik'}/U) \, \vec{S}_i \cdot \sum_{\alpha\beta} (c^\dagger_{k\alpha} \vec{\sigma}_{\alpha\beta} c_{k'\beta})$. In this limit the Anderson lattice problem reduces to the "Kondo lattice".

This problem is clearly more complicated than the Hubbard and t–J problems, and even less is known for sure. Even the problem of a single correlated impurity is highly non-trivial: this is the celebrated Kondo impurity problem that was cracked completely after a concerted effort in the 1970s and 1980s. To give a short summary, the local spin forms a complex, many-body singlet bound state together with an effective spin formed from the conduction electrons, at an energy scale that is exponentially small in the bare coupling $T_K \sim \exp{(-1/(J N_0))}$, where T_K is the Kondo temperature, J the Kondo exchange and N_0 the density of states at E_F. The key to the solution is that this $(0 + 1)$-dimensional problem can be bosonised and the solutions are obtained from the $(1 + 1)$-dimensional boundary conformal field theory [95].

3.2 Mottness, non-Fermi liquids and RVB superconductivity

Departing from the Hubbard or Anderson lattice models, it is obvious that conventional perturbation theory can be mobilised when U is small compared with the bandwidth(s). We will briefly discuss this limit in the next section, and perhaps the surprise is that in weak coupling one finds superconductivity also for repulsive Us, albeit with an unconventional pairing symmetry. However, the physics gets really interesting when U exceeds the band width. Here the question of whether it is at all possible that the system renormalises back to a conventional Fermi liquid characterised by a "large"-Luttinger-volume Fermi surface arises. In the free system the Fermi surface would fill up half the Brillouin zone, and upon hole doping this Luttinger volume would decrease like $1 - x$, where x is the number of holes/dopants. A problem of principle arises with this Fermi surface, although it is easy to argue that there is no fundamental obstruction to the formation of a small-Luttinger-volume Fermi surface $\sim x$.

The best available arguments hinting at a no-go theorem for the Fermi-liquid ground state are of a quantum-statistical origin. A first view of this kind was emphasised by Anderson [86]. This is the theme of spectral weight transfers, expressing that the relationship between the number of particles and the dimensionality of Hilbert space in a doped Mott insulator is very different from that in the corresponding free-fermion system. Let us sketch here the essence of the argument, referring the interested reader to a recent review [96]. We depart from a single-band free-fermion system at a density of $1 - x$. The total number of ways to add or remove an electron in such a system is x and $1 - x$, respectively. Consider next the strongly interacting Mott insulator at half-filling. There will be one electron per site, and upon adding and removing an electron a doubly occupied site and an empty site will be created, respectively, costing a net energy U. One finds accordingly the unoccupied upper and occupied lower Hubbard bands separated by the Mott gap U. Now remove one electron: there are still $N - 1$ sites where one can remove an electron, exciting a state in the occupied lower Hubbard band. However, since an electron has been removed, there are only $N - 1$ states left in the high-energy upper Hubbard band! Instead, there are now two ways to fill the hole site, and these states will end up as unoccupied states right at the chemical potential since it does not cost the energy U to add these electrons. Summarising, one has a state counting $1 - x$ for the occupied states and $2x$ for the low-energy unoccupied states, while the high-energy upper Hubbard loses weight according to $1 - x$. The bottom line is that, very differently from what occurs in a doped band insulator the high-energy states of the upper Hubbard band are "raining down" under the influence of the hole doping into the partly filled lower Hubbard band. Such spectral weight transfers are very pronounced in various spectroscopic measurements on the cuprates [96]. Given the very different way in which the dimensionality of the low-energy Hilbert space changes as a function of density in the doped large-U Mott insulator in comparison with what occurs in the free system, Anderson conjectured that it should be impossible to adiabatically continue the Fermi-liquid ground state to the metal characterised by the microscopic Mottness.

A complementary way to view these quantum statistics aspects is in terms of a first-quantised path-integral representation, to inspect how the exchange statistics works out in a doped Mott insulator (for recent reviews see [97, 98]). At low energies the UV fermions are completely localised in the Mott insulator: these turn into spins, and spins live in a Fock space that is not subjected to anti-symmetrisation! This surely changes when an electron is removed since now fermionic exchanges associated with the moving hole will take place. Surprisingly, these "fermion signs" can be enumerated exactly in terms of the so-called "phase strings". Consider a world history of the configurations of holes and spins. Take arbitrarily the down spins as a background, to find that every time a hole and an up spin exchange by a hopping process ("collision") a minus is added. The overall sign of any world history c in the first-quantised path integral is set by $(-1)^{N_h^\uparrow(c)+N_{ex}^h(c)}$, where $N_h^\uparrow(c)$ is the number of up-spin–hole collisions and $N_{ex}^h(c)$ the number of hole exchanges that one just counts like conventional fermions. One infers immediately that, compared with the case of free fermions, this signals a large "deficit of fermion signs", while in addition the signs are no longer "hard-wired" as in the Fermi gas but become dependent on the dynamical details of the world histories. One also infers that, since the holes are subject to conventional fermion exchanges, they can in principle form small Fermi surfaces $\sim x$. This is, for instance, easily realisable by freezing the spins out by applying a large external staggered field [98]. Some interesting results appeared recently, highlighting the highly non-trivial nature of this phase-string statistics. Using the numerical DMRG method (see section 3.4) one can at will switch off the phase-string signs. It turns out that these spin–hole "collision signs" cause a single hole to "spontaneously" Anderson localise in an otherwise perfectly translationally invariant lattice, lacking any quenched disorder [99]. It also appears that the phase strings are crucial for the Cooper-pairing tendencies revealed by the DMRG computations [100].

Let us now return to 1987, when Anderson suggested the RVB construction [101] as an explanation for high-T_c superconductivity. He relied on the time-tested procedure of guessing an Ansatz wave function. He found here inspiration in an old idea by the quantum chemist Pauling, who had proposed it in 1938 as a theory of the metallic state as such. Pauling was inspired by the Heitler–London theory of the covalent bond in H_2 and accordingly he viewed the origin of the special stability of the benzene molecule as being rooted in the coherent superposition ("resonance") of the two double (π-electron spin-singlet pair)–single (no pair) conformations possible for a ring of six carbon atoms. He subsequently suggested that the metallic state would be of a similar nature: construct all possible configurations of all single and double bonds between the atoms in the metal, with a ground state that is a coherent superposition of all these configurations. From a contemporary perspective this is quite a visionary idea, since this is now understood to be a form of long-range entangled quantum matter.

As early as in 1974 Fazekas and Anderson asserted that this could be a reasonable ground state when one is dealing with frustrated Heisenberg antiferromagnetic spin systems. Now the role of double bonds is taken by pair singlets formed from the spins. Explicitly, in terms of the bare-electron operators such a pair singlet is created by $b_{ij}^\dagger = \left(c_{i\uparrow}^\dagger c_{j\downarrow}^\dagger - c_{i\downarrow}^\dagger c_{j\uparrow}^\dagger \right)$ such that a particular "valence-bond" configuration can be written as

$$|VB, k\rangle = \prod_{(i,j;k)} b_{ij}^\dagger |vac\rangle, \tag{3.4}$$

where k is a particular covering of the entire lattice by the pairs of sites (i, j), such that no site is repeated in the product. RVB states can be written in full generality as

$$|RVB, l\rangle = \sum_k \Psi_k^l |VB, k\rangle, \tag{3.5}$$

where $|\text{VB}, k\rangle$ are coverings of the lattice with spin-singlet pairs that can involve sites that are arbitrarily far apart. Such a "long-range" RVB state of particular interest is

$$|\text{RVB}\rangle = P_{\text{d}}\left[\sum a(r_i - r_j)b_{ij}^{\dagger}\right]^{N/2}|\text{vac}\rangle, \quad P_{\text{d}} = \prod_i\left[1 - n_{i\uparrow}n_{i\downarrow}\right]. \quad (3.6)$$

The function $a(r)$ expresses the range of the singlet pairs as well as the pairing symmetry – d-wave turns out to yield the lowest energy for the reasons explained towards the end of section 3.3. The important entity here is the Gutzwiller projector P_{d}, having the effect that after projection no more than one VB ends at a site in any of the valence-bond configurations. At half-filling such a state is a total spin singlet while it does not break translational symmetry: this was the first example of a *quantum spin liquid*, the study of which has become a thriving subfield in condensed matter, as we will discuss below.

What has this got to do with superconductivity? All that remains is to realise that the wave function can be written in momentum space as

$$|\text{RVB}\rangle = P_{\text{d}}\Pi_k\left[a_k c_{k\uparrow}^{\dagger}c_{-k\downarrow}^{\dagger}\right]^{N/2}|\text{vac}\rangle$$

$$\propto P_{\text{d}}P_{N/2}\exp\left(\sum_k a_k c_{k\uparrow}^{\dagger}c_{-k\downarrow}^{\dagger}\right)|\text{vac}\rangle. \quad (3.7)$$

This is just a Gutzwiller-projected BCS superconductor! Such Gutzwiller-projected states have a long history as variational Ansätze for Hubbard models, and at least the projection faithfully represents the Mottness and there is no conflict between this state and the requirement that doubly occupied states have to be projected out. Away from half-filling the Gutzwiller-projected BCS states represent genuine superconductors, although the superfluid density is strongly reduced since the Gutzwiller projection ensures that the charge condensate completely disappears at half-filling. This is easy to understand qualitatively: in the presence of holes the electrons continue to be bound together in spin singlets – this is the "pairing force". But these effective bosons now delocalise because of the presence of the holes, thereby forming a condensate of charge $2e$. The greater surprise is that this vacuum also carries Bogoliubov excitations, propagating fermionic excitations carrying a quantum number $S = 1/2$. This is easy to see: just break up a pair in the BCS wave function and this propagating Bogoliubov fermion will survive the Gutzwiller projection even at half-filling. This flexibility of the BCS ground state allowing it to survive in conditions that are very adverse to simple Fermi-liquid-based superconductivity has been further explored in the intervening years. The BCS state is now understood as one characterised by topological order, governed by BF topological field theory [102], while for a number of purposes it can also be linked to the topological order associated with the deconfining state of Ising gauge theory [84].

Let us sketch here the essence of the arguments. The BCS superconductor is, in the deep infrared, like a Bose condensate formed from charge-$2e$ bosons. However, differently from a real Bose condensate it also supports Bogoliubov excitations, which are electrons with charge e and spin $1/2$ living in the superconducting vacuum. Obviously, the charge of the electron is completely screened at distances larger than the screening length and it is actually an electrically neutral particle just carrying a spin (a "spinon"). However, when one encircles the vortex/fluxoid of the charge-$2e$ superconductor with a spinon, one finds out that the latter accumulates a topological phase π: the vortex is bound to the flux (or "vison") associated with a Z_2 gauge field. Since these visons are massive, the superconductor corresponds to the deconfining state of the gauge theory [103].

This comes to life when it is combined with the vortex duality in $2+1$ dimensions of section 2.1.2. Upon proliferating and condensing the vortices of the superconductor a charge-$2e$ insulator is realised. However, together with the vortices also the visons proliferate, and therefore the state is also a confining state of the Ising gauge theory. This implies that the spinons are confined in integer-spin gauge-singlet excitations: this insulator is a charge-$2e$ band insulator characterised by integer-spin excitons as spin-full excitations. However, impose now that, for whatever reason, the (anti)vortices bind in pairs when they proliferate. The dual Bose Mott insulator is now characterised by a charge e. However, the visons also proliferate in pairs and, since these neutralise each other, the Z_2 gauge vacuum continues to be deconfining. The result is that one obtains an "electron-like" (charge e) Mott insulator that carries still the propagating Bogoliubov fermions/spinons which are spin-$1/2$ excitations. Departing from a d-wave superconductor these are characterised by massless Dirac spectra: this is the "nodal-liquid" construction devised by Senthil and Fisher [84].

This is in a way the simplest and most transparent construction of a "spin liquid". The "spin solid" is the ordered antiferromagnet, which is characterised by excitations (magnons) that are triplets. In the spin liquid no symmetry is broken, while the triplets fractionalise in spin doublets. This is in turn associated with a long-range entanglement in the ground state that can be enumerated in terms of the deconfinement of the Ising gauge theory. The surprise is that this form of spin-liquid entanglement is the essence rendering the BCS superconductor different from a simple Bose condensate. At no point in this construction does one need as input that the BCS superconductor can be formed from the Fermi-liquid normal state: the BCS fixed point is just more general than what is suggested by the classic BCS theory.

The reader might be surprised: by invoking these seemingly indirect arguments we have suddenly started to use the language of non-perturbative gauge theory.

Differently from the case in the Yang–Mills theory of high-energy physics, these gauge fields simply do not exist in the UV. Instead they emerge in the deep infrared, and they are associated with the collective behaviour of the special form of matter which is realised. These fall within the same category as the topological insulators discussed in section 2.2: this gauge-theoretical structure is just enumerating the special quantum entanglement of the vacuum. In a reversal of the situation of high-energy physics, the *confining* state is associated with the trivial vacuum, while the deconfining state is characterised by long-range entanglement, having as a consequence that it carries excitations characterised by emergent quantum numbers. As in the fractional quantum Hall case, these typically correspond to a fractionalisation of the microscopic quantum numbers. The difference is that we now depart from microscopic spin degrees of freedom, and this part of the topological-order portfolio is therefore called the regime of "quantum spin liquids".

AdS/CFT departs from a UV Yang–Mills theory characterised by a very large gauge symmetry. Accordingly, (de)confinement arises naturally, as we will see in, for instance, chapter 6 dealing with the field theory at zero density. In the finite-density systems of chapter 8 onwards, an explicit understanding of what is happening in the boundary field theory is lacking, and one has to rely completely on the holographic duality. This suggests in turn that at finite density (de)confinement phenomena of a new type occur. The holographic strange metals which are the main actors in this story are somehow related to deconfinement, and are therefore called "fractionalised" phases. One of the main open questions is whether this fractionalisation phenomenon is of the same emergent, long-range-entanglement kind as in the spin liquids, or whether it is just special to the Yang–Mills ultraviolet. An obvious difficulty is that in this particular context only quite primitive emergent gauge symmetries have been identified (Z_2, compact $U(1)$), and one has a long way to go to arrive at the large-N Yang–Mills theories of AdS/CFT. In this regard, the "string-net" construction introduced by Levin and Wen [48, 104] is inspirational. This revolves around quantum liquids formed from microscopic strings defined on a lattice, subjected to particular reconnection rules. The entangled vacua formed from these string nets support a low-energy sector that is in a precise correspondence to non-Abelian Yang–Mills.

To illustrate these ideas further, let us sketch the simplest construction of this kind: the Z_2 spin liquid; for a concise discussion and references see Ref. [105]. This shows in an explicit fashion how the magic of e.g. the Senthil–Fisher nodal liquid is tied to the long-range entanglement of the vacuum. It also highlights the role of the RVB Ansatz in the history of the subject.

We have already alluded to the capacity of Anderson's RVB Ansatz to describe an insulating state that consists of spin-singlet pairs that completely cover the lattice as a state associated with a spin-only Heisenberg-type model. Since this state is not breaking any symmetry, it is also a quantum spin liquid. The Gutzwiller-projected BCS state involves singlet pairs that are of long range, and a first simplification is to insist that these singlet pairs are formed from nearest-neighbour spins on the lattice. As in the Pauling cartoon, these valence bonds form simple "dimers" and the ground state of the insulating system corresponds to a coherent superposition of all possible dimer coverings of the lattice. The simplest way to add dynamics is in terms of a quantum dimer model describing the hopping of pairs of dimers on the lattice [106]. It turns out that the ground state on the triangular lattice of such a dimer model consists of an equal-weight coherent superposition of all dimer configurations [107]. This state is clearly long-range entangled, and it was realised early on (using the slave mean-field language, see the next paragraphs) that this equal-weight superposition implies the topological order associated with the deconfining state of the Ising gauge theory [108, 109]. The first step is to identify the excitations: in the spin liquid one can just break up the dimer into two $S = 1/2$ excitations and these will freely delocalise in the VB quantum liquid: these are the "spinons" [106]. That this relates to deconfinement becomes clear on considering the conventional "confining" states: both in the ordered antiferromagnet and in crystals formed from VBs only $S = 1$ excitations (magnons, triplet excitons) occur, while the spinons are here literally confined.

The subtlety is that this fractionalisation goes hand in hand with the appearance of an extra hidden excitation that is entirely associated with a change in the nature of the long-range entanglement of the RVB ground state. On changing a sign in the coherent superposition of VB configurations, one finds a low-energy excitation that behaves precisely like the vison of the Z_2 gauge theory. It corresponds to a point particle with a gauge seam attached to it that disappears to infinity, but now "formed" entirely from the changes in the long-range entanglement of the excited state (see Fig. 5 in Ref. [105]). The conclusion is that this equal-weight superposition of VB configurations corresponds to the deconfining state of the Ising gauge theory. It includes the gauge fluxes being massive (the visons) and the fractionalisation of the quantum numbers of the matter excitations (spinons). Notice that the same Z_2 topological order underpins the "toric code" of Kitaev, which was invented for topological quantum-computing purposes [110]. This is just the tip of the iceberg: the study of spin liquids is a very active research area, where several promising candidates have been identified in frustrated spin systems as realised in the laboratory [105, 111]. There is a large theoretical literature [48], which is at present rapidly evolving. On the technical side, one should be at least aware of the "slave-particle theory", a mean-field procedure resting on the (vector) large-N limit, which has been very influential in the history of the subject (for an exhaustive review see [112]). It is more of a procedure than a theory, since it never came under real mathematical control, but it has been quite useful in unveiling the topological order in spin liquids [113]. Especially in the early era of high-T_c superconductivity it was popular as a theory for the doped Mott insulator, but due to its lack of empirical success and mathematical control it has remained controversial until the present day. It is also quite influential in the context of the heavy fermions and the Anderson lattice model since it yields a very simple, intuitive interpretation of the formation of the "heavy band structure" [95, 114].

Let us sketch here the basic ideas, taking the t–J model as the context [112]. Crucially, it departs from the assertion that a spin liquid characterised by fermionic spinon excitations is formed in the Mott insulator. One invokes the vector large-N limit to acquire control, by lifting the $SU(2)$ symmetry of the physical spins to $SU(N)$, where N is large, while the microscopic spin S is taken to be small. The $SU(N)$ Heisenberg spin system is parametrised in terms of CP_N fermions on site i as $\vec{S} \propto \sum_{a,\alpha,\beta} \psi^\dagger_{i,a\alpha} (\vec{\sigma})_{\alpha\beta} \psi_{i,a\beta}$, where a is a flavour label. This requires that the number of fermions at every site i is precisely conserved as $\hat{n} = \sum^N_{a=1,\sigma} n_{a\sigma} = 2S$, which can be implemented by e.g. local Lagrange multipliers $\sum_i \lambda_i (\hat{n} - 2S)$. In the infinite-$N$ limit the constraint can be taken to be global ($\lambda_i \rightarrow \mu$, a chemical potential) which is the first step in the mean-field procedure. One is, however, still left with a fermion problem at infinite coupling, characterised by a Hamiltonian of the form $\sum_{\langle ij \rangle} \vec{S}_i \cdot \vec{S}_j \sim \sum_{\langle ij \rangle} \psi^\dagger_i \psi_i \psi^\dagger_j \psi_j$. The next (very hazardous) step is to assume that this can be decoupled in a naive mean-field fashion involving particle–hole VEVs ($\langle \psi^\dagger_i \psi_j \rangle$) and particle–particle VEVs ($\langle \psi^\dagger_i \psi^\dagger_j \rangle$). The first VEV turns the localised fermions into a free Fermi gas, which subsequently turns into a BCS superconductor due to the particle–particle VEVs. The Bogoliubov fermions of this mean-field "solution" describe the spinons representing the fractionalised excitations of the spin problem in the guise of the previous paragraphs. However, at finite N the constraint has to be imposed, and this can be implemented by gauging the spinons with a compact $U(1)$ field ϕ_i: $\psi^\dagger_i \rightarrow e^{i\phi_i} \psi^\dagger_i$. In the bare action this gauge field is infinitely coupled, but it is subsequently asserted that by integrating out the UV fermions an induced Maxwell action will be generated. When this coupling becomes strong enough the compact $U(1)$ gauge theory can get into a deconfining regime, with the caveat that the massless Dirac fermions of the effective d-wave superconductor (on the 2d square lattice) can keep the monopoles out of the vacuum even in $2 + 1$ dimensions. In the deconfining phase the spinons are the real excitations. One is now dealing with the deconfining phase of the compact $U(1)$ gauge theory and this is accordingly a realisation of the $U(1)$ spin liquid.

Up to this point we have been dealing with the spin liquid formed in the Mott insulator itself. How should one deal with the doped system? One writes the electron operator occurring in the hopping term of the t–J model as a composite of the fermionic spinon and a boson representing the charge, the "holon" $b^\dagger_i$ as $c^\dagger_{i\sigma} = \psi^\dagger_{i\sigma} b_i$. One now assumes in addition that the holons will be Bose

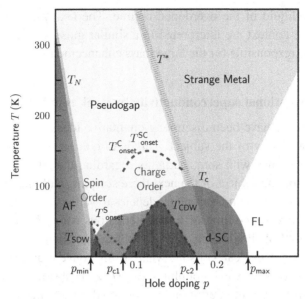

Figure 3.1 The phase diagram of the hole-doped cuprate high-T_c superconductors as a function of doping and temperature. It departs from an antiferromagnetic Mott insulator, turning into a metallic-like system that superconducts at low temperatures; this is the dark grey area starting at a doping p_{min} and ending at a doping p_{max}. At higher temperatures one finds the "pseudogap regime" (lightest grey) characterised by a depletion of states below a characteristic scale (T^*) that decreases with increasing doping. This regime exhibits a plethora of competing static and fluctuating orders (see the main text). With increasing doping the superconducting T_c increases and becomes maximal at an "optimal doping". The pseudogap physics has disappeared and here one finds only the celebrated "strange metal" (medium-grey region). In the overdoped regime the superconductivity deteriorates while the normal state turns into a Fermi liquid that emerges from the higher-temperature strange metal. Figure adapted from [115].

condensed when their density becomes finite in the doped system. According to this "slave mean-field theory" one finds a phase diagram [112] that was early on quite popular as a caricature of the real-life phase diagram of the high-T_c superconductors (see Fig. 3.1). On departing from half-filling the holon condensation temperature increases linearly with the doping, while the spinon BCS gap decreases linearly from a maximum at half-filling. The region where the spinons are condensed and the holons are in the normal state was associated with the pseudogap, which back then was interpreted as just a spin gap, since competing orders had not yet been discovered. The regime where both holons and spinons are condensed is the superconducting state, and the strange metal was ascribed to the regime where both spinons and holons are in their normal states. Finally, the uncondensed spinon Fermi gas together with the condensed holons was associated with the Fermi liquid of the overdoped regime. The heavy Fermi liquids in the Anderson-lattice context are interpreted in a similar guise where the small holon VEVs are made responsible for the large mass enhancements.

3.3 Unconventional superconductivity by weak repulsive interactions

Up to this point we have been discussing qualitative ideas that were put forward during the long history of the subject. This is to quite a degree guess work, but what is actually known with some certainty regarding the Hubbard model and so forth? This will be the subject of the next three sections, with the main conclusion being that invariably one finds strong tendencies towards superconductivity!

Let us start out with the limit of weak interactions where U is small compared with the band width W. Predictably, this is a subject that has generated an extremely large literature, since one can get somewhere using conventional perturbation theory. The claim is that the weak-coupling limit $U \ll W$ has recently been brought under full control using the perturbative functional renormalisation group [116]. The (perhaps) surprising conclusion is that, although one starts out from mere repulsive interactions, this system has generically a *superconducting* ground state, albeit one characterised by an "unconventional" (i.e. non-s-wave) order parameter. There is a vast literature attempting to extend the perturbative procedures up to intermediate couplings $U \simeq W$, but this is invariably based on re-summing subsets of diagrams and therefore lacks rigour (see e.g. Ref. [117] and Ref. [118]). However, in the meantime there is a widespread consensus that superconductivity can be caused by such microscopic repulsions also when the interactions become large. The reason is that, in the wide variety of approaches that we will discuss further below, it shows up in one way or another. Given that the associated energy scales are large, this implies that superconductivity can occur at high temperature: there is no longer a problem in principle with a superconducting order setting in at temperatures well above 100 K.

A crucial ingredient for this to work is that the superconducting order parameter is unconventional: p, d, s±, ... instead of the "phonon" s-wave. A quite sizeable family of such "strongly correlated" superconductors has been identified in the laboratory and all of them share this property [119]. It appears at first sight quite unreasonable that merely repulsive interactions can cause electrons to bind in pairs. This intuition is, however, defeated by a subtlety rooted in a quantum-mechanical phasing effect. These superconductors turn out to work out in a qualitatively similar way, regardless of whether the coupling is weak or strong. A particularly simple cartoon version is the "spin-fluctuation superglue scenario" dating back to the 1960s [120], which is popular among experimentalists, despite its lack of mathematical rigour.

One asserts that the system is a Fermi liquid that is not too far from a transition to an antiferromagnet, with the effect that it is characterised by RPA relaxational spin fluctuations ("paramagnons") of the kind discussed in section 2.4. One now asserts bluntly that these paramagnons act just like phonons in mediating a force between the electrons. Although the basic principle is always the same, which specific pairing symmetry is preferred depends on the details of the crystal structure and the Fermi surface. A simple but typical example is offered by the high-T_c cuprates. The action is in the copper-oxide planes, where the Bravais lattice is a simple two-dimensional square lattice, such that the Brillouin zone (BZ) is also a square. The corners at $\vec{k} = (\pi/a, \pi/a)$ and equivalent points (a is the lattice constant) are called the M-points, and the midpoints of the sides are called the X- ($\vec{k} = (\pi/a, 0)$) and Y- ($\vec{k} = (0, \pi/a)$) points. One departs from the Fermi surface of the non-interacting band structure with a large Luttinger volume. For the sake of the argument one can take a simple circle covering half the BZ. The s-wave gap function would be positive everywhere in the BZ. Given that the orbital momentum is quenched because of the underlying lattice, "d-wave" actually means $d_{x^2-y^2}$: one divides the zone into four pieces by drawing the zone diagonals connecting the M-points (the "nodal lines"), and the gap function changes sign every time one crosses these zone diagonals. Because of this sign change the gap is vanishing along these lines. Differently from the scattering by phonons, which is quite isotropic in momentum space, the spin fluctuations associated with the staggered antiferromagnet are concentrated at momenta $\vec{Q} = (\pi/a, \pi/a)$: they scatter predominantly the electrons near the Fermi surface located at adjacent "antinodal" regions near the X- and Y-points. Although for the s-wave the exchange of these spin fluctuations would cause an effective repulsion, an extra minus is picked up in the expression for the gap function, which is associated with the sign change between adjacent antinodes. This turns into an effectively attractive pairing force causing the gap to be maximum at the antinodal points at the expense of the vanishing of the gap along the nodal lines, explaining why the d-wave state is preferred. Using the same rules, one can deduce the order-parameter symmetry in other cases, depending on the details of the crystal structure, the Fermi surfaces and the magnetic fluctuations: a typical example is the so-called $s_{\pm}$ order parameter predicted for the iron pnictides that has still to be definitively confirmed [119].

The pair wave function in relative *real-space* coordinates has for this d-wave order parameter a node at the origin, which implies that the strong local repulsion which would wreck the s-wave pairing is avoided. In fact, the d-wave pairing is generic, given the square lattice and the dominance of the local repulsions, and is also invariably found in the strong-coupling (non-BCS) theories which will be discussed next. The origin of the d-wave symmetry in this strong-coupling limit is actually quite easy to understand, relying on the way matters work in real space

[121]. In this limit the building blocks are valence-bond-like singlet pairs formed on nearest-neighbour lattice sites, thereby avoiding the strong on-site Hubbard repulsion. The gap symmetry is now determined by the way in which the pair moves around on the square lattice. To move the pair from one nearest-neighbour bond to the next it has to be broken up, by letting one electron unbind by a hop to a next-nearest site. The pair recombines at the adjacent bond by a hop back to the nearest-neighbour site and the net effect is that it tunnels from bond to adjacent bond with a positive definite effective hopping amplitude $\sim |t|^2/V$, where V is the binding energy of the bond pair. This has in turn the consequence that the phasing of the pair wave function upon circling around the four bonds surrounding a site has to be $+, -, +, -$: the d-wave symmetry.

3.4 Numerical machinery I: the density-matrix renormalisation group and tensor-product states

When the coupling becomes strong there is just no reliable mathematical machinery available and one has to fall back on heavily computational approaches. Computers cannot perform miracles: despite the large investments, nothing on the market is foolproof and one should be aware that there are invariably built in limitations and biases. Conventional "brute-force" algorithms just do not deliver. Quantum Monte Carlo becomes useless because of the severe fermion-sign problem, while only tiny systems of 25 sites or so at most can be handled by exact diagonalisation. However, two methods that are generally perceived as delivering reliable information within certain limits have been developed. Both of them are rooted in profound reasoning, making it possible to avoid the sign problem to a degree: the DMFT method, which relies on the limit of large space dimensions of the next section, and the DMRG-type methods, which rest on insights in quantum entanglement of this section.

The "density-matrix renormalisation group" (DMRG) and the recent extensions in the form of the tensor-product states belong to the category of methods based on a variational Ansatz. This tradition started in this context a long time ago using the Gutzwiller Ansatz, which we saw at work in the previous section in the form of the RVB theory. A great leap forward was brought about in the 1990s by the "density-matrix renormalisation-group" (DMRG) construction of Steve White [122]. The original identification with some renormalisation-group scheme turned out to be mistaken. It is actually better to view it as an iterative procedure aimed at truncating the Hilbert space to arrive at a ground state with a built-in variational bias. This bias turns out to correspond to a limitation on the range of the entanglement: as soon as one starts to throw away states there is a maximum length scale beyond which one is actually dealing with an effectively "short-range" entangled product state – this range can just be made quite long. It turns out to be quantitatively very accurate for one (space)-dimensional systems [123], but its use in two and higher dimensions remains problematic up to the present day. The method is particularly suited to dealing with quantum spin problems, as well as the t–J model. The best one can do at present is to study the t–J model on square-lattice "strips" that are infinitely long, with, however, a relatively small width. With an infusion of quantum information insights, the underlying "matrix product" Ansatz was in recent years generalised to the more general category of "tensor-network states", which makes it possible to wire in also long-range entanglements, again in a variational guise [124, 125]. At the time of writing the first results on the t–J model had become available [126], confirming the gross features of the DMRG calculations on the t–J-model strips dating back to the 1990s [127].

The parameter regime of the t–J model which is seen to be of relevance to the cuprates is $J \simeq 0.5t$ and hole dopings typically around 5%–20%. In the DMRG approach, d-wave pairing correlations reminiscent of the RVB mechanism are natural (for the state of the art see [100, 126]), but the surprise which was already revealed in early DMRG calculations on t–J strips [127] is that the superconducting order is in very close competition with another ordering phenomenon called "stripes". Stripe order is quite particular to doped Mott insulators. It was discovered initially in the late 1980s [128, 129] as the generic Hartree–Fock instability of doped Hubbard-type systems at intermediate coupling, $U/W \simeq 1$. Hartree–Fock is controlled by the fluctuations of the order parameter and it is also reliable at large couplings when there is some parameter suppressing the collective quantum fluctuations. For the Mottness physics dominated by magnetism the convenient coupling constant is $1/S$, where S is the size of the microscopic spin [40]. The Hartree–Fock solutions for the Hubbard model on the square lattice turned out to be remarkably complex: they consist of linear Mott insulating and antiferromagnetic domains, separated by domain walls in the antiferromagnet, which in turn bind the excess holes by a mechanism that can be viewed as a generalisation to two dimensions of the Su–Schrieffer–Heeger solitons of one-dimensional physics [130]. However, these mean-field stripe phases acquire their stability from the fact that the holes empty the midgap band associated with the domain walls and the stripe phases are therefore *insulating*. It turns out that such insulating stripes appear to be ubiquitous in non-cuprate doped Mott insulators, which invariably involve spin systems characterised by $S > 1/2$, such as doped nickelates, cobaltates and manganites.

A structurally similar stripe order was discovered in 1995 in a special variety of cuprates characterised by a strongly suppressed superconductivity [131], but these turned out to be metallic and even to a degree superconducting [132]. The surprise revealed by the early DMRG computations [127], which was much later confirmed by the tensor-product state computations [126], is that there is indeed a strong tendency to form the incommensurate magnetism and charge order of the stripe phases. However, this is strongly "entangled" with the tendency to superconduct in a way that is not dissimilar to RVB. It appears that the carriers on the charge stripes form d-wave pairs, but these localise on the domain walls, having a Josephson coupling to adjacent stripes locking in a weak-link two-dimensional superconductor [126]. But there are many other states nearby, such as homogeneous superconductors, insulating stripes and so forth. The gross take-home message is that apparently the physics of the t–J model is in the grip of a new form of frustration, characterised by a variety of orders that are "unreasonably" close in energy, while these ordered states themselves involve a pronounced quantum entanglement. Whether it is a coincidence or whether there is a deep common principle at work, this is strikingly similar to a main message of holography that will unfold in this book. Although it is in detail quite different, we will meet even a holographic incarnation of a stripe phase competing with superconductivity in chapter 12.

3.5 Numerical machinery II: infinitely many dimensions and the dynamical mean-field theory

For Hubbard-type problems there is actually one limit where matters can be brought under full control, even when the interactions are strong: the limit of an infinite number of space dimensions. The surprising physics in this limit was discovered in the late 1980s and it has evolved since then into a large field of activity called "dynamical mean-field theory" (DMFT). This includes both the development of approximate methods marrying DMFT with quantitative band structure theory ("LDA+DMFT") and generalisations aimed at extending the methodology to finite dimensions: the so-called "cluster DMFT" and "dynamical cluster approximation" (DCA). The results of the latter suggest that the qualitative physics found in infinite dimensions might be still quite relevant also in low ($d = 2, 3$) dimensions. The outcomes share some intriguing resemblances with the results of holography. Although "ordered" states are found at the lowest temperatures, an incoherent "quantum soup" takes over rapidly when the temperature increases. Intriguingly, one finds that under conditions where the fermion-sign problem would be particularly severe in a straight quantum Monte Carlo approach the dynamics is in a surprising way "local" in the sense that the dynamics generates long time scales while the length scales stay microscopic. We will see that this "local quantum dynamics" is also a hallmark of the holographic strange metals. To find out whether this resemblance can be made more precise, i.e. whether DMFT can be "mobilised as UV completion" of holography is an outstanding open challenge.

What principle is central to DMFT? When one is dealing with bosonic field theories of the kind discussed in section 2.1, dimensionality is a most powerful control parameter. Whatever the theory, there will be an upper critical dimension above which the critical theory will become a mean-field theory characterised by free field behaviour. When one is considering Hubbard-type models this is no longer the case. Even in infinitely many dimensions one finds that instead the system is described by an effective strongly interacting Anderson impurity problem, living in a host that is self-consistently determined by the solution of the impurity problem [133, 134]. This is quite remarkable: although the spatial dynamics is quenched by the large number of dimensions, and there are no large length scales in the problem, the temporal problem is highly non-trivial and in fact capable of generating large time scales, as embodied by the Kondo effect.

The requirement that one has to find self-consistent solutions adds further structure. The density of states of the host, knowledge of which is required for the solution of the impurity problem, is now set by the spectral function of the impurity fermion itself. In the Kondo regime, this consists of a narrow, small pole strength peak near the Fermi energy with its scale determined by the Kondo temperature (the "Kondo peak") while most of the spectral weight is in the Hubbard bands at energies $\sim \pm U/2$ away from the Fermi energy. Considering the simple Hubbard model at half-filling, this gives rise to the following evolution, when one gradually increases U [135, 136]. When U is small all the spectral weight is near E_F: the "Kondo temperature" is of the order of the band width and the self-consistent solution just looks like a normal band metal. Upon increasing U, the Kondo temperature drops and this transfers the spectral weight away from the Fermi energy, with the effect that the effective coupling $1/(N_0 J)$ increases because N_0 is decreasing: the result is that a narrow band starts to develop near E_F, accompanied by the two Hubbard bands at $\pm U/2$. For increasing U/t this band near the chemical potential further narrows until a self-consistent solution can no longer be found and the system turns, through a first-order transition, into a localised Mott insulator.

This mechanism to produce very narrow bands in infinitely many dimensions works also on departing from an Anderson lattice system, with of course the difference that the system will always stay metallic. The outstanding feature of the empirical heavy-fermion systems is that, when Fermi liquids are formed, they are characterised by an extremely large effective mass, up to a thousand times the electron mass. The DMFT links this to the exponentially small Kondo scale of the single-impurity problem, thereby offering a natural explanation for the mass enhancements. This is then combined with realistic band structure, to arrive at a quantitative and materials-specific description of, for instance, the "volume collapse" observed in plutonium [137]. In addition, since the effective-impurity problems concern serious many-particle systems, these systems do react strongly to temperature by rapidly losing the signatures of quantum coherence such that the spectral functions turn into incoherent continua.

Although it is intuitively appealing, this "single-site DMFT" initially encountered considerable scepticism: why should this limit of infinitely many dimensions have anything to say about the physical systems living in two or three space dimensions? The next step was the development of the "cluster DMFT" and related "dynamical cluster approximation" (DCA) methods [138]. In essence they amount to implementing quantum Monte Carlo for fermions on a small finite-size lattice, augmented, however, with boundary conditions of the DMFT-bath kind. This suffers from the usual fermion-sign problem but the "DMFT bath" has the effect that one can reach much lower temperatures, which are claimed to be so low that meaningful comparison with experiment becomes possible. The size of the clusters is still limited, but the claim is that the results converge surprisingly rapidly as a function of the cluster size. This is in turn taken as evidence for the local (large dynamical critical exponent z) dynamics being not a pathology associated with the use of infinitely many dimensions, but instead a surprising property of the physics of strongly interacting fermions.

With the cluster DMFT one can also address the question of whether there is superconductivity and other types of ordering tendencies. For technical reasons the computations can be done only up to the intermediate coupling regime ($U \simeq W$) of the Hubbard model. One finds clear signs of d-wave superconductivity, actually forming a dome as a function of doping with a maximum T_c around 15% doping [139]. But there is a lot more: this appears to be tied to a quantum critical end point associated with a phase-separation transition involving a non-Fermi liquid phase with a pseudogap at low dopings and a Fermi liquid at high dopings. The normal state at optimal doping appears to show the "conformal" pair susceptibility with a relevant scaling dimension [139] of the kind discussed in section 2.4.

3.6 Quantum matter in the laboratory

We have now arrived at the most hazardous pursuit in this book: trying to summarise in a few pages what one should know about the experiments. Differently from what is possible in high-energy physics, it is relatively easy to do experiments in condensed matter physics that yield information on a highly diverse set of physical properties. The effect is that in the course of time an enormous amount of experimental information accumulates, and from this enormous pile of information one has to filter out what is perceived as significant and reliable. This selection of course depends on the particular theory one has in his or her head, and, especially when matters are not well understood, one can end up focussing on the wrong facts.

Nevertheless, the history of condensed matter physics demonstrates that the art of data analysis has worked miracles in the hands of such masters as Landau, Bardeen and Anderson. As, for instance, the history of the BCS theory demonstrates, without the clues coming from experiment (the presence of an exponentially small gap, the isotope effect and so forth) it would have been impossible to arrive at the right answer. Reality is susceptible to many interpretations, and when one is searching for new physics the best facts are those that violently resist any attempt to fit them into the established wisdom. In this section we will attempt to present a list of some of the most manifest mysteries in correlated electron physics where holography might have the potential to make a difference.

We will focus entirely on the two main families of correlated electron systems: *the heavy-fermion intermetallics* and *the cuprate high-T_c superconductors*. There are many other families of materials that are at present subjects of intense experimental research, such as the pnictide superconductors and graphene. We will ignore these either for the reason that they behave in a conventional way (e.g. graphene) or because the experimental situation is still unsettled to such a degree that it is difficult even to decide whether anything truly strange is going on (e.g. the pnictides).

3.6.1 The heavy fermions in a nutshell

Let us first discuss the heavy-fermion systems. Although the experimental literature is confusing to a degree, with its many different intermetallic compounds, different phase diagrams and so forth, by and large their physics seems to be less complex than that of the cuprates. This field has settled into a relatively mature state and is documented in a well-organised review literature (see e.g. [77, 114, 140, 141, 142]). Up to now, holography has not delivered anything worth mentioning addressing the specific physics in these systems. Let us present a very short list of the most obvious big-picture questions in this field, hoping that this will inspire the holographic model builders to look harder for heavy fermions.

Why does the heavy Fermi liquid exist? There is abundant evidence that the microscopic situation is governed by the physics of the Anderson lattice, with its mix of weakly interacting, wide-band metallic electrons and the strongly interacting electrons of the localised-f-shells. Experiment demonstrates that this can renormalise into a quite well-behaved Fermi liquid, which is, however, characterised by an extremely large mass enhancement (up to a factor of 1,000). Strikingly, it appears that the Fermi surfaces and the dispersions at very low energy are similar to what one would expect from a simple non-interacting band structure, including the hybridisation between the heavy and light bands, except that the band widths of the f-bands are squeezed by a large factor [143, 144]. This heavy-Fermi-liquid state appears to be quite stable at zero temperature: how can this be in the light of the discussion in section 2.3 where we explained that the heavy mass just means that the Fermi surface is on the verge of being destroyed by quantum fluctuations? Upon raising temperature this heavy Fermi liquid falls apart rapidly into a poorly understood incoherent quantum soup, and at a somewhat higher temperature the f electrons resurface in the form of localised Curie–Weiss spins. Except for some hints from DMFT, real theoretical understanding of how the heavy Fermi liquid is born from the higher-temperature incoherent stuff is completely lacking.

What is the origin of the "soft quantum criticality" quantum phase transitions? Upon varying the pressure, magnetic field or chemical doping one finds in a large variety of heavy-fermion intermetallics a quantum phase transition from a non-magnetic heavy Fermi liquid to another heavy Fermi liquid that is accompanied by some form of magnetic order. This magnetism is undoubtedly associated with the f electrons that now acquire their spin-like behaviour at zero temperature. It has been argued that in some of these systems various measured properties are consistent with the expectations from the Hertz–Millis theory (section 2.5), which should behave in a mean-field fashion in these three-dimensional systems [77]. In accord with those expectations, one finds typically that these quantum critical points are "surrounded" by superconducting domes involving unconventional superconductivity [140]. There is surely a need for an explanation for why such systems can rediscover this weakly coupled physics.

How should one describe "hard quantum criticality" quantum phase transitions? There is yet another class of quantum critical heavy fermions where the experiments reveal a remarkably strange and mysterious behaviour, which is directly related to the fermions itself. In Hertz–Millis theory, the fermions are largely spectators: upon moving into the magnetic phase one finds a reconstruction of the Fermi surface according to the Hartree–Fock effective potential recipe, which switches on continuously at the quantum phase transition. However, in this other "hard quantum critical" (or "bad actor") class one finds that the Fermi surfaces jump *discontinuously* in passing through the *continuous* phase transitions [114, 142]. It is as if the delocalised f electrons forming the heavy Fermi liquid in the non-magnetic phase suddenly decide to turn into localised spins that no longer contribute to the Luttinger volume of the Fermi surface. For a first-order transition this would have been easy to understand, but here a genuinely *continuous* quantum phase transition occurs. Everything is supposed to be scale-invariant right at the quantum critical point, including the fermions. But how can one reconcile this with the presence of two very different Fermi surfaces, infinitesimally distant from the critical point? There is evidence from Hall measurements that this Fermi-surface change survives at finite temperature in the form of a crossover line in the middle of the quantum critical wedge. Whatever the explanation is, it must involve a fermionic physics that participates fully in the critical dynamics and has little to do with the stable Fermi liquid.

Local quantum criticality. Among the most striking properties of the holographic strange metals is the property of local quantum criticality (chapter 8). The term local quantum criticality was introduced into condensed matter physics in the 1990s to describe behaviours seen in the measurements. In fact, the most direct evidence for local quantum criticality as found in measurements of collective properties emerged in the heavy-fermion field. Schröder *et al.* [145, 146] found that the magnetic fluctuations in the intermetallic $CeCu_{6-x}Au_x$ as measured by inelastic neutron scattering obey a scaling form $\chi^{-1}(q, \omega) = T^a \left[\Phi(\omega/T) \right] + \chi_0^{-1}(q)$, where $a \simeq 0.75$. This reveals the energy–temperature scaling associated with a strongly interacting critical point, but this appears to be completely momentum-independent and thereby local in space. Notice that inelastic neutron scattering is rather unique in its capacity to measure a dynamical susceptibility in the kinematical regime of relevance to strange-metal behaviour (momentum resolution spanning the whole Brillouin zone, and an energy range from sub-kelvin up to the eV scale in principle). Also notice that the experimentalists have not yet managed to detect the magnetic fluctuations associated with the strange metals realised in optimally doped cuprates using neutron scattering: surely such fluctuations are present, but they are too weak to be detected by the rather insensitive technique of inelastic neutron scattering.

3.6.2 Cuprates in a nutshell

There will frequently be references to experimental results on high-T_c superconductors in this book. This is at least in part related to an awareness of the high-T_c problem in the community which has been developing the condensed matter applications of holography. However, much of this alludes to the physics of the best, optimally doped superconductors. Arguably, this might be the instance where the physics of the cuprate electrons might be most anomalous, but there is a lot more "strange" behaviour that has been observed that eventually might find an explanation in terms of the new principles revealed by holography.

At the time of writing, it appears that rapid progress is being made in the laboratories, especially relating to the physics in the pseudogap regime. Since matters are far from having been settled, there is not much of an overarching and comprehensive review literature available. Here we will just present a short sketch of some of the highlights of this multi-facetted field which should be of particular interest to the holographist. We cannot possibly do justice here to the vast research literature, so we will just focus on a small selection of subjects, referring the reader to reviews that it is hoped will appear in the near future (see Ref. [115] for a concise but authoritative overview).

Any discussion of the cuprates starts with the "phase diagram", which is actually not at all a real phase diagram but rather a way to summarise the different kinds of physics which are found in these very rich systems (Fig. 3.1). The cuprates are layered compounds, with the "active" copper-oxide layers separated by electronically inert insulating layers that also serve the purpose of acting as charge reservoirs: the chemical doping takes place in these insulating layers upon e.g. substituting divalent strontium ions for trivalent lanthanum ions, resulting in a hole doping of the stoichiometric compounds. For instance, consider the simple "214" family, $La_{2-x}Sr_xCuO_4$, where x corresponds to the number of holes in e.g. the $t-J$ model. The undoped compounds form large-gap (2 eV) Mott insulators exhibiting below room temperature a simple two-sublattice antiferromagnetism. As a function of doping and temperature one finds a variety of rather distinct phenomena.

The strange metal. Let us start with the pinnacle in this landscape: the strange metal itself. In fact, not that much has happened since Anderson wrote his book in the mid 1990s, which may still be used as a source of inspiration [86]. Above all, this metal behaves in a way that is "unreasonably simple". The case in point has been all along the famous linear resistivity: $\rho_{DC} \sim T$. In a Fermi-liquid metal, the resistivity is invariably an interesting function of temperature. The momentum dissipation is due to the scattering of the quasiparticles and the scattering mechanism has to be an interesting function of temperature: at low temperature one has electron–electron scattering and at intermediate temperature electron–phonon scattering, while at high temperature the resistivity should saturate when the inelastic mean free path becomes of the order of the lattice constant. It appears that even the residual resistivity associated with zero-temperature elastic scattering has just disappeared in many cuprates despite the fact that these compounds are chemically quite dirty! As we will see in chapter 12, the desire to explain this unreasonable simplicity in terms of the principles governing strongly interacting conformal metals is an important focus of holography. In fact, another simple, but highly anomalous, transport property much emphasised by Anderson is the Hall angle, giving away that the Hall relaxation rate is somehow decoupled from linear momentum relaxation, which is surprisingly difficult to understand (see Ref. [147] for a lucid discussion). Another highly anomalous feature stressed by Anderson is the extreme transport anisotropy. The cuprates are quite reasonable metals in the planar ("a" and "b") directions, while being rather tough insulators along the inter-plane ("c") directions. Optical measurements show that there is considerable spectral weight along the c-axis, but the charge dynamics is completely over-damped. This might be related to the holographic insulators [148] as discussed in chapter 12. Since the mid 1990s not much experimental progress has been made on the strange metals. The optical conductivity measurements showing the "conformal conductivity" in the mid-infrared [149] will figure prominently in chapter 12. The claim in Anderson's book that the strange metal is characterised by some kind of large Fermi surface is not quite right. In discussing the pseudogap we will come back to the mysterious nodal–antinodal dichotomy [150] regarding the rather abrupt change from "quasi-coherent" behaviour near the nodal lines to fully incoherent behaviour near the antinodes as seen in the momentum-dependent electron spectral functions (see Fig. 3.2.)

It turns out that it happens only below the crossover line to the Fermi liquid that these antinodal regions also become quasi-coherent, while in the whole strange-metal regime ARPES shows here only fully incoherent backgrounds [151]. In the nodal regions one finds peaks in the photoemission spectra, but these are surely not real quasiparticles. They can be fitted by assuming that the "marginal Fermi-liquid" electron propagators are of a form indicating that the quasiparticles are on the verge of becoming over-damped [152],

$$G(k, \omega) = \frac{Z}{v_F(k - k_F) - \omega - \Sigma_k(\omega)}, \tag{3.8}$$

where

$$\Sigma_k(\omega) = \lambda \omega \ln\left(\frac{x}{\omega_c}\right) - i\frac{\pi \lambda}{2}x,$$

$x = \max(|\omega|, T)$ and λ is a coupling constant. This is suggestive of the quasiparticles decaying in a continuum of quantum critical excitations while the momentum independence is indicative of a local "$z \to \infty$" dynamics (see chapter 9). In fact, this is the historical origin of the local quantum criticality idea, in an era during which even the notion of quantum critical states had not yet been formulated.

Viewed from the perspective of holography, the experimental information required to decide whether the strange metal is indeed a conformal metal characterised by unusual scaling dimensions (see chapter 8) is still missing. To get a handle on this, one needs to measure the dynamical susceptibilities associated with collective responses, if possible for a large range of kinematical parameters. Although this is hard work, there are surely possibilities to make progress here (e.g. inelastic neutron scattering, the pair susceptibility measurement discussed in section 2.4). The experimental community should be urged to take up this challenge.

The superconducting regime. If one had access solely to data on the low-energy and low-temperature regime of any high-T_c superconductor one would easily be convinced that one was dealing with a meat-and-potatoes BCS d-wave superconductor. It is characterised by impeccable, long-lived Bogoliubov fermions, while the phenomenology of the superconductors behaves according to the textbook, except for the special effects associated with the "soft" flux lines giving rise to the formation of flux liquids realised above H_{c2}, caused by the small

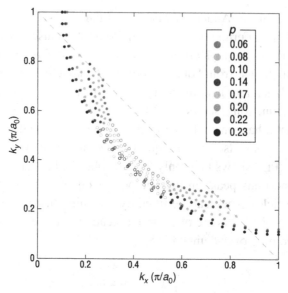

Figure 3.2 The "nodal–antinodal dichotomy". Using the so-called "quasiparticle-interference" technique one can use scanning tunnelling spectroscopy to extract whether quantum-mechanical coherent excitations exist, and in turn reconstruct their dispersions in momentum space. The figure shows how this coherence is distributed in the Brillouin zone at low temperatures deep in the superconducting state as a function of the doping p. The low p values are for the innermost arc, and the value increases successively outwards. Only one quadrant of the Brillouin zone is shown: the diagonal connecting the lower left and upper right corners corresponds to the "nodal line" where the d-wave gap would vanish, while the upper left and lower right corners correspond to the antinodal points where the d-wave gap function would be maximal. The dashed line connecting the antinodal points would be the Umklapp surface associated with a simple translational symmetry-breaking doubling the unit cell in real space. Although a plethora of ordering phenomena can be found in the under-doped "pseudogap regime", none of them have anything to do with such a symmetry-breaking. Remarkably, the locus of the coherent Bogoliubov fermions in momentum space only occurs where the Fermi surface expected for the free band structure "spills out" below the Umklapp surface, where they seem to terminate abruptly. On the other side of this "nodal antinodal wall" one finds rather incoherent excitation spectra that appear to encode for the stripy real-space structures associated with the pseudogap seen in the electronic spectra. In the normal state the BCS-like gap in the nodal regions collapses and one finds the "Fermi arcs" instead in the nodal regime (see the main text). Around optimal doping ($p \simeq 0.24$) it appears that at low temperatures a full underlying Fermi surface is established. Figure source [150]. (Reprinted with permission from the AAAS.)

(20 Ångström) coherence length and the low dimensionality. However, to a degree this is deceptive. The most obvious anomaly is the small *superfluid density*: this has its maximum at optimal doping, where it is still small compared with the BCS value, decreasing both in the over-doped and under-doped regimes. In the

under-doped regime one finds that the superfluid density increases linearly with the doping x, while being proportional to T_c: this "Uemura law" is reminiscent of Bose condensation of preformed pairs. This is consistent with the expectations for RVB-style superconductors which we sketched in section 2.4. As a ramification, one expects severe thermal phase fluctuations, especially in the under-doped regime, and there is considerable evidence for it (diamagnetism, the Nernst effect). A remaining open question is whether this "preformed-pair" physics extends all the way to the pseudogap temperature T^* or whether it terminates at a significantly lower temperature. Another very puzzling feature is the way in which these superconductors react to potential disorder. According to BCS, any form of such disorder should be detrimental for d-wave superconductivity, but the high-T_c variety appears to be oblivious in this regard: the cuprates are chemically quite messy but T_c is rather insensitive. This mystery is further amplified by the fact that a particular form of disorder (planar Cu substituted by Zn or Ni) is extremely destructive for the superconductivity, with the odd effect that these particular impurities behave paramagnetically.

The Fermi-liquid regime. This concerns in the first place the strongly over-doped regime. High-quality quantum oscillations leave no doubt that at low temperatures a real, large-Luttinger-volume Fermi surface is realised, and this contention is further supported by ARPES data. This is perhaps puzzling, given the arguments presented in section 2.4 suggesting the incompatibility of Mottness and the normal Fermi liquid. It has been speculated that this might signal that the Mottness "collapses" near optimal doping [98]. There are still some features that do not seem to make sense. Contrary to expectation, the superfluid density of the superconductor diminishes in the overdoped regime, while also the way in which the resistivity evolves is not at all understood [153]. Another development is the observation of quantum oscillations in the under-doped regime, in very large magnetic fields that suppress the superconductivity. This is indicative of the formation of small-Fermi-surface pockets in this magnetic-field-induced normal state. One problem of principle is associated with the expectation that, at the fields which can be achieved, the system is supposed to form a vortex liquid, and it is not at all clear why here simple Fermi-surface quantum oscillations should occur. Secondly, the size of these pockets is hard to reconcile with the expectations that follow from ARPES and STS measurements, as well as the understanding of the translational symmetry breaking (charge-density wave, stripe) states.

The pseudogap regime: competing orders. This is a "gapping" phenomenon in the sense that one finds that the density of low-energy electronic states is depleted as seen in results from electron spectroscopies, but also in collective properties like the magnetic susceptibility, electric transport and so forth. It became gradually clear that this is related to a *variety* of orders that are closely competing (or "collaborating") with each other and with the superconductivity. This is by itself already quite interesting: in highly itinerant metals this does not happen. Departing from a Fermi liquid it requires much fine tuning to bring even two different instabilities into close competition. One could view this as a consequence of the fact that, since the interactions are strong, the homogenising influence of the quantum kinetic energy is diminished to a degree. This "complex-order" physics would then become reminiscent of classical physics, where complex behaviour is more ubiquitous. However, a development suggesting that the problem is much deeper than this is unfolding. The standard interpretational framework relies on Hartree–Fock, with its central wisdom that order parameters turn into potentials that in turn diffract the electron waves. This rule appears to be grossly violated when one attempts to relate the outcomes of the electron spectroscopies (STS, ARPES) to the information regarding the (competing) order parameters. There is a potential for holography to shed light on these mysteries. The holographic strange metals are, in comparison with the stable Fermi liquids, quite unstable "quantum frustrated" affairs, which can accordingly form the birthplace of a large variety of stable phases that are automatically in close competition, as we will see in this book. We will also find that the relationships between order parameters and dynamical responses are in general quite different from Hartree–Fock, although there is still quite some way to go to establish direct contact with the pseudogap experiments.

Let us first briefly describe the types of order that are suspected to be at work in the pseudogap regime. This discussion is still in flux and it might well be that the picture will change in the near future. The competing-order theme started in the mid 1990s with the discovery of the stripe phases in the 214 system where relatively low-T_c (40 K) superconductors are found. It is believed that these stripes are quite like the stripes which rolled out of the DMRG computations as discussed in section 3.4. These show static incommensurate antiferromagnetism next to their charge order, probably in the form of "crystallised Cooper pairs". Such static magnetism was not found in cuprates with truly high T_cs ("YBCO", "BISCO"), but quite recently definitive evidence was found for a similar static charge order also in the bulk of these systems. These appear to be directly related to the "stripy" patterns which were detected quite a long time ago by real-space scanning tunnelling spectroscopy (STS) measurements on surfaces of under-doped BISCO superconductors. These STS measurements revealed also another form of order that is best called "nematic quantum order" since it lives at zero wave vector, just breaking the rotational symmetry. Recently evidence was presented that this can be stitched together with the stripes, in terms of a special bond-order-density wave with a d-wave form factor [154].

Last but not least, there is evidence for the presence of a completely different type of order: the intra-unit-cell loop-current order. This is believed to correspond to spontaneous orbital currents circulating around the C–O plaquettes inside the unit cell forming a pattern of magnetic fluxes that do not break translational symmetry but instead only violate time reversal. This appears to emerge at a real thermodynamical phase transition at the pseudogap temperature [155]. This orbital magnetism is hard to detect in macroscopic measurements, but it leaves a clear imprint in the neutron scattering, which suggests that it is quite a strong order. This is the first time that such a "current condensation" has been found in a condensed matter system. Intriguingly, we will find out in chapters 8 and 12 that such current orders occur quite naturally in holographic set-ups.

The pseudogap regime: the nodal–antinodal dichotomy. STS, supported by photoemisson, has been quite informative regarding the failure of the Hartree–Fock potentials. The data indicate a highly mysterious connection between the order and the way in which the electron system adjusts to the presence of these electronic VEVs. The essence is illustrated in Fig. 3.2. Using the quasiparticle-scattering interference technique one can measure with STS whether the electrons behave like quantum-mechanical waves that can interfere because of their phase coherence. On measuring at very low temperatures one finds that the Bogoliubov excitations near the nodes are of this kind. However, upon approaching the surface in the Brillouin zone associated with a simple doubling of the unit cell these coherence signals suddenly disappear. Notice that this momentum-space surface does not relate to any of the translational symmetry breakings that have been detected in this doping regime. As a caveat, according to ARPES the "border" between the nodal and antinodal regimes is more of a rapid crossover than a "wall", and it has been argued that this sudden disappearance might be to a degree an artefact of the technique [156].

Upon entering the antinodal regime one finds rather incoherent "backgrounds" that contain the information regarding the "stripy" order. Upon increasing the doping, the coherent nodal regime expands to take over the whole zone around optimal doping [150]. This could be interpreted as the nodal regime being responsible for the superconductivity while the antinodes correspond to the stripy order, but this is too simple. One finds in photoemission that in the pseudogap regime the antinodal regime is completely incoherent, just an energy-independent background with a depleted density of states near E_F. This restructures in the superconducting state, where there emerges a sharp quasiparticle that connects seamlessly with the Bogoliubov excitations near the nodes, having, however, a spectral weight that appears to scale with the superfluid density as a function both of the doping and of the temperature. Moreover, it was also demonstrated recently with STS that the antinodal states react strongly to a magnetic field, indicating that they are involved in the superconducting phase coherence [157]. This becomes even more puzzling in the pseudogap phase itself, above the superconducting transition temperature: the gap stays open in the antinodal regime, but closes in the nodal regime, leaving

"Fermi arcs", pieces of disconnected "Fermi surface" that end at the Umklapp surface of Fig. 3.2. Finally, it appears that the ordering wave vector of the stripy charge order is precisely coincident with the distance between the end points of the arcs [158]. The conclusion is that at present there is not a single theoretical proposal that can explain these observations in any way: they are just plainly mysterious. Is this quantum matter at work?

Quantum critical point, or conformal metal? It is widely believed that the strange metal is related to quantum criticality. The predominant view in the condensed matter community is that Hertz–Millis is somehow at work: it is taken for granted that this is controlled by some competing order or other that undergoes a quantum phase transition at optimal doping, shaking the Fermi liquid "from below" and perhaps causing the superconductivity. There are, however, difficulties with this idea. There appear to be different types of competing orders in the pseudogap phase, all of which seem to disappear at optimal doping: which order parameter is doing the work? How can one explain the non-Hartree–Fock "dichotomy" in the pseudogap phase? Most seriously, Hertz–Millis asserts that the UV consists of a stable Fermi liquid: how can it then be that the resistivity stays linear up to the melting point of the crystal? Why is the Fermi-liquid stability not recovered in the UV, at high energy or high temperature?

An alternative view is suggested by holography, and advocated by Hong Liu, Nabil Iqbal and Mark Mezei [88, 159, 160]. This asserts that at "intermediate" temperatures first a strange metal *phase* of the radical non-Fermi-liquid kind as suggested by holography is realised (see chapter 8). This metal in turn carries the potential for a large variety of cohesive "ordered" states, including charge-ordered states (chapter 12), and superconductors (chapter 10), while even the Fermi liquid can be "born" in the strange metal (chapter 11). Depending on the "small" parameters which are changing as a function of the doping, these various states are then singled out as ground states. Finding out by targeted experimentation whether this basic notion is correct is perhaps the greatest challenge faced both by the holographists and by condensed matter experimentalists.

4

Large-N field theories for holography and condensed matter

The profound puzzles posed by quantum critical metals with Planckian dissipation and long-range entanglement, as observed in cuprates and heavy-fermion systems, cry out for a novel point of view. Holography can provide this new perspective. This book will propose that its concrete manifestation in terms of the AdS/CFT correspondence gives qualitatively new insights into these puzzles. The reason is that holography has to be understood above all as a "weak–strong" duality between two different descriptions of the same physics. In this regard it is qualitatively similar to the Kramers–Wannier or Abelian–Higgs duality we reviewed in chapter 2, but it takes the notion to a new level: it relates quantum field theories to a dual description that includes the gravitational force. For an extremely strongly coupled field theory, the weakly coupled theory is now Einstein's theory of general relativity. Vice versa, a strongly interacting gravitational theory has an equivalent description as a weakly coupled quantum field theory.

General relativity inherently contains the notion of a dynamically fluctuating space-time. The remarkable way in which this emerges in holography is by incorporating the renormalisation-group structure of the quantum field theory into the dualisation. As we previewed in the introduction, the renormalisation-group scale becomes part of the geometrical edifice as an additional space dimension in the gravitational theory.

It is still baffling that a quantitative duality relation can exist between two theories in different space-time dimensions. This paradox is resolved, however, by the *holographic principle* of quantum gravity. This lesson from black-hole physics insists that gravitational systems are less dense in information than conventional quantum field theories in a flat non-dynamical space-time, to the degree that the former can be encoded in a "holographic screen" with one dimension less. The dynamics of this "screen" can be thought of as the dynamics of the dual field theory.

In this chapter we will first provide a brief account of the conceptual and historical background of the holographic principle and in particular its manifestation within string theory (section 4.1). This is where the origin of the AdS/CFT correspondence lies. Fortunately one does not need all this material to understand holographic duality practically. In the remainder of this chapter we approach holography from a constructive angle instead, as was first put together in the excellent review [5]. The anchor here is the large-N limit in quantum field theory as a simplifying tool; this is the subject of section 4.2. We start with the *vector* large-N theories which are familiar both from condensed matter physics and from particle physics. A very natural generalisation, particularly in the context of gauge theories, is to *matrix* large-N models. As first identified by 't Hooft in the 1970s, the matrix large-N limit describes a physics that is completely different from the free-field saddle points of the vector large-N limit. The physics continuous to be very strongly coupled, but at the same time a "master field" mean-field principle of a different kind is at work.

It was realised long before the advent of AdS/CFT that matrix large-N models have a natural connection to string theories. In section 4.3 we introduce the minimal information from string theory needed to provide the final details that ultimately lead to the quantitative formulation of the duality. We then take the reader slowly through the original derivation of Maldacena's canonical example of maximally supersymmetric $U(N)$ Yang–Mills theory dual to so-called type IIB supergravity on $AdS_5 \times S^5$. This highlights precisely the quantitative reason why the correspondence is a strong–weak coupling duality that is closely tied to a matrix large-N limit.

We conclude this chapter with a brief tutorial review of general relativity and the geometry of anti-de Sitter space (box 4.4). This toolkit, together with the explanation of the condensed matter context in the previous chapters and the field-theory background for holography discussed here, will give the reader the proper base to understand the new insights provided by the correspondence in the next and following chapters.

4.1 A short history of the holographic principle, black holes, string theory and the origins of the AdS/CFT correspondence

The origins of holography lie in the seminal work of Bekenstein, Hawking, Penrose and collaborators on black holes. For decades after Schwarzschild's 1916 construction of the first black-hole solution to Einstein's equations of general relativity, the singular nature of the Schwarzschild geometry was often disregarded as an artefact. For any real material the point-like approximation of the mass-density, it was argued, would break down long before the Schwarzschild radius was reached. For elementary particles the quantum nature encoded in the Compton radius is encountered first. This changed with Hawking and Penrose's singularity theorems of the late 1960s and early 1970s. They ruled out the possibility that black holes were just curious solutions that had no bearing on the real world [161, 162]. Instead they showed that ordinary matter would collapse to singular space-times and that the need to address the existence of black holes in general relativity was unavoidable.

From this increased understanding of black holes arose the realisation that the macroscopic properties of black holes show a remarkable resemblance to the laws of thermodynamics. Firstly, so-called black hole "no-hair theorems" proved that a black hole solution in general relativity (in $3+1$ dimensions and when infinitely far from the black-hole space is flat) is uniquely characterised by its mass M, charge Q and angular momentum J [163, 164]. For a black-hole solution they are related as [165]

$$dM = \frac{\kappa_s}{8\pi G} \, dA_H + \Omega \, dJ + \Phi \, dQ, \tag{4.1}$$

where A_H is the area of the black-hole horizon, κ_s is a quantity known as the surface gravity at the horizon, and Ω and Φ are the angular velocity and the electrostatic potential at spatial infinity. Substituting the energy E for the mass, temperature T for the surface gravity and entropy S for the area of the black-hole horizon, this is exactly the first law of thermodynamics. A series of thought experiments by Bekenstein tracing the validity of the second law of thermodynamics – that entropy should always increase, including in the presence of black holes – showed that one should take this resemblance seriously [166]. This was soon followed by Hawking's stunning 1979 discovery that quantum mechanically black holes are not black at all, but in fact radiate at a temperature $T = \kappa_s/(2\pi)$. This proved outright that this resemblance to thermodynamics is not a coincidence.

As remarkable as this discovery was, it also posed an immediate profound puzzle. Since equilibrium thermodynamics is explained by the statistical mechanics of microscopic particles, this implies that *one should also think of the black-hole entropy – measured through the area of the horizon $S_{BH} = A_H/(4G)$ – as a counting of the number of microstates with the same thermodynamical macroscopic properties.* The notion that a black hole should be seen as an ensemble of states stands in direct contrast with the conventional view of black holes. In classical general relativity black holes are inescapable sinks from which nothing can ever escape; hence no distinguishing features can ever be detected by an external probe. The most well-known rephrasing of the conundrum between the notion of black-hole entropy and the classical idea that the black-hole solution can only depend on the mass, charge and angular momentum is Hawking's *information paradox*. As a direct consequence of the no-hair theorems, the Hawking radiation emanating from the black hole must be *exactly* thermal, i.e. a mixed state. But this *necessitates a* loss of information whenever a pure quantum state crosses the horizon and falls into the black hole.

If information would truly be lost whenever an object fell into a black hole, this would pose a formidable challenge to any quantum theory of gravity. It would be a non-unitary theory, and the fundamental precepts of the meaning of the wave function would have to be revisited. A significant number of physicists therefore argued that, rather than giving up the conventional framework of quantum theory, one should investigate the quantum nature of black holes further. When doing so, they argued, one would find that the evolution is in fact fully unitary, i.e. the precise state in the ensemble represented by the black hole would eventually reveal itself in the Hawking radiation by deviations from the exact Planck spectrum.

Trying to actually answer the information paradox, however, required either a novel framework for quantum mechanics or deeper insight into a quantum theory of gravity. These formidable challenges hampered any significant progress, with one exception. In 1993 't Hooft proposed that one should take seriously the most notable aspect of the black-hole entropy: it is not extensive [167]. Unlike in conventional statistical or quantum many-body systems, the black-hole entropy grows geometrically with the area rather than with the volume. Einstein's theory of general relativity already contains the wisdom that lumping a large enough amount of matter together would lead to its collapse to a black hole. 't Hooft and subsequently Susskind argued that this also means that lumping together a large enough amount of states of quantum gravity has to reveal a Bekenstein–Hawking area growth of its macroscopic entropy rather than a geometric volume growth [168]. The exact relation between the horizon area of a black hole and the entropy even shows that one should assign roughly one degree of freedom per unit horizon area in terms of the Planck length squared, the strength of the gravitational force in natural relativistic units: $\ell_P^2 = \hbar G/c^3$. This sounds like a simple idea, but the implications are very deep. The theory of quantum gravity has to be truly very different from any known theory in physics: it has to act like a *hologram* in that the information can be written on the area of the surface of the region whose dynamics it describes [167, 168].

4.1.1 String theory as a theory of quantum gravity

The truly fundamental step forwards had to wait for the input from string theory. Just like the quest for quantum critical metals, the intractability of a quantum theory of gravity using conventional approaches asked for a new point of view. String theory provides this perspective. As with many breakthroughs, this was not its goal when it was developed. In the late 1960s, the precursor of string theory was an attempt to formulate a model to describe the spectrum of elementary particles as measured in high-energy experiments probing the strong interaction. The discovery of the Standard Model in the early 1970s, with as a highlight the asymptotically free theory of quantum chromodynamics (QCD), which is a regular quantum field theory, showed that string theory was not the correct description of the strong interactions. However, during the same era it was realised that a closed string naturally describes a massless spin-2 particle: a graviton, the quantum of the gravitational field. Thus string theory can be used to describe gravity. However, string theory is a fully fledged (first-quantised) relativistic quantum theory. This was therefore not the general relativity of Einstein, but its long-sought quantum extension. It could therefore potentially shed light on the information paradox. Indeed, in 1996 Strominger and Vafa validated the statistical foundation of black-hole thermodynamics by exhibiting the precise counting of the microstates underlying the entropy of a very specific black-hole in string theory [169].

It will be useful to understand qualitatively some of the aspects of string theory to see how it can do this and how it provides the origin of the holographic AdS/CFT correspondence. The characteristic of string theory is that the most elementary object is no longer a point particle as in conventional quantum field theories, but instead a one-dimensional string with a tiny length. Both open and closed strings occur. At scales much larger than the string length, the dynamics of open strings can be shown to reduce to the dynamics of a gauge theory – electromagnetism and its non-Abelian generalisations as in the Standard Model – while the closed strings naturally reduce to the gravitational interactions of Einstein's general relativity plus a selection of massless fields. The fundamentally string-like nature of the elementary excitations has three important consequences [170, 171, 172].

1. To ensure quantum consistency in string theory, i.e. that the quantum theory has the same number of polarisations as the classical theory, one requires an enormous number of additional degrees of freedom at short distances in addition to gravity and gauge interactions. A natural way to group these extra excitations together is to consider the theory to exist in a higher-dimensional world. For the bosonic string theory carrying only bosonic excitations, quantum consistency dictates that it has to live in 26 dimensions.

2. Given that one also wants to use string theory to describe fermionic excitations, one naturally finds that a fermionic string in flat space has a supersymmetric spectrum of excitations. This supersymmetric version of string theory has to live in 10 dimensions for quantum consistency.

3. To get rid of the extra dimensions, one relies on so-called Kaluza–Klein compactification: one rolls up the extra dimensions to a small size. This gaps out states with momentum in the rolled-up directions and it has the added benefit that gauge forces and particles emerge as leftovers from the purely geometrical

theory (we provide some more detail in chapter 13). One can attempt to account for the structure of the Standard Model of particle physics in this way. To do so, particular ingenious ways of compactifying the extra dimensions in the form of so-called Calabi–Yau manifolds with deep connections to mathematics were developed.

It turns out that the quantum theory of strings is even further constrained than its number of space-time dimensions. Requiring the absence of quantum gauge and gravitational anomalies showed that only five superstring theories appear to be consistent, all living in $9 + 1$ space-time dimensions. Besides a single open-string theory with $SO(32)$ gauge symmetry, also known as the type I string, four consistent closed-string theories exist: the type IIA and type IIB closed-superstring theories, and two so-called heterotic-string theories with either $SO(32)$ or $E_8 \times E_8$ gauge symmetry. This discovery that the stringy incarnation of quantum gravity was almost unique triggered the "first string-theory revolution" in pursuit of the idea that a single theory of everything could explain all observed phenomena. In particular, the strict consistency requirements left no room to add other forces, and therefore the known forces of the Standard Model and its matter content had to emerge naturally from the original $(9 + 1)$-dimensional theory. In the course of compactifying these theories to $3 + 1$ dimensions in ways that ever more closely approximated the Standard Model of particle physics, several remarkable discoveries were made.

1. After compactification some string theories turn out to give rise to low-energy effective theories that are *dual* to each other. The theories describe the same physics but from a different perspective, in the same sense as that in which the Abelian–Higgs duality discussed in chapter 2 is dual to a scalar theory.
2. Furthermore, it appears that tuning parameters in the low-energy effective theory changes not only the geometry of the rolled-up extra dimensions but also the topology. This was a small revolution insofar as field-theoretic insights into quantum gravity had hitherto always taken the topology as a fixed attribute of the theory. String theory showed that quantum gravity is much richer than that.
3. Even though there appear to be only five consistent string theories in $9 + 1$ dimensions, many more can exist in lower dimensions. This was not fully appreciated at the time (see, however, [173]).

All these results were derived within string perturbation theory. At the same time, string theory did not – and still does not – have a non-perturbative "field-theoretic" formulation, but only a first-quantised formulation analogous to the Schrödinger equation. This hampered the attempt to gain any insight into non-perturbative

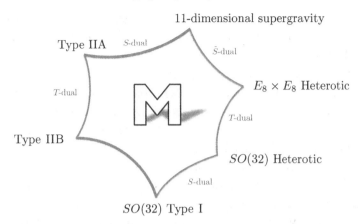

Figure 4.1 The five $(9 + 1)$-dimensional perturbative string theories plus $(10+1)$-dimensional supergravity emerge from an underlying M-theory in various limits. Their common origin is reflected in a sequences of strong–weak Abelian–Higgs-type "S" dualities or stringy "T" dualities that relate any two theories.

phenomena, including the correct ground state of the theory. Moreover, given the strong consistency constraints on the perturbative level, indicating that there is essentially no adjustment possible in any of the five theories, the full non-perturbative extension of the theory has to be consistent on its own in order for the theory to be viable. There is no wriggle room to fix non-perturbative problems. Some in fact worried that not even one of the five string theories, or any of their lower-dimensional cousins, would pass this absolute test.

A talk by Witten at the 1995 annual Strings conference completely changed the general perspective on the meaning of string theory. By trying to discern the non-perturbative dynamics of string theory, he re-identified the existing string theories as semi-classical limits of a single underlying, all-encompassing theory: "M-theory". This triggered the second string revolution: it is emergence at its best. The different semi-classical perturbative string theories are all related to each other by strong–weak "S" (Kramers–Wannier) dualities, or by "T" dualities, a special duality of string theory associated with the exchange of discretised momentum excitations with strings winding around a compact direction. These dualities between the different string theories and, in addition, the unique $(10 + 1)$-dimensional theory of supergravity form a connected "web" of perturbative manifestations of M-theory (Fig. 4.1). This leads to the inescapable conclusion that M-theory has to exist, despite the fact that one is nearly completely in the dark regarding what this theory actually is. One notable aspect is that space-time dimensionality changes throughout the web of dualities: space-time is itself emergent. Partially on the basis of this observation, it is believed that M-theory need not be governed by a local action principle in the usual way. At the same time, nobody knows yet what the principle that controls this theory is instead. This emergence of space-time is a phenomenon that is inherent to holography, as we shall see (see [174] for a deeper discussion on this point).

Instrumental in this non-perturbative extension of string theory to M-theory is the discovery by Polchinski of extra non-perturbative objects that exist in addition to the perturbative strings. Type II string theories have soliton-like solutions, called D-branes. These are spatially extended objects, like strings, but in any number p of spatial dimensions [175]. Their existence in superstring theories is required in order to make the spectrum consistent. Their discovery also shifted the physical perspective of the theory. The previously known open-string theories are better understood as the collective excitations of these Dp-branes. Directly analogously to the way in which the collective excitations of a soliton are tied to its location, the end points of the open string end on such branes. The open strings should really be seen as the vibrational modes of the soliton brane. Open-string end points also act as sources of the gauge fields, as we learned earlier. One therefore naturally has theories with gauge open-string degrees of freedom, confined to a subspace in space-time – the location of the soliton D-brane, while excitations of closed strings are responsible for gravity. The latter live both "on the brane" and "off the brane" and they can probe the full space-time.

Box 4.1 Braneworlds

One interesting consequence of this new perspective is the notion of braneworlds: our universe can be viewed as a three-space-dimensional brane "floating" in the ten-dimensional fundamental space-time of string theory [176, 177] (for a review, see e.g. [178]). The gauge-charged particles of the Standard Model cannot get off the brane, but the neutral gravitons can. Perplexingly, this leaves room for the compactification radius of the extra dimensions to be much "larger" (of the order of micrometres) than had previously been assumed (the Planck/GUT scale $\ell_P \sim 10^{-35}$ m) and still be barely detectable due to the weakness of the gravitational force. In essence, the size of the extra dimension in a braneworld scenario is limited only by our direct observation that no energy "disappears" when a neutral graviton "leaves" the brane to explore the large-extra-dimensional universe. Because the associated true scale of gravity is so much lower in these "large-extra-dimensions" scenarios, one of the observable consequences is the possibility that mini-black holes are created by high-energy collisions. Cosmological constraints do rule this out for the energies achieved at the LHC, but the temptation of such a serendipitous discovery has made large-extra-dimensions scenarios phenomenologically very appealing.

4.1.2 The birth of the AdS/CFT correspondence

AdS/CFT is a direct outcome of the existence of these D-branes. As the crowning touch of the duality discoveries of the second superstring revolution, Maldacena put forward in 1997 the conjecture that type IIB superstring theory Kaluza–Klein compactified on a 5-dimensional sphere, thereby yielding a $(4 + 1)$-dimensional curved anti-de Sitter background for the remaining space-time dimensions, is dual to maximally supersymmetric $\mathcal{N} = 4$ $U(N)$ super-conformal Yang–Mills theory in $3+1$ dimensions. Here $\mathcal{N} = 4$ counts the number of independent supersymmetries the theory possesses. This duality is the prototype of the AdS/CFT correspondence.

We will discuss these roots of the AdS/CFT correspondence in more detail in section 4.3.1, but it is instructive to describe it briefly here. The origin of Maldacena's idea is that there are two ways to describe the dynamics of D-branes. One can view this non-perturbative object either as a soliton in a gravitational field captured by closed strings, or through the prism of its collective excitations. The latter are the open strings that necessarily end on the D-branes. For N coincident D-branes, the two end points of a single open string can reside on two different D-branes, and therefore the excitations of the open strings carry two indices, one for each end, that run from 1 to N. On combining this with the insight that open strings can be viewed as flux-tubes communicating the force between the end points, one finds that the low-energy dynamics of this system of N D-branes, corresponding to the low-energy dynamics of the open strings connecting them, can be described by a $U(N)$ gauge field theory. When the spatially extended D-branes span $p + 1$ dimensions inside a $(d + 1)$-dimensional space-time, this is a $(p + 1)$-dimensional field theory.

Now consider the same physics from the closed-string viewpoint. The fundamental discovery of Polchinski is that these D-branes themselves carry charges with respect to certain closed-string fields [175]. In the weak-coupling limit of the closed-superstring theory, a (super)gravity theory, the charges of these D-branes plus their energy densities source curvature and flux in the closed-string modes, in a way similar to extremal black holes. Maldacena showed that taking the low-energy limit on the open-string side, which thereby turns into the $U(N)$ gauge theory, translates into taking the near-horizon limit of this extremal-black-hole-like configuration in the closed-string/supergravity picture. For the specific case of N coincident $(3 + 1)$-dimensional D-branes in $(9 + 1)$-dimensional type IIB string theory, the near-horizon geometry of these branes turns out to be a $(4 + 1)$-dimensional anti-de Sitter space times a five-sphere: $AdS_5 \times S^5$. The low-energy dynamics on their $d = (3+1)$-dimensional worldvolume in the open-string picture becomes on the other hand $U(N)$ gauge theory with $\mathcal{N} = 4$ supersymmetry.

From the equivalence of these two pictures, Maldacena arrived at the conjecture that the full type IIB superstring theory on $AdS_5 \times S^5$ is dual to $\mathcal{N} = 4$ $U(N)$

super-Yang–Mills field theory in $3 + 1$ dimensions [1]. Crucially, upon taking the low-energy limit of the *open-string* theory so that it turns into the gauge theory, all string-theoretic information is lost. One thereby ends up describing a true field theory. Even though the historical roots of AdS/CFT are in the equivalent open/closed-string descriptions, once the low-energy limit is taken this connection disappears. What makes it work nevertheless is the *holographic principle*: all the information contained in the quantum theory of gravity can be encoded in a regular field theory in one dimension fewer. Shortly after Maldacena's conjecture, Witten, simultaneously with Gubser, Klebanov and Polyakov, showed in an aptly titled article "Anti-de Sitter space and holography" that this is what allows the $(3 + 1)$-dimensional theory to be equivalently described by a $(4 + 1)$-dimensional gravitational theory [2, 3]. Concretely put, AdS/CFT was the first explicit example showing that quantum gravity does indeed obey the postulated holographic principle.

AdS/CFT is formally a conjecture. Nevertheless, it has passed numerous strong tests and it is now taken to be a "theorem". Particularly in supersymmetric versions of the correspondence, there are many protected quantities whose exact answers can be computed on both sides of the correspondence. These answers have been shown to agree in every known case. Since 1998 many further items of non-trivial evidence supporting the conjecture have been found; for a detailed discussion of these checks, see [7] and [9]. As an aside, the surprising progress in solving $(3 + 1)$-dimensional $\mathcal{N} = 4$ super-Yang–Mills field theory [179] has actually given an inkling of hope that AdS/CFT in the form of Maldacena's original conjecture might be proven some day. At present, however, the AdS/CFT conjecture has a similar status to, say, the path integral. Although a strict mathematical proof still has to be delivered, there is no doubt that it is true.

4.1.3 From AdS/CFT to AdS/CMT

Maldacena discovered this holographic AdS/CFT structure as a particular limiting case of the more general open–closed-string dualities of string theory. Taking appropriate limits of other cases, a portfolio of exact AdS/CFT dualities can be extracted with the open-string side as the origin of the field theory living on the holographic screen on the boundary, and the closed-string side accounting for the anti-de Sitter side of the correspondence, i.e. the gravity sector in the bulk. In these "top-down" models, well-defined and transparent rules precisely describe the perturbative limits on both sides. On the gravitational side, it is a compactification of a $(9 + 1)$- or $(10 + 1)$-dimensional classical supergravity to a special background that contains an anti-de Sitter sector, as for instance $AdS_5 \times S^5$. In practice this means that $(d + 1)$-dimensional general relativity is realised in the presence of a negative cosmological constant – responsible for the anti-de Sitter curvature – together with a very particular set of scalar, fermion and gauge fields. In some cases this is complemented by dynamical D-branes: dynamical gravitating degrees of freedom that are restricted to only a subspace of the space-time.

The field-theoretical duals that arise in this way from string theory may initially appear to be very exotic. Their perturbative limits are $U(N)$ Yang–Mills gauge theories with supersymmetry: generalisations of QCD describing the gluons of the strong force, which now include the supersymmetric fermionic partners of the gluons. In addition these theories have a very particular matter content such that the theory flows to a strongly coupled fixed point in the IR. This is the CFT in AdS/CFT. As we have emphasised in the introduction, however, the crucial aspect of the correspondence is that the two descriptions, CFT and AdS, are under perturbative control in *different* regimes. In particular, for the perturbative description on the AdS side, not only does one have $N = 3$ different colours as in QCD, but also one needs to extend it to a nearly arbitrarily large number (large-N). In this limit the actual parameter that governs the effective strength of the interaction is what is known as the 't Hooft coupling, $\lambda \equiv g_{YM}^2 N$, or a variant thereof. *The most important aspect of the holographic AdS/CFT duality is that the regime of perturbative classical general relativity on the gravity side corresponds to taking both this parameter λ and the number of colours N to infinity.* We will discuss precisely how this limit arises in more detail in the next section, but it is this fact that is the foundation of the dream of AdS/CFT. It means that one can use *classical gravity* to study the previously out-of-reach *strongly coupled* physics in the field theory, albeit in the limit of a large number of colours N.

From a deep fundamental viewpoint this is all fine, but a condensed matter physicist can rightly ask how it can be of use to him. What can large-$U(N)$ supersymmetric Yang–Mills theory with a detailed prescribed matter content possibly have to do with correlated electrons even if they are strongly interacting? There is a very good reason why it can be relevant. The articles by Witten [3] and Gubser, Klebanov and Polyakov [2], building on Maldacena's specific examples, argued that the AdS/CFT duality can be formulated on completely general "holographic" grounds that need not appeal to top-down constructions explicitly derived from string theory. They formulated a presumably universal "dictionary": the rules translating the quantities of the boundary field theory into the gravitational bulk and vice versa. Because this general dictionary does not need a string-theoretic origin, it opens up a wide new arena to engineer "bottom-up" the AdS dual for a given strongly coupled CFT of interest. But, *most* importantly, this dictionary confirms the qualitative insight that the extra holographic spatial dimension on the gravity side acts as the RG scale of the theory of interest, as we put forward in chapter 1.

This allows one to deform the CFT with the aim of being able to "engineer" a renormalisation-group flow to different IRs of any strongly coupled QFT of interest. Although one should still seek serious mathematical support by identifying the engineered model with an explicit string construction, it is this bottom-up approach that gives AdS/CFT its status as a breakthrough window into the world of non-perturbative field-theoretical physics.

In the particular context of condensed matter physics, one should view it as follows. At the atomic scale, the physics of electrons is essentially about solving the Schrödinger equation. The art of condensed matter physics is to deduce from this common starting point the emergent universal long-range physics which controls the macroscopic properties, which more often than not completely forget the specifics of the microscopic physics: the "strong emergence" notion. Though AdS/CFT is strikingly novel in its details, the idea is to follow exactly the same programme, but now one replaces the traditional microscopic Hamiltonian by an interacting CFT in the UV.

The ultimate physics question we wish to answer concerns the behaviour of an infinite number of strongly interacting quantum degrees of freedom. What we shall see is that "holography" acts as a "generating functional" that is extremely powerful in revealing the principles controlling the strong-emergence physics. Even though matters are completely intractable on the field-theory side, even if the microscopic theory is known, the dictionary rules are amazingly constraining regarding the construction of the gravitational dual. This is the magic: at least for equilibrium and near-equilibrium problems, one ends up studying minimalistic and very constrained natural GR problems of a kind that are as exciting for professional relativists as their duals are for condensed matter physicists.

This is not surprising insofar as AdS/CFT can be seen as a continuation of the heyday of GR in the 1960s and 1970s with the deep pursuit into the mysteries of black-hole physics, culminating in the discovery of Hawking radiation. Thanks to AdS/CFT, this field has entered a second youth, with new insights into general relativity that are just as remarkable, but now further enhanced by the relations with cutting-edge condensed matter physics. For example, AdS black-holes can un-collapse, as we will see in chapter 6. To a relativist this realisation that black-holes in AdS are not stable states at all has been a stunning surprise. At the same time, AdS/CFT's ability (which will be shown in chapter 7) to nearly effortlessly compute near equilibrium many-body correlations in real time with a seamless transition to the hydrodynamic regime should be nearly as astonishing to a condensed matter physicist.

This does not mean that there are no caveats. The tight gravitational constraints that are reflected in the long-wavelength universality found in the field-theoretical dual ought to dispel most worries that the results obtained with AdS/CFT are particular to large-N supersymmetric Yang–Mills theories. The latter theories generically arise in string constructions, since supersymmetric non-renormalisation theorems (cancellation of bosonic and fermionic quantum radiative corrections) have as a consequence that large classes of supersymmetric Yang–Mills theories are always conformal, i.e. quantum critical, regardless of the value of the coupling constant. However, although supersymmetry is incredibly helpful in constructing the microscopic UV CFT, it is not essential to AdS/CFT: top-down string constructions of field-theory/gravity duals in which supersymmetry is explicitly broken have been identified. Moreover, in the condensed matter context we are typically interested in circumstances where supersymmetry is already explicitly broken by finite temperature, finite density, etc. It appears that the supersymmetric nature of the ultraviolet small-scale theory makes little, if any, essential difference in this regime.

This is not the worry, and neither is the requirement of a large 't Hooft coupling λ. That is in fact the good news. This is what makes the theory strongly interacting. Moreover, one can move away from infinitely strong coupling by the inclusion of so-called α' (string tension) corrections to general relativity. These are specific computable higher-derivative corrections as dictated by string theory. Though it is practically cumbersome, one can add these order by order in perturbation theory in $1/\lambda$, and this is tractable in principle.

The most tenuous requirement with regard to condensed matter applicability is the large-N limit in the field theory: one *can* get away from this limit by studying corrections associated with string *loops*, but this is barely feasible in practice. This is the way in which serious quantum gravity is supposed to show up in string theory, and despite an enormous effort still very little can be done. We shall appeal to the argument that we are interested in highly phenomenological theories describing "strong emergence" physics, which might be rather independent of the specifics of the short-distance physics. To what extent this can be reliably done is an open question because the large-N limit seems to exert a rather undesired influence. Most importantly, it imposes a mean-field attitude of a special kind on everything, even overruling the number of space dimensions.

This special mean-field behaviour could be completely disconnected from real-life low-N systems. Part of the material in this book is a testimony to the fact that this large-N "strong emergence" or "UV independence" works much better than one would expect a priori. By just trying it out, in all cases where we know what to expect for the IR physics on the field-theoretical side, we find the gravitational dual delivers an impeccable description. This is a basis to believe that we can trust the results of holography, even when it predicts states of matter that cannot be described on the basis of available field-theoretical techniques.

4.2 Yang–Mills as a matrix field theory and the ultimate mean field at large-N

The large-N limit, nevertheless, has an honourable history in theoretical physics as a means by which to render an interacting system perturbatively under control. It is in fact a far more accessible inroad into the physics behind AdS/CFT than a full tour through string theory. We therefore leave string theory behind for now, and begin with a review of the large-N limit in conventional field theory. We shall see that both string theory and duality arise very naturally upon inspecting this limit in detail. Although one can be quite ignorant regarding much of string theory, one has to be well informed regarding the general nature of the quantum field theories dealt with by the AdS/CFT correspondence. These are *matrix* large-N field theories. Reflecting the high-energy-physics origin, the ubiquitous example of a matrix field theory is the non-Abelian Yang–Mills theory which is the basis of the Standard Model. The number N here refers to the number of "colours": QCD has $N = 3$, referring to the three colour charges carried by the quarks, but there are N^2 gluons exchanging the colours between the sources. These gluons are naturally encoded in a matrix.

These matrix field theories have their own special traits, and a highlight is what happens in their large-N limit, as was realised first by 't Hooft in the 1970s. In this limit a novel kind of "mean field" or "classical saddle point" is realised, which is, however, of a completely different kind from the conventional "Hartree–Fock" mean fields encountered in the "vector" theories of condensed matter physics. In striking contrast with the latter, in the matrix large-N limit one finds a strongly coupled dynamics, instead of an effectively free theory. For instance, when such a theory becomes scale-invariant one is still dealing with a strongly interacting CFT, while the mean-field equations govern the renormalisation flow between different CFTs in the UV and IR as we will see later (box 5.3). This mean-field structure is at the heart of the relationship between the boundary and the classical limit in the gravitational bulk.

The manifest connection between matrix large-N models and non-Abelian gauge theories has meant that the 't Hooft large-N limit has been applied extensively in the study of QCD. Its application to condensed matter physics until now has been minimal, however. Quantum Hall physics and in particular the composite excitations in the fractional quantum Hall effect do have a natural formulation in terms of a matrix field theory, for which one can probe the large-N limit; for a review see [180]. In addition one of the condensed matter puzzles posed in chapter 3 – a finite density of fermions coupled to a critical boson – has a natural matrix large-N limit [181, 182]. Since it has remained rather unfamiliar in condensed matter physics, we present here a short tutorial on this subject.

4.2.1 Vector large-N models and condensed matter physics

As we already discussed in section 2.1.1, in condensed matter physics one usually considers *vector* field theories. Just as in the semi-classical mean-field theories of the Hartree–Fock type controlled by the smallness of $\hbar$, one finds in the large-N limit order parameters that freeze out. These are typically different from the semi-classical ones, but they share with the Hartree–Fock variety that the mean-field theory is a free theory, which is perturbatively dressed using the $1/N$ expansion [183]. This large-N expansion is particularly powerful in combination with renormalisation-group analysis, where it is complementary to dimensional regularisation (see e.g. [184, 185]).

To set the stage, let us focus in on the zero-density "interacting graphene"-type model, which we introduced in section 2.1.1; in high-energy physics this is known as the Gross–Neveu model (in $d = 1$ dimensions) and the Nambu–Jona–Lasinio model (in $d = 3$ dimensions). This a theory of N species of fermions with a four-fermion self-interaction. Labelling each species by the index $i = 1, \ldots, N$ and using the Einstein summation convention that a repeated upper and lower index implies a summation over its full range, the action is

$$S = \int d^d x \, dt \left(-i\bar{\psi}_i(\gamma^\mu \, \partial_\mu + m)\psi^i + \frac{\lambda}{6}(\bar{\psi}_i \psi^i)^2 \right). \tag{4.2}$$

Here we are using relativistic fermions where the spinor index – ranging from 1 to $2^{d/2}$ in $d = 2n$ and $d = 2n + 1$ dimensions – is suppressed. In $3 + 1$ dimensions these four indices denote the spin-up and spin-down particles and spin-up and spin-down anti-particles. Rotations are generated by the generator $M_{\mu\nu} = \frac{1}{4}[\gamma_\mu, \gamma_\nu]$ with γ_μ the Dirac matrices obeying the Clifford algebra $\{\gamma_\mu, \gamma_\nu\} = 2\eta_{\mu\nu}$. The Dirac conjugate $\bar{\psi}$ is related to the Hermitian conjugate as $\bar{\psi} \equiv \psi^\dagger \gamma^0$.

One can exploit the smallness of λ to describe the weakly interacting limit using conventional perturbation theory. However, when λ becomes large this obviously fails. However, for a large number of species or "flavours" N one can exploit the smallness of $1/N$ to regain control. Using the Hubbard–Stratanovich transformation with a scalar auxiliary/order-parameter field σ, the $U(N)$ four-fermion theory is equal to

$$S = \int d^d x \, dt \left(-i\bar{\psi}_i(\gamma^\mu \, \partial_\mu + m)\psi^i - \frac{3}{2\lambda}\sigma^2 - \sigma \bar{\psi}_i \psi^i \right). \tag{4.3}$$

The N-fermionic fields ψ_i now appear quadratically and can be integrated out to yield

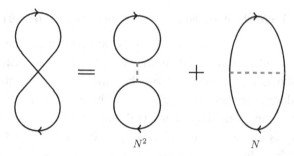

Figure 4.2 The large-N approximation for the four-fermion self-interaction model. On the left-hand side is the standard textbook Feynman diagram of the two-loop free energy where we trace the flow of charge. On the right-hand side is its decomposition where we trace the running of the species/flavour index. This shows that as a function of N there are two terms that contribute with different scaling. In the large-N approximation only the N^2 term is kept.

$$S = \int d^d x \, dt \left(-\frac{3}{2\lambda}\sigma^2 - i\frac{N}{2}\,\mathrm{Tr}\ln(-\gamma^\mu \, \partial_\mu - m - i\sigma) \right)$$

$$= N \int d^d x \, dt \left(-\frac{3}{2\hat{\lambda}}\sigma^2 - i\frac{1}{2}\,\mathrm{Tr}\ln(-\gamma^\mu \, \partial_\mu - m - i\sigma) \right). \qquad (4.4)$$

Here the trace Tr is over the spinor indices. In the last step we have redefined $\lambda = \hat{\lambda}/N$, such that the factor N counting the number of scalar fields now becomes a coefficient in front of the action. This judicious scaling of the coupling constant with N leads to a very powerful result. In the limit $N \to \infty$ with $\hat{\lambda}$ fixed the theory will be dominated by the saddle points of Eq. (4.4), where the action is minimised. The order parameter therefore freezes out, and as in Hartree–Fock it turns into a potential that just scatters the fermions: it becomes straightforward to solve the mean-field equations determining σ.

Notice that each saddle point is still a function of the coupling constant $\hat{\lambda}$. The large-N limit is therefore able to capture non-trivial quantum physics that is quite different from the weak-coupling or semi-classical results, even though one solves saddle-point equations. This "classicalisation" is the essence of the large-N limit.

As in the semi-classical case, the saddle point survives also at large but finite N, and one can explore this regime by perturbative expansion using $1/N$ as a small parameter. Since the large-N mean-field theory describes a free system, the diagrammatic expansion has the structure of conventional weak-coupling perturbation theory. On tracing back one notes that in the large-N limit one sums a subset of diagrams. In terms of the action (4.3), one counts diagrams where fermion loops are linked by the auxiliary field, but not diagrams where the auxiliary field propagates within a loop (Fig. 4.2). Pictorially, one sees that in terms of the original model

Eq. (4.2) we approximate each correlation function by its "maximal loop" decomposition in a way that is identical to the RPA semi-classical mean-field theory discussed in section 2.4

$$\langle \bar\psi_i \psi_j \bar\psi_k \psi_l \rangle \simeq \langle \bar\psi_i \psi_j \rangle \langle \bar\psi_k \psi_l \rangle + \mathcal{O}\left(\frac{1}{N}\right). \tag{4.5}$$

One can now improve on this large-N limit in terms of a $1/N$ expansion, as in Fig. 4.2, in order to capture systems of physical relevance characterised by a finite N.

Box 4.2   The $O(N)$ rotor

This vector large-N mean field is generic and rather independent of the specific theory. One can just as well consider the vector large-N bosonic field theory, as introduced in section 2.1.1:

$$S = \int d^d x \, dt \left[-\frac{1}{2} \partial_\mu \phi_i \, \partial^\mu \phi^i - \frac{m^2}{2} \phi_i \phi^i - \frac{\hat\lambda}{4!N} (\phi_i \phi^i)^2 \right]. \tag{4.6}$$

This theory is also characterised by a global $O(N)$ symmetry where the fields rotate into one another $\delta\phi_i = T_{ij}\phi_j$ with T_{ij} an arbitrary anti-symmetric matrix. After a Hubbard–Stratonovich transformation it is equivalent to

$$S = N \int d^d x \, dt \left(\frac{6}{\hat\lambda} \sigma^2 + \frac{i}{2} \ln(-\Box + m^2 + \sigma) \right), \tag{4.7}$$

which can be analysed in terms of a "classical" large-N saddle point similar to the four-fermion model.

4.2.2 Gauge theories and matrix large-N models

We now turn to matrix field theories, where we will find that the large-N limit is much richer than for vector fields. Although there is a notion of mean-field behaviour at work, this is very different from the conventional saddle points. Explicitly we will see that in the large-N limit the effective theory can remain strongly interacting.

To make a start, let us consider a theory describing many gauge fields, i.e. a vector $U(1)^N$ theory,

$$S = \int d^d x \, dt \left(-\frac{1}{4g^2} F^i_{\mu\nu} F_i^{\mu\nu} \right). \tag{4.8}$$

If one contemplates the physics one is trying to describe for a moment – one is considering a low-energy effective theory with multiple $U(1)$ fields – one realises that a more natural way to think of this theory is as a Higgsed non-Abelian $U(N)$ theory instead of N species of $U(1)$ symmetries. This has a major implication for the large-N limit. In the $U(N)$ theory, the elementary field, in the form of the vector potential $A_\mu = A_\mu^a T_a$, spans the generators T_a of $U(N)$ with T_a the $N \times N$ matrices obeying $[T_a, T_b] = if_{ab}{}^c T_c$. The elementary gauge field A_μ is thus more naturally thought of as an $N \times N$ matrix than as an N-component vector.

The fundamental nature of the limit where we take the rank N of the matrix large is a very different one from the vector large-N limit. A beautiful pictorial representation of the subset of diagrams that survives in the *matrix* large-N limit was discovered by 't Hooft [186]. Keeping A_μ as a matrix, the action for a $U(N)$ gauge theory is

$$S = \int d^d x \, dt \left[-\frac{1}{4g^2} \mathrm{Tr}\left(F_{\mu\nu} F^{\mu\nu} \right) \right] \tag{4.9}$$

with the non-linear field strength

$$F_{\mu\nu} = \partial_\mu A_\nu - \partial_\nu A_\mu - i[A_\mu, A_\nu]. \tag{4.10}$$

The textbook approach to evaluating the theory perturbatively expands the gauge field A_μ in terms of the aforementioned N^2 generators T_a of the gauge group. This yields the standard Yang–Mills Feynman rules (in Lorentz gauge)

$$\langle A_\mu^a(p) A_\nu^b(q) \rangle = g^2 \eta_{\mu\nu} \delta^{ab} \frac{1}{p^2} (2\pi)^4 \delta^4(p+q), \tag{4.11}$$

$$\langle A_\mu^a(p) A_\nu^b(q) A_\rho^c(k) \rangle = \frac{1}{g^2} f^{abc} (2\pi)^4 \delta^4(p+q+k)$$
$$\times \left[(q-k)_\mu \eta_{\nu\rho} + (p-k)_\nu \eta_{\mu\rho} + (p-q)_\rho \eta_{\mu\nu} \right], \tag{4.12}$$

$$\langle A_\mu^a(p) A_\nu^b(q) A_\rho^c(k) A_\sigma^d(s) \rangle = \frac{1}{g^2} f^{eab} f^{cd}{}_e (\eta_{\mu\nu} \eta_{\rho\sigma} - 2\eta_{\mu\rho} \eta_{\nu\sigma} + \eta_{\mu\sigma} \eta_{\nu\rho})$$
$$\times (2\pi)^4 \delta^4(p+q+k+s). \tag{4.13}$$

't Hooft realised that one should keep A_μ as an $N \times N$ matrix instead. Moreover, in the usual Feynman diagram language the pictorial lines either track momentum flow or index flow. When the field is a matrix one can track the charge associated with each index separately. In this double-line notation the Feynman rules are

$$\langle (A_\mu)^i_j(p)(A_\nu)^k_l(q)\rangle = g^2 \delta^{ij}\delta^{kl}\frac{\eta_{\mu\nu}}{p^2}(2\pi)^4\delta^4(p+k),\qquad (4.14)$$

$$\langle (A_\mu)^i_h(p)(A_\nu)^l_j(q)(A_\alpha)^m_n(k)\rangle = \frac{1}{g^2}\delta^i_n\delta^l_h\delta^m_j(p_\mu + k_\nu + q_\alpha)(2\pi)^4\delta^4(p+k+q),$$
$$(4.15)$$

$$\langle (A_\mu)^i_j(p)(A_\nu)^k_l(q)(A_\alpha)^m_n(k)(A_\beta)^h_\zeta(s)\rangle = \frac{1}{g^2}\delta^i_\zeta\delta^j_k\delta^m_l\delta^h_n\eta_{\mu\nu}\eta_{\alpha\beta}$$
$$\times (2\pi)^4\delta^4(p+q+k+s),\quad (4.16)$$

where the indices each run over $i, j, k, l, \ldots = 1, \ldots, N$.

With this double-line representation it is very straightforward to track the power of N of each diagram: each closed loop gives a power of N. It is also easy to see from the pictorial structure of the vertices that each of the Feynman diagrams is a "tiling" together of such closed loops. Each Feynman diagram is thus represented by a two-dimensional tiled surface and comes with a factor N^F, where F is the number of "faces" or "loops" forming the diagram. Combining this with the dependence of the coupling, each Feynman diagram thus scales as

$$\text{Diagram} \sim g^{2P-2V_3-2V_4}N^F,\qquad (4.17)$$

where P is the number of propagators and V_i is the number of vertices with i legs. Euler famously realised that for a two-dimensional surface the combination $P - \sum_i V_i - F = -\chi$, with χ the Euler characteristic $\chi = 2 - 2h - b$, while h is the number of holes in the surface and b is the number of boundaries. This allows us to combine, giving

$$\text{Diagram} \sim (g^2N)^{F-\chi}N^\chi$$
$$\sim (g^2N)^{F-2+2h+b}N^{2-2h-b}.\qquad (4.18)$$

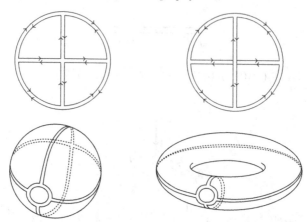

Figure 4.3 Two examples of Feynman diagrams in double-line notation that illustrate the large-N expansion. The left one is planar: it can be drawn on a two-dimensional sphere with no holes. The right one is non-planar: it maps onto a torus. The right one is therefore sub-leading in the 't Hooft large-N expansion. Counting explicit closed index loops, the left diagram scales as $g^6 N^5 = \lambda^3 N^2$ with $\lambda = g^2 N$, whereas the right diagram scales as $g^4 N^2 = \lambda^2$. In the 't Hooft regime where $N \gg \lambda$ the first diagram is therefore more important than the second, despite the fact that it is of higher order in the coupling constant. Figure adapted from [188]. (Reprinted with kind permission from Springer Science and Business Media, © 2013, Springer.)

On defining an effective 't Hooft coupling $\lambda \equiv g^2 N$, one notes that in the limit $N \to \infty$, with λ fixed, the theory reduces to so-called *planar* diagrams with no holes and no boundaries (see Fig. 4.3). Most important for us is again the fact that the complete set of planar diagrams is itself a non-trivial function of the coupling constant. The planar reduction is thus able to retain a large amount of the physics of the full theory. In fact, for the most famous non-Abelian gauge theory, QCD, which is an $SU(3)$ gauge theory, the large-N limit works amazingly well (e.g. [187]).

For any finite value of the 't Hooft coupling, these leading order in "$1/N$" planar diagrams still represent some form of interacting physics. In the matrix large-N limit, differently from the vector large-N case, the theory continues to be (strongly) coupled. Still, there is a different kind of "classicalisation" hidden in the planar-diagram limit. We will soon find out that this requires gravity.

4.2.3 Gauge-invariant operators and their generalisations

Let us now inspect more closely the way in which the various operators in a matrix theory scale as functions of N. The reorganisation of Feynman diagrams in terms of their two-dimensional topology as visible in the double-line notation clearly does not in itself depend on the fact that the fundamental field is a gauge field. We could easily have chosen scalar or fermionic fields as long as they are matrix-valued. In analogy with the case for gauge fields, we do insist that the theory is invariant under the $U(N)$ changes of basis under which the matrix fields Φ^i_j transform as $\Phi \rightarrow U^{-1}\Phi U$. The obvious invariant quantities are traces of products of fields. Although in a generic matrix-valued theory the $U(N)$ similarity transformations are not necessarily gauged, these operators are still often called gauge-invariant.

The trace structure of operators is an essential aspect of matrix large-N theories. For example, consider the matrix generalisation of $\lambda\phi^4$ theory. This is

$$S = \int dt\, d^d x \left(\text{Tr}\left(-\partial_\mu \bar{\Phi} \partial^\mu \Phi - m^2 \bar{\Phi}\Phi \right) - \frac{\lambda_1}{4!} \text{Tr}\left(\bar{\Phi}\Phi\bar{\Phi}\Phi \right) \right.$$
$$\left. - \frac{\lambda_2}{4!} \text{Tr}\left(\bar{\Phi}\Phi \right)\text{Tr}\left(\bar{\Phi}\Phi \right) - \frac{g^2}{4!} \text{Tr}\left([\bar{\Phi},\Phi][\bar{\Phi},\Phi] \right) \right). \quad (4.19)$$

Firstly, the matrix structure allows for the last interaction, which cannot exist for a single scalar field $N = 1$. This is directly analogous to the structure we found for non-Abelian Yang–Mills theory, and we therefore know that a consistent limit arises if the coupling g^2 scales as $\hat{g}^2/N$ with $\hat{g}$ fixed. The index structure of the first single-trace interaction $\lambda_1 \text{Tr}\left(\Phi\Phi\bar{\Phi}\Phi \right)$ is in this regard the same and it will therefore scale the same way: $\hat{\lambda}_1/N$ with $\hat{\lambda}_1$ fixed. The matrix-field action also allows for the second-to-last double-trace interaction $\text{Tr}\left(\bar{\Phi}\Phi \right)\text{Tr}\left(\bar{\Phi}\Phi \right)$. To have a well-defined large-N limit this coupling should also scale appropriately. The graphical representation of the double-trace interaction clearly shows that it is sub-leading compared with the single-trace four-point vertex,

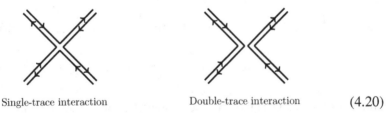

Single-trace interaction   Double-trace interaction   (4.20)

We infer immediately that this double-trace operator is suppressed by a factor of $1/N$ and therefore it will disappear in the large-N limit. To determine its precise behaviour, take the perspective that we can think of a vertex in the Lagrangian as an insertion of a composite operator. Define these operators as

$$\mathcal{O}_{\lambda_1} = \text{Tr}\left(\bar{\Phi}\Phi\bar{\Phi}\Phi \right),$$
$$\mathcal{O}_{\lambda_2} = \text{Tr}\left(\bar{\Phi}\Phi \right)\text{Tr}\left(\bar{\Phi}\Phi \right), \quad (4.21)$$
$$\mathcal{O}_g = \text{Tr}\left([\bar{\Phi},\Phi][\bar{\Phi},\Phi] \right).$$

The scaling of the coupling constant can then be inferred by demanding that the large-N limit is well defined after an appropriate normalisation of any of these composite operators. For single-trace operators to survive in the large-N limit, we demand that their connected two-point function is of order unity. Were it to blow up with N, the theory would be inconsistent; when they diminish they would disappear from the spectrum. Thus we demand that

$$\langle \mathcal{O}_{\lambda_1} \mathcal{O}_{\lambda_1} \rangle_c \sim N^0. \tag{4.22}$$

Note the subscript c for the *connected* two-point function, because this determines the spectrum. To evaluate the diagram there are four propagators connecting the fields,

$$\langle \mathcal{O}_{\lambda_1} \mathcal{O}_{\lambda_1} \rangle = \underbrace{}_{N^4} \tag{4.23}$$

Counting the number of closed loops, this diagram scales as N^4. We readily see that part of the scaling is due to the number of fields. For the single-trace operator $\mathcal{O}_n = \text{Tr}((\bar{\Phi}\Phi)^n)$ the connected parts of $\langle \mathcal{O}_n \mathcal{O}_n \rangle$ scale as N^n. To extract the field dependence we normalise $\Phi \to \hat{\Phi}\sqrt{N}$. To confirm that this is indeed the scaling which 't Hooft determined, note that the action after the rescaling of the field reads

$$S = N \int d^d x\, dt \left(\text{Tr}\left(-\partial_\mu \bar{\hat{\Phi}} \partial^\mu \hat{\Phi} - m^2 \bar{\hat{\Phi}}\hat{\Phi} \right) - \frac{\lambda_1 N}{4!} \text{Tr}\left(\bar{\hat{\Phi}}\hat{\Phi}\bar{\hat{\Phi}}\hat{\Phi} \right) \right.$$
$$\left. - \frac{\lambda_2 N}{4!} \text{Tr}\left(\bar{\hat{\Phi}}\hat{\Phi} \right)\text{Tr}\left(\bar{\hat{\Phi}}\hat{\Phi} \right) - \frac{g^2 N}{4!} \text{Tr}\left([\bar{\hat{\Phi}}, \hat{\Phi}][\bar{\hat{\Phi}}, \hat{\Phi}] \right) \right). \tag{4.24}$$

Thus we do indeed see that, for the single-trace operators to be part of a consistent large-N limit, the combinations $\lambda_1 N$ and $g^2 N$ should be held fixed.

Our goal was to figure out the scaling of the double-trace operator. One can now determine this from the two- or four-point correlation of the single-trace operator $\mathcal{O}_2 = \text{Tr}(\bar{\hat{\Phi}}\hat{\Phi})$. Their connected parts now scale as

$$\langle \mathcal{O}_2 \mathcal{O}_2 \rangle \sim N^0,$$
$$\langle \mathcal{O}_2 \mathcal{O}_2 \mathcal{O}_2 \mathcal{O}_2 \rangle \sim N^{-2}. \tag{4.25}$$

Since the double-trace operator is the normal-ordered product of two single-trace operators there is no freedom in the N dependence of its normalisation. This means that the double-trace operator $\mathcal{O}_{\lambda_2} = c_{\lambda_2}(N)\mathcal{O}_2\mathcal{O}_2$ has a normalisation $c_{\lambda_2} \sim N^0$ that does not scale in the large-N limit. This implies that the consistent coupling in the action carries no N dependence. Thus $\lambda_2 = \hat{\lambda}_2/N^2$, and the double-trace operator therefore decouples from the theory.

Box 4.3 $SU(N)$, $SO(N)$ **and combinations with vector theories**

In the above we have considered only matrix-valued fields that transform in the adjoint of $U(N)$. For completeness let us mention that double-line notation exists for other matrix-valued fields as well, and that one can combine it with a vector-like scaling. For $SU(N)$ theories the main difference is the fact that the adjoint is traceless. That means that from each diagram we need to subtract the trace. For example the propagator becomes,

$$\langle (A_\mu)^i_j(p)(A_\nu)^k_l(q)\rangle = \begin{array}{c} i l \\ \rule{3cm}{0.4pt} \\ j \; \mu \nu \; k \end{array} - \;\; \text{(diagram)} \;\;. \tag{4.26}$$

On closer inspection, one finds that, when the trace is subtracted from the propagator, the trace is subtracted universally both for internal lines and for $SU(N)$-invariant correlations. No additional modification to the Feynman rules is necessary. Notice that the "graphical" index flow of the second term in the $SU(N)$ double-line propagator resembles that of a double-trace operator. It cuts off the colour flow. From the discussion of double-trace operators, it is therefore clear that this term will not survive in the large-N limit. At large-N, $U(N)$ and $SU(N)$ gauge theories are therefore nearly indistinguishable.

For $SO(N)$ or $Sp(N)$ theories the generators are anti-symmetric or symmetric traceless matrices, so we need to (anti-)symmetrise in addition. Thus the propagator becomes

$$\langle (A_\mu)^i_j(p)(A_\nu)^k_l(q)\rangle = \begin{array}{c} i l \\ \rule{3cm}{0.4pt} \\ j \; \mu \nu \; k \end{array} \pm \;\; \text{(diagram)} \;\;. \tag{4.27}$$

Finally, one can add vector-like content to the theory, i.e. fields that have a single index $i = 1, \ldots, N$. From the graphical picture where each closed N-loop is a tile in the Feynman diagram, one readily sees that vector-like flow of the colour charge provides either a boundary or a hole to the two-dimensional surface spanned by the Feynman diagram (see Fig. 4.4).

Box 4.3 (Continued)

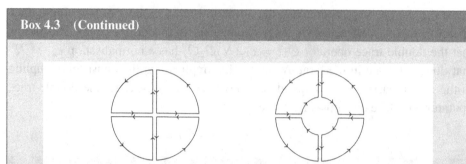

Figure 4.4 Vector-like content in matrix large-N theories in "double-line notation" is denoted by a single line. Feynman diagrams with vector-like content span a two-dimensional surface with topologies that contain a boundary and/or a hole. Counting closed index loops, the left diagram scales as $g^6 N^4 = \lambda^3 N$ with $\lambda = g^2 N$, whereas the the right diagram scales as $g^8 N^4 = \lambda^4$. In the large-N limit, where $\lambda = g^2 N$ is held fixed, the first is therefore sub-leading to a diagram without a boundary, which scales as N^2. The second is even more sub-leading.

4.2.4 Large-N factorisation and the ultimate mean-field theory

We can now explain why an entirely new type of mean-field principle is at work in the large-N limit of strongly coupled gauge theories. In the above we evaluated the connected contribution to the single-trace-operator two-point function $\langle \mathcal{O}_n \mathcal{O}_n \rangle$ with $\mathcal{O}_n = \text{Tr}\big((\bar{\Phi}\Phi)^n\big)$. However, the real power of the large-N limit is exposed by considering the full correlation function including the disconnected diagrams. For the latter there are also four propagators needed, but the index structure is very different.

$$\langle \mathcal{O}_{\lambda_1} \mathcal{O}_{\lambda_1} \rangle = \underbrace{}_{N^6} + \underbrace{}_{N^4} . \tag{4.28}$$

Counting the number of loops, the disconnected diagram scales as N^6 as compared with the N^4 scaling of the connected diagram. It is therefore the disconnected diagram that dominates in the large-N limit. A straightforward "cutting" argument (see Fig. 4.5) shows that this is true for *any* correlation function.

$$N^a \qquad\qquad\qquad\qquad N^{a-1}$$

Figure 4.5 A graphical proof of large-N factorisation. For gauge-invariant operators, the index lines must close somewhere in the blob. If each line closes independently, the left disconnected diagram therefore scales as N^{4+n} with $n \geq 0$. For exactly the same configuration inside the blob, the connected diagram will scale only as N^{3+n}. A disconnected diagram is therefore always leading for N large.

This realisation has a profound consequence. It means that, in the strict large-N limit, correlation functions of gauge-invariant operators *factorise*,

$$\lim_{N \to \infty} \langle \mathcal{O}\mathcal{O} \rangle = \langle \mathcal{O} \rangle \langle \mathcal{O} \rangle + \cdots . \tag{4.29}$$

The expectation value of products of gauge-invariant operators thus becomes equal to the product of their expectation values. Large-N gauge-invariant operators behave essentially as classical variables. This is so in a very strict sense, since factorisation implies that the full variance vanishes,

$$\Delta \mathcal{O}^2 \equiv \langle \mathcal{O}\mathcal{O} \rangle - \langle \mathcal{O} \rangle \langle \mathcal{O} \rangle = 0 + \cdots . \tag{4.30}$$

Note that the vanishing of the variance is a much stronger condition than a mean-field reduction to a Gaussian ensemble. The whole ensemble has collapsed to a *single point*. In this sense one can think of a matrix large-N theory as the ultimate mean-field theory. It is a mean-field theory that is very different from the conventional Hartree–Fock type of field theories defining free field-theoretical systems at the saddle point.

This reduction of the sum over configurations to a single point was first realised by Witten [189], who postulated that it has to mean that there exists a "master-field" formulation of the theory, where the selection of this configuration is manifest. There should be a redefinition of the dynamical variables in the path integral in terms of new degrees of freedom ϕ^{cl}, in terms of which the path integral explicitly localises on a single configuration of these fields. In particular, it should mean that in the large-N limit all expectation values of single-trace operators are simply the evaluation of the operators on this single configuration,

$$\langle \mathcal{O} \rangle = \mathcal{O}(\phi^{\mathrm{cl}}). \tag{4.31}$$

Conversely, the localisation of the path integral on a single classical configuration guarantees factorisation because the (operator) product of classical variables is straightforward multiplication.

Note in particular that factorisation implies the absence of mixing between single- and double-trace operators as we found earlier. The expectation value of any multi-trace operator factorises into a product of single-trace operators,

$$\lim_{N \to \infty} \langle \mathcal{O}\mathcal{O} \rangle = \langle \mathcal{O} \rangle \langle \mathcal{O} \rangle + \cdots,$$

$$\lim_{N \to \infty} \langle \mathcal{O}\mathcal{O}\mathcal{O} \rangle = \langle \mathcal{O} \rangle \langle \mathcal{O} \rangle \langle \mathcal{O} \rangle + \cdots,$$

$$\vdots$$

$$\lim_{N \to \infty} \langle \mathcal{O}^n \rangle = \langle \mathcal{O} \rangle^n + \cdots. \tag{4.32}$$

Multi-trace operators therefore decouple and disappear from the spectrum. This is a more formal way of stating that these multi-trace interactions in the Lagrangian do not contribute to the leading large-N limit.

4.3 The master formulation of large-N matrix models and string theory

Witten's postulation of the existence of a master field led to an intense research effort aimed at identifying the explicit nature of such a theory (see [190]). Starting with the very first paper on the large-N expansion by 't Hooft, it was realised that the matrix large-N limit had something to do with strings. The two-dimensional surface formed by the Feynman diagram in double-line notation starts to show a remarkable resemblance to the two-dimensional worldsheet of a string when the number of faces or plaquettes increases. For growing 't Hooft coupling $\lambda = g^2 N$, this "net" becomes denser and denser, and thereby becomes increasingly string-like; see Fig. 4.6. The master field formulation was therefore most likely a string theory. For $(1+1)$-dimensional gauge theories, lacking dynamical degrees of freedom, one can in fact show directly that this is true [191]. An explicit formulation of such a string-like master field theory remained elusive, however, until the discovery of AdS/CFT.

The true string-like nature of a matrix large-N theory emerges only if we take into consideration another physical concept. This is the notion of duality that we have seen at work in the Abelian–Higgs model in section 2.1.2.

Consider again the prototype matrix large-N theory, a $U(N)$ non-Abelian Yang–Mills gauge theory with action

$$S = \int d^d x \, dt \left(-\frac{1}{4g_{\text{YM}}^2} \, \text{Tr}\big(F_{\mu\nu} F^{\mu\nu}\big) \right). \tag{4.33}$$

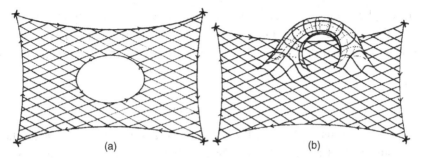

Figure 4.6 For a very large number of tiles the two-dimensional surface representing the Feynman diagram in double-line notation looks more and more like the worldsheet swept out by a dynamical string. This can be either an open string with boundaries (a) (if there is vector-like content in the theory, see box 4.3) or a closed string (b). Figure source [186]. (Reprinted with permission from Elsevier, © 1974.)

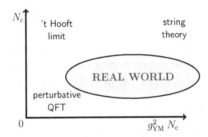

Figure 4.7 The "phases" of a $U(N)$ Yang–Mills theory. Depending on the values of N and $g_{YM}^2 N$ the best description of the theory is as a textbook quantum field theory, a planar diagram expansion or a string theory.

This theory depends on two parameters N and the coupling constant g_{YM}^2. We have just shown that it is more natural to think of the combination $\lambda = g_{YM}^2 N$. For small N and λ the theory is most conveniently analysed via a normal textbook perturbative QFT expansion. For large-N we can use the double-line notation to rearrange the Feynman diagrams in the topological expansion and focus on planar diagrams only, but the planar diagrams themselves are still a perturbative expansion in the coupling constant $\lambda = g_{YM}^2 N$. There is still a loop expansion for each planar diagram. The string-like picture, as we have just seen, however, should emerge as $g_{YM}^2 N$ becomes larger and larger: it corresponds to the strong-coupling regime of the theory. When the diagrams start to look like strings, one is entering a deeply non-perturbative regime. See Fig 4.7 as an illustration.

This gauge-theory/string-theory correspondence, also known as AdS/CFT, is just a more general version of duality, the fact that the most appropriate description of the theory changes as one changes the coupling constant. The subtlety is that it takes place in the large-N regime of the theory. For small 't Hooft coupling λ

one has the planar diagram theory, whereas for large 't Hooft coupling λ AdS/CFT identifies the dual theory as a quantum-gravitational string theory in a curved space with one extra spatial dimension. Let us now show how this identification comes about.

4.3.1 String theory, branes, black-holes

To discuss the origins of AdS/CFT we shall need a few items from string theory. Recall from the first section of this chapter that there are two kinds of strings: open strings, with open ends, and closed strings, which form loops. Whereas the dynamics of a particle is given by its worldline action,

$$S_{\text{particle}} = \frac{1}{2} \int_{\tau_0}^{\tau_1} d\tau \sqrt{-h(\tau)} \left[-h^{-1}(\tau) g_{\mu\nu}(x(\tau)) \frac{dx^\mu(\tau)}{d\tau} \frac{dx^\nu(\tau)}{d\tau} - m^2 \right.$$
$$\left. - iq A_\mu(x(\tau)) \frac{dx^\mu(\tau)}{d\tau} \right], \tag{4.34}$$

with $g_{\mu\nu}$ the space-time metric, m the mass, A_μ a background vector potential and $h(\tau)$ the worldline metric guaranteeing invariance under proper time transformations $\tau \to \tau'(\tau)$, the dynamics of an (open or closed) string is given by the worldsheet action

$$S_{\text{open}} = \frac{1}{\ell_s^2} \int d\tau \int_0^\pi d\sigma \sqrt{-h(\tau)}$$
$$\times \left[-G_{\mu\nu}(X(\sigma,\tau)) h^{\alpha\beta} \partial_\alpha X^\mu(\sigma,\tau) \partial_\beta X^\nu(\sigma,\tau) \right]$$
$$- \oint_{\sigma=0,\pi} d\tau \, A_\mu(X) \partial_\tau X^\mu, \tag{4.35}$$
$$S_{\text{closed}} = \frac{1}{\ell_s^2} \int d\tau \int_0^{2\pi} d\sigma \sqrt{-h(\tau)}$$
$$\times \left[-G_{\mu\nu}(X(\sigma,\tau)) h^{\alpha\beta} \partial_\alpha X^\mu(\sigma,\tau) \partial_\beta X^\nu(\sigma,\tau) \right],$$

with a background space-time metric $G_{\mu\nu}$ and worldsheet metric $h^{\alpha\beta}$ with determinant h. The remarkable aspect of string theory is that the string provides its own background fields. The lowest-energy fluctuations of the open string, those that survive in the limit $\ell_s \to 0$, are the vector-like (gauge-field) fluctuations A_μ which appear in the final term in Eq. (4.34). The lowest-energy fluctuations of the closed string, contain the graviton $G_{\mu\nu}$ which serves as the non-linear factor of the kinetic term in Eq. (4.35). Depending on the precise type of string, these are accompanied by additional fermionic and bosonic massless degrees of freedom.

The crucial part for our discussion is that open strings cannot exist without closed strings. A one-loop diagram of an open string equals a tree-level closed string diagram after a re-parametrisation of the local worldsheet coordinates τ and

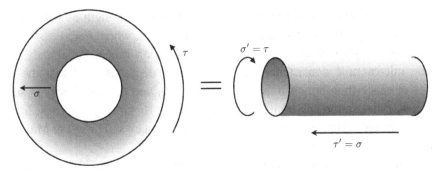

Figure 4.8 An open-string loop diagram equals a closed-string tree diagram after a re-parametrisation.

σ; see Fig. 4.8. A detailed investigation of the required independence with respect to global coordinate changes confirms this fact. Any theory of open strings must therefore contain a closed-string sector, including their mutual interactions. This ability to view certain individual string-theory diagrams and processes from both an open- and a closed-string perspective is colloquially known as *open–closed-string duality*. We will now explain how this is the origin of the duality aspect of AdS/CFT.

The open-string side can be classified in a simple manner. For each spatial direction one can choose boundary conditions corresponding to freely moving end points, such that in the absence of forces their velocity stays fixed, $\partial_\tau X^i(\sigma = 0, \pi) = 0$. Alternatively, one can choose to have the open-string end point "stuck" to a particular location $X^i =$ constant. Because the latter are known as Dirichlet boundary conditions, the hyperplane spanned by all $X^i =$ constant is called a *Dirichlet brane* or D-brane state for short. The energy of such a D-brane state can be shown to scale as e^{-1/g_s} with g_s the string coupling constant. D-branes are therefore non-perturbative solitons of string theory [172].

Recall that solitons are defined as localised finite-energy solutions to the equations, and they therefore always break the translation invariance of the underlying theory. As a corollary, the small excitations around the soliton always contain a zero mode that corresponds to moving the centre of the soliton. The theoretical description of this zero mode is as quantum mechanics on the worldline of the soliton. Similarly, there are two types of low-energy excitations for an open string with Dirichlet boundary conditions: the aforementioned gauge fields, which are now localised on the brane, and the zero modes of the soliton, which are *also* localised on the worldvolume of the brane. For a single $(p + 1)$-dimensional D-brane – a Dp-brane – the low-energy effective action is therefore

$$S_{\text{l.e.}} = \frac{1}{g_s} \int d^p x \; dt \left(-\frac{1}{4} F_{\mu\nu} F^{\mu\nu} - \frac{1}{2} \partial_\mu \Phi^i \, \partial^\mu \Phi_i \right). \tag{4.36}$$

Here g_s is the string-coupling constant, $F_{\mu\nu}$ is the field strength associated with the gauge field and Φ^i are the translational zero modes, where $i = 1, \ldots, d - p$ runs over the number of directions transverse to the Dp-brane. We already noted that the microscopic consistency of string theory demands both that the total number of space-time dimensions is 10, i.e. $d = 9$, and that the theory is super-symmetric. The simplest D-brane is invariant under 16 supersymmetries. In $p = 2n - 1$, $2n$ dimensions there are therefore also $16/2^n$ species of (Weyl or Majorana) fermions.

For instance, the full low-energy effective action of a single D3-brane has four species of Majorana fermions. The symmetry that relates these four species is pre-cisely the $SO(6) = SU(4)$ symmetry that is the remnant rotational symmetry around the D3-brane. This is the same symmetry as that which rotates the scalars. A convenient way to capture that the scalars and the fermions rotate under the same symmetry is to write the **6** of $SO(6)$ as the anti-symmetric representation of $SU(4)$,

$$S_{\text{D3}} = \frac{1}{g_s} \int d^3 x \, dt \left(-\frac{1}{4} F_{\mu\nu} F^{\mu\nu} - \frac{1}{2} \partial_\mu \Phi^{AB} \partial^\mu \Phi_{AB} - i \bar{\Psi}^A \gamma^\mu \partial_\mu \Psi_A \right), \quad (4.37)$$

with $A = 1, \ldots, 4$ and $\Phi_{AB} = -\Phi_{BA}$.

When we consider N Dp-branes on top of each other we get a special enhance-ment of the theory. Each Dp-brane comes with its gauge field, but there are also open strings stretching between each pair of branes i, j. The lowest excitation of such a string has a mass $m_{ij} = (r_i - r_j)/\ell_s^2$, where $r_i - r_j$ is the relative posi-tion of the D-branes in the mutually orthogonal directions. Thus it will become massless as the D-branes are located on top of each other. But this mode is also a vector. Intuitively one can therefore understand that the vector-like modes of these stretched open strings enhance the gauge group from $U(1)^N$ to $U(N)$. These also affect the translational zero-mode sector, however. The relative distance between the branes is also a zero mode, and it can be shown that this combines with the original N translational zero modes to give an $N \times N$ matrix field with a potential that mimics exactly that of the vector fields.

Strictly speaking, the relative distance is only clearly a zero mode to lowest order. For most QFTs this zero mode is lifted at first order: solitons either attract or repel each other. Many D-branes in string theory, however, are "mutually BPS", i.e. one can show using supersymmetry that the relative distance is an exact quantum zero mode.

Moreover, since changes in the labelling of the Dp-branes clearly corresponds to rotations in the gauge group $U(N)$, the scalar zero modes are charged under the $U(N)$ symmetry. The bosonic part of the low-energy effective action is therefore

$$S_{\text{l.e.}} = \frac{1}{g_s} \int d^p x \, dt \left[-\frac{1}{4} \text{Tr}\big(F_{\mu\nu} F^{\mu\nu}\big) - \frac{1}{2} \text{Tr}\big(D_\mu \Phi^{AB} D^\mu \Phi_{AB}\big) \right.$$

$$\left. -\frac{1}{4} \text{Tr}\big([\Phi^{AB}, \Phi^{CD}][\Phi_{AB}, \Phi_{CD}]\big) \right], \qquad (4.38)$$

with $D_\mu \Phi_{AB} = \partial_\mu \Phi_{AB} + i[A_\mu, \Phi_{AB}]$. In the supersymmetric string there are also $16/2^n$ species of $N \times N$ charged fermions with a Yukawa interaction with the scalars. For the D3-brane example above, the full action including the fermions is

$$S_{\text{l.e.}} = \frac{1}{g_s} \int d^p x \, dt \left[-\frac{1}{4} \text{Tr}\big(F_{\mu\nu} F^{\mu\nu}\big) - \frac{1}{2} \text{Tr}\big(D_\mu \Phi^{AB} D^\mu \Phi_{AB}\big) \right.$$

$$-\frac{1}{4} \text{Tr}\big([\Phi^{AB}, \Phi^{CD}][\Phi_{AB}, \Phi_{CD}]\big) - i \, \text{Tr} \, \bar{\Psi}^A \gamma^\mu D_\mu \Psi_A$$

$$\left. - \text{Tr} \, \bar{\Psi}^A [\Phi_{AB}, \Psi^B] \right].$$

This low-energy theory of N D3-branes describes a $(3 + 1)$-dimensional theory that is particularly special. The 16 supersymmetries arrange themselves in four spinor charges. These supersymmetries transform all fields, vectors, spinors and scalars into each other, and together they form a single representation of $\mathcal{N} = 4$, $p = 3 + 1$ supersymmetry, with $\mathcal{N}$ counting the number of independent spinor charges. This theory is the unique theory of $\mathcal{N} = 4$ $U(N)$ super-Yang–Mills. Thanks to the large symmetry, this theory displays a number of remarkable characteristics: among others, it is an interacting theory with the property that electric–magnetic duality is exact. The singular property that will be most important in the present context is that the dimensionless coupling constant g^2 does not renormalise. The theory has no intrinsic scale, even at the quantum level: $\mathcal{N} = 4$ *super-Yang–Mills is a conformal field theory.*

The above is the open-string end of the duality; let us now turn to the closed-string point of view. We noticed that D-branes as solitons have a finite energy. Because the closed-string sector includes gravity, this directly implies that there must be a representation of D-branes on the closed-string side as well. The crucial insight needed to establish that this is indeed the case was Polchinski's discovery that D-branes are charged under one of the additional set of bosonic closed-string fields, the so-called Ramond–Ramond potentials. Not only are they charged, but also their charge equals their mass in the appropriate units. Gravitational solutions that reflect this special charge–mass relation are very well known. They correspond to extremal Reissner–Nordström black-hole solutions. Because in the present case the hypersurface spanned by the soliton is a $(p + 1)$-dimensional membrane, these are more appropriately called "black-brane" solutions. We will encounter the AdS version of these solutions in detail in chapter 8. Here we shall just state their properties.

The N coincident D3-branes of the above can be considered as $(3 + 1)$-dimensional black-hole solitons in a $(9 + 1)$-dimensional space-time. In this case the extremal solution is given by the space-time metric (see box 4.4 at the end of this chapter for the basics of general relativity)

$$ds^2 = H(r)^{-1/2}(-dt^2 + dx_1^2 + dx_2^2 + dx_3^2) + H^{1/2}(r)(dr^2 + r^2 d\Omega_{S^5}^2),$$

$$H(r) = 1 + \frac{4\pi g_s N \ell_s^4}{r^4}.$$

(4.39)

Here $d\Omega_{S^5}^2 = d\theta^2 + \sin^2\theta \, d\Omega_{S^4}^2$ is the metric on the five-dimensional sphere with unit radius, ℓ_s is the $(9+1)$-dimensional Planck length and g_s is the string coupling constant. The latter enters the $(9 + 1)$-dimensional Einstein's equation, as derived from string theory as

$$R_{\mu\nu} - \frac{1}{2}g_{\mu\nu}R = \frac{1}{2\pi}g_s^2(2\pi\ell_s)^8 T_{\mu\nu}.$$

(4.40)

In addition there is a five-fold anti-symmetric tensor generalisation of a $U(1)$ Maxwell field strength that behaves as

$$F_{txyzr} + \frac{1}{5!}\epsilon_{txyzr\alpha\beta\gamma\delta\zeta} F^{\alpha\beta\gamma\delta\zeta} = H^{-2}(r)\frac{16\pi g_s N \ell_s^4}{r^5}.$$

(4.41)

For the informed reader, we note that the coordinates Eq. (4.39) are not the usual black-hole (Schwarzschild) coordinates. In these coordinates, the double horizon, which is characteristic of an extremal black hole, is at $r = 0$.

Now we can analyse both the closed- and the open-string side at the same time. Open–closed-string duality states that the full closed-string theory in the background of this extremal black-brane solution describes the same physics as the full open-plus-closed-string theory, with a low-energy effective action given by $\mathcal{N} = 4$ super-Yang–Mills. Maldacena's deep insight was to attempt to decouple parts of the degrees of freedom on both sides. In particular, since the gravitational coupling constant governing the interaction between the open and closed strings is dimensionful, on taking the string length $\ell_s \to 0$ on the open-string side, the open-string degrees of freedom decouple from the gravitational closed-string sector and reduce to their low-energy sector: pure $\mathcal{N} = 4$ super-Yang–Mills. In order to take the *same* limit on the black-brane closed-string side, Maldacena demanded that the Higgs mass $m_H \sim r/\ell_s^2$ which would arise under a small separation of the N coincident branes should stay fixed. Thus taking $\ell_s \to 0$ meant taking $r \to 0$ simultaneously in the geometry (4.39). The full theory splits into two asymptotically far-away parts: plain closed strings in flat

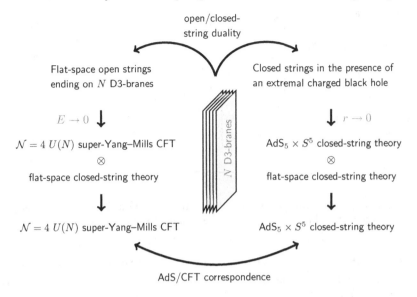

Figure 4.9 A cartoon of the origin of the AdS/CFT correspondence.

space, plus closed strings in this near-horizon limit. Near the horizon the metric becomes

$$ds^2 = \frac{r^2}{L^2}(-dt^2 + dx_1^2 + dx_2^2 + dx_3^2) + \frac{L^2}{r^2}(dr^2 + r^2\,d\Omega_{S^5}^2)$$

$$= \frac{r^2}{L^2}(-dt^2 + d\mathbf{x}_3^2) + \frac{L^2}{r^2}\,dr^2 + L^2\,d\Omega_{S^5}^2, \tag{4.42}$$

with $L^2 = \sqrt{4\pi g_{\rm s} N}\ell_{\rm s}^2$. This is the metric of a $(4+1)$-dimensional anti-de Sitter space times a five-sphere. Thus, in the corresponding limits, both the open- and the closed-string theories result in a decoupled flat-space closed-string sector times another sector. Since the original open–closed-string theories were dual, these two sectors should be equivalent: this implies that the open-string sector of $\mathcal{N} = 4$ super-Yang–Mills should equal closed strings living in the near-horizon limit $AdS_5 \times S^5$ (Fig. 4.9).

4.3.2 Conformal field theories and the boundary of AdS

If these two theories are equal it is a *sine qua non* that their symmetries have to match. $\mathcal{N} = 4$ super-Yang–Mills in fact has an enhanced symmetry group, because it has no intrinsic scale. It is invariant not only under translations, rotations and Lorentz boosts, but also under scale transformations. For a unitary theory it is believed that Lorentz invariance plus scale invariance implies that the theory is in fact invariant under the full conformal group [192, 193, 194].

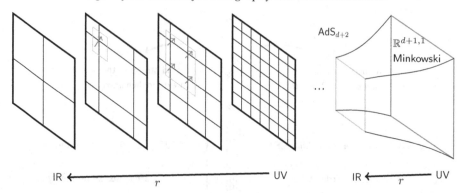

Figure 4.10 We can think of the radial direction of AdS as the successive stages of the conformal field theory under coarse graining. Figure source [5].

This includes in addition the special conformal transformations (see box 4.4). Together these transformations form the group $SO(2, 4)$. One of the early convincing indications that Maldacena's conjecture was right is that the isometry group of five-dimensional anti-de Sitter space is precisely $SO(2, 4)$. This is explained in detail in box 4.4.

The relation is even more precise in that the isometries of anti-de Sitter space reduce exactly to the conformal transformations on the boundary at spatial infinity of anti-de Sitter space $r = \infty$ in the coordinates Eq. (4.42). This, together with the fact that we had to zoom in on the location of the D-branes as the starting point, strongly suggests that the conformal $\mathcal{N} = 4$ Yang–Mills theory "lives" on the boundary of anti-de Sitter space.

The most puzzling feature is the mismatch in dimensions. $\mathcal{N} = 4$ super-Yang–Mills lives in $(3 + 1)$-dimensions, whereas the dual closed-string theory lives in the $(9 + 1)$-dimensional space-time formed by $AdS_5 \times S^5$. The five-sphere is a compact space, and one ought to think of its excitations as an infinite tower of massive modes living on AdS_5 with masses proportional to their angular momentum along the sphere. The actual space-time experienced by the theory is therefore five-dimensional. This is still one dimension more than the super-Yang–Mills theory. The original derivation, which we have just reviewed, explicitly shows us the meaning of this extra dimension. The conformal $\mathcal{N} = 4$ super-Yang–Mills theory has no intrinsic scale. When one Higgses the $U(N)$, however, one introduces a scale in which all other dimensionful quantities are then naturally measured. It is the reference scale that occurs in describing the theory. On the closed-string side the Higgs mass is captured by the position in the radial direction transverse to the boundary. This indicates that one can indeed consider the radial direction as the energy scale at which one probes the system; see Fig. 4.10. We took this deep relation as the starting point of our intuitive guide to AdS/CFT in chapter 1.

4.4 The AdS/CFT correspondence is a field-theory/gravity duality

There is one last issue to confront. As derived, $\mathcal{N} = 4$ $U(N)$ super-Yang–Mills theory has an equivalent description as a *full string theory* on $AdS_5 \times S^5$. Since even the description of string theory in flat space is still a struggle, this is a serious obstacle. Although the description of the full string theory may be unknown, we do know, however, that at low energies closed-string theory just reduces to Einstein's general relativity plus massless fields. Taking the low-energy limit in the background of the anti-de Sitter space means that, in addition to the usual truncation of massive modes, one should also insist on keeping the curvature of AdS very large in units of the string length. The energy-density in the background space-time itself should also be small and to a local observer it should look very flat. This means that the AdS radius $L/\ell_s \gg 1$.

The black-brane solution told us, however, that $L = \ell_s (4\pi g_s N)^{1/4}$. Thus $L/\ell_s \gg 1$ means that $g_s N \gg 1$. But in the Yang–Mills theory as derived from open-string theory Eq. (4.38) the coupling constant g_{YM}^2 is precisely equal to g_s. Thus the low-energy limit where the AdS string theory reduces to classical general relativity precisely corresponds to the large 't Hooft coupling limit in the Yang–Mills theory, which we reviewed in section 4.2.2. Moreover, if we insist that we can neglect string loops, i.e. $g_s \ll 1$, then N must be parametrically larger than $N \gg g_s N \gg 1$. We have to be deeply inside the matrix large-N regime of the gauge theory. *This consideration establishes AdS/CFT as a weak–strong duality.* It exhibits directly that the perturbative weakly coupled classical gravitational side of the theory corresponds to the strong 't Hooft coupling side of the large-N limit of the gauge theory.

String-theoretic open-string/closed-string black-brane duals are the origin of numerous other examples of AdS/CFT correspondences (see chapter 13). The general lesson of *any* duality is that the insights gained should judiciously be viewed as more generally applicable beyond the canonical examples. Many dualities of field theories are qualitatively understood as an exchange between solitons and elementary fields, depending on the value of the coupling, even though there are only a few computable examples. Similarly the insights of AdS/CFT – in particular its concrete dictionary between correlation functions that we will discuss in the next chapter – should carry beyond the few examples that can be derived directly from string theory. This is what validates the bottom-up approach to holography that is the main theme of this book. If a string-theoretic embedding exists, it does ensure that the theory is fully quantum consistent on its own. For the dual field theory this means that it is a UV-complete theory – i.e. we have the absence of a Landau pole/zero, and no other degrees of freedom are needed. This is an important condition, but not a necessary one. Certainly in condensed matter physics one is familiar with plenty of IR theories that are not UV-complete. Nevertheless, these are often the ones which are of relevance to the description of reality.

With the promise that this outlook holds, we now turn to the quantitative details of the AdS/CFT correspondence in the next chapter.

Box 4.4   Basics of general relativity

Here we give a brief overview of general relativity to refamiliarise the reader with the necessary concepts. For more detail we refer the reader to the standard textbooks, e.g. [195, 196, 197].

In special relativity, the near-equal footing with which time and space are considered is made manifest by using four-vector notation,

$$V^\mu = \begin{pmatrix} V^0 \\ V^i \end{pmatrix}. \tag{4.43}$$

The Lorentz-invariant inner product $V^2 \equiv -(V^0)^2 + (V^i)^2$ can be considered as a conventional length of this four-vector in a space-time where distances are measured according to the line element

$$ds^2 = \eta_{\mu\nu}\, dx^\mu\, dx^\nu, \quad \eta_{\mu\nu} = \begin{pmatrix} -1 & 0 & 0 & 0 \\ 0 & 1 & 0 & 0 \\ 0 & 0 & 1 & 0 \\ 0 & 0 & 0 & 1 \end{pmatrix}. \tag{4.44}$$

Einstein's insight was that special relativity together with the equivalence of inertial mass with gravitational mass indicated that the relativistic version of the gravitational response to an object of mass M,

$$\frac{d^2r}{dt^2} = \frac{GM}{r} + \cdots, \tag{4.45}$$

had to be universal for all objects. This universality could be guaranteed if the space-time geometry accounted for the change. The formalism for this was provided by Riemann. For a curved space-time with locally changing line element $ds^2 = g_{\mu\nu}(x)dx^\mu\, dx^\nu$, a freely falling object follows the trajectory labelled by τ which is defined by

$$\frac{d^2x^\mu(\tau)}{d\tau^2} + \Gamma^\mu_{\nu\rho}\frac{dx^\nu(\tau)}{d\tau}\frac{dx^\rho(\tau)}{d\tau} = 0, \tag{4.46}$$

where the Christoffel symbol $\Gamma^{\mu}_{\nu\rho}$ is a function of the derivatives of the metric $g_{\mu\nu}$

$$\Gamma^{\mu}_{\nu\rho} = \frac{1}{2} g^{\mu\sigma} \left(\partial_{\nu} g_{\sigma\rho} + \partial_{\rho} g_{\sigma\nu} - \partial_{\sigma} g_{\nu\rho} \right). \tag{4.47}$$

Here τ is taken as the proper time of the object. Indeed, the geodesic equation of motion follows from the action for a point particle that is proportional to the length of its worldline

$$S = -m \int_{\tau_i}^{\tau_f} ds(\tau) = -m \int_{\tau_i}^{\tau_f} d\tau \sqrt{-g_{\mu\nu} \frac{dx^{\mu}}{d\tau} \frac{dx^{\nu}}{d\tau}}. \tag{4.48}$$

This is the same action as Eq. (4.34).

If space-time were to adjust itself to the presence of matter, then probe particles would all feel the same adjustment to their trajectories encoded in the Christoffel term. Fundamentally this meant a change in the metric of the space-time. The metric description, however, also changes when one changes coordinates, but clearly this should not have a physical effect. It is the Riemann tensor

$$R_{\mu\nu\rho}^{\sigma} = \partial_{\nu} \Gamma^{\sigma}_{\mu\rho} + \Gamma^{\sigma}_{\nu\tau} \Gamma^{\tau}_{\mu\rho} - (\mu \leftrightarrow \nu) \tag{4.49}$$

that captures the change in geometry irrespective of the coordinates. By combining this with energy–momentum conservation, Einstein showed that the way space-time responds to geometry is given by

$$R_{\mu\nu} - \frac{1}{2} g_{\mu\nu} R = 8\pi G T_{\mu\nu}, \tag{4.50}$$

where $R_{\mu\nu} = R_{\mu\rho\nu}^{\rho}$, $R = g^{\mu\nu} R_{\mu\nu}$ are the Ricci tensor and scalar, respectively, and $T_{\mu\nu}$ is the stress-energy tensor of the matter configuration.

This equation follows as the equation of motion from the Einstein–Hilbert action

$$S = \frac{1}{16\pi G} \int d^{d+2}x \sqrt{-g} \left[R - \mathcal{L}_{\text{matter}} \right], \tag{4.51}$$

with $\mathcal{L}_{\text{matter}}$ the matter Lagrangian density and g the determinant of the metric.

From a modern quantum point of view the Einstein–Hilbert action is the low-energy effective action of quantum-gravity/string theory. From this perspective there is an additional constant term that is allowed up to two-derivatives.

$$S = \frac{1}{16\pi G} \int d^{d+2}x \sqrt{-g} \left[R - 2\Lambda - \mathcal{L}_{\text{matter}} \right]. \tag{4.52}$$

Box 4.4   (Continued)

The sign of this cosmological constant Λ is chosen such that positive vacuum energy corresponds to a positive cosmological constant. The unique space-time with only a positive cosmological constant/vacuum energy and no other source of energy–momentum is called de Sitter space. It is the Lorentzian analogue of the sphere. The unique space-time with only a negative cosmological constant/vacuum energy is called anti-de Sitter space. It is the Lorentzian analogue of the hyperboloid; see below.

Energy–momentum tensor for matter

The action (4.52) is the one we will study again and again in this book with different matter fields. The corresponding equation of motion for the metric field is

$$R_{\mu\nu} - \frac{1}{2}g_{\mu\nu}(R - 2\Lambda) = 8\pi G T_{\mu\nu}, \tag{4.53}$$

where

$$T_{\mu\nu} = -\frac{2}{\sqrt{-g}}\frac{\delta S_{\text{matter}}}{\delta g^{\mu\nu}}. \tag{4.54}$$

For a free scalar and vector matter with the actions

$$\mathcal{L}_{\text{real scalar}} = -\frac{1}{2}\nabla_\mu\phi\,\nabla^\mu\phi - V(\phi), \tag{4.55}$$

$$\mathcal{L}_{\text{gauge}} = -\frac{1}{4}F_{\mu\nu}F^{\mu\nu}, \tag{4.56}$$

where ∇_μ is a standard convention for the covariant derivative

$$\nabla_\mu\psi = \partial_\mu\phi, \tag{4.57}$$

$$\nabla_\mu A_\nu = \partial_\mu A_\nu - \Gamma^\alpha_{\mu\nu}A_\alpha. \tag{4.58}$$

The corresponding energy–momentum tensors are

$$T^{\text{real scalar}}_{\mu\nu} = \partial_\mu\phi\,\partial_\nu\phi - \frac{1}{2}g_{\mu\nu}((\partial\phi)^2 + V(\phi)), \tag{4.59}$$

$$T^{\text{gauge}}_{\mu\nu} = F_{\mu\alpha}F_\nu{}^\alpha - \frac{1}{4}g_{\mu\nu}F^2. \tag{4.60}$$

For a charged complex scalar the action is

$$\mathcal{L}_{\text{charged scalar}} = -D_\mu\Phi(D^\mu\Phi)^* - V(\Phi), \tag{4.61}$$

where

$$D_\mu = \nabla_\mu - iqA_\mu, \tag{4.62}$$

and the corresponding energy–momentum tensor is

$$T_{\mu\nu}^{\text{charged scalar}} = D_\mu \Phi (D_\nu \Phi)^* + D_\nu \Phi (D_\mu \Phi)^* + g_{\mu\nu} \mathcal{L}_{\text{charged scalar}}. \tag{4.63}$$

Tangent space, spinor

Fermions in curved space-time require an additional notion. Through the spin-statistics theorem, we know that fermions are associated with half-integer-spin representations of the Lorentz group. A generic space-time is not invariant under Lorentz boosts, however. Locally, however, it should be. This leads to the introduction of a local tangent space, where the local Lorentz transformations are manifest. This is done through the introduction of a $D = (d + 1)$-dimensional *orthonormal basis* $\theta^a = e_\mu^a(x)dx^\mu$ such that $ds^2 = \eta_{ab}\theta^a\theta^b$, with η_{ab} the Minkowski metric. The $D \times D$ invertible matrix e_μ^a is called the tetrad or vielbein, and, together with its inverse matrix e_a^μ, it obeys the following relations:

$$e_a^\mu e_b^\nu g_{\mu\nu} = \eta_{ab}, \quad \eta_{ab} e_\mu^a e_\nu^b = g_{\mu\nu}. \tag{4.64}$$

For the vielbein e_μ^a, the tangent-space index a is lowered or raised by η_{ab}, while the coordinate space index μ is acted on by $g_{\mu\nu}$. Any tensor in a coordinate basis can in this way be mapped to a tensor in the orthonormal basis $V_b^a = e_\mu^a e_b^\nu V_\nu^\mu$.

As we move along the manifold, the local orthonormal basis may change. Infinitesimally this is captured by a spin connection

$$\nabla_\mu V_b^a = \partial_\mu V_b^a + \omega_{\mu\ c}^{\ a} V_b^c - \omega_{\mu\ b}^{\ c} V_c^a. \tag{4.65}$$

This is equivalent to the normal way a tensor transforms under infinitesimal motion,

$$\nabla_\mu V_\sigma^\rho = \partial_\mu V_\sigma^\rho + \Gamma_{\rho\alpha}^\mu V_\sigma^\alpha - \Gamma_{\mu\sigma}^\alpha V_\alpha^\rho, \tag{4.66}$$

if the vielbein is covariantly constant,

$$\nabla_\mu e_a^\nu = \partial_\mu e_a^\nu + \Gamma_{\mu\rho}^\nu e_a^\rho - e_b^\nu \omega_{\mu\ a}^{\ b} = 0. \tag{4.67}$$

This condition fixes the spin connection in terms of the metric and the vielbein

$$\omega_{\mu\ a}^{\ b} = \Gamma_{\mu\rho}^\nu e_a^\rho e_\nu^b - e_a^\nu \partial_\mu e_\nu^b. \tag{4.68}$$

Note that, from $\nabla_\mu \eta_{ab} = 0$, we have $\omega_{\mu ab} = -\omega_{\mu ba}$.

Box 4.4   (Continued)

What Eq. (4.65) shows is that one can write tensors also in terms of representations of the local Lorentz group, where the spin connection $\omega_{\mu ab}$ acts as the gauge field. This shows us how to write down a Dirac equation in a curved space-time. We extend the conventional derivative to a covariant derivative with a gauge field (the spin connection) times the generator of the representation. For spin-1/2 fields the generator of Lorentz transformations is

$$M^{bc} = \frac{1}{4}[\Gamma^b, \Gamma^c] \equiv \frac{1}{2}\Gamma^{bc}. \tag{4.69}$$

The curved-space Dirac equation is therefore

$$D_\mu \Psi = \partial_\mu \Psi + \frac{1}{4}\omega_{\mu bc}\Gamma^{bc}\Psi, \tag{4.70}$$

which follows from the action

$$S = \int dt\, d^d x \sqrt{-g}\left[-i\bar{\Psi}\left(e_a^\mu \Gamma^a \left(\partial_\mu + \frac{1}{4}\omega_{\mu bc}\Gamma^{bc}\right) - m\right)\Psi\right], \tag{4.71}$$

where $\bar{\Psi} = \Psi^\dagger \Gamma^0$.

Induced metric, extrinsic curvature

For reference, we will introduce the notions of extrinsic geometry which will be used when we compute the thermal quantities and D-brane physics in this book. Extrinsic geometry is the more intuitive geometry of the way a surface curves in a larger space-time. A surface can be intrinsically flat (a piece of paper), but still have extrinsic curvature (the paper is bent into an S-shape). More abstractly, to characterise a $(D-1)$ dimensional hypersurface Σ that is embedded in a higher D-dimensional space-time with a metric $g_{\mu\nu}$ ($\mu, \nu = 0, 1, \ldots, D-1$), one uses its normal unit vector n^ν. When $n^\nu n_\nu = 1$ (or -1), Σ is spacelike (or timelike). This vector is constructed from the way in which the surface is embedded in the space-time. Most generally, if we give the surface its own set of coordinates ξ^α, $\alpha = 0, 1, \ldots, D-1$, then for each point ξ^α there is a point $X^\mu(\xi)$ that tells where the surface is located. Infinitesimal motion along the surface is given by the $(D-1)$-vectors $\partial_{\xi^\alpha} X^\mu\, d\xi^\alpha$. The normal vector n^μ is the orthogonal complement to ξ^α, $n_\mu \partial_{\xi^\alpha} X^\mu = 0$.

Two important quantities for Σ are the induced metric and the extrinsic curvature. The induced metric $h_{\mu\nu}$ on Σ by $g_{\mu\nu}$ is defined as

$$h_{\mu\nu} = g_{\mu\nu} - n_\mu n_\nu. \tag{4.72}$$

Its indices are raised and lowered by $g_{\mu\nu}$, and by construction $h_{\mu\nu} n^\nu = 0$. In terms of the coordinates on Σ the induced metric equals $h_{\alpha\beta} = g_{\mu\nu} \partial_{\xi^\alpha} X^\mu \partial_{\xi^\beta} X^\nu$. The extrinsic curvature of Σ is characterised by the variation of the normal unit vector,

$$K_{\mu\nu} = h_\mu^\rho h_\nu^\sigma (\nabla_\rho n_\sigma + \nabla_\beta n_\alpha). \tag{4.73}$$

The trace of the extrinsic curvature is defined by

$$K = h^{\mu\nu} K_{\mu\nu}. \tag{4.74}$$

Anti-de Sitter space

The AdS in AdS/CFT stands for anti-de Sitter space. Here we give a brief introduction to this space-time. As always in physics, the best way to classify space-times is through their symmetries. For space-times, symmetries are the isometries of the space-time manifold, i.e. coordinate transformations that leave the metric invariant. To each such symmetry we can assign a distinct *Killing vector*. Recall that under general coordinate transformations $\delta x^\nu = \xi^\nu(x)$ the metric transforms as

$$\delta g_{\mu\nu} = \nabla_\mu \xi_\nu + \nabla_\nu \xi_\mu. \tag{4.75}$$

A Killing vector is a vector such that $\nabla_\mu \xi_\nu + \nabla_\nu \xi_\mu = 0$.

The simplest space-times are those with the most symmetries. These will have the most Killing vectors. Since a D-dimensional symmetric metric has $D(D+1)/2$ components, there can be at most $D(D+1)/2$ independent Killing equations and thus at most $D(D+1)/2$ Killing vectors. Space-times with this many Killing vectors are called maximally symmetric. These Killing vectors will form a group. The simplest group that contains this number of generators is $SO(D+1)$, but this would not account for the Lorentzian signature. It turns out there are only three distinct maximally symmetric space-times. They are classified by the group formed by their Killing vectors. They are as follows:

Group	Space-time
$SO(1, D)$	D-dimensional de Sitter space
$SO(2, D-1)$	D-dimensional anti-de Sitter space
$ISO(1, D-1) = SO(1, D-1)$	
$\propto \text{Translations}_D$	D-dimensional Minkowski space

The simplicity of maximally symmetric space-times is reflected in their curvature. Since for a maximally symmetric space-time essentially every point

Box 4.4   (Continued)

is similar to every other point, the curvature cannot have a derivative dependence. Using conventions where $[\nabla_\mu, \nabla_\nu]V^\rho = R_{\mu\nu}{}^\rho{}_\sigma V^\sigma$ and $g_{\mu\nu}$ is mostly plus, the curvature of a maximally symmetric space-time takes the form

$$R_{\mu\nu\rho\sigma} = c(g_{\mu\rho}g_{\nu\sigma} - g_{\mu\sigma}g_{\nu\rho}),$$
$$R_{\mu\nu} = c(D-1)g_{\mu\nu},$$
$$R = c(D-1)D, \tag{4.76}$$

with c a constant. All these space-times are in fact solutions to the vacuum dynamical Einstein equations in D dimensions supplemented by a cosmological constant,

$$R_{\mu\nu} - \frac{1}{2}g_{\mu\nu}(R - 2\Lambda) = 8\pi G T_{\mu\nu}. \tag{4.77}$$

In the vacuum $T_{\mu\nu} = 0$, and the contraction with the metric gives

$$R = \frac{2D}{D-2}\Lambda \tag{4.78}$$

and thus $c = 2\Lambda/((D-2)(D-1))$.

It turns out that $\Lambda > 0$ corresponds to de Sitter space, $\Lambda = 0$ to Minkowski space and $\Lambda < 0$ to anti-de Sitter space. In other words, anti-de Sitter space is the maximally symmetric space which is the unique solution to the vacuum Einstein equations with a *negative* cosmological constant.

This may not say much, but maximally symmetric Euclidean spaces are in fact very familiar. The maximally symmetric Euclidean spaces are the sphere ($\Lambda > 0$), flat space ($\Lambda = 0$) and the hyperboloid ($\Lambda < 0$). De Sitter space is thus the Lorentzian generalisation of the sphere, and anti-de Sitter space is the Lorentzian generalisation of the hyperboloid. Just as the D-dimensional sphere S^D is defined as the solution in Euclidean flat space-time to the constraint

$$X_1^2 + X_2^2 + \cdots + X_{D+1}^2 = L^2, \tag{4.79}$$

the D-dimensional hyperboloid is defined as the solution to the constraint

$$-X_{-1}^2 + X_1^2 + \cdots + X_D^2 = -L^2 \tag{4.80}$$

in Euclidean flat space-time $\mathbb{R}^{d+3}$. In this way D-dimensional anti-de Sitter space (AdS$_{d+2}$) can be defined as the surface

$$-X_{-1}^2 - X_0^2 + X_1^2 + \cdots + X_{d+1}^2 = -L^2 \tag{4.81}$$

in a flat space with two time directions $\mathbb{R}^{2,D-1}$. We immediately see the $SO(2, D-1)$ symmetry. Note that, despite the two timelike dimensions in $\mathbb{R}^{2,D-1}$, the AdS surface only has one time direction. A solution to this defining relation is

$$X_1 = L_+ \cos \theta_1,$$

$$X_2 = L_+ \sin \theta_1 \cos \theta_2,$$

$$\vdots$$

$$X_{D-2} = L_+ \sin \theta_1 \sin \theta_2 \ldots \cos \theta_d,$$

$$X_{D-1} = L_+ \sin \theta_1 \sin \theta_2 \ldots \sin \theta_d,$$

$$X_0 = L_- \cos \tau,$$

$$X_{-1} = L_- \sin \tau, \tag{4.82}$$

where $L_+ = L \sinh \rho$, $L_- = L \cosh \rho$, $0 \le \tau < 2\pi$, $0 \le \theta_i < \pi$ $(i = 1, \ldots, d-1)$ and $0 \le \theta_d < 2\pi$. The induced metric one gets is the AdS_{d+2} metric

$$\begin{aligned} ds^2 &= -dX_{-1}^2 - dX_0^2 + dX_1^2 + \cdots dX_{d+1}^2 \\ &= L^2(d\rho^2 - \cosh^2 \rho \, d\tau^2 + \sinh^2 \rho \, d\Omega_d^2), \end{aligned} \tag{4.83}$$

where $d\Omega_d^2$ is the metric on the d-dimensional sphere S^d. Anti-de Sitter space is the universal cover of this, where we unroll the periodic timelike coordinate τ to the range $-\infty < \tau < \infty$. This coordinate system is quite special because it is a *global* coordinate system. It covers the whole of AdS (recall that generically one needs multiple coordinate patches to cover the whole space). The Euclidean AdS space-time is a hyperbolic space as mentioned before.

Now consider the coordinate transformation $\rho = \text{arcsinh} \tan \theta$ with $0 \le \theta \le \pi/2$. (For AdS_2, $-\pi/2 \le \theta \le \pi/2$.) This maps the hyperbolic direction ρ to a finite range. In these coordinates the metric is

$$ds^2 = \frac{L^2}{\cos^2 \theta} \left(d\theta^2 - d\tau^2 + \sin^2 \theta \, d\Omega_d^2 \right). \tag{4.84}$$

This metric allows us to understand the topology of AdS. For this we can ignore the overall conformal factor $L^2/\cos^2 \theta$. The topologically equivalent space-time

$$ds_{\text{teq}}^2 \sim \left(d\theta^2 - d\tau^2 + \sin^2 \theta \, d\Omega_d^2 \right) \tag{4.85}$$

describes a cylinder with radial direction θ and longitudinal direction τ, and each point in (θ, τ) is an S^d. See Fig. 4.11. We thus see that AdS has a (conformal) boundary at $\theta = \pi/2$. Note that in physical units this is an

Box 4.4 (Continued)

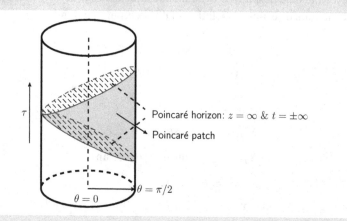

Figure 4.11 The topology of anti-de Sitter space is as a solid cylinder with boundary $\mathbb{R} \times S^d$. The Poincaré patch (grey regime) covers only a portion of the global AdS.

infinite distance away. This conformal boundary will play an important role in AdS/CFT.

There is a third convenient coordinate system for AdS_{d+2}. This solves the defining relation (4.81) with the relations

$$X_1 = \frac{L}{z} x_1,$$

$$X_2 = \frac{L}{z} x_2,$$

$$\vdots$$

$$X_{D-2} = \frac{L}{z} x_d,$$

$$X_{D-1} = \frac{1}{2z} \left(L^2 - z^2 + t^2 - \sum_{i=1}^{d} x_i^2 \right),$$

$$X_0 = \frac{L}{z} t,$$

$$X_{-1} = \frac{1}{2z} \left(L^2 + z^2 - t^2 + \sum_{i=1}^{d} x_i^2 \right). \tag{4.86}$$

The metric of AdS$_{d+2}$ becomes

$$ds^2 = \frac{L^2}{z^2}\left(dz^2 - dt^2 + dx_1^2 + \cdots + dx_d^2\right). \qquad (4.87)$$

This coordinate system, the so-called *Poincaré patch*, covers only a portion of AdS. The boundary of the Poincaré patch is $\mathbb{R}^{1,d}$, thus the Poincaré coordinates are convenient when the dual CFT is defined in Minkowski space $\mathbb{R}^{1,d}$. Relating it back to the global coordinate system through the defining relation (4.81), one can show that it covers the grey area depicted in Fig. 4.11. For a review see [198, 199].

Anti-de Sitter isometries and conformal symmetry

We will briefly discuss a special aspect of the boundary of anti-de Sitter space. This (conformal) boundary inherits from anti-de Sitter space not just translation and rotation invariance, but in fact a full invariance under *conformal transformations*. These conformal symmetries are precisely the manifestion of the full $SO(2, d+1)$ isometries of AdS$_{d+2}$ on the boundary.

Let us briefly recall some aspects of $SO(2, d+1)$ as the conformal group. The generating algebra of $SO(2, d+1)$ is

$$[M_{ij}, M_{kl}] = \eta_{jk}M_{il} - \eta_{ik}M_{jl} - \eta_{jl}M_{ik} + \eta_{il}M_{jk}, \qquad (4.88)$$

where $M_{ij} = -M_{ji}$ and η_{jk} is the $SO(2, d+1)$ metric. This directly follows from the fact that rotations in a flat space-time with two time directions, $\mathbb{R}^{2,d+1}$, are generated by $i(\hat{x}_i\hat{p}_j - \hat{p}_i\hat{x}_j)$. To interpret this algebra as the conformal group in d dimensions we isolate the Lorentz group $SO(2, d+1) = SO(1,1) \times SO(1,d)$. Using lightcone coordinates for the additional $(1,1)$ directions and identifying $M_{+-} = D$, $M_{i-} = P_i$ and $M_{i+} = K_i$, one has the generating algebra of the conformal group. Here D is the generator of *dilatations*, P_i is the generator of translations (one gets this at no cost) and K_i is the generator of special conformal transformations. The first two are familiar, but the nature of the last one is less clear. A convenient way to think about K_i is as a generator that "flips scales". The conformal group can also be thought of as the combination of the Poincaré group with *inversions*, where *inversion* is the operation

$$I : x^\mu \to \frac{x^\mu}{x^2}. \qquad (4.89)$$

The subtlety about inversion is that as a group element it is *not* connected to the identity. It does not have determinant unity. Therefore it does not appear in the algebra. Its manifestion in the algebra is through the generator K_i, which one can show to be equal to $K_i = I P_i I$. Since the inversion operator appears twice, this element does have determinant unity and its infinitesimal version is an element of the algebra.

Box 4.4   (Continued)

We can now argue why the boundary inherits the full conformal group from the isometries of AdS. In the Poincaré coordinates translations and boosts are directly present. Also the scaling symmetry is easy to see. The Poincaré metric

$$ds^2 = \frac{L^2}{z^2}(\eta_{\mu\nu}\, dx^\mu\, dx^\nu + dz^2) \tag{4.90}$$

is clearly invariant under $x^\mu \to \lambda x^\mu$, $z \to \lambda z$. The non-trivial transformation is the generalisation of this CFT inversion symmetry. It is not so hard to see that the Poincaré coordinates are also invariant under

$$x^\mu \to \frac{x^\mu}{z^2 + \eta_{\mu\nu}x^\mu x^\nu}, \quad z \to \frac{z}{z^2 + \eta_{\mu\nu}x^\mu x^\nu}. \tag{4.91}$$

On the boundary $z = 0$ this reduces to the inversion symmetry and thus the boundary is invariant under the full conformal group.

5

The AdS/CFT correspondence as computational device: the dictionary

Having both understood the condensed matter context where AdS/CFT might prove a unique perspective and the field-theory context in which it should be considered, we are now in place to get to the heart of the matter. This is the "Gubser–Klebanov–Polyakov–Witten" (GKPW) rule that provides the precise dictionary between properties of the field theory and the dual gravity theory. Building on Maldacena's discovery, Witten and independently Gubser, Klebanov and Polyakov deduced this general rule relating the bulk and the boundary in 1998 [2, 3]. Just like everything else in holography, it still has a conjectural status. The previous summaries are part of the intuition behind this educated guess, but, as we discussed in the previous chapter, the true support follows from string theory [7]. We will just state the "GKPW" rule as coming from a divine authority. Its essence is the equality of the partition functions on the two sides of the correspondence

$$Z_{\text{CFT}}(N) = \int \mathcal{D}\phi \, e^{iN^2 S_{\text{AdS}}(\phi)}. \tag{5.1}$$

On the left-hand side one has the full quantum partition sum or "vacuum amplitude" of the field theory, Z_{CFT}. On the right-hand side one has the path integral over all fields in the gravity theory. The most crucial aspect is that in the large-N limit the right-hand side reduces to the saddle point. In particular, it means that the free energy of the CFT at large-N coincides with the least (Euclidean) action of the gravitational theory, which is equal to $S_{\text{AdS}}^{\text{on-shell}}$, evaluated for a solution to the equations of motion.

In this book we will address only the equilibrium physics of the CFT, and an implication of this rule is that this corresponds to *stationary* solutions to the bulk equations of motion. Part of the equilibrium characteristics is constituted by the dynamical susceptibilities in linear response that provide us with information on the excitations living in the undisturbed system. The second great power of AdS/CFT will be on display here. Unlike in conventional approaches, where one must Euclideanise the theory and confront the ill-controlled Wick rotation of discrete Matsubara frequencies back to real time, in AdS/CFT one can compute in real time directly. A finite temperature can be encoded in a black-hole geometry, and this automatically controls the temperature dependence, in that its real-time geometry effectively contains the periodically identified imaginary-time direction. We will elaborate on this in chapter 6.

When one knows how to compute a partition sum, the computation of the linear response functions is straightforward using generating functionals. One perturbs the system with an external field ("source") $J_I(x)$ coupling to a local operator $\mathcal{O}_I(x)$ of the field theory (the "response"), by adding a term to the Lagrangian of the field theory,

$$\mathcal{L}(x) \to \mathcal{L}(x) - i \sum_I J_I(x)\mathcal{O}_I(x). \tag{5.2}$$

In full generality, any n-point function can then be computed by varying the log of the partition sum including the source term and taking the limit of vanishing source strength:

$$\langle \mathcal{O}_1(x_1)\mathcal{O}_2(x_2)\dots\mathcal{O}_n(x_n)\rangle = \prod_{i=1}^{n} \frac{\delta}{\delta J_i(x_i)} \ln Z_{\text{QFT}}\bigg|_{J=0}. \tag{5.3}$$

The insight of GKPW is how to incorporate this recipe on the right-hand side of Eq. (5.1). One cannot simply introduce a corresponding local source everywhere in the bulk. This will not work. Following the insights from string theory that open strings are sources for closed strings and vice versa, Gubser, Klebanov, Polyakov and Witten [2, 3] discovered the right procedure. The *source* in the CFT should be encoded in a *field* in the gravity theory. Then, following the second dictionary rule, if the CFT can be thought to live on the boundary of the AdS space-time, then on the gravitational side the source is restricted to this boundary. The sources act therefore as boundary conditions for the classical fields propagating in the bulk gravity theory. In mathematical form, taking the N^2 dependence implicit in the definition of the action:

Rule 5   The GKPW rule, the mathematical backbone of AdS/CFT, identifies the partition function of a QFT in the presence of a source J with the partition function of a bulk gravitational theory where the asymptotically anti-de Sitter boundary value of the fields ϕ are equated with the sources J:

$$\langle e^{\int d^{d+1}x J(x)\mathcal{O}(x)}\rangle_{\text{QFT}} = \int \mathcal{D}\phi \, e^{iS_{\text{bulk}}(\phi(x,r))}|_{\phi(x,r=\infty)=J(x)}. \tag{5.4}$$

On thinking this through, one realises that this does justice to the notion that the renormalisation group of the CFT is "geometrised" by the AdS dynamics. The operators $\mathcal{O}_I(x)$ are the bare operators of the theory which have a meaning in the deep UV, and one wants to know their fate in the IR by tracking the energy dependence of the associated propagator. Think photoemission: this operator would be the literal electron injected into the interacting electron system from the outside. In the solid the electron remains the bare electron at very short times, since there has been no time yet for it to realise that there are other electrons to interact with. It is in the deep UV. But as time evolves the electron becomes increasingly dressed, and after a very long time it will have "fallen apart" in the true, highly collective low-energy excitations associated with the strongly interacting vacuum. When these are the quasi-electrons of the strongly interacting Fermi liquid, these will show up as tiny but very sharp peaks in the bare electron spectral function at very low energy: the "quasiparticle peaks" seen routinely in strongly interacting Fermi liquids by observing their photoemission. In chapter 9 we will find out that such quasiparticle poles can quite literally show up in holographic computations.

Paying tribute to the idea that the radial direction in AdS corresponds to the scaling direction in the CFT, it had to be the case that the information of the bare UV operators should be "transferred" to the bulk right *at* the boundary. This is the deep insight underlying the GKPW rule. Starting in section 5.1, we will gradually expose its powers. As a first step, we will show that it validates in quantitative detail the intuitive construction of the "RG = GR" rule for the CFT correlators we presented at the beginning (section 1.3), emphasising that we can do so in real time. We can not only set up the full formalism for retarded, advanced and mixed Green functions, but also compute quantum expectation values directly. Here we will focus on zero temperature, and complete this discussion at the end of the next chapter for the finite-temperature cases (section 6.4). Section 5.1 really serves as the foundation for the remainder of the book. It will also present the algorithm by means of which to compute holographically higher-order correlation functions and explain how to include quantum corrections in the bulk in principle. In section 5.2 we will focus on how the renormalisation group in the boundary is encoded in the bulk, elucidating the relations between the mass of the bulk field and the scaling dimensions of the boundary operators in the case of the CFTs. We will pay special attention to the notion of an aspect of holography called alternative quantisation and how this relates to the mean-field nature of the double-trace deformations. In the final section, section 5.3, we will complete the dictionary by focussing on the role of symmetry, discussing why fields governed by global symmetries in the boundary are dual to fields in the bulk governed by local symmetries. This adds enough information that we can present the "dictionary" in the form of Table 5.2 at the very end of this technical introduction to the correspondence.

5.1 The GKPW rule in action: computing correlation functions

Let us focus on how the GKPW rule precisely validates the ad-hoc construction introduced in section 1.3, Eq. (1.20), stating that the two-point correlation function can be read off from the near-boundary asymptotics of AdS waves. Using the "saddle-point reduction" in the large-N limit, the GKPW rule instructs us to consider the *on-shell* bulk action with the boundary value of the field equal to the source in the dual-field theory. For the simple real-scalar theory the bulk action can be written in terms of a "bulk" contribution and a "boundary" contribution as

$$
S = \int_{AdS} dr \, d^{d+1}x \sqrt{-g} \left[-\frac{1}{2} g^{\mu\nu} \, \partial_\mu \phi \, \partial_\nu \phi - \frac{1}{2} m^2 \phi^2 \right]
$$
$$
= \int_{AdS} dr \, d^{d+1}x \sqrt{-g} \, \frac{1}{2} \phi (\Box - m^2) \phi - \frac{1}{2} \oint_{\partial AdS} d^{d+1}x \sqrt{-h} \, \phi \, \partial_n \phi. \tag{5.5}
$$

Here h is the determinant of the "induced metric" on the boundary of AdS. For the commonly used AdS metrics of the type

$$
ds^2 = \frac{L^2}{r^2} \, dr^2 + h_{\mu\nu}(r, x) dx^\mu \, dx^\nu, \tag{5.6}
$$

it is just the tensor $h_{\mu\nu}(r, x)$ evaluated at the boundary at $r = \infty$. The formal definition can be found in the review in box 4.4 (at the end of chapter 4) of the basics of general relativity and the Riemannian mathematics of curved space-times. The normal derivative $\partial_n \equiv n^\mu \partial_\mu$ in the boundary term is outward directed from the boundary along the unit normal vector n^μ. For the metric Eq. (5.6) $n^\mu = (r/L, 0, \ldots, 0)$, and, taking for simplicity the pure AdS metric, $h_{\mu\nu} = (r^2/L^2)\eta_{\mu\nu}$.

This boundary term does not contribute to the equations of motion, but it is crucial in the correspondence. On solving the equations of motion and substituting a solution for $\phi(x)$ into the action, the "bulk" term vanishes. We learned in section 1.3 that this solution, after Fourier transformation along the boundary directions, has a universal asymptotic behaviour near the boundary $r = \infty$ as

$$
\phi_{sol}(\omega, \mathbf{k}, r) = A(\omega, \mathbf{k}) \left(\frac{r}{L} \right)^{-(d+1-\Delta)} + B(\omega, \mathbf{k}) \left(\frac{r}{L} \right)^{-\Delta} + \cdots , \tag{5.7}
$$

where $\Delta = (d + 1)/2 + \sqrt{(d + 1)^2/4 + m^2 L^2}$. According to the GKPW rule, the boundary value of $\phi(\omega, \mathbf{k}, r = \infty)$ should be taken *fixed* and equal to the source $J(\omega, \mathbf{k})$. We now face a problem since the power of $d + 1 - \Delta$ will be generically negative and thus the boundary value of ϕ is not well defined. However, we know what the meaning of this divergence is in the boundary field theory. Approaching the boundary is like increasing the renormalisation scale to infinity, and here one typically encounters UV divergences. In other words, the theory has to be regulated, and this can be done in a particularly elegant way using the geometrical bulk language.

GKPW proposed that one should compute at an infinitesimal distance $r = \epsilon^{-1}$ away from the formal boundary, and then modify the theory such that one can take an appropriate limit $\epsilon \to 0$. Using that $h_{\mu\nu} = (r^2/L^2)\,\eta_{\mu\nu}$ and again Fourier transforming along the boundary, the "regulated" on-shell action equals

$$
S_{\text{on-shell}}(\epsilon) = \frac{1}{2L} \oint_{r=\epsilon^{-1}} \frac{d\omega\, d^d \mathbf{k}}{(2\pi)^{d+1}} \left(\frac{r}{L}\right)^d \left((d + 1 - \Delta) A^2 \left(\frac{r}{L}\right)^{-(2d+1-2\Delta)} \right.
$$
$$
\left. + (d + 1) A B \left(\frac{r}{L}\right)^{-d} + \cdots \right). \quad (5.8)
$$

The first term is the problematic formally divergent term. The key to making the action well defined is that adding an arbitrary boundary term to the action never changes the equation of motion. However, such an extra boundary term can be used to remove this divergence. Adding a boundary counter-term of the form

$$
S_{\text{counter}}(\epsilon) = -\frac{1}{2L}(d + 1 - \Delta) \oint_{r=\epsilon^{-1}} \frac{d\omega\, d^d \mathbf{k}}{(2\pi)^{d+1}} \sqrt{-h}\phi^2
$$
$$
= -\frac{1}{2L}(d + 1 - \Delta) \oint_{r=\epsilon^{-1}} \frac{d\omega\, d^d \mathbf{k}}{(2\pi)^{d+1}} \left(\frac{r}{L}\right)^{d+1}
$$
$$
\times \left(A^2 \left(\frac{r}{L}\right)^{-(2d+2-2\Delta)} + 2 A B \left(\frac{r}{L}\right)^{-(d+1)} + \cdots \right) \quad (5.9)
$$

yields, in combination with Eq. (5.8),

$$
S_{\text{on-shell}}(\epsilon) + S_{\text{counter}}(\epsilon) = \frac{1}{2L} \oint_{r=\epsilon^{-1}} \frac{d\omega\, d^d \mathbf{k}}{(2\pi)^{d+1}} \left(\frac{r}{L}\right)^{d+1}
$$
$$
\times \left((2\Delta - d - 1) A B \left(\frac{r}{L}\right)^{-(d+1)} + \cdots \right). \quad (5.10)
$$

This is now all finite. The leading behaviour (the coefficient $A(\omega, \mathbf{k})$) of the solution $\phi_{\text{sol}}(\omega, \mathbf{k}, z)$ can then be equated to the local source $J(\omega, \mathbf{k})$. Given that the above should coincide with the combination $S_{\text{source}} = -i \int d^{d+1}x\, J O$ in the field theory, taking the single derivative with respect to J yields the expectation value $\langle O \rangle_J$ of the field-theory operator (sourced by J) in the presence of the source. It is given by

$$
\langle O(\omega, \mathbf{k}) \rangle_J = \frac{2\Delta - d - 1}{2L} B(\omega, \mathbf{k}). \quad (5.11)
$$

Thus we see that equating the leading near-boundary behaviour $A(\omega, \mathbf{k})$ of the solution $\phi_{\text{sol}}(\omega, \mathbf{k}, r)$ to the source $J(\omega, \mathbf{k})$ implies that the *sub-leading* near-boundary behaviour $B(\omega, \mathbf{k})$ is the corresponding *response*.

Linear-response theory then tells us that the response $B(\omega, \mathbf{k})$ ought to be linearly proportional to the source $A(\omega, \mathbf{k})$. Dividing out this proportionality, together with a combinatorial factor 2, precisely gives the CFT correlation function

$$\langle \mathcal{O}(-\omega, -\mathbf{k})\mathcal{O}(\omega, \mathbf{k})\rangle = \frac{2\Delta - d - 1}{L}\frac{B(\omega, \mathbf{k})}{A(\omega, \mathbf{k})} \tag{5.12}$$

and we have demonstrated that the propagator rule Eq. (5.12) is indeed a consequence of the fundamental GKPW rule. In box 5.1 we will show this answer directly from the GKPW rule by taking an additional derivative w.r.t. J and setting it to vanish.

For a generic value of Δ the UV divergence reflects the fact that the solution with $A \neq 0$, $B = 0$ is not a normalisable solution on an AdS space, whereas the solution with $A = 0$, $B \neq 0$ is. The leading and sub-leading solutions are therefore also often called the *non-normalisable* and *normalisable* solutions, respectively. Note, however, that there are values of Δ for which *both* solutions are normalisable, and therefore leading and sub-leading is a more accurate labelling. We will discuss the physics associated with the regime where both solutions are normalisable in section 5.2.

Box 5.1   Two-point correlation functions in AdS/CFT

To compute the linear response two-point correlation function directly from the GKPW rule by differentiating w.r.t. the source J, one has to realise that one is in fact solving a simple Dirichlet boundary-value problem. Recall that the equation of motion for ϕ is a second-order differential equation with two linearly independent solutions. Let us denote the solution with $A = 0$ as ϕ_B with boundary behaviour $\phi_B(r) = Br^{-\Delta}(1 + \sum_n c_n r^n)$. This is the appropriate Dirichlet solution which vanishes at the boundary. For the other independent solution, we will choose a solution that is regular in the interior; we denote this solution as $\phi_{\text{int}}(r)$. For computational simplicity, we set $L = 1$ in this box. Note that this can always be realised by rescaling the coordinate to dimensionless quantities, i.e. $(t, x_i, r) \rightarrow L(t, x_i, r)$, thus L plays a role only as an overall coefficient in front of the action.

On writing the equation of motion for ϕ abstractly as

$$\frac{\partial^2}{\partial r^2}\phi(r) + P(r)\frac{\partial}{\partial r}\phi(r) + Q(r)\phi(r) = 0, \tag{5.13}$$

the Dirichlet AdS Green function obeying

$$(\nabla^\mu \nabla_\mu - m^2)\mathcal{G}^{\text{AdS}}(r, r') = e^{-\int^r P}\delta(r - r'), \quad \lim_{r \to \infty}\mathcal{G}^{\text{AdS}}(r, r') = 0 \tag{5.14}$$

with $\int^r P \equiv \int^r d\mathsf{r}\, P(\mathsf{r})$, where the sans-serif r is an integration variable, then equals

$$\mathcal{G}^{\text{AdS}}(r, r') = \frac{\phi_B(r)\phi_{\text{int}}(r')\theta(r - r') + \phi_{\text{int}}(r)\phi_B(r')\theta(r' - r)}{W(\phi_{\text{int}}, \phi_B)} \tag{5.15}$$

with

$$W(\phi_{\text{int}}, \phi_B) \equiv e^{\int^r P}(\phi_{\text{int}}\, \partial_r \phi_B - \phi_B\, \partial_r \phi_{\text{int}}). \tag{5.16}$$

The Wronskian $W(\phi_{\text{int}}, \phi_B)$ in the denominator ensures that we have the correct normalisation and that it is independent of r, i.e. it may be evaluated for any preferred r. Then for a boundary source $J(\omega, \mathbf{k})$ the solution to the equation of motion is

$$\phi_{\text{sol}}(\omega_1, \mathbf{k}_1, r) = \lim_{\epsilon \to 0} \oint_{r'=\epsilon^{-1}} \frac{d\omega\, d^d\mathbf{k}}{(2\pi)^{d+1}} e^{\int^{r'} P}\, \partial_{r'}\mathcal{G}(r, \omega_1, \mathbf{k}_1; r', \omega, \mathbf{k})J(\omega, \mathbf{k})$$

$$= \lim_{\epsilon \to 0} \int \frac{d\omega\, d^d\mathbf{k} \cdot e^{\int^{r'} P}\, \partial_{r'}\phi_B(r')\phi_{\text{int}}(r)|_{r'=\epsilon^{-1}}}{(2\pi)^{d+1}e^{\int^r P}(\phi_{\text{int}}\, \partial_r \phi_B - \phi_B\, \partial_r \phi_{\text{int}})}J(\omega, \mathbf{k}). \tag{5.17}$$

By construction this obeys $\lim_{r \to \infty} \phi_{\text{sol}}(\omega, \mathbf{k}, r) = J(\omega, \mathbf{k})$, which can be seen by noting that the Wronskian reduces to $e^{\int^r P}\phi_{\text{int}}\, \partial_r \phi_B$ for $r \to \infty$. The normal derivative of the solution follows straightforwardly

$$\partial_r\phi_{\text{sol}}(\omega_1, \mathbf{k}_1, r)$$

$$= \lim_{\epsilon \to 0} \oint_{r'=\epsilon^{-1}} \frac{d\omega\, d^d\mathbf{k}}{(2\pi)^{d+1}} e^{\int^{r'} P}\, \partial_r \partial_{r'}\mathcal{G}(r, \omega_1, \mathbf{k}_1; r', \omega, \mathbf{k})J(\omega, \mathbf{k})$$

$$= \lim_{\epsilon \to 0} \int \frac{d\omega\, d^d\mathbf{k}}{(2\pi)^{d+1}} \frac{\partial_r\phi_{\text{int}}(r)}{\phi_{\text{int}}(\epsilon^{-1})}J(\omega, \mathbf{k}). \tag{5.18}$$

On substituting this into the action one finds

$$S_{\text{on-shell}} + S_{\text{counter}}$$

$$= \lim_{r \to \infty} \left(-\frac{1}{2}\int d^{d+1}x\, r^{d+2}\phi_{\text{sol}}\, \partial_r\phi_{\text{sol}} - \frac{1}{2}(d + 1 - \Delta)\int d^{d+1}x\, r^{d+1}\phi_{\text{sol}}^2 \right)$$

$$= \lim_{\epsilon \to 0} \int \frac{d\omega\, d^d\mathbf{k}}{(2\pi)^{d+1}} \left[\frac{1}{2}J(-\omega, -\mathbf{k})\left(-\epsilon^{-d-2}\frac{\partial_r\phi_{\text{int}}(r)}{\phi_{\text{int}}(r)}\bigg|_{r=\epsilon^{-1}} \right.\right.$$

$$\left.\left. - (d + 1 - \Delta)\epsilon^{-d-1} \right)J(\omega, \mathbf{k}) \right]. \tag{5.19}$$

On recalling that the generic behaviour of the solution ϕ_{int} near the boundary equals $\phi_{\text{int}} = A_{\text{int}}r^{-(d+1-\Delta)} + B_{\text{int}}r^{-\Delta}$ and by taking two derivatives w.r.t. the source J, one finds

---

Box 5.1   (Continued)

$$\langle \mathcal{O}(-\omega, -\mathbf{k}) \mathcal{O}(\omega, \mathbf{k}) \rangle = \lim_{\epsilon \to 0} \left[-\epsilon^{-d-2} \frac{\partial_\epsilon^{-1} \phi_{\text{int}}(\epsilon^{-1})}{\phi_{\text{int}}(\epsilon^{-1})} - (d+1-\Delta)\epsilon^{-d-1} \right]$$

$$= \lim_{\epsilon \to 0} \epsilon^{1-\Delta} \frac{\partial_\epsilon (\epsilon^{-(d+1-\Delta)} \phi_{\text{int}}(\epsilon^{-1}))}{\phi_{\text{int}}(\epsilon^{-1})}$$

$$= \lim_{\epsilon \to 0} \epsilon^{2\Delta-2d-2} (2\Delta - d - 1) \frac{B(\omega, \mathbf{k})}{A(\omega, \mathbf{k})}. \tag{5.20}$$

The last step is that one simply drops the overall dependence on the regulator, and one is left with the same linear response answer as we found in Eq. (5.12). This step is justified because there is an ambiguity in identifying $J(\omega, \mathbf{k})$ with the boundary source. Owing to the conformal symmetry, one can only identify it up to a scale transformation [3]. Since J has scaling dimension $d + 1 - \Delta$, this precisely accounts for the factor $\epsilon^{2(\Delta-d-1)}$. Put more precisely, one has to renormalise the boundary source $J(\omega, \mathbf{k})$ to match it to the field source $J_{bdy}(\omega, \mathbf{k}; \epsilon) = \epsilon^{d+1-\Delta} J_{\text{CFT}}(\omega, \mathbf{k})$. Using this, the overall regulator dependence will be immediately accounted for.

5.1.1 Example: real-time correlation functions in a CFT

We have shown that our engineering rule is correct. Let us now show in detail how this exact application of the GKPW rule does indeed reproduce the correct two-point correlation functions. As an example we take a scalar operator $\mathcal{O}$ in a generic CFT with scaling dimension Δ for which we know the answer by scaling. The dual of this scalar operator is a scalar field with the action Eq. (5.5) and the background metric is that of pure AdS,

$$ds^2 \equiv g_{\mu\nu} dx^\mu dx^\nu = \frac{r^2}{L^2} \left(-dt^2 + \sum_{i=1}^{d} dx_i^2 \right) + \frac{L^2}{r^2} dr^2, \tag{5.21}$$

since we are studying a CFT. The equation of motion (EOM) for ϕ is therefore

$$(\nabla^\mu \nabla_\mu - m^2)\phi = 0, \tag{5.22}$$

with ∇_μ the covariant derivative $\nabla_\mu V^\nu = \partial_\mu V^\nu + \Gamma^\nu_{\mu\sigma} V^\sigma$ and $\Gamma^\nu_{\mu\sigma}$ equal to the Christoffel connection for the metric Eq. (5.21) $\Gamma^\nu_{\mu\sigma} = \frac{1}{2} g^{\nu\rho} (\partial_\mu g_{\rho\sigma} + \partial_\sigma g_{\rho\mu} - \partial_\rho g_{\mu\sigma})$. After the Fourier transformation to frequency–momentum space with regard to directions shared with the boundary, the equation of motion we obtain is

$$\left[\frac{\partial^2}{\partial r^2} + \frac{d+2}{r} \frac{\partial}{\partial r} - \left(\frac{k^2 L^4}{r^4} + \frac{m^2 L^2}{r^2} \right) \right] \phi(\omega, \mathbf{k}; r) = 0, \tag{5.23}$$

where $k^2 = -\omega^2 + \mathbf{k}^2$. Given that we started from a second-order differential equation, we need two boundary conditions. One is the normalisation (which will drop out of the final answer), but the second is the boundary condition in the interior. The exact choice here is a crucial step to obtain the exact answers for correlation functions. The available choices will precisely determine the *type* of Green function we compute. In the Euclidean case there is no ambiguity regarding the interior boundary condition. In this case $k^2 = \omega_E^2 + \mathbf{k}^2 > 0$, and the solutions to this second-order ODE are

$$\phi(r) = a_K \left(\frac{r}{L^2} \right)^{-(d+1)/2} K_\nu \left(\frac{kL^2}{r} \right) + a_I \left(\frac{r}{L^2} \right)^{-(d+1)/2} I_\nu \left(\frac{kL^2}{r} \right), \tag{5.24}$$

$$\nu = \Delta - \frac{d+1}{2} = \sqrt{\frac{(d+1)^2}{4} + m^2 L^2}, \tag{5.25}$$

where $K_\nu(x)$ and $I_\nu(x)$ are modified Bessel functions of the second kind and $a_{K,I}$ are constants independent of k. Given the behaviour of the Bessel functions in the interior,

$$K_\nu \left(\frac{kL^2}{r} \right) \sim e^{-kL^2/r}, \quad I_\nu \left(\frac{kL^2}{r} \right) \sim e^{kL^2/r}, \tag{5.26}$$

a regularity condition in the interior ($r \to 0$), i.e. $a_I = 0$, uniquely fixes the solution.

In Lorentzian (real-time) signature causality matters: whether the four-momentum is timelike or spacelike distinguishes different Green functions. For spacelike k, i.e. $k^2 = -\omega^2 + \mathbf{k}^2 > 0$, the independent solutions are the same as Eq. (5.24). Thus one must demand that ϕ is regular in the interior of AdS. For timelike k, i.e. $k^2 = -\omega^2 + \mathbf{k}^2 < 0$, the independent solutions are

$$\phi(r) = a_+ \left(\frac{r}{L^2} \right)^{-(d+1)/2} H_\nu^{(1)} \left(\sqrt{\omega^2 - \mathbf{k}^2} L^2/r \right)$$

$$+ a_- \left(\frac{r}{L^2} \right)^{-(d+1)/2} H_\nu^{(2)} \left(\sqrt{\omega^2 - \mathbf{k}^2} L^2/r \right)$$

$$\sim a_+ e^{-i\sqrt{\omega^2 - \mathbf{k}^2} L^2/r} + a_- e^{i\sqrt{\omega^2 - \mathbf{k}^2} L^2/r} \text{ at } r \to 0, \tag{5.27}$$

with $H_\nu^{(1)}$, $H_\nu^{(2)}$ equal to the Hankel functions. Unlike in the previous case, we cannot impose a regularity condition. Instead we are dealing with truly fluctuating fields. We have two choices. The first one is to impose in-falling $e^{i\sqrt{\omega^2 - \mathbf{k}^2} L^2/r}$ boundary conditions with $a_+ = 0$. Note that, combined with the standard $e^{-i\omega t}$, this describes a wavefront moving into the interior at small r for positive ω. The other choice is to impose out-going $\sim e^{-i\sqrt{\omega^2 - \mathbf{k}^2} L^2/r}$ boundary conditions with $a_- = 0$

for positive ω. These choices correspond to the calculation of real-time *retarded* and *advanced* Green functions for positive ω, respectively. It is straightforward to generalise this to the negative-ω case. This intuitively sensible connection – the causal response to a boundary source moves inward – can be proven through the singularity structure of the resulting Green functions both at zero and at finite temperature [201, 202]. In the next chapter, chapter 6, we will complete the dictionary and elucidate how to generalise the holographic recipe to compute the real-time propagators in the finite-temperature case.

We can now straightforwardly write down the results for the two-point function using Eq. (5.12). Upon expanding the modified Bessel function (for the spacelike case) and Hankel function (for the timelike case) near the boundary ($r \to \infty$),

$$K_\nu\left(\frac{kL^2}{r}\right) = \left(\frac{kL^2}{2r}\right)^{-\nu}\frac{\Gamma(\nu)}{2} + \left(\frac{kL^2}{2r}\right)^{\nu}\frac{\Gamma(-\nu)}{2} + \cdots, \tag{5.28}$$

$$H_\nu^{(1)}\left(\frac{qL^2}{r}\right) = \left(\frac{qL^2}{2r}\right)^{-\nu}\left(-\frac{i\Gamma(\nu)}{\pi}\right) + \left(\frac{qL^2}{2r}\right)^{\nu}\left(-\frac{i\Gamma(-\nu)}{\pi}e^{-i\pi\nu}\right) + \cdots, \tag{5.29}$$

$$H_\nu^{(2)}\left(\frac{qL^2}{r}\right) = \left(\frac{qL^2}{2r}\right)^{-\nu}\left(\frac{i\Gamma(\nu)}{\pi}\right) + \left(\frac{qL^2}{2r}\right)^{\nu}\left(\frac{i\Gamma(-\nu)}{\pi}e^{i\pi\nu}\right) + \cdots \tag{5.30}$$

with $q = ik = \sqrt{\omega^2 - \mathbf{k}^2}$ we obtain

$$\langle\mathcal{O}_\Delta(-k)\mathcal{O}_\Delta(k)\rangle$$
$$= \begin{cases} 2\nu(\Gamma(-\nu)/\Gamma(\nu))(k/2)^{2\Delta-d-1}, & k^2 > 0, \\ 2\nu e^{i\pi\,\mathrm{vsgn}(\omega)}(\Gamma(-\nu)/\Gamma(\nu))(ik/2)^{2\Delta-d-1}, & k^2 < 0, \text{in}, \\ -2\nu e^{-i\pi\,\mathrm{vsgn}(\omega)}(\Gamma(-\nu)/\Gamma(\nu))(ik/2)^{2\Delta-d-1}, & k^2 < 0, \text{out}. \end{cases} \tag{5.31}$$

An overall prefactor of L is left out, since it can be absorbed into the global renormalisation constant in front of the action for the scalar field Eq. (5.5). On Fourier transforming back to position space, we find precisely the scaling dependence on Δ expected of a two-point correlation function:

$$\langle\mathcal{O}_\Delta(x)\mathcal{O}_\Delta(y)\rangle = \begin{cases} 2\nu\Gamma(\Delta)/\left(\pi^{(d+1)/2}\Gamma(\nu)|x-y|^{2\Delta}\right), & k^2 > 0, \\ i\theta(x_0 - y_0)\dfrac{2\nu\Gamma(\Delta)}{\pi^{(d+1)/2}\Gamma(\nu)|x-y|^{2\Delta}}, & k^2 < 0, \text{in}, \\ -i\theta(y_0 - x_0)\dfrac{2\nu\Gamma(\Delta)}{\pi^{(d+1)/2}\Gamma(\nu)|x-y|^{2\Delta}}, & k^2 < 0, \text{out}. \end{cases} \tag{5.32}$$

The normalisation of these propagators is unconventional, but it is convenient since it is the natural choice in AdS/CFT. For two-point correlation functions it plays no further role, but for multi-point correlation functions the normalisation relative to

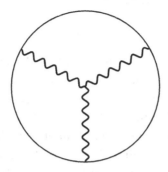

Figure 5.1 The Witten diagram for the leading-order contribution to the three-point correlation function $\langle \mathcal{O}_\Delta \mathcal{O}_\Delta \mathcal{O}_\Delta \rangle$ due to a cubic interaction $S_{\text{AdS}}^{(3)} = \int d^{d+2}x \sqrt{-g}\left(-(\lambda/3!)\phi^3\right)$ for the field ϕ dual to $\mathcal{O}_\Delta$.

the lower-point ones does matter [203]. This normalisation depends sensitively on the regularisation procedure. As long as one always places the boundary at $r = \epsilon^{-1}$ as the very first step in the calculation and takes the limit $\epsilon \to 0$ at the very end, one is guaranteed that all relative normalisation factors are correct.

5.1.2 Three-point correlations and beyond

Given that the partition function can be used as generating functional for correlation functions up to all orders, the GKPW rule allows us to compute in principle higher-order correlation functions in the boundary as well. These can be computed perturbatively in the weakly coupled bulk gravity theory, and the task at hand is to derive the appropriate Feynman rules for such a bulk computation. This goes through exactly as in standard quantum field theory, with the difference that the external fields are sourced at the boundary of AdS.

Because AdS space (at a fixed moment in time) can be thought of as a ball whose boundary is a sphere (box 4.4), and because the boundary plays such an important role, Witten proposed a pictorial representation of these rules to emphasise this fact [3]. For example, suppose the field dual to a scalar operator has a cubic interaction in the bulk,

$$S_{\text{AdS}}^{\text{int}} = \int d^{d+2}x \sqrt{-g}\left(-\frac{\lambda}{3!}\phi^3\right). \tag{5.33}$$

The "Witten diagram" for the leading-order contribution to the three-point function of a scalar operator then looks as in Fig. 5.1.

The one new aspect in the Feynman rules corresponding to this diagram is the expression assigned to the external lines extending to the boundary. For internal lines between two points in the interior, one uses the conventional "bulk-to-bulk" Green function. For a massive scalar this is the function which inverts the kinetic operator in the curved space-time background,

$$(\nabla^\mu \nabla_\mu - m^2)\mathcal{G}(x, r; y, r') = \frac{1}{\sqrt{-g}}\delta^{d+1}(x - y)\delta(r - r'). \qquad (5.34)$$

Note the extra factor of $1/\sqrt{-g}$ in the definition. This Green function is by construction normalisable, however. It therefore falls off faster than $r^{-(d+2)}$ at the boundary, $r = \infty$. This implies that the Green function itself cannot be the propagator for a boundary disturbance: it does not feel the boundary. This is precisely analogous to the electrostatics problem of an electric field created by a charged, perfectly conducting object within which the electrostatic potential vanishes. Outside of the conductor the electrostatic potential which determines the field is given by the normal derivative of the Green function integrated over the boundary of the conductor. It is therefore a boundary-value problem for the electric field, rather than the potential.

The same Dirichlet problem is at play here. The normal derivative of the Green function w.r.t. boundary does respond to a local source on the boundary (see also box 5.1 above). It is straightforward to show that this "bulk-to-boundary propagator" $\mathcal{K}(x; y, r') \equiv n^\mu \partial_\mu \mathcal{G}(x, r; y, r')|_{r=\infty}$ with n^μ a unit normal vector to the boundary (i.e. in the r direction) obeys

$$(\nabla^{y\mu} \nabla_{y\mu} - m^2)\mathcal{K}(x; y, r') = 0, \quad \lim_{\epsilon \to 0} \mathcal{K}(x; y, \epsilon^{-1}) = \epsilon^{d+1}\delta^{d+1}(x - y). \quad (5.35)$$

This "bulk-to-boundary" Green function is the object which is assigned to each external line in the Witten diagram (Table 5.1). In principle any n-point correlation function to any order can be computed this way. In accordance with the normalisation $Z_{CFT} = \int \mathcal{D}\phi \exp(iN^2 S_{bulk})$ in the GKPW rule, tree-level diagrams on the gravity side correspond to the large-N limit of CFT. In this limit one need only compute the classical field theory in the bulk. Following the dictionary, loop corrections in the bulk correspond to $1/N$ corrections to the boundary correlation functions. At lowest order in derivatives, these are still evaluated at infinitely strong 't Hooft coupling λ. Higher-derivative terms in the AdS action account for the effects of λ becoming finite.

An aspect that should be dealt with carefully in this context of multi-point functions is associated with the UV divergences in the CFT which are related to placing the AdS boundary at infinity. As we have already announced, the straightforward way to deal with them is to compute first with a regulated boundary at a distance $r = \epsilon^{-1}$, in order to take the limit $\epsilon \to 0$ at the end of the computations. To do so consistently, one uses the regulated bulk-to-boundary propagator instead

$$(\nabla^{y\mu} \nabla_{y\mu} - m^2)\mathcal{K}^{reg}(x; y, r') = 0, \quad \lim_{r' \to \epsilon^{-1}} \mathcal{K}^{reg}(x; y, r') = \delta^{d+1}(x - y). \quad (5.36)$$

From the insight that the radial direction corresponds to the RG scale, this procedure is the dual of regulating the UV boundary theory. As we saw at the

Table 5.1 *Real-space Feynman rules for AdS Witten diagrams*

Each internal line	A bulk-to-bulk propagator $\mathcal{G}(x, r; y, r')$
Each vertex	$\int dr\, d^{d+1}x\sqrt{-g}\lambda$
Symmetry factor	As usual
Each external line	A (regulated) bulk-to-boundary propagator
(connected to the boundary)	$\mathcal{K}^{\text{reg}}(x; y, r') \equiv n^\mu\, \partial_{x^\mu}\mathcal{G}(x, r; y, r')|_{r\to\infty}$
Conversion to the CFT Green function	Renormalisation with a factor $\epsilon^{-\Delta}$ for each external line (see box 5.2), take the limit $\epsilon \to 0$

beginning of this section, to obtain a well-defined limit where we can remove the cut-off ϵ, this may call for the introduction of a boundary counter-term. This procedure of *holographic renormalisation* can be consistently implemented at all orders [204].

As an example, we compute in box 5.2 the three-point functions of a pure CFT in the large-N limit. Just like for the two-point functions, the conformal invariance fixes the form of such a three-point correlator completely, and we demonstrate that this is reproduced correctly by a tree-level bulk computation of the corresponding Witten diagram for the case of simple scalar fields.

Box 5.2   CFT three-point function from Witten diagrams

In a Lorentz-invariant CFT, conformal invariance also fixes the three-point function up to an overall constant. For three scalar fields of scaling dimensions Δ_1, Δ_2 and Δ_3, the only possible form of the three-point correlation function compatible with the symmetries is

$$\langle \mathcal{O}_{\Delta_1}(x_1)\mathcal{O}_{\Delta_2}(x_2)\mathcal{O}_{\Delta_3}(x_3)\rangle$$
$$= \frac{C_{123}}{|x_1 - x_2|^{\Delta_1+\Delta_2-\Delta_3}|x_1 - x_3|^{\Delta_1+\Delta_3-\Delta_2}|x_2 - x_3|^{\Delta_2+\Delta_3-\Delta_1}}, \tag{5.37}$$

where $|x| = \sqrt{\sum_{i=1}^d x_i^2 - t^2}$ is the invariant distance. Let us now show how the AdS/CFT Feynman rules reproduce this result. We will work in the Euclidean signature and set $L = 1$. For simplicity we shall take an AdS theory with a single massive scalar field with cubic interaction

$$S_{\text{AdS}}^{\text{int}} = \int d^{d+2}x\sqrt{-g}\left(-\frac{\lambda}{3!}\phi^3\right). \tag{5.38}$$

The Witten diagram is the one given in Fig. 5.1, and we can now use the rules of Table 5.1. There are three external lines corresponding to

Box 5.2   (Continued)

bulk-to-boundary propagators integrated over a common vertex, and the symmetry factor is unity. The expression we need to evaluate is

$$\langle \mathcal{O}(x_1)\mathcal{O}(x_2)\mathcal{O}(x_3)\rangle$$

$$= \lim_{\epsilon \to 0} \epsilon^{-3\Delta}\lambda \int dr\, d^d x \sqrt{-g}\, \mathcal{K}^{\mathrm{reg}}(x_1; x, r)\mathcal{K}^{\mathrm{reg}}(x_2; x, r)\mathcal{K}^{\mathrm{reg}}(x_3; x, r). \quad (5.39)$$

We therefore need the expression for the bulk-to-boundary propagator. From the homogeneous solutions to the wave equation Eq. (5.24), we can immediately write down the bulk-to-bulk Green function

$$\mathcal{G}(x, r; y, r') = -\int \frac{d^{d+1}k}{(2\pi)^{d+1}}\left[(rr')^{-(d+1)/2}\left(K_\nu\left(\frac{k}{r}\right)I_\nu\left(\frac{k}{r'}\right)\theta(r'-r)\right.\right.$$

$$\left.\left. + K_\nu\left(\frac{k}{r'}\right)I_\nu\left(\frac{k}{r}\right)\theta(r-r')\right)e^{ik(x-y)}\right]. \quad (5.40)$$

The easiest way to obtain the regularised bulk-to-boundary propagator is to regularise the bulk-to-bulk propagator first. Instead of vanishing at $r = \infty$, we "regularise" by requiring that it vanishes at $r = \epsilon^{-1}$,

$$\mathcal{G}^{\mathrm{reg}}(x, r; y, r') = \mathcal{G}(x, r; y, r') + \int \frac{d^{d+1}k}{(2\pi)^{d+1}}(rr')^{-(d+1)/2}$$

$$\times K_\nu\left(\frac{k}{r}\right)K_\nu\left(\frac{k}{r'}\right)\frac{I_\nu(k\epsilon)}{K_\nu(k\epsilon)}e^{ik(x-y)}. \quad (5.41)$$

The normal derivative at $r = \epsilon^{-1}$ of the regularised bulk-to-bulk Green function then gives us the regularised bulk-to-boundary propagator

$$\mathcal{K}^{\mathrm{reg}}(x; y, r) = \left(\frac{\epsilon}{r}\right)^{(d+1)/2}\int \frac{d^{d+1}k}{(2\pi)^{d+1}}\frac{K_\nu(k/r)}{K_\nu(k\epsilon)}e^{ik(x-y)}. \quad (5.42)$$

Using the integral representation of the Bessel function,

$$K_\nu(ks) = \frac{1}{2}\left(\frac{2s}{k}\right)^\nu \int_0^\infty d\tau\, \tau^{\nu-1}e^{-\tau s^2 - k^2/4\tau}, \quad (5.43)$$

one can give a closed-form expression for the bulk-to-boundary Green function

$$\mathcal{K}^{\mathrm{reg}}(x_1; x, r) = \frac{\Gamma(\Delta)}{\pi^{(d+1)/2}\Gamma(\nu)}\epsilon^\Delta\left(\frac{r^{-1}}{r^{-2} + |x_1 - x|^2}\right)^\Delta. \quad (5.44)$$

On substituting this expression for the bulk-to-boundary propagator into the formal holographic answer for the three-point function, we find

$$\langle \mathcal{O}(x_1) \mathcal{O}(x_2) \mathcal{O}(x_3) \rangle$$

$$= \lim_{\epsilon \to 0} \left(\frac{\Gamma(\Delta)}{(\pi^{(d+1)/2} \Gamma(\nu))} \right)^3 \lambda \int dr \, d^d x$$

$$\times r^d \frac{r^{-3\Delta}}{[(r^{-2} + |x - x_1|^2)(r^{-2} + |x - x_2|^2)(r^{-2} + |x - x_3|^2)]^\Delta}. \tag{5.45}$$

Using the identity

$$\frac{1}{A^\alpha} = \frac{1}{\Gamma(\alpha)} \int_0^\infty d\tau \, \tau^{\alpha-1} e^{-\tau A}, \tag{5.46}$$

the integral over the radial coordinate r can immediately be done. Shifting the common bulk coordinate x such that only relative distances $|x_i - x_j|$ remain allows this integral to be evaluated. The remaining three auxiliary τ integrals are straightforward, and one finds

$$\langle \mathcal{O}(x_1) \mathcal{O}(x_2) \mathcal{O}(x_3) \rangle = \frac{C}{|x_1 - x_2|^\Delta |x_1 - x_3|^\Delta |x_2 - x_3|^\Delta} \tag{5.47}$$

with

$$C = \lambda \frac{\Gamma(\frac{1}{2}\Delta + \nu)}{2\pi^{(d+1)}} \left[\frac{\Gamma(\Delta/2)}{\Gamma(\nu)} \right]^3. \tag{5.48}$$

In pure AdS one can arrive at the closed-form position-space expressions for the propagators in a much more direct way using the discrete inversion symmetry in AdS [205]. The derivation here is more general and can be applied in situations where the bulk geometry is not pure AdS.

5.2 Correlations, scaling and RG flows

The calculations above are performed in pure AdS backgrounds dual to exact CFTs. The majority of this book, however, will involve backgrounds where we deform away from the exact CFT. Intuitively the most straightforward way to do so is to turn on a relevant operator. This will trigger an RG flow to a new fixed point. One of the beautiful aspects of holography is that we will be able to follow this flow completely in the bulk space-time geometry. The asymptotic boundary at $r \to \infty$ corresponds to the UV of the theory, and it will reflect the original conformal invariance of the UV fixed point. As we move into the interior, we are lowering the energy scale, and the geometry will change to reflect both the loss of conformal invariance and the emergence of new IR-specific characteristics. Eventually the flow ends in the interior in a new fixed point. As we have stressed repeatedly, this "geometrisation of the RG" is at the very heart of the appeal of AdS/CFT, and allows for new types of renormalisation flows towards novel IR fixed points that encode for new states of quantum critical matter.

In order to proceed, we need to identify fields in the dual gravity that are dual to relevant operators. This information is contained in the dictionary rule that relates scaling dimensions to masses. A relevant operator is any scalar operator with dimension $\Delta < d + 1$. From the calculation of the two-point function of an operator dual to a massive scalar field, we know that the scaling dimension of the field in terms of the mass equals

$$\Delta = \frac{d+1}{2} + \sqrt{\frac{(d+1)^2}{4} + m^2 L^2}. \tag{5.49}$$

Any normal scalar field with positive $m^2 > 0$ therefore has $\Delta > d + 1$. It is dual to an *irrelevant operator*. Similarly, a massless scalar field is dual to a marginal operator with dimension $\Delta = d+1$ – as in standard field theory this operator dimension will get corrections in general. Our objective, however, is to find the relevant operators. The bizarre observation is that we would need an operator with *negative* mass squared, $m^2 L^2 < 0$. In a flat space-time background, such an operator would imply a linear ("tachyonic") instability. The scalar mode is fluctuation around a background, where one sits at the top of a local potential maximum. However, in curved AdS spaces such negative mass-squared fields can be perfectly consistent, be it only within a small window. This is precisely consistent with the window for which the scaling dimension of the operator in the dual field theory stays real. Specifically, for scalar fields the mass squared in the bulk can be negative, but not more negative than

$$m^2 L^2 > -\frac{(d+1)^2}{4}. \tag{5.50}$$

How is the instability signalled by negative mass in flat space-time countered in a curved geometry? Breitenlohner and Freedman discovered in the 1970s [6] that in anti-de Sitter space there is an additional contribution to the potential from the gravitational background. The effective potential felt by the scalar field is therefore not equal to the potential in the Lagrangian. It is precisely for negative mass squared in the window Eq. (5.50) that the effective potential remains stable. It is straightforward to see this from the equation of motion Eq. (5.23). By redefining $\phi(k, r) = r^{-(d+2)/2}\Phi(k, r)$ one finds an equivalent Schrödinger-like equation in an effective flat space-time,

$$\frac{\partial^2}{\partial r^2}\Phi - \left(\frac{k^2 L^4}{r^4} + \frac{m^2 L^2 + d(d+2)/4}{r^2}\right)\Phi = 0, \tag{5.51}$$

which shows that due to the gravitational background the mass term picks up an additional factor $d(d+2)/4$. It is well known that, for a potential that behaves like

$1/r^2$ as $r \to \infty$, the quantum theory will be stable if $m^2L^2 + d(d+2)/4 > -\frac{1}{4}$ which is exactly the Breitenlohner–Freedman (BF) bound (5.50). In other words, for values of the negative mass squared that are more negative than this BF bound the background solution is yet again perturbatively unstable. These are excluded on principle; it is well understood in relativistic field theory that perturbatively unstable fluctuations break unitarity [206]. An AdS theory with a mass below the BF bound is therefore a non-unitary theory, and we will not consider such theories. Note that in later chapters we shall encounter theories where the effective mass runs between the UV AdS boundary and an emergent geometry in the IR. In the IR the bound can and will be violated. The exclusion here refers to the UV value of the mass.

> **Rule 6** Irrelevant scalar operators in the field theory are dual to massive scalar fields in the gravity theory. Marginal scalar operators are dual to massless scalar fields, while relevant scalar operators are dual to scalar fields with a negative mass squared. There is a small window of negative-mass-squared fields that are nevertheless stable in AdS backgrounds. A similar rule applies to fields with spin.

The precise way in which the unitarity constraint manifests itself in holography is quite instructive. Unitary bounds in conformal field theories are a well-studied subject (see e.g. [207]). A simple exercise reveals that the scaling dimensions of operators are all bounded from below by their free-field values: for a relativistic scalar operator $\Delta > (d-1)/2$, with d the number of spatial dimensions. If we use the mass-scaling dimension relation derived earlier, however, we find that the lowest possible scaling dimension equal to the BF bound is $\Delta = (d+1)/2$. How do we account for unitary operators with dimension $(d-1)/2 < \Delta < (d+1)/2$? They observed that precisely for this range of values of Δ *both* asymptotic fall-offs of the solution to the linearised scalar wave equation are normalisable. Recall that the solution behaves as

$$\phi(r) = Ar^{-(d+1-\Delta)} + Br^{-\Delta} + \cdots. \tag{5.52}$$

Thus in an AdS background with metric

$$ds^2 = \frac{L^2}{r^2} dr^2 + \frac{r^2}{L^2} \eta_{\mu\nu} dx^\mu dx^\nu, \tag{5.53}$$

the boundary contribution from the norm behaves as

$$\int d^{d+1}x \sqrt{-h}\phi^2 \sim \lim_{r\to\infty} A^2 r^{-(d+1-2\Delta)} \tag{5.54}$$

and is *finite* for $\Delta < (d+1)/2$. In the standard approach to correlation functions in holography we treated the solution to the wave equation with leading behaviour near the boundary as the non-normalisable source at infinity, and the sub-leading normalisable solution as the quantised fluctuating mode. This analysis shows that for $\Delta < (d+1)/2$ there is a different way to quantise the theory, where we take the leading solution as the fluctuation. Following the usual quantisation rules, its canonical conjugate, i.e. the sub-leading contribution, should then be considered as the source. In this *alternative quantisation* the two-point correlation function is therefore equal to

$$G_{\text{alt}}(\omega, \mathbf{k}) = \frac{A(\omega, \mathbf{k})}{B(\omega, \mathbf{k})} \sim k^{d+1-2\Delta_{\text{std}}}, \tag{5.55}$$

where $\Delta_{\text{std}} = (d+1)/2 + \sqrt{(d+1)^2/4 + m^2 L^2}$ is the scaling dimension of the operator in standard quantisation. The alternatively quantised AdS theory is, however, dual to a CFT that is fundamentally different from the standard quantised AdS theory. The spectrum of operators differs between the two CFTs, and we should interpret this correlation function as a two-point function of an operator with scaling dimension Δ_{alt},

$$G_{\text{alt}} = \langle \mathcal{O}_{\Delta_{\text{alt}}} \mathcal{O}_{\Delta_{\text{alt}}} \rangle \equiv k^{2\Delta_{\text{alt}}-d-1}. \tag{5.56}$$

Upon equating the scaling behaviour we see that $\Delta_{\text{alt}} = d + 1 - \Delta_{\text{std}}$, which precisely mimics the exchange of roles between leading and sub-leading coefficients of the solution to the equation of motion.

In the window between the unitarity bound and the BF bound as applied to standard quantisation $(d-1)/2 < \Delta < (d+1)/2$, one therefore has to use an alternative quantisation to describe the theory. The perspective that one takes the scaling dimensions of the operators as input also makes clear that the choice between standard and alternative quantisation is therefore an explicit choice one has to make. There is, however, a very interesting relation between the two CFTs. The alternatively quantised theory has a natural relevant operator, in the form of the square of this operator: the composite operator $\mathcal{O}_{\Delta_{\text{alt}}} \mathcal{O}_{\Delta_{\text{alt}}}$ has the naive scaling dimension $2\Delta_{\text{alt}} < d + 1$. With the canonical example of a matrix large-N field theory for the CFT in mind, where the operators are $U(N)$-invariant single-trace operators, the composite squared operator is a double-trace operator; see section 4.3. At leading order in large N, a double-trace deformation is naturally suppressed.

As we will elucidate in detail in box 5.3, due to $1/N$ suppression the RG flow triggered by this relevant operator can easily be tracked in the matrix large-N theory. Under deformation by a relevant operator generically everything will mix and change as one flows to a new theory in the IR. We will see and use this many times in the remainder of the book. With the identification of the radial extra dimension as the RG scale, the primary way in which this flow manifests itself is through a geometry that changes as we move from the UV conformal fixed point on the boundary to the new emergent IR. For a double-trace deformation, however, the geometry remains unchanged. It is a very particular RG flow that affects only a subset of the couplings/anomalous dimensions (due to the large-N limit). The only coupling it affects is the scaling dimension of the operator itself, and it flows precisely from alternative ($\Delta_{\text{UV}} = \Delta_{\text{alt}}$) to standard quantisation, i.e. $\Delta_{\text{IR}} = \Delta_{\text{std}}$.

As we will show explicitly in box 5.3 using holographic means, this takes a form that is very familiar in condensed matter theory. In terms of the correlation function of the operator $\mathcal{O}$, the consequence of deforming the original theory by its square

$$S = S_{\text{CFT}} + \int d^{d+1}x \, \frac{f}{2} \mathcal{O}^2 \tag{5.57}$$

is that the Green function takes the form

$$\langle \mathcal{O}(-k)\mathcal{O}(k) \rangle = \frac{G(k)}{1 + fG(k)}. \tag{5.58}$$

This is precisely coincident with the time-dependent Hartree–Fock/RPA result which we discussed in section 2.4 dealing with the conventional mean-field theory describing spontaneous symmetry breaking. This is a signal of the mean-field nature associated with the matrix large-N limit. We discussed the factorisation property emerging in the matrix large-N limit which is responsible for this mean-field behaviour in section 4.3. However, one also infers immediately from the above discussion that this mean field has a *very different* meaning from the conventional Hartree–Fock mean-field theory. This matrix theory mean field *ties together different strongly interacting quantum critical states* in this specific conformally invariant context – the previous paragraphs may have been quite confusing in this regard for the condensed matter readership. Although we will find out, especially in chapter 10, that it can also have consequences that are more reminiscent of a conventional mean-field behaviour (such as driving BCS-like instabilities, suppressing thermal critical fluctuations), one should be very aware that the matrix large-N mean field is fundamentally very different. Let us conclude this section by highlighting an entertaining technical subtlety. Even in the straightforward holographic computations for exact conformal field theories we have seen various instances supporting the identification of the radial AdS direction with the energy scale of the boundary theory. There is one aspect, however, which is strikingly different from the conventional renormalisation-group theory. The radial evolution equations in AdS characterising the flow along the RG scale are actually second-order differential equations. This is in apparent conflict with the "diffusive" nature of the renormalisation group, as characterised by first-order equations for the flow of coupling constants.

The resolution of this apparent paradox is as follows. The actual problem one has to solve is a boundary-value problem for the on-shell action as a function of the boundary values of the fields; these are dual to the sources of the operators in the boundary field theory. A standard component of classical mechanics is that, given a set of abstract boundary values, one can solve the equations of motion implicitly in terms of these boundary values. In this well-known Hamilton–Jacobi formalism the equations of motion for the boundary values become first order [208]. Since the boundary values of the fields in AdS are precisely the couplings in the dual CFT on the boundary, one recovers the fact that the RG evolution in the boundary is governed by the first-order differential equations.

> **Box 5.3   Double-trace deformations and the holographic emergence of large-N mean-field behaviour**
>
> In section 4.3 we emphasised that one of the most important aspects of matrix large-N theories is factorisation: the property that the higher-point correlation function of a product of $U(N)$-(gauge)-invariant single-trace operators equals the product of the lowest non-vanishing correlation function at leading order in large-N, generically the one-point function of the single-trace operator itself. This factorisation property is the aspect that these theories share with the conventional mean fields of statistical and condensed matter physics, but we have already emphasised that the consequences of this factorisation can be quite different.
>
> This has as a consequence that a deformation of the theory by a product of single operators is tractable [209]. Here we show how this is encoded in AdS/CFT. We take the field theory of interest with a source for the single-trace operator $\mathcal{O}$ and deform it by an arbitrary multi-trace potential. Then the partition function becomes
>
> $$Z_{\text{QFT}} = \int \mathcal{D}\phi \, e^{i \int d^{d+1}x \left(\mathcal{L}(\mathcal{O}) - iJ\mathcal{O} - iW(\mathcal{O}) \right)}. \tag{5.59}$$
>
> This shows that, in the presence of an expectation value for $\mathcal{O}$, the source for the fluctuating part of the gravity field dual to $\mathcal{O}$ should then correspond to $J + \partial W / \partial \mathcal{O}$. In other words the function $W(\mathcal{O})$ affects the boundary conditions of the field ϕ dual to the operator $\mathcal{O}$.

How it does so can be explicitly derived in a Hamiltonian formalism [210], with an outcome that is quite simple. If we are in a regime where the relevant fluctuations are those of the operator itself, the saddle point of the action is given by

$$J + \left\langle \frac{\partial W(\mathcal{O})}{\partial \mathcal{O}} \right\rangle = 0. \tag{5.60}$$

Factorisation in the large-N limit means that we can simplify this to

$$J + \frac{\partial W(\langle \mathcal{O} \rangle)}{\partial \langle \mathcal{O} \rangle} = 0. \tag{5.61}$$

Now we know that we should identify the source J for the operator with the coefficient A of the leading solution to the equation of motion for the dual field ϕ, Eq. (5.7), and the expectation value $\langle \mathcal{O} \rangle$ with the coefficient B of the sub-leading solution. This factorised saddle-point equation thus translates into the relation between these coefficients as

$$A + \frac{\partial W(B)}{\partial B} = 0. \tag{5.62}$$

Note that, for a linear function $W(\mathcal{O}) = \beta \mathcal{O}$, this precisely recovers the identification of the leading coefficient with the source.

Consider now a double-trace deformation where $W(\mathcal{O}) = \beta \mathcal{O} + (f/2)\mathcal{O}^2$ is a quadratic function. Using the insight above, this means that on the AdS side we should consider the solution to the equation of motion for ϕ with the relation between the leading and sub-leading coefficients as

$$A = \beta + fB. \tag{5.63}$$

We know that in the absence of such a relation the on-shell action equals Eq. (5.10) (for simplicity the prefactor is ignored here):

$$S = \oint_{r=\epsilon^{-1}} \frac{d\omega \, d\mathbf{k}}{(2\pi)^{d+1}} \frac{1}{2} AB + \cdots . \tag{5.64}$$

Earlier we showed that, for fixed A, regularity in the bulk determines that $B = GA$, where G is the conformal two-point function for the operator $\mathcal{O}$. Inverting this relation means that for a given response B one must have the source $A = G^{-1}B$. It is now convenient to rewrite the action in terms of the dynamical variable B, which gives the on-shell action

$$S = \oint_{r=\epsilon^{-1}} \frac{d\omega \, d\mathbf{k}}{(2\pi)^{d+1}} \frac{1}{2} BG^{-1}B + \cdots . \tag{5.65}$$

Box 5.3   (Continued)

The reason is that we can now add $W(B)$ directly to this action. The deformed theory can thus be characterised by

$$S_{\mathrm{def}} = \oint_{r=\epsilon^{-1}} \frac{d\omega\, d\mathbf{k}}{(2\pi)^{d+1}} \left(\frac{1}{2} BG^{-1}B + \beta B + \frac{f}{2}B^2 \right). \qquad (5.66)$$

On solving for the dynamical variable B in terms of the source β, we find

$$B = -\frac{G}{1+fG}\beta. \qquad (5.67)$$

On substituting this into the action and differentiating twice w.r.t. the source β, we thus find the two-point function

$$\langle \mathcal{O}\mathcal{O} \rangle = \frac{G}{1+fG} = G\sum_{n=0}^{\infty}(-fG)^n. \qquad (5.68)$$

This is immediately recognised as the RPA form of the Green function that follows from a mean-field Dyson re-summation. Note that this becomes exact in the large-N limit.

5.3 The identification of symmetries and the dictionary table

To complete the dictionary there are two more clear aspects to be discussed. With regard to the physics of the boundary theory, we have focussed entirely on how the GKPW rule is used to compute boundary propagators, and how the renormalisation group of the boundary theory is geometrised in the bulk. We used a simple neutral scalar operator to illustrate this. What if the operator we wish to consider has charge and/or spin? In particular, an operator in any relativistic field theory is the spin-2 energy–momentum tensor $T_{\mu\nu}$. This operator is always there as the generator of time- and space-translation symmetries. Energy–momentum is the quantum number under this symmetry. The aspects we still need to address are symmetries and quantum numbers.

We will take the ubiquitous energy–momentum tensor as an example. As a symmetry generator, it obeys a conservation law $\partial_\mu T^{\mu\nu} = 0$. This conservation law is the heart of the relativistic hydrodynamics as we will discuss at length in chapter 7. In addition, the energy–momentum tensor also generates rescalings. When one considers a conformal field theory, these rescalings are also symmetries. The associated conservation law is that the trace of the stress tensor must vanish identically: $\eta_{\mu\nu}T^{\mu\nu} = 0$. At the quantum level the exact scaling may be broken –

in a $(1 + 1)$-dimensional theory this is precisely what is captured by the central charge. Nevertheless, in a flat space-time background the one-point function still identically vanishes for a CFT: $\eta_{\mu\nu}\langle T^{\mu\nu}\rangle = 0$.

For fields with spin or internal quantum numbers, the GKPW rule goes through without modification in principle. From the identification of the field-theory source with the gravity field, it is clear that the numbers and properties have to be equal on both sides. However, the fact that a physical operator such as the energy–momentum tensor is associated with a symmetry has an important consequence for the tensorial nature of the operator. In the case of the energy–momentum tensor, the conservation law (Lorentz invariance) projects out the spin-1 component of the tensor, and the vanishing of the trace (conformal invariance) projects out the spin-0 part. The dynamical part of the stress tensor is therefore a pure spin-2 degree of freedom. These constraints already have an immediate effect on the two-point function for the stress tensor. Together with the symmetries, they impose that the stress tensor two-point function in a conformal field theory must be of the form

$$\langle T_{\mu\nu}(x)T_{\rho\sigma}(y)\rangle = \frac{c_T}{s^{2(d+1)}} J_{\mu\alpha}(s) J_{\nu\beta}(s) P_{\alpha\beta,\rho\sigma}, \tag{5.69}$$

where $s = x - y$,

$$J_{\mu\nu}(x) = \delta_{\mu\nu} - 2\frac{x_\mu x_\nu}{|x|^2},$$

$$P_{\mu\nu,\rho\sigma} = \frac{1}{2}(\delta_{\mu\rho}\delta_{\nu\sigma} + \delta_{\mu\sigma}\delta_{\nu\rho}) - \frac{1}{d+1}\delta_{\mu\nu}\delta_{\rho\sigma} \tag{5.70}$$

and c_T is a constant. One notices that the conformal dimension of the energy–momentum tensor is fixed to be $\Delta_T = d + 1$. This follows directly from the simple fact that the energy is subjected to "engineering scaling", because it is set by the volume.

We should now use the rule we learned in section 5.2 to translate the scaling dimension of the operator into the mass of the dual field in the bulk: the engineering scaling implies that the former is marginal, $\Delta = d + 1$. For a spin-2 field there is a different relation between masses and dimensions from that for a spin-0 scalar field. Nevertheless, for a spin-2 field marginality also means that the bulk field has to be massless. Since the spin characterises how the operator transforms under rotations, and since the space-time of the field theory is identified with the space-time at the boundary of AdS, it follows that the dual field should transform in the same way under rotations in the bulk space-time. This field must therefore be a spin-2 field as well. The only consistent theory of massless spin-2 fields we know is Einstein's theory of general relativity: *the bulk field dual to the boundary energy–momentum tensor must therefore be the graviton!*

In the introductory chapter we argued that the insight that the radial direction of AdS is dual to the energy scale implies that the space-time must be dynamical, since it must respond to deformations of the boundary field theory that set up a non-trivial RG flow. Here we see the precise way in which the dynamics of space-time must come out. Space-time fluctuations encode the energy–momentum fluctuations in the dual-field theory. On closer inspection a subtlety arises; since the gravity theory in the bulk is characterised by one extra dimension, the indices of the graviton $g_{\mu\nu}$, $\mu = 0, 1, \ldots, d, r$ run over one more value than the indices of the stress tensor. The graviton is the gauge field associated with space-time diffeomorphisms, and one can use the independent coordinate transformations in the $(d + 2)$-dimensional bulk to precisely choose a gauge where $g_{rr} = 0$, $g_{r\mu} = 0$, with the effect that everything precisely matches.

It is clear that this is not a coincidence but a consequence of the symmetries. Note that, as foretold in the introduction, the global symmetry of translations on the boundary is related to the local symmetry of coordinate transformations in the bulk. Exactly the same global–local dual correspondence applies to the global internal symmetries characterising the physics of the boundary system. If the conformal theory has a global internal (non-space-time) symmetry, with an associated conserved (spin-1) current J^μ, the field in the gravity theory that is dual to this operator is a (spin-1) massless vector field A_μ with a gauged version of the symmetry in the field theory. The radial component A_r, which has no clear boundary part, can again be set to vanish thanks to the gauge invariance in the bulk.

We can now summarise what we have learned. As always in quantum physics, symmetries and their associated currents play a central role throughout in the construction. They are supplemented by an abstract set of operators. In that sense AdS/CFT, especially in the bottom-up construction, works very much like an advanced version of phenomenological Landau–Ginzburg theory. The elementary ingredients, a set of operators $\mathcal{O}$ with quantum numbers q and spin s, are dual to a set of fields ϕ with the same quantum numbers q and spin s. One then has to solve a boundary-value problem in the gravity theory, where the boundary value of the field is equated with the source for the operator $\mathcal{O}$. In linear response the sub-leading part of the solution to the equation of motion for the field is the expectation value of the operator $\mathcal{O}$. Higher-order effects can be perturbatively included by computing Witten–Feynman diagrams for the boundary-value problem. To present these items of wisdom conveniently for the novice

Table 5.2 *The basic dictionary for AdS/CFT correspondence.*

Boundary: field theory	Bulk: gravity
Partition function	Partition function
Scalar operator/order parameter $\mathcal{O}$	Scalar field ϕ
Source of the operator	Boundary value of the field (leading coefficient of the non-normalisable solution)
VEV of the operator	Boundary value of the radial momentum of the field (leading coefficient of the normalisable solution; sub-leading to the non-normalisable solution)
Conformal dimension of the operator	Mass of the field
Spin/charge of the operator	Spin/charge of the field
Energy–momentum tensor T^{ab}	Metric field g_{ab}
Global internal symmetry current J^a	Maxwell field A_a
Fermionic operator $\mathcal{O}_\psi$	Dirac field ψ
Two-point correlation function	Ratio of normalisable to non-normalisable solution evaluated at the boundary
Higher-point correlation functions	Witten diagram computation
Double-trace deformation (RPA)	Mixed boundary conditions
RG flow	Evolution in the radial AdS direction
Number of degrees of freedom	Radius of AdS space
Global space-time symmetry	Local isometry
Global internal symmetry	Local gauge symmetry
Finite temperature	Black-hole Hawking temperature or radius of compact Euclidean time circle
Chemical potential/charge density	Boundary values of the electrostatic potential A_t (radial electric field)
Free energy	On-shell value of the action
Entropy	Area of black-hole horizon
Phase transition	Instability of black holes
Wilson line along $\mathcal{C}$	String worldsheet with end points on $\mathcal{C}$
Entanglement entropy of area A	Minimal surface with boundary equal to A
Quantum anomalies	Chern–Simons terms

holographist, we have summarised them in Table 5.2. In this table we include also the novel entries to the dictionary for finite temperatures, finite density and collective aspects which we will address in later chapters. This is to whet the reader's appetite!

6

Finite-temperature magic: black holes and holographic thermodynamics

Already shortly after the discovery of the AdS/CFT correspondence it became clear that holography can deal remarkably easily with the finite-temperature physics of the boundary system. The key is that one can account for the thermal physics of the strongly interacting critical state in the boundary by adding a black hole in the deep interior in the AdS bulk. As we will explain in detail in the first section, the temperature of the boundary system turns out to be identical to the Hawking temperature of the black hole living in an AdS space-time. This involves an intriguing and non-trivial twist of the "classic" consideration explaining Hawking radiation. In Hawking's computation one is dealing with *quantised* fields living in the classical black-hole space-time, whereas in the AdS bulk everything is strictly classical and zero temperature. Instead, via a remarkably elegant construction it is easy to understand that the black-hole bulk geometry "projects" onto the boundary system a finite temperature that is coincident with the Hawking temperature one would find in a bulk with quantised fields.

Having identified the dictionary rule that finite temperature is encoded by the bulk black-hole geometry, it turns out that these black holes also encode in an impeccable way for all the thermodynamics principles governing thermal equilibrium physics. This direct map of the "rules of black holes" to the thermodynamics of a *real physical system with microscopic degrees of freedom* is why the AdS/CFT correspondence manifests the holographic principle explained in the preceding chapters. The most poignant aspect hereof, as we will also discuss in the first section, is the identification of the Bekenstein–Hawking black-hole entropy with the entropy of microscopic configurations of the boundary field theory.

However, holographic thermodynamics is a lot more powerful than these classic black-hole thermodynamics notions. In later chapters we will show that AdS black holes are very different from their more familiar featureless all-engorging flat-space cousins. AdS black holes are actually able to describe rather rich, real-life phase diagrams of the matter in the boundary. In section 6.2 we will highlight a historically important example of such a phase diagram: the "Hawking–Page" confinement–deconfinement phase transition in a finite volume as discovered early on by Witten [211]. This is a mandatory exercise for the holography student, since it highlights the role played by the gauge-theoretical (de)confinement phenomenon in holography, which will be a recurrent theme in this book.

In section 6.3 we will drill a bit deeper into this (de)confinement theme by summarising some of the basic constructions behind a development that to a degree preceded AdS/CMT. This is AdS/QCD: the use of holography to address specifics of real-life quantum chromodynamics of relevance to particle physics. Differently from the super-conformal Yang–Mills theory of the original AdS/CFT construction, at zero temperature the QCD vacuum is confining, with a dynamically generated mass gap. One now faces the question of how to encode this confinement scale and the associated gap in the bulk geometry. This can be accomplished by "capping off the geometry": upon descending along the radial direction towards the interior, at some distance along the radial direction the bulk geometry is forced to end in a hard wall. Given the identification of the radial direction with the scaling direction in the boundary, this translates into an energy scale in the field theory that coincides with the confinement scale. The probe fields in the bulk perceive this capped-off geometry as a box, and this gives rise to quantisation of the radial modes. These quantised modes translate into a tower of massive particle excitations in the boundary, which are identified as the gauge singlets ("mesons") of the confining boundary theory. At finite temperature a generalised version of the prototype Hawking–Page transition takes place. Upon raising temperature the black-hole horizon increases along the radial direction and at some point crosses the "capping-off" point. This signals a transition/crossover from a confining meson state to a deconfining high-temperature state of many quarks and gluons. It is therefore identified with the quark–gluon plasma.

This "capping off" of the geometry in the bulk can be described on various levels of sophistication. Given that this confinement theme will play an important role in the story that follows, we will introduce not only the very simple "hard-wall" construction, but also the more sophisticated "soft walls" and "AdS solitons"; the former require the addition of a scalar degree of freedom whose dynamics is encoded in a more involved Einstein-dilaton theory to describe the bulk. Also for future use, we will explain the dictionary entry for Wilson loops and how these are used to detect (de)confinement in the boundary.

We end this chapter with a dictionary entry: the recipe to compute the real-time propagators of the boundary field theory when the temperature is finite (section 6.4). This is part of the general dictionary which was the subject of the previous chapter, but we have delayed its introduction until this chapter dealing with finite temperature. Given the ambiguities which accompany the Wick rotation from the

Euclidean to real-time correlation functions, this has the reputation of being notoriously difficult in direct field-theoretical approaches. But it is not so in holography: it is easy to carry out computations in a Minkowski-signature bulk that is dual to real time in the boundary. All one needs to know in addition is how to apply boundary conditions at the black-hole horizon. The outcome is elegant and intuitive: for retarded propagators of relevance to spectral functions one has to choose in-falling boundary conditions. The stuff that disappears behind the horizon encodes for the dissipative aspects of the finite-temperature physics in the boundary. Similarly, the advanced Green functions are associated with out-going boundary conditions. In fact, the full Schwinger–Keldysh propagators of non-equilibrium physics are elegantly encoded in the causal structure of the bulk space-time.

6.1 Black holes in the bulk and finite temperature in the boundary

6.1.1 Black holes in flat space-time

Before delving into finite-temperature holography, let us first briefly summarise some of the highlights of black-hole physics which act as a reference frame for what follows.

About a month after Einstein had proposed his famous theory of general relativity, Schwarzschild found the first non-trivial exact solution for Einstein's equation. This was much later recognised as *the* iconic object of general relativity – the Schwarzschild black hole. In $3 + 1$ dimensions the Schwarzschild metric is

$$ds^2 = -f(r)dt^2 + \frac{dr^2}{f(r)} + r^2(d\theta^2 + \sin^2\theta \, d\phi^2). \tag{6.1}$$

Here $f(r)$ is the so-called emblackening factor

$$f(r) = 1 - \frac{2GM}{r}, \tag{6.2}$$

where G is equal to Newton's constant and M is a parameter that can be interpreted as the mass of the black hole. When $M = 0$, the above geometry (6.1) reduces back to Minkowski space-time in polar coordinates. Also far away from the black hole, when $r \rightarrow \infty$, the geometry (6.1) becomes a Minkowski flat space-time. This property is known as asymptotic flatness. The location $r_0 = 2GM$, where $g_{tt} = 0$, appears to be special. This is the horizon; beyond it, when $r < 2GM$, the time and the radial direction appear to exchange roles with each other.

The horizon is in fact not a physical object: a different observer will not see the horizon in the same place. The horizon is a coordinate singularity – for another choice of coordinates, such as the Lemaître coordinates (describing a free-falling observer), the metric will become smooth at the horizon, lacking any sign of a singularity. Nevertheless, to an external observer at $r = \infty$ any object located beyond $r_0 = 2GM$ needs an escape velocity larger than the speed of light. Thus nothing can escape from this horizon: it is the "boundary" of a black hole. The horizon is the extremal edge of the gravitational potential well created by the black hole, and the location of extreme redshift and time dilatations. To an observer at $r = \infty$, objects that fall towards the horizon in fact never cross, since to them time comes to a standstill at the horizon. This infinite time dilatation also applies to the case of a collapsing star: to an observer at $r = \infty$ the collapse of the star seems to freeze at the horizon.

A free-falling observer, on the other hand, will not notice anything upon crossing the horizon. However, his or her future fate is sealed: not only can he or she never escape back to outer space once he or she has crossed the horizon, but also he or she will get "spaghettified" by the extreme tidal forces as he or she gets closer and closer to the black hole and its real singularity at $r = 0$.

This story acquired extra impetus in the 1970s when the ramifications of such a classical space-time as a background for quantum physics were explored. Hawking famously discovered that the space-time of a Schwarzschild black hole "tears apart" the coherent vacuum of free quantum fields, rendering virtual vacuum fluctuations real, turning into radiation of a black body with a Hawking temperature inversely proportional to its mass. Departing from (free) quantised fields in the classical black-hole space-time, the Hawking temperature is given by

$$T_{\mathrm{H}} = \frac{\hbar c^3}{8\pi G M k_{\mathrm{B}}} \tag{6.3}$$

or, in natural units ($\hbar = c = k_{\mathrm{B}} = 1$),

$$T_{\mathrm{H}} = \frac{1}{8\pi G M}. \tag{6.4}$$

The Hawking temperature can also be derived from the Gibbons–Hawking formulation of Euclidean quantum gravity [212]. In fact, general relativity can be formulated regardless of the choice of signature, i.e. whether time is chosen to be real (Minkowski) instead of imaginary (Euclidean). The Euclidean formulation is a hazardous affair given that the analytic continuation to imaginary time erases the direct connection between light-like propagation and causality. However, for the stationary space-times one is perfectly safe. These will be most important for us, because, in the context of AdS/CFT, stationary space-times are associated with equilibrium configurations in the dual boundary theory. To obtain the Euclidean black-hole metric we can therefore simply analytically continue the Schwarzschild metric in Minkowski space-time (6.1) to imaginary time by substituting $\tau = it$ such that τ becomes a spacelike dimension. We will discuss this procedure in more detail in box 6.1. The outcome is that, in order to avoid a conical singularity at the horizon, τ has to be compactified on a circle with a periodicity at radial infinity,

$$\tau \sim \tau + \beta,\tag{6.5}$$

where $\beta = \hbar/(k_B T_H)$ and T_H is precisely equal to the Hawking temperature. Quantum fields at radial infinity are therefore described by a *thermal* equilibrium field theory characterised by an imaginary-time axis that is compactified on a circle with a radius $R_\tau = \hbar/(k_B 2\pi T_H)$. This means that their apparent temperature is "just" T_H and one recovers the black-body radiation without needing to do any further computation – all one needs to know is the geometry of the imaginary-time axis.

Before Hawking's discovery Bekenstein had already realised that thermodynamic insights indicated that black holes could not be completely featureless. Given the process of absorption of macroscopic matter by a black hole and the validity of the second law, a Bekenstein entropy had to be associated with the black hole. This entropy is equal to

$$S_{BH} = \frac{k_B}{c\hbar} 4\pi G M^2 = 4\pi G M^2,\tag{6.6}$$

again in explicit and natural units. This turned out to be in perfect synchrony with the explicit computation of the entropy on the basis of Hawking's considerations.

The salient feature of this Bekenstein–Hawking (or black-hole) entropy is that it can also be expressed as

$$S_{BH} = k_B \frac{A}{4\ell_P^2}\tag{6.7}$$

where A is the horizon area and ℓ_P is the four-dimensional Planck length $\ell_P = \sqrt{G\hbar/c^3}$. Any "normal" material system has an extensive entropy that grows as the volume of the space it occupies, but the Bekenstein–Hawking entropy tells us, however, that for black holes one should associate an entropy with the enveloping "horizon" area, divided into cells with a size set by the Planck length, where every cell contains one bit of information. This is the counterintuitive area law at the basis of the holographic principle [167, 168] discussed in chapter 4: *the number of degrees of freedom of a gravitational system in $D + 1$ space-time dimensions can be counted in a non-gravitational field-theoretical system living in D dimensions.* It is believed to be a generic property of quantum gravity, rather than being merely a specific characteristic of a black hole.

6.1.2 Black holes in AdS and AdS/CFT thermodynamics

After these preliminaries, let us now turn to the subject of this chapter: how does holography deal with finite-temperature physics in the boundary field theory? The punch line of the formalism is that the outcomes for the thermal physics in the boundary are precisely coincident with the rules of "classic" black-hole physics outlined above, but now for a black hole in an AdS space-time.

Although these insights from 1970s black-hole physics played an important guiding role in the discovery of the AdS/CFT correspondence, it is in a way remarkable that it works so well. After all, to a degree one is comparing apples and pears. In the "classic" black-hole physics one considers the effects of a classical black-hole space-time on the vacuum of a quantised field theory living in that space-time. In the AdS/CFT correspondence, the bulk is not only a classical space-time but the fields living in this space-time are classical as well (at least, in the large-N limit of the field theory). In the strict classical limit, there is no such thing as a Hawking temperature in this bulk, but the projection of the bulk geometry on the boundary makes the field theory on the "holographic screen" behave precisely as it should at a finite temperature coincident with the one computed by Hawking.

We will now show that this follows directly from the fundamentals introduced in chapter 5 and present explicit recipes for computing the temperature (box 6.1) and free energy/entropy (box 6.2) using the holographic dictionary. The GKPW rule can be effortlessly combined with the requirements of thermal field theory. For matter in equilibrium in the boundary, one can address the stationary gravity problem in the bulk in Euclidean signature. The temperature of the boundary field theory is then coincident with the Hawking temperature of the black hole in the bulk (box 6.1). In the final section, section 6.4, we will show that in Minkowski space real-time finite-temperature correlation functions of the field theory are impeccably reproduced by the GKPW rule, by choosing appropriate boundary conditions at the black-hole horizon.

In the same way, the entropy at the boundary equals the Bekenstein entropy of the black hole, and one finds this correspondence for the whole of the black-hole thermodynamics. The only difference is that one has to pay tribute to the AdS asymptotics upon computing the geometry in the bulk, which does alter matters significantly, compared with the flat asymptotics of the "classic" black-hole physics.

Similarly to the Minkowski-space example, the Hawking temperature of an AdS black hole can be computed from the Euclidean gravity by demanding that the Euclidean black-hole solution should be smooth, as is explained in box 6.1. As for flat Minkowski asymptotics, one finds a time circle with a radius that increases as one moves away from the horizon along the radial direction, as shown in Fig. 6.1. On making the adjustment that the boundary field theory has the natural metric $d\tilde{s}^2 = \eta_{\mu\nu} \, dx^\mu \, dx^\nu$, whereas a section of AdS at constant r has metric $ds^2 = (r^2/L^2)\eta_{\mu\nu} \, dx^\mu \, dx^\nu$, at *conformal infinity* ($r \rightarrow \infty$, where one divides out the overall r^2/L^2 factor), the Euclidean time of the black-hole bulk is "shared" with

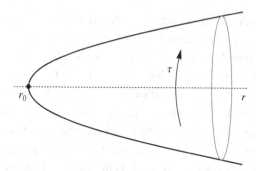

Figure 6.1 The Euclidean geometry of a black-hole space-time $ds^2 = g_{tt}(r)d\tau^2 + g_{rr}(r)dr^2$, where g_{tt} vanishes at the horizon $r = r_0$. In order that the geometry is smooth at $r = r_0$, τ has period $4\pi \sqrt{g'_{tt}(r_0)g^{rr\prime}(r_0)}$, where a prime denotes a derivative w.r.t. r and $g^{rr} = 1/g_{rr}$. The proper length of the perimeter of the circle at r is $\ell_{\text{perimeter}} = \sqrt{g_{tt}(r)}4\pi/\sqrt{g'_{tt}(r_0)g^{rr\prime}(r_0)}$. In the case in which the original Lorentzian space-time asymptotically approaches flat Minkowski space, the proper length approaches a constant and the above geometry looks like a cigar. In the asymptotically AdS case, the proper length diverges as $\ell_{\text{perimeter}} \sim r$. This is not in conflict with AdS/CFT since the boundary of AdS is conformal only to the flat space-time on which the boundary field theory lives.

that of the boundary and here the radius of the time circle is still finite, implying that the boundary field theory lives at a finite temperature. However, the change in the asymptotic behaviour of the space-time plays a very important role. The classic Hawking result for the Schwarzschild black hole for *flat* asymptotics shows that the temperature is inversely proportional to the horizon radius: small black holes are very hot and large black holes are cold. For anti-de Sitter asymptotics the opposite holds. The explicit computation (box 6.1) demonstrates that the temperature in the boundary equals, in explicit units,

$$T = \frac{\hbar c}{k_{\text{B}}} \frac{(d+1)r_0}{4\pi L^2} \tag{6.8}$$

in terms of the horizon radius r_0, the number of space-time dimensions $d + 2$ and the AdS radius L. We thus learn the following:

Rule 7 A black hole in the deep interior of the AdS bulk encodes finite temperature in the field theory with the temperature equal to the Hawking temperature of this black hole. In AdS this Hawking temperature increases proportionally to the radius of the horizon.

This growth of the AdS black-hole horizon with temperature is elegantly in synchrony with the notion that the radial direction is coincident with the scaling direction of the field theory. The perfect scale invariance in the field theory at zero temperature is embodied by the pure AdS metric. At finite temperature the "infrared" part of the geometry in the deep interior is disturbed by the black-hole horizon, and this effect increases, moving to higher and higher energy scales when the temperature rises. This encodes the finite-size scaling effect of temperature in the field theory in a geometrical language. Moreover, as we will discuss here and in the next chapter, the point-of-no-return nature of the black-hole horizon gives it precisely the distinguishing character of thermal physics. The irreversible nature of the dynamics of the finite-temperature boundary system obeying the second law of hydrodynamics is encoded in the fact that the modes of the classical fields propagating in the bulk will eventually fall through the horizon. These bulk modes thereby acquire a finite lifetime (the "quasinormal modes [213]"), which in turn encodes for the origin of dissipation.

Box 6.1   The Hawking temperature of the AdS–Schwarzschild black hole from Euclidean gravity

To compute the temperature of AdS black holes, we need to know the AdS metric. The Einstein equation for a $(d+2)$-dimensional Minkowski-signature space-time with a negative cosmological constant is [214]

$$R_{\mu\nu} - \frac{1}{2} g_{\mu\nu} R - \frac{d(d+1)}{2L^2} g_{\mu\nu} = 0. \qquad (6.9)$$

This is solved by an AdS–Schwarzschild black-hole background with the metric

$$ds^2 = \frac{r^2}{L^2}(-f(r)dt^2 + d\Sigma_k^2) + \frac{L^2}{r^2 f(r)} dr^2, \quad i = 1, \ldots, d, \qquad (6.10)$$

where L is the AdS radius and r is the radial direction, while the emblackening factor is

$$f(r) = 1 - \frac{m}{r^{d+1}} + \frac{kL^2}{r^2} \qquad (6.11)$$

and

$$d\Sigma_k^2 = \begin{cases} L^2 d\Omega_d^2, & \text{for } k = 1, \text{ spherical horizon,} \\ \sum_{i=1}^{d} dx_i^2, & \text{for } k = 0, \text{ flat, planar horizon,} \\ L^2 dH_d^2, & \text{for } k = -1, \text{ hyperbolic horizon,} \end{cases} \qquad (6.12)$$

Box 6.1   (Continued)

where $d\Omega_d^2$ is the unit metric on the sphere S^d and dH_d^2 is the unit metric on the d-dimensional hyperbolic space $\mathbb{H}^d$.

The solution to $f(r) = 0$ gives us the black horizon, and we will pick up the outer horizon $r = r_0$; thus

$$m = r_0^{d+1} + kL^2 r_0^{d-1}. \tag{6.13}$$

Note that the horizon spanned by the space at a fixed t and $r = r_0$ is a flat d-dimensional Euclidean space $\mathbb{R}^d$ for $k = 0$, a sphere for $k = 1$ and hyperbolic space for $k = -1$. To compute the temperature, we use the Euclidean-gravity insight from Gibbons and Hawking. Consider a general static black-hole metric,

$$ds^2 = -g_{tt}(r)dt^2 + \frac{dr^2}{g^{rr}(r)} + g_{xx}(r)d\vec{x}^2, \tag{6.14}$$

where $g_{tt}(r)$ and $g^{rr}(r)$ have a single zero at the horizon r_0. Upon Wick rotating to Euclidean signature $\tau = it$, one finds

$$ds_E^2 = g_{tt}(r)d\tau^2 + \frac{dr^2}{g^{rr}(r)} + g_{xx}(r)d\vec{x}^2. \tag{6.15}$$

Let's make the natural assumption that the properties of the black hole are reflected in the geometry near the horizon where g_{tt} and g^{rr} are vanishing (because they are proportional to the emblackening factor). To focus on this region we expand the metric near $r = r_0$, giving $g_{tt}(r) = g'_{tt}(r_0)(r - r_0) + \cdots$, $g^{rr}(r) = g^{rr\prime}(r_0)(r - r_0) + \cdots$ and $g_{xx}(r) = g_{xx}(r_0) + \cdots$, where the prefactors, the radial derivatives $g'_{tt}(r_0)$ and $g^{rr\prime}(r_0)$ evaluated at the horizon, are now just numbers. Thus the near-horizon (Euclidean) metric equals

$$ds_E^2 = g'_{tt}(r_0)(r - r_0)d\tau^2 + \frac{dr^2}{g^{rr\prime}(r_0)(r - r_0)} + g_{xx}(r_0)d\vec{x}^2 + \cdots. \tag{6.16}$$

It is now convenient to re-parametrise the radial direction in terms of a new variable $R_0 = 2\sqrt{r - r_0}/\sqrt{g^{rr\prime}(r_0)}$. The metric becomes

$$ds_E^2 = \frac{1}{4}R_0^2 g'_{tt}(r_0)g^{rr\prime}(r_0)d\tau^2 + dR_0^2 + g_{xx}(r_0)d\vec{x}^2 + \cdots. \tag{6.17}$$

What matters is the plane spanned by this R_0 and the imaginary-time direction τ. This is just the metric of a plane in polar coordinates with τ acting as the compact angular direction. Upon approaching the horizon $R_0 \to 0$, one sees that the prefactor of $d\tau^2$ vanishes: this means that the Euclidean time direction shrinks to a point. However, since the horizon is not a special point, we should not allow this point to be singular. Smoothness at the horizon can be achieved by insisting that $R_0 = 0$ is the centre of a Euclidean polar coordinate system, and this implies that τ is periodic with period $4\pi/\sqrt{g'_{tt}(r_0)g^{rr'}(r_0)}$. Figure 6.1 illustrates what this Euclidean space-time with its periodic imaginary-time direction looks like as a function of the radial direction. This periodicity is directly identified with the inverse temperature of the black hole because the time coordinate is identified as the one in dual field theory.

For the AdS–Schwarzschild black hole (6.10), we thus find

$$
T = \frac{\sqrt{g'_{tt}g^{rr'}}}{4\pi}\bigg|_{r=r_0} = \frac{r_0^2}{4\pi L^2}\frac{df(r)}{dr}\bigg|_{r=r_0}
$$

$$
= \frac{1}{4\pi L^2}\left((d+1)r_0 + \frac{kL^2(d-1)}{r_0}\right). \tag{6.18}
$$

In this book we will mostly focus on the case $k = 0$, and we study the $k = 1$ case solely in the next section (section 6.2). For the $k = 0$ case, it is more appropriate to call it a *black-(mem)brane* solution rather than a black hole. One can also call it a black hole with a planar horizon. We will call it a "black hole" without causing any confusion. For convenience, we write out the case of a black hole with a planar horizon explicitly for future references. The metric is

$$
ds^2 = \frac{r^2}{L_i^2}\left(-f(r)dt^2 + dx_i^2\right) + \frac{L^2}{r^2 f(r)}dr^2, \quad i = 1, \ldots, d, \tag{6.19}
$$

with

$$
f(r) = 1 - \frac{r_0^{d+1}}{r^{d+1}}, \tag{6.20}
$$

and the corresponding Hawking temperature is

$$
T = \frac{(d+1)r_0}{4\pi L^2}. \tag{6.21}
$$

AdS/CFT thermodynamics

The next obvious quantity to compute in the thermal ensemble is the free energy. The GKPW rule suggests that this should be very simple since the partition function of the CFT at large-N is directly related to the on-shell action of the gravitational bulk. This is true, but with a subtlety that we have already seen in the previous chapter: given that AdS can be considered as a space with a boundary, one has to introduce a boundary contribution to the action. In the context of gravitational dynamics this boundary action is determined by the requirement that one has a well-defined variational principle where the boundary metric is held fixed. The free energy of the field theory also receives a contribution from these so-called "Gibbons–Hawking–York" boundary terms when they are evaluated for the particular geometry realised in the bulk. We will go through this procedure in detail in box 6.2.

The AdS path integral in the presence of a black-hole background computes the finite-temperature physics of a boundary conformal field theory. This is special in that in the conformal field theory there are no internal energy scales. The temperature dependence of the thermodynamic variables is therefore completely determined by (hyper)scaling. In particular,

$$F = \alpha \frac{1}{d+1} T^{d+1}, \quad U = \alpha \frac{d}{d+1} T^{d+1}, \quad S = \alpha T^d. \tag{6.22}$$

On computing the holographic free energy departing from the AdS–Schwarzschild black hole this is indeed confirmed, and it is found that the entropy density equals (box 6.2)

$$s = \frac{S}{\text{Vol}_d} = \frac{1}{4G} \frac{k_B c^3}{\hbar} \frac{r_0^d}{L^d} \tag{6.23}$$

in explicit units in terms of the quantities G (Newton's constant), L (the AdS radius) and r_0 (the horizon radius). Using the relation Eq. (6.8) between the horizon radius and the temperature, this translates into

$$s = \frac{1}{4G} \frac{k_B c^3}{\hbar} \frac{k_B^d}{\hbar^d c^d} \left(\frac{4\pi}{d+1} \right)^d L^d T^d = \frac{1}{4G} \left(\frac{4\pi}{d+1} \right)^d L^d T^d \tag{6.24}$$

in both explicit and natural units.

Now note that, for AdS black-hole metrics of the form (6.19), the horizon area can be directly computed as,

$$A = \int d^d \mathbf{x} \sqrt{\det g_{ij}} \Big|_{r=r_0}, \tag{6.25}$$

where i, j run over all directions except t and r and the integral is over the surface located at the horizon, $r = r_0$. For an AdS–Schwarzschild black hole the answer is simply

$$A_{\text{AdS–Schwarzschild}} = \frac{r_0^d}{L^d} \text{Vol}_d. \tag{6.26}$$

On combining the result for the area with the entropy density derived from the free energy (6.23), it follows that the entropy itself can be written as

$$S = \frac{1}{4G}\frac{k_B c^3}{\hbar}A = k_B \frac{1}{4}\frac{A}{\ell_P^d}.$$ (6.27)

This is exactly the famous Bekenstein–Hawking black-hole entropy, including the factor $1/4$. This demonstrates how the AdS/CFT correspondence holographically gives substance to the microscopic foundation of black-hole thermodynamics.

We do still have to express the entropy in terms of the quantities natural to the field theory. We need to translate Newton's constant G and the AdS radius L specifying the gravitational bulk into properties of the boundary field theory. This part of the dictionary requires an explicit top-down model. For Maldacena's original example of $(3+1)$-dimensional, $\mathcal{N}=4$ super-Yang–Mills theory, the following relation between the gravitational quantities and those of the field theory holds:

$$\frac{c^3}{\hbar}\frac{L^3}{8\pi G} = \frac{N^2}{4\pi^2}\left(1+\mathcal{O}\left(\frac{1}{N}\right)\right).$$ (6.28)

As discussed in chapter 4, this identification is at the heart of the duality nature of AdS/CFT. In order for us to be able to trust classical gravity, the curvature in the bulk has to be small $(L^3/G \gg 1)$ in natural units, and this goes hand in hand with $N^2 \gg 1$ for the boundary field theory.

On combining Eqs. (6.8), (6.23) and (6.28) we arrive at the answer for the entropy of the large-N CFT. Explicitly the entropy density in large-N $\mathcal{N}=4$ super-Yang–Mills at strong coupling equals

$$s = \frac{\pi^2}{2}N^2\frac{k_B^4}{\hbar^3 c^3}T^3.$$ (6.29)

We thus arrive at the next dictionary entry:

Rule 8   The entropy of the boundary field theory is equal to the Bekenstein–Hawking entropy of the black hole in the bulk and is determined by the horizon area of this black hole. To establish its absolute magnitude, the gravitational units have to be converted into field-theoretical quantities and this requires top-down information.

In the particular example we have computed, the physics information is fully contained in the prefactor. We have already noted that for a conformal field theory the dependence on temperature follows directly from scaling in a manner that is similar to the T^d behaviour of the Debye entropy associated with e.g. acoustic phonons, or the Stefan–Boltzmann law for radiation. Furthermore, as we shall see in the next section, the N^2 factor is characteristic of the deconfined phase. Despite being a number, this prefactor does contain a very interesting result. It reveals the truly interacting nature of the dual AdS description. The theory of which we just computed the entropy using the canonical top-down AdS/CFT duality – maximally supersymmetric Yang–Mills theory at strong 't Hooft coupling $\lambda = g^2 N$ – is surely among the most strongly coupled CFTs that can be imagined. Now compare this entropy with the result of the *free*-field theory. This is trivial since it simply counts the number of massless modes. The free-field content of $\mathcal{N} = 4$ super-Yang–Mills in $d = 3 + 1$ dimensions is one vector field, four real fermions and six real scalars, each in the adjoint representation of $U(N)$, i.e. with multiplicity N^2. The result is $s_{\lambda=0} = \frac{4}{3}(\pi^2/2)N^2T^3$ in natural units [215]. Both the N^2 dependence and the T^3 dependence arise for the same qualitative reasons as before. Note that the result is the usual relation for a free gas of massless particles $s = (n_b + \frac{7}{8}n_f)\frac{1}{6}\sigma_B T^3$ times the number of physical polarisations $n_b = n_f = 8$ and a degeneracy factor of N^2, where σ_B is the Stefan–Boltzmann constant. This gives a different answer from the holographic computation, showing that the holographic answer is not that of a free-field theory. Remarkably, the prefactor in the interacting case is nearly the same: $s_{\lambda=0}/s_{\lambda=\infty} = 4/3$. Apparently the thermodynamic potentials of such zero-density quantum critical states vary quite slowly as functions of the coupling strength as illustrated in Fig. 6.2. Historically this comparison of entropies was done even

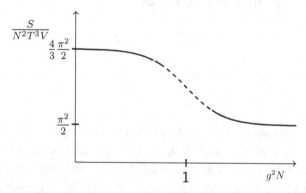

Figure 6.2 The entropy of the gauge-theory plasmas of $\mathcal{N} = 4$ super-Yang–Mills as a function of the 't Hooft coupling in the large-N limit. Owing to the symmetries, the entropy must be proportional to N^2T^3V. The numerical prefactor does change as a function of the coupling. It interpolates between $4\pi^2/6$ in the free theory and $\pi^2/2$ in the strongly interacting theory. (Figure source [13].)

before the full dictionary was established at a time when the role of λ was not yet fully understood [216] and played an important guiding role in constructing the duality.

Box 6.2   Computing the holographic free energy and entropy of the CFT

The starting point is the defining AdS/CFT relation Eq. (5.1) between the partition sum of the large-N strongly coupled CFT and the on-shell action of the bulk (classical) gravity. This identity holds equally well for equilibrium field theory at a finite-temperature dual to Euclidean Einstein theory for the black hole in the deep interior of AdS. The free energy of the field theory is thus given by the dictionary entry

$$F = -k_B T \ln \mathcal{Z}_{CFT} = k_B T S_E^{AdS}[g_E], \qquad (6.30)$$

where g_E is the Euclidean metric (6.15) and $S_E^{AdS}[g_E]$ is the on-shell Euclidean Einstein–Hilbert action: notice that temperature is now entirely encoded in the bulk through the Euclidean black-hole geometry. Because AdS should be considered as a space-time with a boundary, even though this boundary is infinitely far away, there are *two* subtleties one must take into account.

- To have a well-defined variational problem for fluctuations of the bulk metric, given the fixed boundary metric, we must insist that after integration by parts there are no boundary terms proportional to $\partial_r \delta g_{\mu\nu}$ (we are imposing Dirichlet boundary conditions on the metric). Gibbons, Hawking and York showed that to do so one has to supplement the action with a "Gibbons–Hawking–York" boundary term [217, 218].
- Even with this Gibbons–Hawking–York term the on-shell action is divergent in the sense that, if one regulates by locating the boundary at a fixed location $\bar{r}$, the on-shell action scales as a positive power of $\bar{r}$. This is in fact the same divergence as that which appears when one computes the two-point function of the stress–energy tensor as outlined in the previous chapter. Likewise this divergence can be cancelled out by a local boundary counter-term. Rather than performing the fluctuation analysis, there is a more straightforward way to determine it in this case by demanding that the on-shell action vanishes for pure AdS [219].

The last set of counter-terms can be expanded in terms of the intrinsic curvature of the boundary. For the AdS–Schwarzschild black brane considered

Box 6.2   (Continued)

here, only the lowest-order term is relevant. The result is that for our purposes the correct gravitational action to consider equals

$$S = S_{\text{bulk}} + S_{\text{GHY}} + S_{\text{ct}}, \tag{6.31}$$

where

$$S_{\text{bulk}} = -\frac{1}{2\kappa^2} \int_0^\beta d\tau \int_{r_0}^\infty dr \int_{-\infty}^\infty d^d x_i \sqrt{g_{\text{E}}} \left(R + \frac{d(d+1)}{L^2} \right),$$

$$S_{\text{GHY}} = \frac{1}{2\kappa^2} \int_0^\beta d\tau \int_{r\to\infty} d^d x_i \sqrt{h}(-2K), \tag{6.32}$$

$$S_{\text{ct}} = S_{\text{ct}}^{(0)} + \cdots,$$

in which

$$S_{\text{ct}}^{(0)} = \frac{1}{2\kappa^2} \int_0^\beta d\tau \int_{r\to\infty} d^d x_i \sqrt{h} \frac{2d}{L}, \tag{6.33}$$

with r_0 the location of the horizon. In these formal expressions, $h_{\mu\nu}$ is the induced metric on the boundary at $r \to \infty$. It is defined as $h_{\mu\nu} = g_{\mu\nu} - n_\mu n_\nu$, with n^ν an outward unit vector normal to the boundary. n^μ is thus a null eigenvector of $h_{\mu\nu}$ and its determinant is by definition taken only in directions orthogonal to n^μ. The quantity $K = h^{\mu\nu} \nabla_\mu n_\nu$ is the trace of the extrinsic curvature of the induced metric. These quantities were introduced in box 4.4, and here we will take the AdS–Schwarzschild black-hole geometry of Eqs. (6.19) and (6.20) as an example to illustrate the computation. The Euclidean AdS–Schwarzschild black-hole geometry is

$$ds_{\text{E}}^2 = \frac{r^2}{L^2} \left(f(r)d\tau^2 + dx_i^2 \right) + \frac{L^2}{r^2 f(r)} dr^2, \quad \iota = 1, \ldots, d. \tag{6.34}$$

The determinant and Ricci scalar for this metric are

$$g_{\text{E}} = \left(\frac{r^2}{L^2} \right)^d, \quad R = -\frac{(d+2)(d+1)}{L^2}. \tag{6.35}$$

The outward-pointing normal vector n^μ is clearly in the radial direction. To ensure that it has unit length, i.e. $n^\mu n^\nu g_{\mu\nu} = 1$, one must have $n^\mu = (0, \ldots, 0, r\sqrt{f}/L)$, with n^r the only non-zero component. The induced metric is then

$$h_{\mu\nu} = \text{diag}\left\{ \frac{r^2 f}{L^2}, \frac{r^2}{L^2}, \ldots, \frac{r^2}{L^2}, 0 \right\}, \tag{6.36}$$

with (reduced) determinant $h \equiv \det h_{ij\ (i,j \neq r)} = (r^2/L^2)^{d+1} f$. Since the induced metric is orthogonal to the normal vector, the trace of the extrinsic curvature reduces to

$$K = -h^{\mu\nu} \gamma^{\alpha}_{\mu\nu} n_{\alpha},$$

(6.37)

where $\gamma^{\alpha}_{\mu\nu} \equiv \frac{1}{2} g^{\alpha\beta} (\partial_{\mu} g_{\beta\nu} + \partial_{\nu} g_{\beta\mu} - \partial_{\beta} g_{\mu\nu})$ is the Christoffel connection. On substituting the AdS–Schwarzschild metric the trace of the extrinsic curvature of the boundary is thus

$$K = n_r \left(-h^{\tau\tau} \gamma^r_{\tau\tau} - \sum_{i=1}^{d} h^{x_i x_i} \Gamma^r_{x_i x_i} \right)$$

$$= \frac{\sqrt{f}}{L} \left[(d+1) + \frac{r}{2} \frac{f'}{f} \right].$$

(6.38)

By substituting Eqs. (6.20), (6.35) and (6.37) into Eqs. (6.32) and (6.40) one finds a remarkably simple result for the free energy, namely

$$F = \frac{1}{\beta} S_{\mathrm{E}}[g_{\mathrm{E}}] = -\frac{1}{2\kappa^2} \frac{r_0^{d+1}}{L^{d+2}} \mathrm{Vol}_d,$$

(6.39)

where $\mathrm{Vol}_d = \int d^d x_i$ is the volume of the d-spatial-dimensional volume of the boundary field theory. Using the relation Eq. (6.8) between the horizon radius and the temperature, this becomes

$$F = -\frac{2\pi L^d}{\kappa^2} \left(\frac{4\pi}{d+1} \right)^d \frac{T^{d+1}}{d+1} \mathrm{Vol}_d.$$

(6.40)

The internal energy and the entropy for the dual field theory follow immediately.

The first law of black-hole thermodynamics, energies and Fefferman–Graham coordinates

This computation also serves as an illustration of the first law of black-hole thermodynamics, which is the foundation both of holography and of the AdS/CFT correspondence. We can independently compute the entropy from the area of the horizon,

$$S_{\mathrm{Area}} = \frac{2\pi}{\kappa^2} \frac{r_0^d}{L^d} \mathrm{Vol}_d = \frac{2\pi}{\kappa^2} \frac{(4\pi L)^d}{(d+1)^d} T^d \mathrm{Vol}_d,$$

(6.41)

which matches the entropy obtained from the free energy, and we can compute the energy using the GKPW rule from chapter 5. The expectation value of the energy should be the sub-leading part of the g_{tt} component of the

Box 6.2   (Continued)

metric. The subtle part here is that g_{tt} is not gauge-invariant: it depends on the choice of coordinates. There is a specific set of coordinates, called Fefferman–Graham coordinates – denoted by r, essentially the gauge $g_{r\mu} = 0$ – where the identification of the sub-leading part of g_{ij} with the boundary expectation value of T_{ij} is directly true [200, 204, 220]. These coordinates are therefore of the form

$$ds^2_{\text{FG}} = \frac{L^2}{r^2} dr^2 + \frac{r^2}{L^2} \tilde{g}_{ij}(x, r) dx^i \, dx^j, \tag{6.42}$$

with the metric $\tilde{g}_{ij}$ having the asymptotic form

$$\tilde{g}_{ij}(x, r) = \eta_{ij} + \frac{\tilde{g}^{(2)}_{ij}}{r^2} + \cdots + \frac{\tilde{g}^{(d+1)}_{ij}}{r^{d+1}} + \frac{\tilde{h}^{(d+1)}_{ij} \ln r^2}{r^{d+1}} + \cdots, \tag{6.43}$$

where the logarithmic terms appear only when d is even and "$\cdots$" are terms of higher order. The terms with power higher than r^{d+1} are all accounted for by the counter-terms in the boundary action. The term from which the stress tensor can be read off is the term with power r^{d+1} corresponding to the appropriate dimension of the stress tensor. One has

$$\langle T_{ij} \rangle = \frac{d+1}{2\kappa^2} \tilde{g}^{(d+1)}_{ij}. \tag{6.44}$$

On transforming the AdS–Schwarzschild black-hole metric Eq. (6.19) into these coordinates through $r(r) = r[(1 + \sqrt{f(r)})/2]^{2/(d+1)}$ (equivalently $r(r) = r[1 + r_0^{d+1}/(4r^{d+1})]^{2/(d+1)}$), one finds

$$ds^2_{\text{FG–AdS–BH}} = \frac{L^2}{r^2} dr^2 + \frac{r^2}{L^2} \left[\left(1 + \frac{r_0^{d+1}}{4r^{d+1}}\right)^{4/(d+1)} dx^2 \right.$$

$$\left. - \frac{(1 - r_0^{d+1}/r^{d+1})^2}{[1 + r_0^{d+1}/(4r^{d+1})]^{2(d-1)/(d+1)}} dt^2 \right]. \tag{6.45}$$

The energy density thus equals

$$\epsilon = \langle T^{00} \rangle = \frac{d}{2\kappa^2} r_0^{(d+1)} = \frac{d}{2\kappa^2} \frac{(4\pi)^{d+1} L^{2(d+1)}}{(d+1)^{d+1}} T^{d+1}. \tag{6.46}$$

One readily verifies the (integrated) first law,

$$F = E - TS, \tag{6.47}$$

where $E = \epsilon \, \text{Vol}_d$ and $dE = T \, dS$.

6.2 Holographic thermodynamics: the Hawking–Page transition

Conformal invariance is an extremely powerful symmetry that strongly constrains the physics: all two-point correlation functions are fixed by their scaling dimensions, and so forth. Even at finite temperature the only meaningful thermodynamic property is the prefactor of the entropy (a universal amplitude) as discussed in the previous section. However as soon as we add another scale – in particular when we turn on a finite density – this "remnant" of conformal invariance at finite temperature will be broken. In chapter 8 onwards we will use the fact that, with such a second scale, holography can be used to compute rather rich phase diagrams containing both known stable phases of matter (such as superconductors and Fermi liquids), in addition to novel emergent quantum critical phases, and phase transitions between them.

Here we will make a start on addressing this theme, relying on the simplest way one can imagine to break the conformal invariance with an additional scale: let the boundary field theory live in a finite volume. Naively this just gaps the theory, but, as discovered by Witten [211], the story is much more intricate. It reveals a gravitational *Hawking–Page phase transition* in the gravitational dual, which describes a thermal *confinement–deconfinement* transition in the finite-volume boundary theory. The confinement phenomenon, which will constitute an important theme in the chapters to come, is familiar from a Yang–Mills theory like QCD. At low temperature and energies the quarks and gluons do not appear as asymptotic states but are confined in gauge singlets (baryons, mesons). Only by heating up the QCD vacuum to temperatures above the confinement scale (this is achieved in relativistic heavy-ion collisions that produce the quark–gluon plasma) or by high-energy collisions (in more conventional particle accelerators) can one discern that elementary quarks and gluons exist; at sufficiently high energy they behave more and more like nearly free particles due to asymptotic freedom of the coupling. But which is the physics in the context of the strongly interacting super-conformal Yang–Mills theories described by the present form of holography?

A first clue follows from the entropy computed in the previous section. We found that this scales with N^2, and, from the discussion in chapter 4, it is obvious that this counts the size of the Yang–Mills matrices. These in turn count the number of gluons, and the entropy can know about these degrees of freedom only when they are deconfined. It follows immediately that any finite-temperature state that is described by a Schwarzschild black hole with a finite horizon area corresponds to a deconfining state of the field theory.

A confining state, on the other hand, has an entropy of order unity: the order-N^2 "partons" combine in a single gauge-singlet, confined degree of freedom. Given that we rely on classical gravity that captures only the leading large-N contribution, such order-one contributions to the entropy should be invisible, and a finite-temperature confining state should therefore exhibit a vanishing holographic entropy. We thus face the following puzzle. Is there an alternative finite-temperature state in the boundary that is different from a black hole in the bulk, since the latter comes with a horizon representing an order N^2 entropy?

It is immediately clear that one needs a scale for confinement, and conformal invariance somehow has to be broken. This scale should act in the infrared. The easiest way to introduce such an IR scale is to put the system into a box. Up to this point we have always assumed that the boundary theory lives in an infinitely large flat space; given that at finite temperature the topology of the Euclidean time circle is S^1, this gives us a total topology of space-time of $S^1 \times \mathbb{R}^d$. It is in principle straightforward, however, to put the boundary theory in a finite volume. Geometrically it will be easier for us to choose a sphere rather than a box, but the qualitative physics will be the same. We therefore constrain the spatial directions to a d-dimensional sphere, yielding an overall topology of $S^1 \times S^d$. This is completely compatible with the AdS geometry of the bulk, if the radius of the sphere is the same as the radius of AdS itself. Recall from section 4.4 that in so-called global coordinates the boundary of AdS has precisely this topology.

Given this compact boundary geometry, what are the possible solutions in the Euclidean-signature Einstein equations for a negative cosmological constant? A famous result for "global AdS" goes back to the heyday of general relativity. In the 1980s Hawking and Page proved that there are exactly two (isotropic) solutions [221]. One is the finite Euclidean time-circle black hole but now with a spherical horizon – a proper black hole, therefore, rather than the planar flat $\mathbb{R}^d$ horizon black hole or black "brane" we have discussed until now. The other solution is just *pristine* global AdS with its compact spatial topology but imaginary time compactified to a circle. This is called "thermal AdS": it represents a finite-temperature state in the finite volume boundary, without the presence of a black-hole horizon. Since there is no horizon, there is no macroscopic order-N^2 entropy suggesting that thermal AdS is dual to a finite-temperature confining state in the boundary.

Which solution should one choose? One should compare the boundary free energies of the two competing states as a function of the spatial compactification radius and temperature to find out which state is thermodynamically preferred. Employing the machinery of box 6.2, we illustrate in detail how this works for the Hawking–Page system in box 6.3. The result is shown in Fig. 6.3. At high temperatures, when the radius of the imaginary time circle is small compared with the radius of the compact spatial directions of the boundary, the Schwarzschild-type solutions are thermodynamically preferred. This confirms the intuition that at high temperature the field theory is deconfined. On the other hand, when the time circle becomes larger than the spatial radius, the pristine AdS geometry extending all the way to the deep interior is the preferred solution. Since the time circle is still finite, this represents a confining thermal state in the field theory.

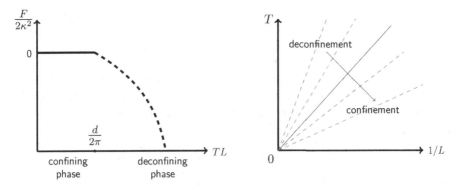

Figure 6.3 The Hawking–Page phase transition. One can consider the boundary field theory in a finite volume by placing it on a spatial sphere with radius L. Within AdS/CFT this can be naturally accounted for by using "global AdS space" in the bulk instead of the Poincaré patch. It can be proven that there are only two static homogeneous solutions to Einstein's equations in global AdS: either the Schwarzschild black-hole-type solution or a thermal AdS geometry with a finite imaginary-time circle. By computing the boundary/black-hole free energy Hawking and Page showed already in the 1970s that thermal AdS is stable at low temperature (left-hand panel), and undergoes a first-order phase transition to a Schwarzschild black-hole state at higher temperature. The physics behind this transition can be made clear with AdS/CFT. Holography shows both from the entropy and from a Wilson-loop computation that thermal AdS describes a confining state that undergoes a first-order transition to a high-temperature deconfining state at a temperature $T \propto 1/L$ (right-hand panel).

Box 6.3   The Hawking–Page thermal phase transition between the Schwarzschild black hole and thermal AdS space-time

As we emphasised, to introduce a second scale in the boundary field theory in addition to the temperature, the spatial dimensions are compactified on a sphere. The bulk black-hole solution with a spherical horizon which we are considering is Eq. (6.10) with $k = 1$,

$$ds^2 = V(r)d\tau^2 + \frac{dr^2}{V(r)} + r^2 \, d\Omega_d^2. \qquad (6.48)$$

The difference from the AdS–Schwarzschild black-brane metric Eq. (6.19) is the last term. Instead of flat space we have $d\Omega_d^2$, the line element of a d-dimensional sphere with unit radius; for convenience we have set the AdS radius to $L = 1$. The asymptotic AdS solution can be written as

$$V(r) = 1 + \frac{r^2}{L^2} - \omega_d \frac{M}{r^{d-1}}, \qquad \omega_d = \frac{2\kappa^2}{d\,\mathrm{Vol}(S^d)}. \qquad (6.49)$$

Box 6.3   (Continued)

For $M = 0$ this is the pristine global AdS solution of chapter 4; "thermal AdS" just corresponds to choosing Euclidean time to be compact. For $M \neq 0$ this is the global AdS black hole, whose Euclidean periodicity is determined by demanding that the horizon, the largest zero of $f(r)$, is a coordinate singularity. It is worth pointing out that, in the scaling limit

$$t = \lambda t, r = \lambda^{-1} r, \theta = \lambda \theta \text{ for } d\Omega_d^2 \equiv d\theta^2 + \sin^2\theta \, d\Omega_{d-1}^2 \text{ with } \lambda \to 0, \quad (6.50)$$

the metric (6.48) changes to AdS_{d+2} in Poincaré coordinates with a flat $\mathbb{R}^d$ boundary. If we scale $M \to \lambda^{-d+1} M$ at the same time, we get the planar AdS–Schwarzschild black hole (6.19).

Note that the emblackening factor $V(r)$ for the global AdS black hole has multiple zeros. The outermost one, i.e. the larger solution of the equation

$$1 + \frac{r^2}{L^2} - \omega_d \frac{M}{r^{d-1}} = 0, \quad (6.51)$$

is the horizon r_0. Following the near-horizon prescription of box 6.1, one can deduce the periodicity in the time direction as

$$\beta = \frac{4\pi L^2 r_0}{(d+1)r_0^2 + (d-1)L^2}, \quad (6.52)$$

and its inverse gives the temperature of the black hole $T = 1/\beta$.

We can evaluate which solution is thermodynamically favoured by using the algorithm explained in box 6.2 to compute the free energies of the states of the boundary theory corresponding to these two different bulk geometries. A crucial extra ingredient is that, in order to compare these free energies, one has to insist that the bulk geometries describe precisely the same boundary space-time at the same temperature. This is manifestly not the same, since, due to the warping of space-time, energies can be blue/redshifted between different locations. On choosing a reference location in the form of a cut-off in the radial direction at $r = R$, one can express the imaginary-time periodicity of the thermal AdS β' in terms of the black-hole inverse temperature β such that both systems live in the same boundary space, with a boundary time circle of the same proper length. At the cut-off radius R the Euclidean time periodicities in each geometry are

$$\beta' \left(1 + \frac{R^2}{L^2}\right)^{1/2} = \beta \left(1 + \frac{R^2}{L^2} - \frac{\omega_d M}{R^{d-1}}\right)^{1/2}. \quad (6.53)$$

Taking the black-hole temperature β as the reference, we solve for β' in terms of β and R. One now computes the boundary field-theory free-energy duals to a black hole straightforwardly from the difference in the Euclidean actions

$$\beta F_{BH} = I_{ThAdS} - I_{BH}$$

$$= \frac{d}{\kappa^2 L^2} \lim_{R \to \infty} \left(\int_0^\beta dt \int_{r_0}^R dr \int_{S^d} d\Omega \, r^d - \int_0^{\beta'} dt \int_0^R dr \int_{S^d} d\Omega \, r^d \right)$$

$$= \frac{4\pi \, \mathrm{Vol}(S^d) r_0^d (L^2 - r_0^2)}{2\kappa^2 ((d+1)r_0^2 + (d-1)L^2)}, \tag{6.54}$$

where I_{ThAdS} and I_{BH} are the on-shell Euclidean actions for the thermal AdS and black hole solutions. Note that the free energy for the field theory dual to thermal AdS is zero. Plotting this difference as a function of T for fixed L shows a first-order transition from the deconfined phase at $T > d/(2\pi L)$ to a confined phase at $T < d/(2\pi L)$. See Fig. 6.3.

Long before AdS/CFT Hawking and Page had already noted that the free energy of the preferred state, which is obtained from the saddle-point approximation, changes as a function of the black-hole radius r_0 [221]. At low temperature – which is also the temperature in the field theory – the black hole has a radius smaller than the radius of the curvature of AdS ($r_0 < L$) and the thermal AdS has a lower free energy than the black-hole solution. As the temperature rises a sign change occurs in the difference between the free energies when r_0 becomes larger than L. This signals a first-order transition from the thermal AdS to a high-temperature state associated with the black hole geometry in the bulk: the Hawking–Page transition, that takes place at a critical temperature $T_c = d/(2\pi L)$. With the advent of AdS/CFT and the holographic meaning of the extra radial direction as the energy scale, the occurrence of this transition is even clearer. The horizon radius r_0 of the black hole can be literally thought of as exploring the energy landscape in the radial direction.

The real contribution of AdS/CFT is that we now understand what this phase transition means. From the purely gravitational perspective it appeared as a mystery. However, by computing the expectation value of a Wilson–loop through AdS/CFT, Witten showed that it unambiguously encodes for a confinement/deconfinement transition in the field theory [211]. Although Wilson loops have not yet played an important role in the finite-density physics discussed in the later chapters, their holographic encoding in the bulk is an interesting affair; it connects with the underlying string theory, while providing a beautiful exercise in the geometrical calculations of general relativity [222, 223, 224]. This is elaborated upon in further detail in box 6.4. Let us here summarise some qualitative features.

The confinement/deconfinement nature of a gauge-theory vacuum is measured through the expectation value of the Wilson-loop operator. The Wilson loop is defined by the trace of the path-ordered exponential of a gauge field A_μ,

$$W(C) = \text{Tr}\left[\mathcal{P}\exp\left(i\oint_C A_\mu\,dx^\mu\right)\right],\tag{6.55}$$

where C is a closed curve and $\mathcal{P}$ denotes the path-ordering operator. The trace is taken in the fundamental vector representation of the gauge field A_μ. The standard zero-temperature field-theory approach is as follows. In order to discern the potential between the quarks (sources of the gauge fields), one takes for the loop C a rectangle with one time direction and one space direction, with lengths L_t and L_s, respectively. In the limit $L_t \to \infty$ this zero-temperature Euclidean field-theory computation should give

$$\lim_{L_t\to\infty} \langle W(C)\rangle \sim e^{-L_t V(L_s)},\tag{6.56}$$

where $V(L_s)$ is the static potential for an infinite heavy-quark–antiquark pair, separated by a spatial distance L_s. If, in the limit $L_s \to \infty$, one finds a potential that behaves as $V(L_s) \sim$ constant or decays with L_s, one is in the deconfining phase; if the potential grows linearly with L_s ($V(L_s) \sim L_s$) or faster, the gauge-theory vacuum is in the confining state. Equivalently, for a large but arbitrary spatial loop C, one finds a perimeter law in the deconfining vacuum with $V(L_s) \sim$ constant,

$$\ln\langle W(C)\rangle \propto L(C),\tag{6.57}$$

while confinement with a linear potential $V(L_s) \sim L_s$ is signalled by an area law,

$$\ln\langle W(C)\rangle \propto A(C).\tag{6.58}$$

The fact that one of the legs of the Wilson loop is along the Euclidean time direction is in fact immaterial in a relativistic gauge theory at zero temperature. This changes at finite temperature. A purely spatial Wilson loop should still exhibit an area law in the confining phase, but a more direct measurement is a *temporal* Wilson–Polyakov loop,

$$P(\mathbf{x}) = \text{Tr}\left[\mathcal{P}\exp\left(i\int_0^{1/T} A_0(\mathbf{x})dt\right)\right].\tag{6.59}$$

This object is a proper order parameter for confinement (see e.g. [211, 225]). The expectation value of the temporal Wilson–Polyakov loop directly measures the change of free energy of the system induced by the presence of a probe quark: $\langle P(\mathbf{x})\rangle \propto \exp(-F(T)/T)$. In the confining phase this free energy should be infinite ($\langle P(\mathbf{x})\rangle = 0$), whereas in the deconfining phase of the thermal gauge theory it is finite and hence $\langle P(\mathbf{x})\rangle \neq 0$. It thereby plays the role of the order parameter of the deconfining phase – the symmetry associated with it is the centre of the gauge group [211].

How does one compute this temporal Wilson loop in AdS/CFT: what is its dictionary entry? Given that it is a non-local operator it cannot be directly deduced from the GKPW rule. One can, however, use the string-theoretical origins of the correspondence. In the top-down realisations of AdS/CFT Wilson loops are very natural objects. A quark in the fundamental representation corresponds to an end point of an (open) string on the brane, and the string accounts for the force exerted on the quark. The Wilson loop corresponds in turn to the propagation of a quark–antiquark pair along the loop. Upon attaching a string stretching from the quark to the antiquark, whose end points trace the Wilson loop, one therefore expects that its expectation value in the boundary is given by [222, 223]

$$\langle W(C) \rangle \sim \exp(-S_{\text{NG}}), \tag{6.60}$$

where S_{NG} is the on-shell value of the so-called Nambu–Goto action of the string

$$S_{\text{NG}} = \frac{1}{\ell_s^2} \int d\tau \int_0^{2\pi} d\sigma \sqrt{-\det G_{\mu\nu}(X(\sigma,\tau))\partial_\alpha X^\mu(\sigma,\tau)\partial_\beta X^\nu(\sigma,\tau)}. \tag{6.61}$$

This is Eq. (4.35), after integrating out the worldsheet metric $h_{\mu\nu}$, and directly measures the area of the worldsheet. In the AdS description, however, this string can now extend into the extra direction in the bulk. This has a simple geometrical interpretation: in the same way as one extends a curve in space into a worldsheet in space-time, the curve C in the boundary extends from the boundary at radial infinity to a worldsheet in the radial direction. To find out how it does so, one just has to minimise the action of the worldsheet in the bulk geometry (proportional to its area), translating this in turn into the behaviour of the Wilson–Polyakov loop in the boundary.

In box 6.4 this computation is presented in more detail. We can readily illustrate the result (Fig. 6.4). In the absence of a black-hole horizon, as in the case of the thermal AdS, the worldsheet has to form a connected surface in the bulk. This means that its (properly renormalised) area is finite, and the Wilson–Polyakov loop will exhibit the area-law behaviour in the boundary, signalling that the gauge theory is in a confining phase. However, when a black-hole horizon is present this will "cut open" the string worldsheet. Such a disconnected worldsheet formally has vanishing area and thus the phase described by the black hole is deconfining. Using the radial-direction/scaling-energy translation, this elegantly describes the scaling behaviour of the Wilson–Polyakov loop: when the loop is small and probes only the short-distance physics, the worldsheet in the bulk does not extend far. Effectively it therefore sees a "zero-temperature"-like state because it does not reach far enough along in the radial direction to probe the horizon. Only when the size of the Wilson

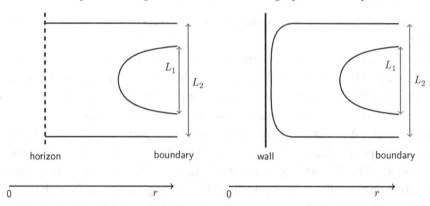

horizon   boundary   wall   boundary

0   r 0 r

Figure 6.4 A cartoon illustrating the geometry of the string extending in the bulk corresponding to the Wilson loop in the boundary. The left and right panels illustrate the situation in the presence of a black hole (the deconfining state), and in the case in which either the geometry is capped off (here literally the hard wall introduced in section 6.3) or the thermal AdS applies, corresponding to confinement. Wilson loops of short length L_1 probe only the short-scale structure of the theory. The gravitational dual description here is the same in both situations. One has a connected worldsheet, which translates into an area law for the Wilson–Polyakov loop in the boundary signalling confinement. Upon increasing the size of the Wilson loop L_2, one finds that in the black-hole case there is a critical length L_* beyond which the minimal worldsheet consists of two strings that drop straight into the black-hole horizon. The worldsheet becomes disconnected at the horizon: this gives rise to a perimeter law in the boundary, signalling that the black hole describes a thermal deconfining state. Upon increasing the size of the Wilson loop L_2 in the hard-wall (and thermal AdS) case, on the other hand, the worldsheet of the string remains connected. It will just extend itself along the wall where the geometry terminates (right panel). The area-law scaling therefore persists into the deep infrared, signalling that the gauge theory in the boundary is in the confining phase.

loop reaches a critical length L_*, such that the worldsheet extends all the way to the horizon, does it feel the change. As its size grows beyond L_* the worldsheet ruptures at the horizon and the Wilson loop signals that the system "enters" the deconfining thermal infrared.

Rule 9 Wilson loops in the boundary field theory can be computed with the semi-classical action of a string in AdS that ends on the location of the Wilson loop in the boundary. If the string extends all the way to the horizon of a black hole, it can break. This signals the transition from a confining to a deconfining phase in the boundary field theory.

In summary, this Hawking–Page transition is a first example of the important moral of this book that Einsteinian theory in AdS can switch suddenly between qualitatively different stationary space-times. This switching literally translates into a phase transition between different phases of the boundary field theory. This specific example has a special appeal because it just involves space-time itself and nothing else, while it displays a remarkable lesson. It may seem obvious that a "hot" space-time has a preference to form a black hole. It is, however, quite surprising that a cold black-hole space-time may wish to "un-collapse". We see that, even though naively a black hole is the most stable object possible, in a more complete theory this turns out to be no longer the case. We will see this astonishing *instability* of black holes again and again in later chapters. It is one of the pillars of the connections of holography to condensed matter physics.

In this particular example we are dealing with a Yang–Mills theory at zero density, where the general expectation is that it can be in either a confining or a deconfining phase, and this is beautifully confirmed by holography. We learn that the large-N super-conformal Yang–Mills theories that are the primary example of these boundary field theories as they arise in explicit top-down constructions of holographic duals are always on the verge of confinement: in infinite volume deconfinement occurs at any finite temperature, but breaking the conformal invariance by imposing a finite volume is sufficient to stabilise a confining ground state. Given the dictionary entry for Wilson loops, this follows directly from the observation that the presence of a Schwarzschild horizon in the bulk implies that the field theory is in a deconfining state. When we turn to holographic metals realised at finite density this theme of deconfinement will play an important role.

The alert reader will have noticed immediately that the Hawking–Page claim that a real thermodynamic phase transition can be enforced by placing the system in a *finite* volume seems to violate a basic thermodynamical principle: true thermodynamical singularities can occur only in the thermodynamic limit, i.e. in an *infinite* volume. The reason why one does still find a genuine phase transition is the matrix large-N limit. In this limit the thermal fluctuations are suppressed in a way that is very similar to what happens in the case of infinitely many dimensions, and the effect is that the thermal transitions behave in a mean-field way. This unavoidable large-N mean-field behaviour will become more evident on dealing with the symmetry-breaking phase transitions of the later chapters, beginning with the holographic superconductivity of chapter 10. This is, however, perceived as a problem of principle because it turns out that the thermal fluctuations are restored by tractable quantum corrections in the bulk.

Box 6.4   Holographic Wilson loop and (de)confinement

This dictionary rule for Wilson loops follows naturally from the interpretation of open-string end points as quarks [222, 223]. In AdS/CFT the body of this string is free to explore the full bulk geometry, and its action is the AdS quantity dual to the Wilson-loop expectation value. Specifically,

$$\langle W(C) \rangle \sim e^{-S_{\mathrm{NG}}}, \tag{6.62}$$

where S_{NG} is the on-shell value for the (Euclidean) Nambu–Goto action for the string worldsheet whose boundary is exactly the curve C on the dual field theory,

$$S_{\mathrm{NG}} = \frac{1}{2\pi\alpha'} \int d^2\sigma \sqrt{\det\left[G_{\mu\nu}(\partial_m X^\mu)(\partial_n X^\nu) \right]}, \tag{6.63}$$

with $\mu, \nu = 0, 1, \ldots, d+1$, which refers to the coordinate X^μ of the string in the AdS space index, and $m, n = 0, 1$, which refers to the coordinate σ^m used to parametrise the worldsheet; $G_{\mu\nu}$ is the metric for the bulk geometry. By choosing the specific contour C above, i.e. τ from $-L_\tau/2$ to $L_\tau/2$ and x from $-L_s/2$ to $L_s/2$ with $L_\tau \to \infty$, the quark–antiquark potential can be obtained by computing

$$V(L) = \lim_{L_\tau \to \infty} \frac{S_{\mathrm{NG}}}{L_\tau}. \tag{6.64}$$

This computation is straightforward, with one subtlety. As we have encountered before, since the boundary of AdS is formally infinitely far away, the area of the worldsheet S_{NG} is actually divergent. This divergence can be subtracted by a consistent regularisation procedure, and a final finite result can be obtained.

Let us first consider the computation of the quark–antiquark static potential dual to the pure Euclidean AdS case [222, 223]. Note that we are computing not the temporal Wilson–Polyakov loop, which is the order parameter for confinement, but the direct quark–antiquark potential. For simplicity, we will choose a specific dimension $d = 3$. This is the natural dimension in which to consider Yang–Mills theories. Setting the AdS radius to unity for convenience, the metric is thus simply that of Euclidean AdS$_5$,

$$ds^2 = r^2 \left(d\tau^2 + dx^2 \right) + \frac{dr^2}{r^2}. \tag{6.65}$$

For the Wilson loop we choose the standard rectangle $x \in (-L_s/2, L_s/2)$, $\tau \in (-L_\tau/2, L_\tau/2)$ at the boundary of AdS$_5$. The natural parametrisation for the string worldsheet is to choose $\sigma^0 = \tau$, $\sigma^1 = x$. For the configuration fields of the string $X^\mu(\tau, x)$, we can choose a physical gauge where $X^0 = \tau$, $X^1 = x$. If we then look for a static solution, all other configuration fields are independent of τ. Moreover, as we move along the worldsheet we stay at a fixed point in the other two spatial directions. Thus $X^2(\tau, x) = \text{constant}$, $X^3(\tau, x) = \text{constant}$. On the other hand, as we move along the worldsheet, we do expect the string to move in the radial AdS direction. Thus X^r is a function of $X^r(x) \equiv R(x)$. In addition we have the boundary conditions that the end points of the string should be fixed to the location of the Wilson loop, i.e. the value of the worldsheet parameter x, t ranges only from $-L_s/2$ to $L_s/2$, and $-L_t/2$, to $L_t/2$, respectively.

With these choices and restrictions, and using the metric Eq. (6.65), the Nambu–Goto action in Euclidean AdS becomes

$$S_{\text{NG}} = \frac{L_\tau}{2\pi\alpha'} \int_{-L_s/2}^{L_s/2} dx \, R^2 \sqrt{1 + \frac{R'^2}{R^4}}, \tag{6.66}$$

with $R' = \partial_x R$. The system has become effectively one-dimensional, and for one-dimensional systems the Hamiltonian $H = P\dot{X} - \mathcal{L}$ is always conserved:

$$\frac{R^2}{\sqrt{1 + R'^2/R^4}} = \text{constant} \,. \tag{6.67}$$

We can now do a simple turning-point analysis. The "maximal potential energy" turning point ($R'(x) = 0$) for the string is, by symmetry, at the midpoint $R(x = 0) = R_{\text{max}}$, thus we have

$$\frac{R^2}{\sqrt{1 + R'^2/R^4}} = R_{\text{max}}^2 \quad \Rightarrow \quad R' = \pm R^2 \sqrt{\frac{R^4}{R_{\text{max}}^4} - 1}. \tag{6.68}$$

Solving the differential equation on the RHS of Eq. (6.68) immediately gives us a relation between L_s and R_{max},

$$L_s = 2 \int_{R_{\text{max}}}^{\infty} dR \, \frac{1}{R^2 \sqrt{R^4/R_{\text{max}}^4 - 1}} = \frac{2\sqrt{\pi}\Gamma(3/4)}{\Gamma(1/4)R_{\text{max}}}. \tag{6.69}$$

For the value of the on-shell action, we change integration variables from x to R and obtain,

$$S_{\text{NG}} = \frac{L_\tau}{\pi\alpha'} \int_{R_{\text{max}}}^{\infty} dR \, \frac{R^2}{R_{\text{max}}^2 \sqrt{R^4/R_{\text{max}}^4 - 1}} = \frac{L_\tau R_{\text{max}}}{\pi\alpha'} \int_1^{\infty} dy \, \frac{y^2}{\sqrt{y^4 - 1}}. \tag{6.70}$$

Note that the prefactor has doubled. This is to account for both the string in the region $(-L_s/2, 0)$ and the string in the region $(0, L_s/2)$.

Box 6.4 (Continued)

It is easy to see that (6.70) is divergent. To regularise this, note that we are interested in the quark–antiquark potential. A bare static quark in finite-temperature Yang–Mills also contains energy density. In the AdS string prescription this bare rest energy corresponds to a single string just falling straight into the horizon of the black hole. On subtracting this bare single-string energy both for the quark and for the antiquark, the regularised S_{NG} is

$$S_{NG} = \frac{L_\tau R_{max}}{\pi \alpha'} \left[\int_1^\infty dy \left(\frac{y^2}{\sqrt{y^4 - 1}} - 1 \right) - 1 \right]. \tag{6.71}$$

Thus the regularised potential for the quark–antiquark pair can be obtained as

$$V(L_s) = -\frac{c}{L_s}, \tag{6.72}$$

with a positive constant c [222]. It is a Coulomb-like potential, which is a consequence of conformal symmetry.

At finite temperature [226, 227] we consider the behaviour of the string in the following background:

$$ds^2 = r^2 \left(f(r)d\tau^2 + d\mathbf{x}^2 \right) + \frac{dr^2}{r^2 f(r)}. \tag{6.73}$$

It is straightforward to repeat the above computation. Recall that there are two solutions to the equations of motion. There is the U-shaped string trajectory we have just considered. In this case, we have the static quark–antiquark potential,

$$V = \frac{R_{max}}{\pi \alpha} \left[\left(\int_1^\infty dy \, \frac{\sqrt{y^4 - r_0^4/R_{max}^4}}{\sqrt{y^4 - 1}} \right) - 1 \right], \tag{6.74}$$

with R_{max} the turning point, r_0 the location of the horizon and

$$L(R_{max}) = 2 \frac{L_s^2}{R_{max}} \sqrt{1 - \frac{r_0^4}{R_{max}^4}} \int_1^\infty dy \, \frac{1}{\sqrt{(y^4 - r_0^4/R_{max}^4)(y^4 - 1)}}. \tag{6.75}$$

The other solution is just the bare quarks, whose dual description is two straight strings dropping into the horizon. It is easy to see that, in this case, the total area of the "disconnected" strings (compared with the zero-temperature solution) is

$$V = -\frac{r_0}{\pi \alpha'}. \tag{6.76}$$

On plotting both potentials as a function of L_s, we clearly see the existence of a special scale L_*. For $L_s < L_*$ the U-shaped string has the minimal area, and the quarks experience a Coulomb-like attraction, whereas for $L_s > L_*$ the disconnected string solution is the energetically favourable solution (see Fig. 6.4). The free energy is now independent of the separation and this clearly denotes a deconfined phase.

6.3 A brief introduction to AdS/QCD

Confinement is an essential part of the manifestation of quantum chromodynamics (QCD) in the real world. Nevertheless, with strong-coupling physics as its underpinning, a thorough quantitative understanding is still actively sought. Strong coupling is, of course, what holography is good at. Indeed, the study of QCD using holographic duality ("AdS/QCD") is a vast subject in itself. The fact that large-N versions of the same non-Abelian Yang–Mills theories as those which govern the Standard Model of particle physics are the canonical examples of holography made this the natural arena for applications. Early in this century, holography acquired considerable credibility in the context of QCD through its successful description of the nearly perfect-fluid behaviour of the quark–gluon plasma (QGP) produced in heavy-ion collisions, as we will discuss in the next chapter. This evolved into a substantial research effort to use holography to address various aspects of the physics of QCD. We will review this briefly in this section. Our focus will be on confinement aspects in particular, since these will play a role in the applications to CMT discussed in subsequent chapters.

The outlook of QCD has traditionally been on the physics of non-Abelian Yang–Mills theory at zero or extremely low ("nuclear") density under equilibrium conditions. In this setting the fermion signs can be handled thanks to the success of the "lattice", i.e. quantum Monte Carlo computations. The gross qualitative principles governing the physics in this regime appear to be quite well understood. Here one is dealing with the transition at the QCD scale to a confining phase from the high-energy asymptotically free deconfined phase, and associated with this is the phenomenon of chiral symmetry breaking. What remains are quantitative issues like the spectra of mesons and baryons at zero temperature, and macroscopic questions such as the equation of state as a function of temperature and (low) density [11, 12].

The really difficult problem of strongly interacting quantum matter at finite density which is the main subject of this book is also of relevance for QCD. Although at very high quark densities QCD becomes weakly coupled due to the asymptotic freedom and the set of conventional condensed matter tools (e.g. BCS theory) can be applied, at intermediate densities there is room for surprises. In this regime it becomes a prototypical quantum matter problem, where finite-density fermions combine with the strong colour forces to turn it into a problem that cannot be handled by established methods.

Another issue is the high-temperature regime of QCD dynamics at nuclear densities which is currently being explored in particle accelerator heavy-ion collisions. The primary objective here is the QGP, a new state of matter where quarks and gluons form a primary hot soup. Somewhat unexpectedly, the physical manifestation of the QGP found in these experiments at Brookhaven and CERN is not a simple state of moderately weakly coupled quarks and gluons, but instead behaves as a strongly interacting highly collective plasma. One can make the case that holography captures features of this strongly coupled QGP (sQGP) better than any other method. In particular, due to its unique ability to describe real-time non-equilibrium physics in strongly interacting quantum systems at zero or finite density, there are questions for which at present holography is the only method that is available. The details of this rapidly evolving subject can be found in a book dedicated to the field [12].

The transition between the QGP and the zero-temperature vacuum is what confinement is all about. The sQGP at high temperature can be quite reasonably captured by the holography of the super-conformal Yang–Mills theory of the basic AdS/CFT correspondence. To describe the (zero-density) confined phase of QCD to a greater or lesser degree in a realistic fashion, various holographic set-ups were proposed. The finite-volume confinement of the previous section is indicative for how it can work, but is clearly quite different from the mechanism in real QCD. Some more intricate holographic dualities that are more QCD-like in this regard have been uncovered (see e.g. Ref. [228]). They include a number of top-down constructions, which break conformal invariance explicitly and naturally have confining ground states [229, 230, 231]; especially the "brane intersection" top-down model (see chapter 13) constructed by Sakai and Sugimoto [232, 233] mimics QCD quite well. These models, but also direct engineering, have generated a variety of bottom-up constructions for holographic confinement. These are of interest to us, since we will use them later as a diagnostic device in finite-density holographic systems, where confinement will simplify matters significantly.

Confinement comes with a scale, and this implies that at the location in the radial direction corresponding to this scale the AdS geometry in the bulk must change. The bottom-up constructions turn this around. The bulk geometry is altered by hand, in such a way that the dual field theory becomes confining. There are various ways of accomplishing this, all of which share the basic ingredient that the geometry in the deep interior (encoding for the IR physics) is "capped off". This is based on the intuition that confinement comes with a mass gap. Below the gap there should no longer be dynamical degrees of freedom present. This means that beyond the corresponding location in the radial direction there ought to be no space-time left in the bulk since there should be no room for gravitons (which are dual to energy) to fluctuate. The simplest and crudest way to accomplish this is by imposing a "hard wall" [234, 235], where the geometry is suddenly terminated by hand at some radial coordinate. Writing the metric in Poincaré coordinates,

$$ds^2 = r^2 \left(-dt^2 + dx^2\right) + \frac{dr^2}{r^2}, \tag{6.77}$$

one incorporates the wall by limiting the range of the radial coordinate to $\infty > r > r_c$. From the discussion of the Wilson loop in the previous section, one can deduce that this describes a confining state with an area law that behaves like that of QCD. Compared with the computation in box 6.4, one now has to add explicit boundary conditions at the wall. When the contour C is large enough that the string worldsheet starts to touch the wall, from this point onwards it just "spreads" along the wall without breaking up, and for large enough L_s the on-shell string action just increases precisely by the area enclosed by C (see Fig. 6.4).

Capped-off geometries such as the hard-wall one have a profound influence on the nature of the excitations. Instead of the "un-particle" branch-cut spectra associated with conformal invariance as discussed in chapter 4, one now finds distinct particle excitations. The way this works in the bulk is very simple. Both the boundary and the wall near the deep interior now act as mirrors for the modes of the classical fields in the bulk. This means that these fields are literally living in a box in the radial direction, with the associated radial quantisation of these modes. In the boundary field theory these radially quantised modes correspond to perfectly undamped propagating particle poles, whose masses are functions of the radial quantum number. These are identified as the spectrum of gauge-singlet "mesons" associated with the confining gauge theory. In box 6.5 we illustrate this simple computation by studying a particular vector field in the bulk and its dual boundary spectrum.

To turn this into a more realistic meson spectrum for QCD one has to pay tribute to the flavour sector as well. The confinement in QCD goes hand in hand with chiral symmetry breaking of the global $SU(N_{\text{flavour}})_L \times SU(N_{\text{flavour}})_R$ to diagonal $SU(N_{\text{flavour}})$. Using the translation of global field theory symmetries to local symmetries in the AdS dual, this can be encoded in $SU(N)$ gauge fields in the bulk. The chiral symmetry breaking is encoded by introducing a chiral order parameter in the hard-wall model [234, 236] (see box 6.5). On combining all these ingredients in a confining hard-wall AdS geometry, one is left with three free parameters that have to be fitted with experimental data: the confinement scale (the location of the hard wall), the strength of the vector-current two-point function (the coupling strength of the $SU(3)$ flavour currents) and the VEV of the chiral-symmetry-breaking order-parameter matrix (the derivative of the chiral order-parameter field at the boundary). The masses, decay rates and couplings of the lightest mesons are then completely fixed and turn out to be in remarkable agreement with the experimental (and lattice) results for the mesons [234, 236].

The hard-wall model also yields quite decent phenomenological results [235] with regard to the finite-temperature phase diagram. One follows the same procedure as for the Hawking–Page transition; that is, one compares the free energy for different configurations. To describe the finite-temperature confining state, one turns the time axis of the Euclidean version of the hard-wall metric (6.77) into the usual circle. The finite-temperature deconfining state is the hard-wall cut-off AdS with the emblackening factor of the planar Schwarzschild black-hole solution in AdS$_5$,

$$ds^2 = \frac{r^2}{L^2}(-f(r)dt^2 + d\mathbf{x}^2) + \frac{L^2}{r^2 f(r)}\,dr^2, \quad f(r) = 1 - \frac{r_0^4}{r^4}, \tag{6.78}$$

with r_0 the location of the horizon while $r \geq \max(r_c, r_0)$. From the difference in free energy between the two solutions one finds out that the hard-wall geometry wins at low temperature, while a first-order transition to the high-temperature deconfining (black-hole) state follows at a temperature set by r_c.

In box 6.5 we discuss the holographic computation of the spectrum of gauge-singlet "meson" bound states in the confining regime of the theory. This is conceptually a simple affair: in the presence of the hard-wall AdS turns quite literally into a box and the solution of the equations of motion in the bulk corresponds to simple standing waves along the radial direction. These dualise in infinitely long-lived particle poles in the boundary, with a mass spectrum set by the harmonic eigenvalues in the bulk. The experimental meson masses obey the famous "Regge trajectory" relation $m_n^2 \propto n$. A main shortcoming of the hard-wall set-up is that it does not quite reproduce the correct Regge-trajectory behaviour. One finds instead the relation $m_n^2 \sim n^2$, where n is the radial excitation number. This result is hard-wired into the hard-wall geometry, indicating that this bulk geometry does not quite capture the true confinement behaviour observed in experiment. There are several other problematic features with the hard-wall construction that are correlated with the abrupt termination of the geometry [235, 237]. To improve on these Regge trajectories, Karch, Katz, Son and Stephanov [238] modified the bulk space-time by changing the sudden hard-wall IR cut-off into a smooth cut-off: the "soft wall". In this construction one insists that the bulk geometry is the usual unaltered AdS, but the action is modified by the introduction of a so-called "dilaton field" $\Phi(r)$. This is simply incorporated by changing the bulk action from

$$S = \int dt\, dr\, d^d x \sqrt{-g} \mathcal{L}$$

to

$$S = \int dt\, dr\, d^d x \sqrt{-g} e^{-\Phi(r)} \mathcal{L}. \tag{6.79}$$

Such dilaton fields may be unfamiliar to the reader without a string-theoretical background. They are, however, quite ubiquitous in the low-energy theories generated by Kaluza–Klein compactifications of string theories (see chapter 13). We will encounter them later (especially in chapter 8) as dynamical fields with their own potentials. However, for the purpose of a soft wall, $\Phi(r)$ is a non-dynamical field whose profile just serves to cut off the dynamics at small values of r. A hard wall would correspond to a step-function cut-off $\exp\{-\Phi(r)\} = \theta(r - r_c)$. Taking a smooth profile of the form $\Phi(r) = c/r^2$ instead ameliorates much of the confounding behaviour of the hard wall and in addition tunes the Regge trajectory to acquire the correct form.

A further construction to encode confinement in the bulk geometry is called the "AdS soliton" [211, 239]. To obtain the scale necessary for confinement, one considers a $(d + 1)$-dimensional boundary field theory with *one* of its spatial dimensions compactified on a circle, instead of all directions compactified to a sphere. The Euclidean finite-temperature version of this theory therefore lives in $S^1 \times \mathbb{R}^{d-1} \times S^1$, with the first S^1 representing the Euclidean time circle and the second S^1 representing an extra compact dimension. This might look complicated, but the AdS-soliton geometry is easy to find. One doubly Wick rotates the AdS–Schwarzschild solution Eq. (6.19), which has the same topology but with the roles of the S^1 reversed. By Wick rotating both time and the compact spatial direction, one obtains the desired AdS-soliton geometry, e.g. in AdS$_5$,

$$ds^2 = \frac{L^2}{r^2 f(r)} dr^2 + \frac{r^2}{L^2}(-dt^2 + dx^2 + dy^2) + \frac{r^2 f(r)}{L^2} d\phi^2, \tag{6.80}$$

with

$$f(r) = 1 - \frac{r_0^4}{r^4},$$

where ϕ must obey $\phi \sim \phi + \pi L/r_0$ in order to have a smooth geometry. Because it is just a doubly Wick-rotated black hole, the soliton geometry Eq. (6.80) is a solution of Einsteinian gravity with a negative cosmological constant. At zero temperature, one can easily see that r_0 plays the same role as r_c in the hard-wall model, and one anticipates that this geometry describes a confining state. At finite temperature, one finds two Euclidean solutions with the same boundary condition, which are described by the black hole Eq. (6.19) and the confining soliton

Eq. (6.80). A first-order phase transition occurs at $T_c = r_0/\pi$, corresponding to the confinement–deconfinement transition. The balance between the extra scale introduced by the radius of the compact spatial dimension and the temperature works therefore roughly in the same way as in the Hawking–Page transition in global AdS.

By and large these AdS/QCD constructions confirm an intuition that is widespread in condensed matter physics: confining vacua are featureless entities like the Bose Mott insulators discussed in chapter 2. Since these are short-ranged entangled product states they typically support particle excitations. This seems to be the message of the gravitational dual as well: there is nothing that can be encoded in a deep interior geometry beyond the confinement scale as in the above "wall" constructions. This in turn goes hand in hand with the radial "finite-volume quantisation", which gives rise to simple particle poles in the spectrum of the boundary field theory. In fact, as we will see in later chapters, when we force these confining systems to a finite density they continue to behave in a very simple way. At densities near the confinement scale, they just describe systems formed from a finite density of weakly (or even non-)interacting particles. These describe the conventional states of condensed matter, superfluids, Fermi gasses, etc. In this emergent IR of excitations of weakly interacting particles, some of the strong–weak duality power of holography is lost. At low energies the collective physics can be expressed in terms of these emergent excitations on either side of the duality in the same weakly coupled way. This makes perfect sense, since we know from our own experience that gauge singlets such as mesons turn into weakly interacting particles in the confining phase. However, the bright side of confinement is that it acts like a sanity check, and helps us to get a clearer view of the nature of the truly "strange" holographic matter.

Box 6.5   The QCD meson spectrum as the excitations in the hard-wall model

In this box we will show how to compute the discrete particle spectrum of the hard-wall AdS/QCD model [234]. These excitations are naturally interpreted as the mesons – bound massive states of quark–antiquark pairs – of the QCD-like theory. Since massive states break the independent left-and right-helicity symmetries of massless relativistic fermions, the existence of a meson excitation serves as an order parameter for chiral symmetry breaking. In QCD there are three flavours of quarks, and chiral symmetry breaking is therefore characterised by the breaking of $SU(3)_L \times SU(3)_R$ to $SU(3)_{\text{diag}}$. The scalar order-parameter field transforms as a $(\bar{3}, 3)$ under the $SU(3)_L \times SU(3)_R$. Using the AdS/CFT dictionary we can therefore readily write down a simple realisation of chiral symmetry breaking in the bulk: for each current $J_L = \bar{q}_L \gamma^\mu t^a q_L$ and $J_R = \bar{q}_R \gamma^\mu t^a q_R$ in QCD in the field theory, we introduce gauge fields $A_{L,R}$; the dual to the order parameter $\bar{q}_R q_L$ should be a scalar field X. Since the order parameter has classical dimension $\Delta = 3$ at the boundary – we are considering a $(3 + 1)$-dimensional boundary – we choose the mass of X to equal $m_X^2 L^2 = \Delta(\Delta - 4) = -3$ in Eq. (6.81). The minimal Lagrangian in AdS is therefore

$$\mathcal{L} = \text{Tr}\left(|DX|^2 + 3|X|^2 - \frac{1}{4g_5^2}(F_L^2 + F_R^2)\right), \tag{6.81}$$

where

$$D_\mu X = \partial_\mu X - i A_{L\mu} X + i X A_{R\mu}, \tag{6.82}$$

with

$$A_{L,R} = A_{L,R}^a t^a, \quad F_{\mu\nu} = \partial_\mu A_\nu - \partial_\nu A_\mu - i[A_\mu, A_\nu].$$

The gauge coupling g_5 is determined by matching the vector current correlators with the result from QCD. One finds $g_5 = 12\pi^2/N_c$ if the gauge group of the QCD-like theory is $SU(N_c)$.

We now construct the background solution which corresponds to the broken ground state. From chapter 5 we know that the sub-leading component of the field X near the boundary characterises its expectation value. From the general behaviour of a scalar field, we know that the asymptotic behaviour for a field with dimension $\Delta = 3$ will be

$$X_0 = \frac{M}{2}\frac{1}{r} + \frac{\Sigma}{2}\frac{1}{r^3} + \cdots. \tag{6.83}$$

Insisting that the expectation value of $\langle X \rangle = \Sigma/2$ is non-vanishing means that we are in the broken state. Choosing the chiral condensate $\Sigma = \sigma \delta^{ab}$ to be diagonal means that the diagonal subgroup $SU(3)_{\text{diag}}$ will remain unbroken as desired. In this case the leading term of the field X is naturally interpreted as the quark mass matrix $M = m_q \delta^{ab}$. The leading term is the source of a deformation of the action of the boundary CFT, and one can view the quark mass term as this deformation.

In the absence of gauge fields, it turns out that

$$X_0 = \frac{M}{2}\frac{1}{r} + \frac{\Sigma}{2}\frac{1}{r^3} \tag{6.84}$$

with no further terms is an exact solution to the equations of motion. Now we study the fluctuations around this background. To do so, it is useful to rewrite

Box 6.5 (Continued)

$X = X_0 \exp\{2i\pi^a t^a\}$ and vector and axial fields as $V_\mu = (A_{L\mu} + A_{R\mu})/2$ and $A_\mu = (A_{L\mu} - A_{R\mu})/2$. In the radial gauge $V_r = A_r = 0$, the vector mesons ρ are identified as the excitations dual to normalisable modes of $V_{\mu \neq r}$ and the axial mesons a_1 are those of $A_{\mu \neq r}$; π^a corresponds to pseudo-scalar mesons π, and the fluctuations of X_0 correspond to the scalar meson. On substituting these redefinitions as well as (6.83) into Eq. (6.81), the action becomes

$$S = \int d^5x \sqrt{-g} \left[-\frac{1}{8g_5^2} \left((\partial_\mu V_\nu^a - \partial_\nu V_\mu^a)^2 + (\partial_\mu A_\nu^a - \partial_\nu A_\mu^a)^2 \right) \right.$$
$$\left. + \frac{X_0^2}{2} (\partial_\mu \pi^a - A_\mu^a)^2 \right], \tag{6.85}$$

where $\mathrm{Tr}(t^a t^b) = \delta^{ab}/2$ is used.

One can find that at quadratic order the gauge field V_μ decouples from A_μ and π. For simplicity we will consider only the spectrum for the vector meson in this box. In the radial gauge, the equation of motion for the transverse part of V_μ obeying $\partial^\mu V_\mu = 0$ ($\mu = 0, 1, 2, 3$) is

$$\partial_r^2 V_\mu + \frac{3}{r} \partial_r V_\mu + \frac{1}{r^4} (-\partial_t^2 + \partial_\nu^2) V_\mu = 0. \tag{6.86}$$

To solve the system, we need to impose suitable boundary conditions. At the hard wall $r = r_c$, we impose the gauge-invariant boundary condition $F_{r\mu}^{(V)} = 0$. At the boundary we impose normalisable boundary condition $V_\mu(r \to \infty) = 0$. Fourier transforming $V_\mu(r, x) = f_\mu(r)e^{-ikx}$ with $k^2 = -m^2$ and redefining $f_\mu = r^{-3/2} \tilde{f}_\mu$, the equation of motion for $\tilde{f}_\mu(r)$ is

$$-\tilde{f}_\mu'' + \left(\frac{3}{4r^2} - \frac{m^2}{r^4} \right) \tilde{f}_\mu = 0. \tag{6.87}$$

It has two independent solutions. The sector which satisfies the condition $f_\mu(\infty) = r^{-3/2} \tilde{f}_\mu = 0$ is $\tilde{f}_\mu = c_\mu \sqrt{r} J_1[m/r]$, with c_μ the constant and J_1 the Bessel function of the first kind. Thus

$$f_\mu = c_\mu \frac{1}{r} J_1 \left[m/r \right]. \tag{6.88}$$

We should impose the boundary condition $\partial_r f_\mu |_{r=r_c} = 0$ at the hard wall. Using the fact that $\partial_x(x J_1[x]) = x J_0[x]$, the boundary condition at the hard wall is then equivalent to

$$J_0 \left[m/r_c \right] = 0. \tag{6.89}$$

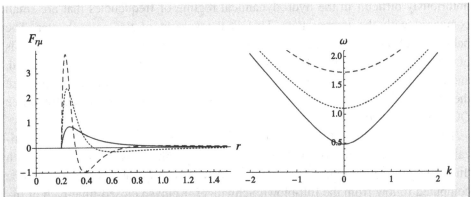

Figure 6.5 The left panel shows the quantised normalisable radial excitations in a hard-wall AdS/QCD model. The quantisation condition translates into the dispersion relation for the modes. The right panel shows the spectrum of meson states in the dual field theory that follows.

For a fixed value of r_c, one can satisfy this condition only for discrete values of m. For large m/r_c, we have $J_0[m/r_c] \simeq \sqrt{2r_c/(\pi m)} \cos(m/r_c - \pi/4)$. The higher excited modes of the vector meson are therefore found to have masses

$$m_n \simeq (n - 1/4)\pi r_c. \tag{6.90}$$

We clearly see that normalisable solutions to the equation (6.87) exist only for discrete value of m^2; see Fig. 6.5. This discretisation is directly due to the boundary conditions, and it is therefore completely analogous to the quantisation of a particle in a box. The hard-wall model has many extensions beyond the meson spectrum, e.g. one can study the glueball spectrum which corresponds to a scalar field in the bulk [240] or the spin-1/2 nucleon spectrum by introducing fermions [241].

6.4 The GKPW rule at finite temperature: thermal correlation functions and the Schwinger–Keldysh formalism

We will now use the GKPW rule to highlight one of the most powerful aspects of the AdS/CFT correspondence: the ability to compute real-time thermal correlation functions in the boundary field theory, even under non-equilibrium conditions. Real-time thermal correlation functions in field theory are notoriously cumbersome. This is already so for the linear response of equilibrium systems. In principle one can compute the real-time response by Wick rotation of the Euclidean correlation functions. However, here one faces an infamous information-loss problem: after analytical continuation to real time, the smooth Euclidean correlation functions turn into very "bumpy" real-time propagators and one has to know the former with an extreme precision in order for this to be successful. This becomes notoriously difficult in the hydrodynamical regime of frequencies that are small compared with the temperature. This is the subject of the next chapter. The behaviour of the classical fluid realised in this regime is buried in the fine details of the long imaginary tails of the Euclidean propagators. Any noise is detrimental, and it has proven impossible to the present day to compute a simple quantity like the viscosity of the quark–gluon plasma using the otherwise so successful lattice QCD quantum Monte Carlo approach.

This becomes much harder for fully non-equilibrium quantum field-theoretical physics. There are no good field-theoretical methods available to deal with this problem in a systematic and controlled way. The quantities that have to be computed follow from the full-blown Schwinger–Keldysh closed-time-path or in–in formalism in field theory, which computes probabilities directly, instead of amplitudes. However, except for the very special cases of integrable field theories in $1 + 1$ dimensions, this can be handled only in a perturbative expansion around free-field theory. The non-perturbative physics of field-theoretical systems away from equilibrium is beyond the present capacity of the methods of field theory.

In holography this situation changes drastically. We have already encountered the ease with which the dictionary handles the Wick rotation. The Minkowski or Euclidean signature in the boundary is in one-to-one correspondence to the signature in the bulk. We used the Euclidean signature all along in this chapter since it is the most convenient means to compute equilibrium thermodynamics. For equilibrium systems one can proceed equally well by computing the boundary Euclidean propagators holographically, to subsequently Wick rotate [201, 202, 242, 243]: this now works since the bulk equations of motion allow one to compute the boundary propagators with arbitrary precision [244]. However, with similar ease one can directly compute the real-time boundary propagators directly by using a Minkowski signature in the bulk. We used this strategy repeatedly in our discussion of the GKPW rule in chapters 1 and 5. As we will discuss in this section, this can straightforwardly be extended to finite-temperature equilibrium systems. The Minkowski advanced and retarded Green functions are simply computed by choosing appropriate "boundary conditions" at the horizon of the black hole that encodes the finite-temperature state [201]. The outcome is quite intuitive; the black-hole horizon in the bulk represents the finite-temperature scale in the boundary, and in order to address the effects of dissipation in the retarded propagator, which is of primary physical interest, one has to choose in-falling boundary conditions at the horizon. The stuff in the bulk that disappears in the black hole represents the effects of dissipation in the boundary, and one finds that the propagators describe seamlessly the crossover from the coherent regime at high frequency to the hydrodynamical responses at small frequencies [24]. Similarly, to compute the advanced Green function, one chooses out-going boundary conditions [201]. Since all "normal" time evolution should result in disappearance into the black hole, these choices inherently reflect the correct nature of the retarded and advanced Green functions.

This is not yet the complete story. Given the identification of real time in the boundary with the Minkowski signature in the bulk, it is immediately obvious that a non-stationary bulk corresponds to non-equilibrium physics in the boundary. As we have already emphasised, the study of non-equilibrium systems by holography is at present a flourishing research subject that is so rich that it deserved its own comprehensive treatise [23]. We will only touch on it – in section 7.2 dealing with the derivation of the (near-equilibrium) Navier–Stokes equations we will drill down a bit deeper in this subject. The other exception is the remainder of this section. In non-equilibrium systems one needs the full Schwinger–Keldysh formalism, and we will discuss this here in its full generality.

The way that the black-hole boundary conditions follow from the GKPW rule is one of the most remarkable aspects of holography [202, 245]. The crucial characteristic of the GKPW rule is that one should read off the properties of the field theory from the boundary behaviour. The *complete* space-time of a black hole has *two* boundaries, however. This is revealed by the well-known Kruskal extension of the metric. Technically the standard AdS–Schwarzschild metric Eq. (6.19) is not geodesically complete. In this metric trajectories of particles fall into the black hole in a finite proper time. Beyond that moment, the Schwarzschild metric is unreliable, but one can construct a different set of coordinates based on in-falling and out-going light rays that better illustrates what happens to such particles. These are the Kruskal coordinates U, V that cover the whole of the AdS black-hole space-time with the metric

$$ds^2 = -\frac{r^2 f(r) e^{-4\pi T r_*}}{(2\pi T)^2} \, dU \, dV + r^2 \, dx_i^2, \tag{6.91}$$

with $r_* = \int_{r_0}^r dy/(y^2 f(y))$. The original AdS–Schwarzschild metric (6.19) is reobtained after the transformation

$$U = -e^{-2\pi T(t-r_*)}, \quad V = e^{-2\pi T(t+r_*)}. \tag{6.92}$$

These Kruskal coordinates show that each static black hole is *always* accompanied by a mirror white hole in the infinite past, and there is a complete hidden region of space-time connected by a (zero-size) wormhole to the original Schwarzschild region (Fig. 6.6). This means that the true *full* AdS–Schwarzschild geometry in fact has two boundaries, one in the original region and one in the hidden region. The

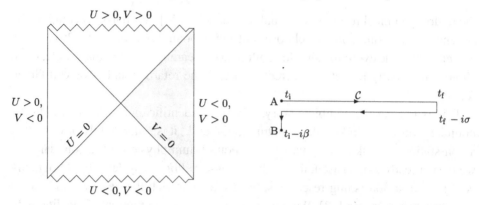

Figure 6.6 Left: the Penrose diagram of an AdS–Schwarzschild black hole. The original Schwarzschild coordinates cover only the shaded area. The solid lines at 45° are the horizon. The top wiggly line is the black-hole singularity; the bottom wiggly line is the white-hole singularity that, together with the hidden region in the left quadrant, is revealed by the Kruskal extension. Right: the Schwinger–Keldysh contour. The parameter σ can be arbitrarily chosen to be any value $0 < \sigma < \beta$. The most convenient choice is $\sigma = \beta/2$.

beauty is that this matches precisely with the Schwinger–Keldysh prescription for computing probabilities. The probability is the square of the absolute value of the transition amplitude, and the time evolution of each amplitude separately,

$$
\begin{aligned}
P_{\text{in}\to\text{out}} &= \langle\text{in}|e^{i\int_{t_i}^{t_f} H}|\text{out}\rangle\langle\text{out}|e^{-i\int_{t_i}^{t_f} H}|\text{in}\rangle \\
&= \langle\text{in}_B|e^{i\int_{t_i}^{t_f} H}|\text{out}\rangle\langle\text{out}|e^{-i\int_{t_i}^{t_f} H}|\text{in}_A\rangle\Big|_{\text{in}_A=\text{in}_B},
\end{aligned}
\tag{6.93}
$$

automatically gives us a "doubled" initial condition, one evolving forwards in time and one evolving backwards. In the path integral this double evolution is combined to form the well-known complex Schwinger–Keldysh contour (Fig. 6.6). This doubled initial condition in the Schwinger–Keldysh formalism is translated through the GKPW rule into the bulk requirement that one must choose boundary conditions both in the original region and in the hidden region, which mirror each other. From a detailed inspection it follows that this match between black holes and real-time finite-temperature field theory is truly perfect owing to the fact that the Schwarzschild time runs "backwards" in the hidden region. This captures precisely the complex-conjugate evolution of the amplitude.

By carefully solving the evolution on the full extended Kruskal space-time with two boundaries, one can show that the advanced and retarded Green functions of the boundary field theory can also be obtained by taking the ratio of subleading to leading terms of the solution to the bulk equation of motion in the original Schwarzschild region only, with out-going and in-falling boundary conditions respectively (box 6.6). Especially for two-point correlation functions, there is therefore no necessity to employ the full Kruskal extended space-time, although it should be clear that if there is any ambiguity regarding how to compute a boundary quantity one should resort to the original GKPW construction in the full extended space-time.

Rule 10 The real-time propagators of the field theory at finite temperature can be computed directly in Minkowski signature in the bulk. Retarded and advanced Green functions require in-falling and out-going boundary conditions at the black-hole horizon, respectively, while the full structure of the Schwinger–Keldysh formalism is in one-to-one correspondence to the causal structure of the bulk geometry.

We will use this powerful way to compute real-time properties in the field theory repeatedly in the remaining chapters.

Box 6.6   Real-time thermal Green functions in holography

Let us first recall how retarded and advanced Green functions in the Schwinger–Keldysh formalism follow from a doubled source: one on the forward time direction and one on the backward time direction. In the complex-time path integral that computes the probabilities Eq. (6.93) directly, the time integral of the Lagrangian is defined over the contour B in Fig. 6.6 on the right,

$$
\begin{aligned}
S &= \int_C dt\, L(t) \\
&= \int_{t_i}^{t_f} dt\, L(t) + \int_{t_f}^{t_f - i\sigma} dt\, L(t) - \int_{t_i - i\sigma}^{t_f - i\sigma} dt\, L(t) - \int_{t_i - i\sigma}^{t_i - i\beta} dt\, L(t),
\end{aligned} \tag{6.94}
$$

where $L(t) = \int d^d x\, \mathcal{L}[\phi(t,x), \partial_\mu \phi(t,x)]$ and $\mathcal{L}$ is the Lagrangian density. We will set the free parameter $\sigma = \beta/2$ in the remainder. By coupling to sources on both of the Minkowski parts of the contour,

$$
S_{\text{total}} = S + \int_{t_i}^{t_f} dt \int d^d x\, J_1 \mathcal{O}_1 - \int_{t_i - \beta/2}^{t_f - \beta/2} dt \int d^d x\, J_2 \mathcal{O}_2, \tag{6.95}
$$

one can calculate the propagator matrix,

$$
i G_{ab}(x - y) = i \begin{pmatrix} G_{11} & -G_{12} \\ -G_{21} & G_{22} \end{pmatrix}, \tag{6.96}
$$

Box 6.6 (Continued)

where

$$iG_{11}(t,x) = \langle T\mathcal{O}_1(t,x)\mathcal{O}_1(0)\rangle, \quad iG_{12}(t,x) = \langle \mathcal{O}_2(0)\mathcal{O}_1(t,x)\rangle, \qquad (6.97)$$

$$iG_{21}(t,x) = \langle \mathcal{O}_2(t,x)\mathcal{O}_1(0)\rangle, \quad iG_{22}(t,x) = \langle \bar{T}\mathcal{O}_2(t,x)\mathcal{O}_2(0)\rangle \qquad (6.98)$$

and T denotes the time ordering while $\bar{T}$ denotes the reversed time ordering. Recalling the definition of the retarded and advanced Green functions,

$$G_R(k) = -i\int d^{d+1}x\, e^{-ikx}\theta(t)\langle [\mathcal{O}(x), \mathcal{O}(0)]\rangle, \qquad (6.99)$$

$$G_A(k) = i\int d^{d+1}x\, e^{-ikx}\theta(-t)\langle [\mathcal{O}(x), \mathcal{O}(0)]\rangle, \qquad (6.100)$$

one can show that

$$G_{11}(k) = \text{Re}\, G_R(k) + i\coth\left(\frac{\beta\omega}{2}\right)\text{Im}\, G_R,$$

$$G_{12}(k) = G_{21}(k) = \frac{2ie^{-\frac{\beta}{2}\omega}}{1 - e^{-\beta\omega}}\text{Im}\, G_R, \qquad (6.101)$$

$$G_{22}(k) = -\text{Re}\, G_R(k) + i\coth\left(\frac{\beta\omega}{2}\right)\text{Im}\, G_R.$$

Using that $G_R(k)^* = G_A(k)$, these identities can also be written as

$$G_{11}(k) = (1+n)G_R - nG_A,$$
$$G_{12}(k) = \sqrt{n(n+1)}(G_R - G_A), \qquad (6.102)$$
$$G_{22}(k) = nG_R(k) - (1+n)G_A,$$

with $n = 1/(e^{\beta\omega} - 1)$ the Bose–Einstein distribution.

Let us now reproduce these relations from holography [202]. In the Kruskal coordinates for the AdS black hole, it is convenient to define a Kruskal time $t_K = U + V$ and Kruskal radial coordinate $x_K = U - V$. Plane waves in these Kruskal coordinates can be separated into four classes. For $\omega > 0$ these are

$$\text{out-going: } e^{-i\omega U} = e^{-i\omega(t_K - x_K)/2},$$

$$\text{out-going: } e^{i\omega U} = e^{i\omega(t_K - x_K)/2},$$

$$\text{in-falling: } e^{-i\omega V} = e^{-i\omega(t_K + x_K)/2},$$

$$\text{in-falling: } e^{i\omega V} = e^{-i\omega(t_K + x_K)/2}. \qquad (6.103)$$

The notions of "out-going" and "in-falling" refer to the viewpoint from the original Schwarzschild region. An observer in this region would have solved the (scalar) wave equation in terms of functions

$$u_{k,\text{Schw},\pm} = \begin{cases} e^{-i\omega t + k \cdot x} f_{\pm k}(r), & \text{in the Schwarzschild region,} \\ 0, & \text{in the hidden region.} \end{cases} \quad (6.104)$$

An observer in the mirrored hidden region would instead have solved the wave equation in terms of functions

$$u_{k,\text{Hid},\pm} = \begin{cases} 0, & \text{in the Schwarzschild region,} \\ e^{-i\omega t + k \cdot x} f_{\pm k}(r), & \text{in the hidden region.} \end{cases} \quad (6.105)$$

By symmetry the functions $f_{\pm k}(r)$ are the same, and clearly $f_k(r) = f^*_{-k}(r)$. It is easy to show that, near the horizon, $f_{\pm k}(r) = e^{i\omega r_*}$, with $r_* = \ln(-UV/(4\pi T))$. This shows that $u_{k,\text{Schw},+}$ and $u_{k,\text{Hid},+}$ are out-going modes and $u_{k,\text{Schw},-}$ and $u_{k,\text{Hid},-}$ are in-falling modes. The solutions u_k mix positive and negative frequencies, however. By combining solutions from the Schwarzschild region with those for the hidden region, one can separate them out:

$$u_{\text{out},+} = u_{k,\text{Schw},+} + e^{-\frac{\beta}{2}\omega} u_{k,\text{Hid},+},$$

$$u_{\text{out},-} = u_{k,\text{Schw},+} + e^{\frac{\beta}{2}\omega} u_{k,\text{Hid},+},$$

$$u_{\text{in},+} = u_{k,\text{Schw},-} + e^{-\frac{\beta}{2}\omega} u_{k,\text{Hid},-}, \quad (6.106)$$

$$u_{\text{in},-} = u_{k,\text{Schw},-} + e^{\frac{\beta}{2}\omega} u_{k,\text{Hid},-}.$$

These are the basis functions in terms of which the solution to the scalar equation of motion can be written. Note that we have solved the system by solving the wave equation independently in two disconnected regions. The unique solution is therefore determined by two continuity conditions at the "interface" – the horizon – and two boundary conditions. The continuity conditions are natural and intuitive: positive-frequency modes should be in-falling, whereas negative-frequency modes should be out-going. Thus the appropriate solution for a scalar wave behaves as

$$\phi_k(r) = C_{-,k} u_{\text{out},-} + C_{+,k} u_{\text{in},+}. \quad (6.107)$$

The values of the two coefficients are determined by the boundary conditions on the two boundaries $\phi = J_1|_{\partial\text{Schw}}$ and $\phi = J_2|_{\partial\text{Hid}}$. The explicit solution becomes

$$\phi_k(r)\big|_{r \in \text{Schw}} = \big((n+1) f^*_k(r) - n f_k(r)\big) J_{1,k}$$
$$+ \sqrt{n(n+1)} \big(f_k(r) - f^*_k(r)\big) J_{2,k}, \quad (6.108)$$

$$\phi_k(r)\big|_{r \in \text{Hid}} = \sqrt{n(n+1)} \big(f^*_k(r) - f_k(r)\big) J_{1,k}$$
$$+ \big((n+1) f_k(r) - n f^*_k(r)\big) J_{2,k},$$

Box 6.6   (Continued)

where $n = 1/(e^{\beta \omega} - 1)$ is again the Bose–Einstein distribution. We have used here the freedom to normalise $f_k(r)$ to unity at the boundary,

$$f_k(r)\big|_{\partial \text{Schw}} = 1 = f_k(r)\big|_{\partial \text{Hid}}; \tag{6.109}$$

this is obviously the leading term as we approach the boundary on the RHS of Eq. (6.108).

With the explicit solution in hand, we can now straightforwardly apply the GKPW rule. The one point requiring care is that there are two boundaries. The bulk on-shell action vanishes as before (in section 5.1) and the only terms are

$$S_{\text{on-shell}} = \frac{1}{2} \int_{\partial \text{Schw}} \frac{d^{d+1}k}{(2\pi)^{d+1}} \sqrt{-g} g^{rr} \phi_{-k} \, \partial_r \phi_k$$

$$- \frac{1}{2} \int_{\partial \text{Hid}} \frac{d^{d+1}k}{(2\pi)^{d+1}} \sqrt{-g} g^{rr} \phi_{-k} \, \partial_r \phi_k. \tag{6.110}$$

On substituting the explicit solution, we find

$$S_{\text{on-shell}} = \frac{1}{2} \int \frac{d^{d+1}k}{(2\pi)^{d+1}} \sqrt{-g} g^{rr}$$

$$\times \Bigg[\phi_{1,-k} \big((1+n) f_k \, \partial_r f_k^* - n f_k^* \, \partial_r f_k \big) \phi_{1,k}$$

$$+ \phi_{1,-k} \sqrt{n(1+n)} \big(-f_k \, \partial_r f_k^* + f_k^* \, \partial_r f_k \big) \phi_{2,k}$$

$$+ \phi_{2,-k} \sqrt{n(1+n)} \big(-f_k \, \partial_r f_k^* + f_k^* \, \partial_r f_k \big) \phi_{1,k}$$

$$+ \phi_{2,-k} \big(n f_k \, \partial_r f_k^* - (1+n) f_k^* \, \partial_r f_k \big) \phi_{2,k} \Bigg]. \tag{6.111}$$

After identifying

$$G_R = -\sqrt{-g} g^{rr} f_k \, \partial_r f_k^* \big|_{\partial \text{AdS}} \, , \quad G_A = -\sqrt{-g} g^{rr} f_k^* \, \partial_r f_k \big|_{\partial \text{AdS}}, \tag{6.112}$$

one sees immediately that the functional derivatives with respect to the sources J_1 and J_2 do exactly reproduce the Schwinger–Keldysh Green functions. Were $f_k(r)$ not normalised to unity, the appropriate expression for G_R would be

$$G_R = -\sqrt{-g} g^{rr} \frac{1}{f_k f_k^*} f_k \, \partial_r f_k^* \big|_{\partial \text{AdS}} = -\sqrt{-g} g^{rr} \frac{\partial_r f_k^*}{f_k^*} \bigg|_{\partial \text{AdS}}, \tag{6.113}$$

and we recognise the standard ratio of sub-leading to leading coefficients of the near-boundary solution to the equation of motion, where the other boundary condition on $f_k^*(r)$ is now set by in-falling behaviour at the horizon. It can be shown that the derivation presented here for the scalar fields straightforwardly generalises to the case of fermion fields [246].

7

Holographic hydrodynamics

Any system at a finite temperature in the macroscopic realms of large distances and long times that does not break translational or rotational symmetry has to turn into a fluid that is governed by the nineteenth century Navier–Stokes theory of hydrodynamics. The structure of the hydrodynamical equations is universal, rooted in a strong emergence principle associated with globally conserved quantities. However, one needs "microscopic" data in order to assign values to the hydrodynamical parameters. These include the equation of state, but also the phenomenologically important transport coefficients, such as the shear and bulk viscosities, which account for the dissipative properties of the fluid. For a weakly interacting "particle-physics" system it is well established how to compute these numbers through Boltzmann's kinetic equations. To be practical as a computational device, it relies on a perspective where the transport of macroscopic quantities is the consequence of microscopic collision processes between nearly free particles. Even for purely classical fluids this procedure fails when these are dense and strongly interacting. A case in point is that the computation of the viscosity of liquid water, departing from the microscopic physics of water molecules, is to a degree still a challenge.

This procedure completely breaks down in a strongly interacting quantum critical state. As we discussed at length in chapter 2, at zero temperature this "un-particle" theory has no nearly free particle-like excitations in its spectrum. Nevertheless, at long distances and late times, the conservation laws which are the foundation of fluid mechanics still hold, and a fluid description should emerge, including its dissipative properties. In section 2.1 we presented elementary scaling considerations leading to the general notion of Planckian dissipation, which is applicable to any strongly interacting quantum critical state. This amounts to the statement that the energy relaxation is governed by a time $\tau_E \simeq \hbar/(k_B T)$, which can be argued to be the shortest possible "entropy-production time" admissible by the general principles of thermal field theory. This requires only temporal scale invariance of the quantum dynamics and therefore it should be a very general notion. However, there are still questions of principle on a quantitative level. This includes the values of the prefactors that set the absolute magnitudes of the characteristic relaxation times, but also the precise form of the universal scaling functions of $\omega/(k_B T)$ governing the behaviour of dynamical linear response quantities.

These questions also arise in the context of the "bosonic" (zero-density) CFTs, which can in principle be "solved" numerically using quantum Monte Carlo [244]. Although thermodynamical quantities can be obtained to any desired accuracy in this way, numerics fails to address the long-time dynamical properties at finite temperature, even dealing with the simplest field theories. The reason lies in the "information-loss problem" associated with the Wick rotation. One computes in Euclidean space-time, but any noise is detrimental to the analytic continuation to real time in the hydrodynamical regime. As a consequence, even for the simple complex scalar CFT associated with e.g. the superfluid to Bose Mott-insulator transition in $2+1$ dimensions as discussed in section 2.1, the forms of the $\omega/(k_B T)$ scaling functions are not known accurately. This is also a challenge for "lattice QCD" dealing with the regime of nuclear densities. Although the proton mass, equation of state and so forth can be computed with quite high precision, the hydrodynamical properties of e.g. the quark–gluon plasma are "shrouded behind the Wick rotation".

There is a crucial observation to be made here. As we discussed in chapter 2, the Planckian dissipation principle was already fully recognised some twenty years ago [247] in the condensed matter community [25]. Its power is pre-eminent if the hydrodynamic description applies. However, translational invariance is broken in the typical condensed matter systems of interest at the microscopic scale, and this motive plays a key role in the transport theory of the conventional electron systems realised in solids (Fermi liquids, band and even Mott insulators and so forth). The role of translational symmetry breaking in the quantum critical systems is a very interesting affair, which will be discussed in great detail in chapter 12. The conclusion will be that it is a highly delicate affair to discern the truly hydrodynamical nature of the finite-temperature quantum critical liquid directly in measurable transport properties. For this reason, the question regarding the role of Planckian dissipation in the context of hydrodynamics was just not considered. Otherwise it would have already been common wisdom in this community that the ratio of viscosity and entropy density should have been of order $\eta/s = T\tau_E \simeq \hbar/k_B$ according to the simple dimensional analysis presented in section 2.1.

Why does holography make a great difference in this regard? The reason is simply that the Wick rotation in the boundary field theory is effortlessly accomplished in the bulk by switching from Euclidean to Minkowski signature, as we explained in detail in chapter 5, and in the previous chapter, chapter 6, we learned how to encode finite temperature in terms of a bulk black hole. Hydrodynamics describes the collective response of the system to small deviations from equilibrium, and in particular the response associated with the collective energy and momentum of the many-body state. We have also seen that the energy–momentum tensor in the boundary dualises in the propagation of gravitons in the bulk. As we highlighted, it is among the most elementary dynamical quantities to consider in holographic duality.

These insights together show that holography is almost tailor-made for the computation of transport in the finite-temperature quantum critical systems. Historically, the holographic computation by Policastro, Son and Starinets [46, 248] of the viscosity in a critical quantum fluid was the birth of the subject of this whole book. In 2001 they showed that according to holography the viscosity/entropy density ratio of the hydrodynamical fluid formed from the $\mathcal{N} = 4$ supersymmetric Yang–Mills theory in $3 + 1$ dimensions equals $\eta/s = (1/(4\pi))\hbar/k_\mathrm{B}$. Nearly simultaneously, it was realised that the quark–gluon plasma as produced in the heavy-ion collisions at the RHIC accelarator in Brookhaven showed a much less viscous behaviour than expected from perturbative computations: it appears to behave as an almost perfect fluid [249]. In 2003/2004 the news broke that the QGP produced at the RHIC displayed an anomalously small η/s ratio, which turned out to be very close to the holographic result [250, 251, 252]. This was the first direct confrontation of a prediction by holography with experiment. Also the direct contact between condensed matter physics and holography achieved in 2007 was established through this theme of quantum critical transport [24, 253]. This was inspired by Nernst-effect experiments in strongly under-doped cuprate superconductors. The Nernst coefficient is a complicated transport coefficient associated with the electric current in the y direction induced by a temperature gradient in the x direction in the presence of a magnetic field in the z direction. In the empirical context of under-doped cuprates it is not unreasonable to assert that this relates to a Mott insulator–superconductor quantum critical point in $2 + 1$ dimensions forming a natural arena for quantum critical hydrodynamics. However, even in hydrodynamics the computation of a magneto-electric–thermal transport phenomenon like the Nernst effect is quite a hairy affair, and the holographic gravity calculation in the bulk was employed to nail down the general structure of the linear-response theory that was subsequently re-derived directly in the hydrodynamical theory.

Initially holography was used to investigate the linear-response properties of this hydrodynamical regime. These are relatively easy to compute. After reviewing the basics of (relativistic) hydrodynamics, we will explain the holographic linear-response computation for the viscosity at length in section 7.1. Besides its far-reaching consequences for physics, it is also a good playground to exercise holographic techniques. Hydrodynamics, however, is more than linear response, since it is able to fully address quite finite departures from equilibrium: the theory is solely controlled by the requirement that the length and time scales associated with the flow are large compared with any microscopic scale. In section 7.2 we will present an iconic achievement of holographic duality: the explicit derivation of the Navier–Stokes equations in the boundary from the gravity in the bulk [254]. This result by Bhattacharyya, Hubeny, Minwalla and Rangamani is a stunning demonstration of the powers of holography as a "generating functional of phenomenological theories governed by strong emergence". Quantum field theory, general relativity and Navier–Stokes hydrodynamics are theories with very distinct mathematical structures, seemingly alluding to very different forms of physical reality. Holographic duality unifies these three theories into a single coherent whole. Much of what will follow after this chapter is based on the hope that holography exerts a similar magic when dealing with the unknowns of the zero-temperature quantum matter systems. Just like for "hydrodynamics", what is at stake is whether the *mathematical structure* of the holographic phenomenological theories is also in this context generic.

The conventional strategy used to derive hydrodynamical equations is disciplined; it rests on symmetry principles and the gradient expansion. However, when the circumstances become more complicated due to the presence of magnetic fields, thermal gradients or other environmental forces, this can turn into quite a hazardous affair, where it is easy to overlook delicate but crucial points. The discovery of the gravitational dual has, surprisingly, triggered a revival of the study of such forms of complicated hydrodynamics, with some remarkable results already. In the gravitational dual, the "messy" hydrodynamical derivations turn into a transparent set of algebraic equations that one can solve rather mechanically, albeit with sufficient proficiency. The dualisation to the boundary is straightforward with the GKPW rule and, knowing the answer, one can reconstruct matters following the conventional strategy. To illustrate the power of holography to address such new questions in hydrodynamics, we include a discussion at the end of this chapter of the effect on hydrodynamics of quantum anomalies of the underlying quantum field theory. It turns out that there are parity-violating non-dissipative terms in the hydrodynamical evolution equations that are completely consistent with all principles, but were simply overlooked in the long history of the subject.

In the penultimate section, section 7.3, we will change gear, and we will present the first preparations for the holographic description of matter in a proper sense. The term matter refers to "stuff" that does not disappear: we have to deal with a conserved global charge. The first topic of interest is matter that is characterised by just a simple density associated with a $U(1)$ charge. Adhering to the dictionary, we begin by introducing a Maxwell gauge field in the bulk, which is now minimally described by Einstein–Maxwell theory. The dynamics of this gauge field is dual to that of a conserved global $U(1)$ charge in the boundary field theory. This will allow us to compute the frequency-dependent "optical" conductivity, a quantity that is of central interest for experiments in real systems. The optical conductivity in the boundary turns out to be dual to the propagation of electromagnetic waves ("photons") in the bulk. Here we consider as a warm-up the holographic system described by an AdS–Schwarzschild black-hole background dual to finite-temperature zero-density CFT. Specifically in $2 + 1$ dimensions, one faces an interesting question of principle regarding the scaling behaviour of the optical conductivity that cannot be addressed directly because of the Wick-rotation issue. However, holography turns out to shed interesting light on this matter.

7.1 Quantum criticality and the minimal viscosity

We will present here the computation of the shear viscosity of a quantum critical system. This has been the first example of the use of the GKPW formula to compute a dissipative linear-response quantity using the propagation of classical fields in the bulk. It is remarkably straightforward. Shear viscosity encodes the dissipation of a transverse motion in the fluid, due to the relaxation of its momentum. From the dictionary we know that the momentum operator is encoded in the fluctuations of the metric itself and we therefore need only the pure AdS–Einstein theory in the bulk with action

$$S = \frac{1}{16\pi G} \int d^{d+2}x \sqrt{-g}\Big[R - 2\Lambda\Big]. \tag{7.1}$$

In terms of the description of the field theory, it should be noted that one considers here the hydrodynamics of a relativistic fluid, which looks a bit different from the familiar non-relativistic Navier–Stokes theory. The requirement of Lorentz invariance actually simplifies the structure of relativistic hydrodynamics compared with the non-relativistic limit. Recall the non-relativistic Navier–Stokes equation [255, 256],

$$\rho\frac{D\vec{v}}{Dt} = -\nabla p + \nabla \cdot \mathbb{T} + \vec{f}, \tag{7.2}$$

where ρ is the mass density, $\vec{v}$ is the fluid velocity and

$$\frac{D\vec{v}}{Dt} = \frac{\partial\vec{v}}{\partial t} + (\vec{v} \cdot \nabla)\vec{v} \tag{7.3}$$

is the "material" derivative. This equation just expresses that the mass density times the acceleration of the fluid is equal to the forces exerted on the fluid, as collected on the right-hand side. This includes the gradient of the pressure p, the external forces $\vec{f}$ and the divergence of the stress tensor $\mathbb{T}$. The latter describes the viscous stresses which are in turn proportional to the gradients of the velocity field,

$$\mathbb{T}_{ij} = \eta \left(\partial_i v_j + \partial_j v_i - \frac{2}{3}\delta_{ij}\,\partial_k v_k \right) + \zeta \delta_{ij}\,\partial_k v_k. \qquad (7.4)$$

The constants of proportionality in this equation are the shear (η) and bulk (ζ) viscosities associated with the components transverse and parallel to the flow. For *incompressible* flows ($\partial_i v_i = 0$ and typical velocities small compared with the sound velocity) only the shear viscosity plays a role.

The Navier–Stokes equation is a clear generalisation of Newton's second law. Fundamentally Newton's second law is a consequence of the conservation of linear momentum in the Galilean continuum. Together with the conservation of energy, this implies that in a relativistic system the Navier–Stokes equation can be written in a very compact form,

$$\partial_\mu T^{\mu\nu} = 0. \qquad (7.5)$$

It simply expresses the conservation of the energy–momentum tensor $T^{\mu\nu}$.

The structure of the relativistic hydrodynamics is now obtained straightforwardly, by assuming that the stress tensor is that of fluid with a local four-velocity $u^\mu(\mathbf{x}, t)$, conventionally normalised to $u^\mu u_\mu = -1$ (Minkowski signature). Using the assumption of local equilibrium, one then recasts the local energy and pressure/momentum densities encoded in the stress tensor in terms of the local temperature $T(x)$ and the d components of fluid velocities through a gradient expansion [257]. With these constitutive relations which reduce the number of components of the stress tensor from $(d+1)(d+2)/2$ to $d+1$, the number of equations and unknowns matches, and the fluid behaviour can be determined. To zeroth order in gradients, the stress tensor is that of a "perfect fluid" as characterised by vanishing viscosity. According to the textbook [256],

$$T^{\mu\nu}_{(0)} = (\epsilon + P)u^\mu u^\nu + Pg^{\mu\nu}, \qquad (7.6)$$

where $g^{\mu\nu}$ is the background metric of the space-time (in the examples below we shall consider a flat space-time: $g^{\mu\nu} = \eta^{\mu\nu}$), while ϵ and P are the free-energy density and pressure, respectively, associated with the local equilibrium temperature $T(x)$. In addition, the thermodynamical relations $d\epsilon = T\,ds$, $dP = s\,dT$, and $\epsilon + P = Ts$ are required, where s is the entropy density. Organising the expansion in terms of the rotational symmetry of the fluid at rest, the next-to-leading-order gradient terms of the relativistic Navier–Stokes theory have the form

$$T^{\mu\nu} = T^{\mu\nu}_{(0)} + T^{\mu\nu}_{(1)}, \qquad (7.7)$$

$$T^{\mu\nu}_{(1)} = -P^{\mu\kappa}P^{\nu\lambda}\left[\eta \left(\partial_\kappa u_\lambda + \partial_\lambda u_\kappa - \frac{2}{d}g_{\kappa\lambda}\,\partial_\alpha u^\alpha \right) + \zeta g_{\kappa\lambda}\,\partial_\alpha u^\alpha \right], \qquad (7.8)$$

where $P^{\mu\nu} = g^{\mu\nu} + u^\mu u^\nu$ is the projection operator onto directions perpendicular to u^ν. The coefficients η and ζ are the shear and bulk viscosities. Just as the pressure and energy – the coefficients of the zeroth-order expansion – the viscosities are functions of the local temperature. Meanwhile, the hydrodynamics should follow the physical requirements imposed by the existence of entropy current with non-negative divergence which can constrain transport coefficients η, ζ to be non-negative. The usual non-relativistic Navier–Stokes equation is just the special limit obtained from Eq. (7.5) by taking the limit $\varepsilon \to 0$, after the rescalings

$$\partial_i \to \varepsilon\, \partial_i, \quad \partial_\tau \to \varepsilon^2\, \partial_\tau, \quad v_i \to \varepsilon v_i, \quad P \to \bar{p} + \varepsilon^2 P. \tag{7.9}$$

In almost all examples we shall consider the fluid to be associated with the finite-temperature state of the quantum critical fluid described by the conformal field theory. For such special fluids the conformal symmetry imposes the additional constraint that the stress tensor is traceless: $g_{\mu\nu} T^{\mu\nu} = 0$. This reduces the number of parameters in the constitutive relations. At zeroth order this imposes that the pressure is directly related to the energy density $P = (1/d)\epsilon$, while at first order it has the consequence that the bulk viscosity must vanish identically $\zeta = 0$. To this order a conformal fluid is thus completely specified by just two parameters: its energy and shear viscosity.

If there is a $U(1)$ conserved current associated with the fluid system, one has to deal with an additional hydrodynamical equation, the continuity equation:

$$\partial_\mu J^\mu = 0. \tag{7.10}$$

Similarly, we have the constituent relation

$$J^\mu = n u^\mu - \sigma T P^{\mu\nu} \partial_\nu(\mu/T), \tag{7.11}$$

with $n(T, \mu)$ equal to the charge density and $\sigma(T, \mu)$ the conductivity. In addition, a new thermodynamic parameter is required: the local chemical potential μ.

The objective is now to determine these parameters for a specific fluid. The zeroth-order energy, pressure and charge density follow from textbook calculations. The transport coefficients are defined by the response of the system to an infinitesimal perturbation away from equilibrium, and are the subject of linear-response theory. Because of the fluctuation–dissipation theorem, each transport coefficient can be extracted from a two-point correlation function through a Kubo relation [258]. The shear viscosity follows from the absorptive part of the spatially transverse energy–momentum (T_{xy}) propagator through

$$\eta = \lim_{\omega \to 0} \frac{1}{\omega} \operatorname{Im} G^R_{xy,xy}(\omega, \mathbf{k} = 0), \tag{7.12}$$

where $G^R_{xy,xy}(\omega, 0)$ is the *retarded* Green function of the xy component of the energy–momentum tensor, defined as

$$G^R_{\mu\nu,\alpha\beta}(\omega, \mathbf{k}) = -i \int dt \, d\mathbf{x} \, e^{i\omega t - i\mathbf{k}\cdot\mathbf{x}} \theta(t) \langle [T_{\mu\nu}(t, \mathbf{x}), T_{\alpha\beta}(0, \mathbf{0})] \rangle. \qquad (7.13)$$

We now see why holography is ideally suited to perform this computation; all the ingredients are at hand. In our introduction to the AdS/CFT basics in chapter 5, we considered precisely the energy–momentum tensor in the boundary, concluding that this is dual to metric perturbations in the bulk. The excitations of the transverse components T_{xy} dualise literally to transverse gravitational waves in the bulk. The difference from chapter 5 is that we are now considering the system at a finite T. The dual gravitational description is not a gravitational wave in pure AdS, but, as we explained in chapter 6, we have to consider instead what happens in the presence of an AdS black hole. In addition, the causality structure of the propagators in the field theory is carefully reflected in the behaviour of the wave functions near the horizon. The linear-response dynamical susceptibilities are associated with the retarded propagators of the real-time formalism, and we learned in section 6.4 that these require *in-falling* boundary conditions at the horizon in the bulk. The stuff that disappears behind the horizon in the bulk encodes for the absorption (dissipation) in the boundary field theory.

Formally the computation is thus straightforward. Technically, the presence of the black hole complicates the bulk equations of motions of the gravitons – they are readily solved numerically, but the analytic computations require some insight (box 7.1). This exploits the conservation of the flux density upon moving inwards along the radial direction to relate the long-time hydrodynamical regime of the field theory directly to the near-horizon geometry in the bulk [237]. This happens to be a very particular property related to the fact that the shear viscosity does not renormalise [259]. As such it is in general not shared by other transport coefficients, but it again shows the beautiful way RG field-theory properties are encoded in the radial AdS geometry.

The outcome can be understood on the basis of a very simple argument, resting on the fact that the Kubo relation involves the absorptive part of the correlation function. As we have already emphasised, the long-time limit of the finite-temperature boundary is eventually described by the near-horizon physics in the bulk. Given the rule that the absorptions in the boundary are associated with stuff falling through the horizon in the bulk, while the energy–momentum in the boundary is dual to gravitons, it should be the case that the viscosity is proportional to the absorption cross section of gravitons by the black hole. This turns out to be correct. The exact computation gives [46, 252],

$$\eta = \frac{\sigma_{abs}(0)}{16\pi G},$$ (7.14)

where $\sigma_{abs}(0)$ is the graviton absorption cross section of the black hole in the $\omega \to 0$ limit. In full generality, a zero-frequency absorption cross section measures the size of an object times its density, regardless of the nature of the probe. The gravitons are not an exception, and indeed

$$\sigma_{abs}(0) = \frac{A_{hor}}{\text{Vol}_d},$$ (7.15)

where A_{hor} is the horizon area of the black hole. With Hawking and Bekenstein's insight, this is now readily related to quantities in the field theory. The horizon area gives the entropy of the system, Eq. (6.27) in units of $4G$. The absorption cross section is therefore precisely the entropy density s in units of $4G$. We thus find [46]

$$\eta = \frac{1}{4\pi}s.$$ (7.16)

We will present the full calculation in box 7.1 at the end of this section.

This is by now a famous result. Note that the ratio of the viscosity and the entropy density becomes completely independent of anything other than geometrical factors. Upon restoring explicit units, one has

$$\frac{\eta}{s} = \frac{1}{4\pi}\frac{\hbar}{k_B}.$$ (7.17)

That this result has a special sense of universality is clearly evident. The uniqueness of the solutions of Einstein gravity, together with the dictionary, insists that there is no other outcome possible, and the result is therefore generic for any translationally invariant field-theoretical plasma with an Einstein-gravitational dual [46, 252, 260, 261]. It turns out even to remain true in systems at a finite chemical potential [262, 263]. In addition to its universality, the notable aspect is that, compared with typical values found in normal fluids, the numerical value of Eq. (7.17) is extremely small [252]. We stress that this is in the first instance due to the underlying Planckian dissipation principle, rendering its dimension of order $\hbar/k_B$. In addition, the prefactor $1/(4\pi)$ is very small. The smallness reflects the extremely strong coupling of the system. The value $1/(4\pi)$ is so small that it has in fact been conjectured to be a "minimal viscosity" – the KSS bound [260] – the closest approach any many-body system can make to the realisation of the perfect, non-viscous fluid. This would validate a long-held thought that quantum effects guarantee a minimal amount of dissipation.

The holographically computed value from classical gravity is of course associated with the large-N and large 't Hooft coupling of the CFT. Upon closer inspection, it appears that one can in fact approach the perfect fluid even more closely by moving away from the large 't Hooft coupling limit. In the explicit example of the $AdS_5 \times S^5$ gravity dual to $\mathcal{N} = 4$ super-Yang–Mills, finite $\lambda = g_{YM}^2 N$ corrections do increase the ratio [264]. However, more generic theories allow for a richer set of higher-order derivative correction in the bulk (Einstein–Gauss–Bonnet gravity) corresponding to effects of finite 't Hooft coupling in the dual gauge theory that do show that the prefactor can be smaller than $1/(4\pi)$ [265]. Violation of the KSS bound was also found in an anisotropic system (e.g. [266, 267]), as well as in some particular top-down models [268]. More detailed arguments based on causality constraints do suggest that a new lower bound exists, but its details depend on bulk physics [265, 269, 270].

Summary 1 The viscosity-to-entropy-density ratio η/s is a natural measure for the "Planckian dissipation". According to AdS/CFT the viscosity of the field theory is proportional to the absorption cross section of gravitons by a black hole. For any isotropic field theory dual to classical Einstein gravity this yields the universal ratio $\eta/s = (1/(4\pi))\hbar/k_B$. It is believed that the factor $1/(4\pi)$ is very close to an absolute lower bound for this ratio. The "minimal viscosity" is an upper bound for the dissipative power of this hydrodynamical liquid.

This ratio of viscosity and entropy turns out to be a natural observable quantity in experiments on relativistic plasmas. The holographic results caused a big splash when they were compared with estimates for this quantity in the quark–gluon plasma derived from experiments at the RHIC in Brookhaven [250]. The purpose of this machine was to create and study the quark–gluon plasma (QGP): at sufficiently high density and temperatures nuclear matter is supposed to undergo a phase transition to this deconfined plasma. By smashing heavy nuclei together at very high energies, such a high-density state can ever so briefly be created. Phenomenological estimates of the viscosity-to-entropy-density ratio did exist, but, until the advent of AdS/CFT, the only way to address the quantity η/s directly in QCD was by weak-coupling perturbation theory departing from the asymptotically free UV regime. Lattice calculations are of no help for time-dependent quantities because of the problems with the Wick rotation. According to the perturbative computations, the ratio is of the form (see e.g. [11, 271, 272])

$$\frac{\eta}{s}\bigg|_{\lambda \to 0} = \frac{A}{\lambda^2 \ln(B/\sqrt{\lambda})}, \tag{7.18}$$

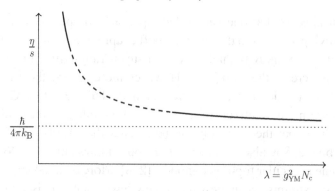

Figure 7.1 The qualitative behaviour of the ratio of the shear viscosity to the entropy density as a function of the 't Hooft coupling $\lambda = g^2 N_c$. In the perturbative regime the viscosity always scales inversely with the coupling, since it measures the relaxation rate. The solid line for small λ is based on perturbative computations of the first correction around perturbatively free Yang–Mills [11]. In the holographic dual of the strongly coupled theory one naturally finds values of order unity. The solid line for large λ is based on a holographic computation of the large-N, large-g^2N holographic dual of $\mathcal{N} = 4$ super-Yang–Mills [264]. The monotonic interpolation is a guess. Figure adapted from [252]. (Reprinted with permission from the American Physical Society, © 2005.)

where A, B are constants of order 1 and theory-dependent. In the weak-coupling case, these have a strong dependence on the 't Hooft coupling λ, with the ratio diverging when λ flows to 0 in the deep UV (Fig. 7.1). This just reflects the somewhat confusing fact that the viscosity of a near-ideal gas is very high because the momentum is carried around by the microscopic degrees of freedom over long distances between the collisions. It completely agrees with the qualitative insight that perturbative computations of the viscosity-to-entropy ratio will always result in ratios greater than unity. It was therefore a big surprise when the RHIC found that the viscosity-to-entropy ratio of the QGP is within the error bar consistent with the AdS/CFT result [11, 250, 251]. Remarkably, this indicated that, despite the transition to a deconfined state, the plasma remained strongly coupled. Though earlier phenomenological estimates argued that just such a thing might happen, it was a shock nevertheless. For these reasons the observed strongly coupled QGP is often denoted as sQGP to distinguish it from its naively weakly coupled cousin.

To end this physics story, let us clarify why the designation "strongly coupled" is, strictly speaking, not specific enough to explain the small-viscosity phenomenon. At extremely strong coupling, one would rather expect that the system reorganises itself into a confined state that is a stable state, with the confinement energy as the characteristic scale and weakly interacting quasiparticles (mesons and so forth) as excitations below this scale. Instead, AdS/CFT is describing the "unparticle physics" associated with the strongly interacting *quantum critical* state lacking particle-like excitations. It is subject to debate why and to what degree the quark–gluon plasma under the experimental conditions can be described as a quantum critical state with its holographic AdS dual [11, 273]. Let us briefly mention that this theme has been becoming quite prominent also in the cold-atom community. By tuning a Feshbach resonance, it appears possible to create a "unitary Fermi gas" that is strongly interacting but scale-free due to the diverging scattering length. Various claims that the minimal viscosity has been observed also in these systems have appeared, e.g. see Refs. [274, 275].

Box 7.1 The viscosity from linearised gravity in a Schwarzschild black-hole geometry

Let us present the explicit holographic computation for the shear viscosity of a strongly interacting plasma [201]. This plasma is the deconfined finite-temperature state whose dual description is the AdS–Schwarzschild black hole. We already introduced the quantity which has to be computed: the viscosity is given by the Kubo formula (7.12), where $G^R_{x_1x_2,x_1x_2}(\omega, \vec{0})$ is the retarded Green function of the "xy" component (we use "x_1, x_2" here) of the energy–momentum tensor as defined by Eq. (7.13). The task is to compute the retarded Green function using the GKPW formula. The power of AdS/CFT is that we can do so directly in real time following the prescription in chapter 6. Holographically the stress-tensor perturbations are encoded in fluctuations of the metric. As usual, one decomposes the metric as $g_{\mu\nu} = \bar{g}_{\mu\nu} + h_{\mu\nu}$, where $\bar{g}_{\mu\nu}$ is the background metric which solves for the AdS–Schwarzschild black hole and $h_{\mu\nu}$ is the infinitesimal metric fluctuation: the gravitational wave or graviton.

The counting of the number of degrees of freedom (DOF) for the massless graviton is a bit intricate. For concreteness, and because historically this was the first choice made, we will consider a $d + 2 = (4 + 1)$-dimensional bulk dual to a $d + 1 = (3 + 1)$-dimensional strongly interacting field theory. The $(4 + 1)$-dimensional graviton modes can be classified as follows [276]. Owing to rotation invariance and without loss of generality, the spatial momentum is taken in the x_3 direction $\mathbf{k} = (0, 0, k)$ and the perturbations correspond to $h_{\mu\nu}(t, r, x_3)$. The system has an $O(2)$ symmetry in the $x_1 - x_2$ plane and, given the way in which the graviton modes transform under this symmetry, these decouple into three sets:

(i) the tensor mode (transverse mode): $h_{x_1x_2}$;
(ii) the vector mode (shear mode): $h_{tx_1}, h_{x_3x_1}, h_{rx_1}, h_{tx_2}, h_{x_3x_2}$ and h_{rx_2};
(iii) the scalar (sound) mode: $h_{tt}, h_{tx_3}, h_{x_3x_3}, h_{x_1x_1} + h_{x_2x_2}, h_{rr}, h_{tr}$ and h_{rx_3}.

This classification extends to other dimensions as well in an obvious way.

Box 7.1   (Continued)

The tensor mode is the one of relevance to the viscosity computation. It follows directly from the linearised coordinate transformations on $h_{\mu\nu}$,

$$\delta h_{\mu\nu} = \bar{\nabla}_\mu \xi_\nu + \bar{\nabla}_\nu \xi_\mu, \tag{7.19}$$

that coordinate transformations consistent with the choice $h_{\mu\nu}(t, r, x_3)$, i.e. $\xi_\mu = \xi_\mu(r)e^{-i\omega t + ikx_3}$, are compatible with the decomposition. Thus $h_{x_1 x_2}$ in this momentum configuration is a gauge-invariant term and thereby a physical mode.

The next step is to perturb the Einstein equation Eq. (6.9) around the solution describing the AdS–Schwarzschild black-hole geometry Eq. (6.19). In other words, we insert the metric $g_{\mu\nu} = \bar{g}_{\mu\nu}^{\text{AdS–Schw}} + h_{\mu\nu}$ into Einstein's equation for the AdS background and expand to linear order in $h_{\mu\nu}$. On performing a partial Fourier transform of the perturbation $h_{x_2}^{x_1}(t, r, x_3) \equiv g^{x_1 \mu} h_{\mu x_2}$ in the x_3 direction [265],

$$h_{x_2}^{x_1}(t, r, x_3) = \int \frac{d\omega\, dk}{(2\pi)^2} \phi(r; \omega, k) e^{-i\omega t + ikx_3}, \tag{7.20}$$

one obtains a linearised equation of motion for the Fourier components $\phi(r; \omega, k)$. One finds that this isolated tensor mode propagates in the same way as the massless scalar discussed in box 5.1 and Eq. (5.23), with equation of motion

$$\frac{1}{\sqrt{-g}} \partial_\mu \left[\sqrt{-g} g^{\mu\nu} \partial_\nu h_{x_2}^{x_1}(r; t, x_3) \right] = 0. \tag{7.21}$$

The difference with the problem discussed in chapter 5 is that there is now a black hole present, and substituting the background metric Eq. (6.19) we find after a Fourier transformation,

$$\phi''(r; \omega, k) + \left(\frac{d+2}{r} + \frac{f'}{f} \right) \phi'(r; \omega, k) + \frac{(\omega^2 - k^2 f)L^4}{r^4 f^2} \phi = 0 \tag{7.22}$$

with the emblackening factor for the $(d+2)$-dimensional AdS black hole $f = 1 - r_0^{d+1}/r^{d+1}$. Note that we are using coordinates r, where $r = 0$ is the interior and $r = \infty$ is the boundary.

To obtain the retarded Green function in the CFT we must now solve this equation with in-falling boundary conditions at the horizon. According to the GKPW rule, the retarded Green function is then given by the ratio of the sub-leading coefficient to the leading coefficient of the solution at the boundary:

$$\phi_{\text{sol}}(u, \omega, k) = A(\omega, k) r^{\Delta - d - 1} + B(\omega, k) r^{-\Delta} + \cdots, \tag{7.23}$$

with $\Delta = d + 1$ since the graviton is massless, and

$$G_{\text{R},xy,xy}^{\text{CFT}}(\omega, k) = (2\Delta - d - 1)\frac{B(\omega, k)}{A(\omega, k)}. \tag{7.24}$$

Differently from the pure AdS case discussed in chapter 5, closed solutions are no longer available for $A(\omega, k)$ and $B(\omega, k)$. Although it is relatively straightforward to obtain these numerically, analytical solutions can be obtained for small frequencies and small momenta of interest here [277]. These can be used in turn to compute the viscosity via the Kubo formula (7.12) in closed form. This is a multi-step computation. We will describe the more general methods in chapter 9. With only a minor extra investment it is possible to evaluate the viscosity Kubo formula in a much more elegant way.

The hydrodynamical long-distance and late-time regime in the boundary should be governed by the near-horizon geometry in the bulk, and the computation which now follows [237, 278] exploits this fact to maximal effect. The key property which makes it work is the fact that the flux of the gravitational radiation is conserved along the radial direction. The first step is to rewrite the GKPW formula for the propagator in terms of the Wronskian which encodes this flux.

For this purpose, we first rewrite the Green function Eq. (7.24) directly in terms of the solution $\phi_{\text{ns}} = r^{d+1-\Delta}\phi_{\text{sol}}$ following the GKPW construction:

$$G_{\text{R}}^{\text{CFT}}(\omega, k) = -\frac{1}{2\kappa^2} \lim_{r \to \infty} r^{-2(d+1-\Delta)} \sqrt{-g}g^{rr} \frac{\partial_r \phi_{\text{ns}}(r)}{\phi_{\text{ns}}(r)}. \tag{7.25}$$

The overall factor $1/(2\kappa^2)$ follows from the normalisation of the Einstein–Hilbert action. The next key observation is that the computation of the viscosity (and all other dissipative transport coefficients) through a Kubo formula requires only the imaginary part of the retarded Green function. Upon inserting $1 = \phi_{\text{ns}}^*(r)/\phi_{\text{ns}}^*(r)$ inside the limit, one obtains an expression for the imaginary part of Eq. (7.25) only,

$$\text{Im}\, G_{\text{R}}^{\text{CFT}}(\omega, k) = -\frac{1}{2\kappa^2} \lim_{r \to \infty} r^{-2(d+1-\Delta)} \sqrt{-g}g^{rr}$$

$$\times \frac{\phi_{\text{ns}}^*(r)\partial_r \phi_{\text{ns}}(r) - \phi_{\text{ns}}(r)\partial_r \phi_{\text{ns}}^*(r)}{2i\phi_{\text{ns}}^*(r)\phi_{\text{ns}}(r)}. \tag{7.26}$$

The numerator is readily recognised as the Wronskian which measures the flux density through a surface at fixed r. This can be seen by acting on the numerator with ∂_r and using the equation of motion (7.22). Note that the combination $\sqrt{-g}g^{rr}$ in Eq. (7.26) is precisely of the form $e^{\int^r P(r)}$ as used in Eq. (5.16).

Box 7.1   (Continued)

By rewriting the imaginary part of the retarded Green function (7.26) directly in terms of the conserved Wronskian we find

$$\text{Im } G_R^{\text{CFT}}(\omega, k) = \frac{1}{2\kappa^2} \lim_{r \to \infty} r^{-2(d+1-\Delta)} \frac{W(r)}{2i\phi_{\text{ns}}^*(r)\phi_{\text{ns}}(r)}. \qquad (7.27)$$

For our special case of $\Delta = d + 1$, we have $\phi_{\text{ns}} = \phi_{\text{sol}}$. In the following we will not distinguish between these two terms. The point of this rewriting is now immediately evident. We can use the conservation of the Wronskian to evaluate the numerator at any point r. The most convenient radial coordinate is the horizon itself, where the in-falling boundary conditions are set. We can then "pull up" the result to the boundary, where it translates into the imaginary part of the field-theory Green function.

Near the single zero of $f(r)$ associated with the horizon r_0, the equation of motion Eq. (6.24) reduces to

$$\phi'' + \frac{f'}{f}\phi' + \frac{L^4\omega^2}{r^4 f^2}\phi + \cdots = 0. \qquad (7.28)$$

Upon expanding $f(r)$ near the horizon, $f(r) = (r - r_0)(d + 1)/r_0 + \mathcal{O}((r - r_0)^2)$,

$$\phi'' + \frac{1}{r - r_0}\phi' + \frac{L^4\omega^2}{(d + 1)^2 r_0^2 (r - r_0)^2}\phi + \cdots = 0. \qquad (7.29)$$

We can deduce the power-law dependence of the solution near the horizon by substituting the Ansatz

$$\phi_{\text{sol}}(r; \omega, k) = (r - r_0)^\alpha (1 + \cdots), \qquad (7.30)$$

with "$\cdots$" denoting higher-order terms, into $r - r_0$. One finds

$$\alpha(\alpha - 1) + \alpha + \frac{L^4\omega^2}{(d + 1)^2 r_0^2} = 0, \qquad (7.31)$$

with solutions $\alpha = \pm i\omega L^2/((d + 1)r_0) = \pm i\omega/(4\pi T)$. In the last step we used the relation between the horizon location and the black-hole temperature derived in Eq. (6.8). The choice $\alpha = -i\omega/(4\pi T)$ corresponds to the in-falling solution. Thus, near the horizon, we may parametrise

$$\phi_{\text{sol}}(r; \omega, k) = (r - r_0)^{-i\omega/(4\pi T)} F(r; \omega, k), \qquad (7.32)$$

where $F(r; \omega, k)$ is regular at the horizon $r = r_0$ and can be power expanded according to the order of ω. Upon evaluating the conserved Wronskian near the horizon, one now finds

$$
\begin{aligned}
W(r_0) &= -\lim_{r \to r_0} \sqrt{-g} \, g^{rr} \phi_{\text{sol}}^* \overset{\leftrightarrow}{\partial_r} \phi_{\text{sol}} \\
&= -\lim_{r \to r_0} \frac{r^{d+2}}{L^{d+2}} \left(1 - \frac{r_0^{d+1}}{r^{d+1}} \right) \left[\left(\frac{-2i\omega}{4\pi T} (r - r_0)^{-1} \right) F^*(r) F(r) + \cdots \right] \\
&= -\frac{r_0^{d+2}}{L^{d+2}} \frac{d+1}{r_0} \left(\frac{-2i\omega}{4\pi T} F^*(r_0) F(r_0) \right) \\
&= \left(\frac{4}{d+1} \pi T L \right)^d (2i\omega) F^*(r_0) F(r_0).
\end{aligned}
\tag{7.33}
$$

The ellipsis in the second line are terms of regular functions of r thus it plays no role when we perform the limit $r \to r_0$. In the last line, we have again used the definition of the temperature Eq. (6.8): $4\pi T = (d+1) r_0 / L^2$.

We now simply substitute these results into the customised GKPW-rule expression for the imaginary part of the boundary Green function:

$$
\text{Im} \, G_R^{\text{CFT}}(\omega) = \frac{1}{2\kappa^2} \lim_{r \to \infty} \left(\frac{4}{d+1} \pi T L \right)^d \omega \frac{F^*(r_0) F(r_0)}{F_{\text{sol}}^*(r) F_{\text{sol}}(r)}.
\tag{7.34}
$$

The remaining unknown is the ratio of the absolute value of $|F(\dot{r}_0)|$ at the horizon to $\lim_{r \to \infty} |F(r)|$ at the boundary. Formally one still needs to solve for $F(r)$ to determine this. Here we encounter the final simplification that makes it possible to pull the near-horizon solution directly to the boundary: in the limit $\omega \to 0, k = 0$, the leading contribution will be the ω-independent solution for the remaining function F. From Eq. (7.22), this is readily seen to be the trivial constant function. This is the leading solution $\phi \sim A r^{\Delta - d - 1}$ near $r \to \infty$ with our special case of $\Delta = d + 1$. Hence, the leading order of the imaginary part of the retarded Green function in an expansion in ω equals

$$
\text{Im} \, G_R^{\text{CFT}}(\omega, 0) = \frac{1}{2\kappa^2} \left(\frac{4}{d+1} \pi T L \right)^d (\omega) + \mathcal{O}(\omega^2).
\tag{7.35}
$$

Using now the Kubo relation (7.12)

$$
\eta = \lim_{\omega \to 0} \frac{1}{\omega} \text{Im} \langle T_{x_1 x_2}(-\omega) T_{x_1 x_2}(\omega) \rangle = \lim_{\omega \to 0} \frac{1}{\omega} \text{Im} \, G_R^{\text{CFT}}(\omega, \mathbf{0}),
\tag{7.36}
$$

we find that the shear viscosity equals

$$
\eta = \frac{1}{2\kappa^2} \left(\frac{4}{d+1} \pi T L \right)^d.
\tag{7.37}
$$

Box 7.1   (Continued)

On recalling that $2\kappa^2 \equiv 16\pi G$ and comparing this with the entropy density $s = (4\pi/(2\kappa^2))(4\pi T L/(d+1))^d$ for a $(d+2)$-dimensional AdS–Schwarzschild black hole as derived in Eq. (6.24), we find the famous ratio

$$\frac{\eta}{s} = \frac{1}{4\pi}\frac{\hbar}{k_{\mathrm{B}}}. \tag{7.38}$$

7.2 Deriving the Navier–Stokes fluid from the bulk dynamical gravity

The remarkable capacity of AdS/CFT to encode classical hydrodynamical properties was beautifully exposed by the explicit demonstration by Bhattacharyya, Hubeny, Minwalla and Rangamani [254] that in a particular long-wavelength limit the bulk *dynamical* gravity is in a one-to-one [279] relation with the Navier–Stokes equations in the boundary. This is an astonishing realisation given that classical hydrodynamics is a theory that has a very different structure from both quantum field theory and classical Einstein theory, but the dictionary fuses it together into a single mathematical entity. This perhaps pleases the theoretical physicist's soul more than any other accomplishment of holography! After all, we are dealing here with the three monumental theories of physics which turn out to be just different sides of the same "holographic coin".

Of course, holography purports only to describe special matrix large-N field theories. This is reflected in the fact that the computation reveals very particular values for the hydrodynamical parameters – the equation of state, the minimal viscosity, etc. This, however, is all that remains from the microscopic "UV". On going to the IR at long distances, one can completely rely on classical gravity to get the *structure* of the hydrodynamical equations right. One can then promote the parameters to *free* parameters at the end of the calculations, and in this sense one achieves a completely UV-independent description of the physics of the finite-temperature fluid. It is remarkable and a bit mysterious that this works so well, and is very suggestive that it might extend beyond the literal large-N realm of classical AdS/CFT. In fact, the idea that there is a deep connection between the dynamics of the horizon in black-hole physics and the Navier–Stokes equation in hydrodynamics has a long history, dating back to the 1980s, according to which the black hole can be replaced by a fictitious fluid living on the horizon with dynamics governed by Einstein's equations (the "membrane paradigm" [280]). The crucial new addition of AdS/CFT is that in the fluid/gravity correspondence the fluid lives on the *boundary*.

The use of Einsteinian gravity as a "generating functional" to determine hydro-dynamical equations also has more practical benefits. As we have already stressed, when the questions of interest become more complicated it can be hard work to determine the higher-order equations using the conventional strategy resting on symmetry principles and the gradient expansion. Using the holographic machin-ery, on the other hand, it is just a matter of "mechanically" solving equations where the algebra serves as an infallible compass. This has led to a recent revival of "foundational" hydrodynamics [281]. A first example is the recent clas-sification of the two derivative terms in relativistic hydrodynamics [254, 257] correcting the conventional result derived by Israel and Stewart in the late 1970s. As a second example, the two-fluid description of the finite-temperature super-fluid was improved [282, 283, 284]. We will highlight these matters in the final section where we will sketch recent developments in a third example con-cerning parity violation in hydrodynamics [285] and the effects of quantum anomalies [286].

One can use the fluid/gravity version of AdS/CFT in reverse to ask the follow-ing question: what can we learn regarding gravity from the hydrodynamical dual? Depending on the Reynolds number, hydrodynamics is characterised by a regime of smooth laminar flows or a turbulent regime. The traditional view on dynamical gravity is associated with low-Reynolds-number smooth hydrodynamical flows. But if fluid/gravity is correct there must also be a turbulent regime. Turbulence is notoriously hard to model, and often is accessible only in numerical simulations. Fortunately a technical breakthrough amounting to the realisation that numerical GR is much more straightforward in asymptotically AdS spaces than conventional flat space-time has opened this door [23, 287]. The first highlight has been the dis-covery of highly turbulent near-horizon dynamical geometries characterised by a fractal geometry reflecting the Kolmogorov scaling of the turbulence in the bound-ary [288]. The next accomplishment has been the holographic study of superfluid turbulence [288], which we will discuss in more detail in chapter 10.

Summary 2   The Navier–Stokes equations of a macroscopic classical fluid, describing the universal long-range and late-time collective evolution in the finite-temperature boundary field theory, are in a fully non-linear and non-equilibrium sense encoded in the dynamical gravity of the black hole in the bulk, as described by Einsteinian gravity.

The construction of the "fluid/gravity duality", which is summarised in box 7.2, is conceptually quite straightforward. The hydrodynamics is controlled by a gradient expansion that is based on the requirement that the length and time scales associated with the flow are very large compared with any microscopic scale. Given the holographic correspondence between the long-distance hydrodynamical regime of the boundary and the gravity in the bulk, which we highlighted in box 7.1, it has to be the case that the gradient expansion in the boundary should translate into a gradient expansion of the dynamical gravity in the near-horizon gravity of the bulk. The latter can be "uplifted" to the boundary, and converted into the boundary hydrodynamics using the GKPW rule.

Box 7.2   A brief introduction to fluid/gravity duality

We will merely sketch the derivation. We have seen that the dynamics of a relativistic fluid is elegantly captured in the single conservation equation for the energy–momentum tensor $\partial_\mu T^{\mu\nu} = 0$. This conservation is clearly preserved in the correspondence. The issue in establishing the fluid/gravity duality is the constitutive relations of the energy–momentum tensor in terms of the macroscopic properties, namely the temperature and the fluid velocity of a classical relativistic fluid. A key insight emphasised by AdS/CFT is the recognition that matters on both sides of the duality should be organised in terms of a gradient expansion [277].

Since hydrodynamics is a dynamical theory of the stress–energy tensor at long times and lengths, we have to study the corresponding metric fluctuations in the bulk. We shall give a short description of how the procedure of Bhattacharyya, Hubeny, Minwalla and Rangamani [254] works, focussing on the steps of relevance to the bulk gravity side (see also [257, 289] and the reviews [290, 291]). We start from the thermodynamic equilibrium configuration dual to the AdS–Schwarzschild black hole. The first step is to perform a coordinate transformation to so-called Eddington–Finkelstein coordinates and subsequently Lorentz boost along the Minkowski boundary with a velocity v. In units where the speed of light is $c = 1$ and the AdS radius is set to $L = 1$, one obtains

$$ds^2 = -2u_\mu \, dx^\mu \, dr - r^2 f(r/T) u_\mu u_\nu \, dx^\mu \, dx^\nu + r^2 (u_\mu u_\nu + \eta_{\mu\nu}) dx^\mu \, dx^\nu, \quad (7.39)$$

where x^μ with $\mu, \nu = 0, \ldots, d$ are the coordinates of the boundary field theory, $f(r/T) = 1 - (4\pi T/((d+1)r))^{d+1}$ and $(u^0, u^i) = (1, v_i)/\sqrt{1 - v^i v_i}$ with the constant velocities v_i. Eddington–Finkelstein coordinates emphasise that the horizon is a coordinate singularity, not a physical entity. In these coordinates an in-falling object manifestly encounters no singular behaviour at $r_0 = 4\pi T/(d+1)$. For in-falling objects the coordinate system is completely regular for all $r > 0$.

Suppose we now disturb the fluid slightly by imposing a slowly varying temperature profile, i.e. we change the horizon location as a function of the coordinates x^μ. It is immediately obvious that we must then also promote the other components $u_\mu(x)$ to slowly varying functions in order that the solution remains a solution to Einstein's equations. Upon implementing this, the metric takes the formal form

$$ds^2 = -2u_\mu(x)dx^\mu\, dr - r^2 f\left(\frac{r}{T(x)}\right)u_\mu(x)u_\nu(x)dx^\mu\, dx^\nu$$
$$+ r^2\big(u_\mu(x)u_\nu(x) + \eta_{\mu\nu}\big)dx^\mu\, dx^\nu. \tag{7.40}$$

Note that we have fixed a gauge by setting $g_{rr} = 0$ and $g_{r\mu} = -u_\mu$. This will lead to constraint equations. For constant T and u_μ, the metric (7.40) solves the Einstein equations by construction. For generic but slowly varying functions $u_\mu(x)$ and $T(x)$, however, one can now expand the u_μ and T in a gradient expansion with regard to the coordinates x^μ and obtain perturbative solutions order by order, just as in the derivation of fluid dynamics. Identifying the orders in this expansion in terms of a small parameter ε after rescaling the coordinate x to εx as $\partial_\mu \sim \varepsilon$, $\partial_r \sim \varepsilon^0$, $u_\mu \sim \varepsilon^0$, $p \sim \varepsilon^0$, this gradient expansion relative to the stationary solution takes the form

$$g_{\mu\nu} = \sum_{n=0}^{\infty} \varepsilon^n g_{\mu\nu}^{(n)}(T(\varepsilon x), u(\varepsilon x)),$$

$$u^\mu = \sum_{n=0}^{\infty} \varepsilon^n u^{\mu\,(n)}(\varepsilon x), \tag{7.41}$$

$$T = \sum_{n=0}^{\infty} \varepsilon^n T^{(n)}(\varepsilon x).$$

We can now solve the bulk Einstein equations Eq. (6.9) iteratively, in principle up to arbitrary order. Upon imposing the boundary conditions of normalisability at infinity and *regularity in the interior for all $r > 0$*, the solution is uniquely determined. The standard "single-derivative" hydrodynamics at the beginning of this section corresponds to the expansion up to first order in ε. Let us illustrate this.

To first order one finds the gravitational solution

$$ds^2 = ds_0^2 + ds_1^2 + \cdots,$$
$$ds_0^2 = -2u_\mu\, dx^\mu\, dr - r^2\tilde{f}(\beta r)u_\mu u_\nu\, dx^\mu\, dx^\nu$$
$$+ r^2\big(u_\mu u_\nu + \eta_{\mu\nu}\big)dx^\mu\, dx^\nu, \tag{7.42}$$
$$ds_1^2 = 2r^2\beta F(\beta r)\sigma_{\mu\nu}\, dx^\mu\, dx^\nu + \frac{2}{d}ru_\mu u_\nu\, \partial_\alpha u^\alpha\, dx^\mu\, dx^\nu$$
$$- ru^\alpha\, \partial_\alpha(u_\mu u_\nu)dx^\mu\, dx^\nu,$$

Box 7.2   (Continued)

where

$$\tilde{f}(r) = 1 - \frac{1}{r^{d+1}}, \quad \beta = \frac{d+1}{4\pi T}, \quad F(r) = \int_r^\infty dy \left(\frac{y^d - 1}{y(y^{d+1} - 1)} \right) \tag{7.43}$$

and

$$\sigma^{\mu\nu} = P^{\mu\alpha} P^{\nu\beta} \left[\frac{1}{2} \left(\nabla_\alpha u_\beta + \nabla_\beta u_\alpha \right) - \frac{1}{d} P_{\alpha\beta} \nabla_\rho u^\rho \right]. \tag{7.44}$$

This time-dependent, inhomogeneous solution describes the metric throughout the bulk, including the region near the boundary. This information must now be dualised to the boundary field theory. Recall that our goal is to find the constitutive relations expressing the stress–energy tensor in terms of the temperature T and velocities u_μ. Using the GKPW formula, the expectation value of the energy–momentum tensor of the boundary theory should be given by the variation of the on-shell bulk action with respect to the source. This source is the boundary metric itself. The *boundary stress–energy tensor* thus follows from the variation of the bulk gravitational action with respect to the boundary metric $h^{\mu\nu}$ [219, 292, 293],

$$\langle T_{\mu\nu} \rangle = \frac{2}{\sqrt{-h}} \frac{\delta S_{\text{grav}}}{\delta h^{\mu\nu}}$$

$$= \lim_{r \to \infty} \frac{-r^{d+1}}{\kappa^2} \left[K_{\mu\nu} - K h_{\mu\nu} + d h_{\mu\nu} - \frac{1}{d-1} \left({}^h R_{\mu\nu} - \frac{1}{2} {}^h R h_{\mu\nu} \right) \right],$$

$$\tag{7.45}$$

where $h_{\mu\nu}$ is the induced metric near the boundary surface $r \to \infty$, ${}^h R_{\mu\nu}$ and ${}^h R$ are the corresponding Ricci tensor and scalar, and $K_{\mu\nu}$ and K are the extrinsic curvature and its trace (see section 4.4). This combination was already known in the GR community as the Brown–York tensor [292]. By substituting (7.42) into (7.45) we obtain

$$T^{\mu\nu} = (\epsilon + P) u^\mu u^\nu + P \eta^{\mu\nu} - 2\eta \left(\frac{1}{2} P^{\mu\alpha} P^{\nu\beta} [\partial_\alpha u_\beta + \partial_\beta u_\alpha] - \frac{1}{d} P^{\mu\nu} \partial_\alpha u^\alpha \right), \tag{7.46}$$

which corresponds precisely to the general energy–momentum tensor of a relativistic Navier–Stokes fluid. The Navier–Stokes equation $\partial_\mu T^{\mu\nu} = 0$ is simply the constraint equation related to the gauge choice. The detailed AdS-gravity answer has very special values for the parameters: in $3 + 1$ dimensions, one finds the standard equation of state for a conformal fluid, $\epsilon = 3P$. The pressure equals $P = (\pi^2/8)N^2T^4$, which can be easily understood because the temperature remains the only scale in a CFT. Finally, the shear viscosity equals $\eta = (\pi/8)N^2T^3$. Given the expression for the entropy (6.29) this turns into the "minimal viscosity ratio" $\eta/s = 1/(4\pi)$. This could have been anticipated since this is the "dissipation" way to compute the viscosity as opposed to the fluctuation linear-response approach of the previous section.

The real power of this "fluid/AdS-gravity" approach is illustrated at the next orders. In fact, this gravitational procedure has turned out to be quite useful to derive the higher-derivative generalisations of constitutive relations [254, 257]. Instead of constructing these terms by hand relying on symmetries, the second law and unitarity, using GR this exercise just turns into straightforward algebra and solving ordinary differential equations.

7.3 The conductivity: conserved currents as photons in the bulk

Up to this point the focus has been on the pure gravity in the bulk. The connection between bulk isometry and boundary global (space-time) symmetry shows that the pure gravity in the bulk encodes for everything associated with the global space-time symmetries in the boundary field theory. This governs the (free) energy/thermodynamics and the hydrodynamics of the previous sections, given their roots in energy and momentum conservation.

Matter is more, of course. One also needs a "conserved charge". This is just the usual stuff of condensed matter physics: in water we know that water molecules do not disappear by themselves, and the number of electrons in a piece of solid is not changing with time. The number of conserved electrons, or their total electrical charge, is governed by a global $U(1)$ symmetry: this is familiar to condensed matter physicists as the symmetry which is spontaneously broken in a superconductor. In the AdS/CMT context this is the most important global symmetry. One can address larger, non-Abelian "flavour" symmetries such as $SU(2)$ associated e.g. with the spin currents of condensed matter, but we will focus our attention mainly on $U(1)$ theories.

As we have already discussed in chapter 5, through the general global–local duality structure of the AdS/CFT correspondence, a global internal symmetry in the boundary is dual to its gauged version in the bulk. The global $U(1)$ "counting the matter" in the boundary is therefore represented by a local $U(1)$ symmetry in the bulk. In the bottom-up setting, the next rule is the weak–strong duality which insists that the bulk theory should have a minimal number of gradients, while it should be considered in the classical limit associated with the matrix large-N limit of the "colour" degrees of freedom in the boundary. The minimal derivative classical $U(1)$ gauge theory is of course nothing other than classical Maxwell electrodynamics. Upon adding this theory to the Einstein theory in the bulk, one deduces that the strongly coupled large-N physics of material systems characterised by a conserved number is described by the classical AdS–Einstein–Maxwell system in the bulk,

$$S = \int d^{d+2}x\sqrt{-g}\left[\frac{1}{2\kappa^2}\left(R + \frac{d(d+1)}{L^2}\right) - \frac{1}{4g_F^2}F_{\mu\nu}F^{\mu\nu}\right]. \qquad (7.47)$$

Summary 3   A conserved current associated with a global symmetry dualises in a gauge field with the same symmetry in the bulk. For a simple conserved number in the boundary (global $U(1)$), the minimal implementation is in terms of AdS–Einstein–Maxwell theory in the bulk.

This simple system is the point of departure for a large part of the recent AdS/CMT discoveries. Similarly to encoding a finite T through a black hole as a finite energy source, one can address the physics of the field theory at finite density by sourcing the bulk Maxwell fields with an electrical charge residing in a black hole. Starting in the next chapter, much of the remainder of this book is dedicated to the rich appearances, births and deaths of such charged black holes which translate into many revealing details and surprises regarding the nature of the finite-density matter of the boundary dual. In this chapter our focus is on macroscopic properties and the collective response in the hydrodynamical limit. As a first introduction to how these global internal symmetries are handled by the correspondence, we will focus here on the linear response of the finite-temperature but zero-density fluid of the pure AdS–Schwarzschild black hole to a small local distribution of charge. This will give us the *optical conductivity* of the zero-density but finite-temperature conformal state.

Differently from e.g. the viscosity, by virtue of their very definition conductivities are associated with the conserved charge. Literally, the electrical conductivity σ of an electron system in a solid quantifies the way that the electrical charge is transported through the system in response to an external electric field

$$J^i = \sigma E^i. \qquad (7.48)$$

According to the dictionary rule we just introduced, the current response J is dual to electromagnetic waves in the bulk. The interest is now in computing the frequency-dependent "optical" conductivity, typically at zero momentum, since that is of practical interest for condensed matter applications. In the laboratory one can measure currents only at very small momenta because the velocity of light is very large compared with the typical velocities of the material (electron) system. One is, in particular, interested in the dependence of the conductivity on temperature, since this quantifies the way the spectrum of charged excitations re-arranges itself. Especially the temperature dependence of the zero-frequency DC conductivity is of special interest.

In full generality, in systems at a finite density such currents are protected both by charge conservation *and* momentum conservation. In Galilean-invariant systems the momentum conservation imposes that any finite-density system turns into *a perfect conductor*. This implies that the real part of the optical conductivity consists of a "diamagnetic" delta-function peak at zero frequency. As we will show in the next chapter, this is also reproduced by holography. However, due to the "equalising" effect of momentum conservation there is not that much one can learn regarding the details of the system. Conductivities become more informative when translational symmetry is broken, and chapter 12 is dedicated to this rich subject in holography.

At zero density, as we consider here, the situation is different: because of charge conjugation the currents are no longer protected by momentum conservation, with the effect that the DC conductivity can be finite even in the Galilean continuum. This general wisdom is easy to understand considering simple free Dirac fermions. Applying a constant electric field will induce a current at zero momentum composed of excited "electrons/particles" moving to the right, but also of "positrons/holes" moving to the left. Because the positron will have exactly the opposite momentum, there is a net electrical current although there is no transport of momentum.

When the system is quantum critical it is easy to deduce the general scaling form of the conductivity, using the fact that the charge is conserved. This implies that the conductivity is governed by engineering dimensions (recall the discussion in chapter 5). In a relativistic system at zero temperature the real part of the optical conductivity σ_1 has to behave according to $\sigma_1(\omega, T = 0) \propto \omega^{2-d}$, whereas at zero frequency and finite temperature it must scale as $\sigma_1(\omega = 0, T) \propto T^{2-d}$. For intermediate values one has an interpolating function $\sigma_1(\omega, T) = T^{2-d} f(\omega/T)$, which should depend only on the universality class of the theory. We immediately see that in two spatial dimensions the conductivity becomes a constant. Indeed, for $d = 2$ the conductivity is dimensionless and one finds both a temperature-independent DC conductivity and a frequency-independent high-frequency behaviour: $\sigma(\omega = 0, T) = \sigma_0$ and $\sigma(\omega, T = 0) = \sigma_\infty$. A typical example is the electron system in graphene, where the electronic band structure realises at low energies a zero-density massless Dirac system that does indeed exhibit this behaviour. This is supposedly a free system in the scaling limit, and one can correct for the influence of interactions by perturbation theory in the usual guise of particles scattering from each other: see Ref. [294] and references therein.

However, as realised by Sachdev and collaborators [24, 295], one encounters an interesting problem of principle concerning the optical conductivity of strongly interacting zero-density quantum critical states in two-dimensional systems. We refer to the "critical graphene" model, i.e. zero-density electrons coupled to a critical boson, and especially the "bare-bones" Bose–Hubbard model (2.3) which we introduced in chapter 2. Focussing in on the latter, we argued that at the superfluid to Bose–Mott-insulator transition a relativistic, strongly interacting zero-density CFT, which is manifestly associated with a conserved $U(1)$ charge, is realised. The optical conductivity at the quantum critical point is of the form outlined above and fully described by a universal scaling function $\sigma(\omega, T) = F_\sigma(\omega/T)$, asymptoting to a constant σ_∞ and a constant σ_0 at large and small ω/T, respectively. The precise form of this crossover function is, however, a remarkable conundrum. Despite the simple nature of the quantum critical state – it is just the XY model in space-time – the explicit form of the optical conductivity is not known with certainty. Owing to the usual troubles with the Wick rotation, even numerical quantum Monte Carlo methods fail; see, however, [244].

It is also an instance where the "particle paradigm" fails spectacularly in dealing with the quantum critical state. One can attempt to compute the conductivity perturbatively departing from the particle excitations of e.g. the stable superfluid. In the hydrodynamical regime ($\omega/T \ll 0$) one should now use the perturbative (relative to the particle excitations of the stable state) quantum Boltzmann approach, and this yields a relaxational Drude-like response, translating into the usual peak in the optical conductivity centred at $\omega = 0$. However, as was first realised by Sachdev, one runs into a paradox in this way. The $d = 2$ Abelian–Higgs duality of section 2.1.2 has as a consequence that the *conductivity* of the particles should be identical to the *resistivity* associated with the dual system of vortices. On the other hand, on treating the vortex "pancakes" perturbatively in a quantum Boltzmann setting as if they were scattering particles, one finds that they should too give rise to a Drude peak in the vortex conductivity. But this vortex conductivity peak translates into a resistivity dip in the particle dual: one obtains the opposite behaviour depending on whether one computes the conductivity in a particle or vortex language.

Holography, as we know by now, is precisely able to address this conundrum. It naturally encodes an "un-particle" theory. For the reasons discussed above, we are particularly interested in the $(2 + 1)$-dimensional zero-density system. The outcome will be very simple, but at the same time quite satisfactory: the conductivity is just a constant, regardless of the energy and the temperature, i.e. $\sigma_0 = \sigma_\infty = \sigma(\omega, T)$. In a way this is a perfect solution of Sachdev's paradox. When the conductivity is completely featureless as a function of energy and temperature, it is automatically coincident with its dual vortex conductivity. In fact, it can be shown that this holographic result is rooted in the special property of $(3 + 1)$-dimensional Einstein–Maxwell bulk theory that its electromagnetic sector is self-dual. This forces the conductivity to be a constant [24]. This bulk electromagnetic self-duality is the holographic encoding of the particle–vortex duality in the boundary theory. Thus particle–vortex duality in these strongly coupled critical large-N theories has the remarkable consequence that the σ_0 and σ_∞ constants are the same, showing an unanticipated connection between the coherent and dissipative regimes of quantum critical transport.

Summary 4   The holographic computation of the optical conductivity associated with the conserved current in the zero-density, strongly coupled large-N boundary conformal field theory in $2 + 1$ dimensions shows that it is completely independent of both frequency and temperature. This resolves Sachdev's "particle–vortex paradox", which is encountered when the problem is addressed with the quantum Boltzmann approach.

The strongly interacting matrix large-N holographic computation is of course not quite applicable to the simple $U(1)$ quantum critical state, but it might well be that the "maximally interacting" CFT of holography is much closer to the desired answer than anything that can be accomplished with perturbation theory. At the same time the Abelian–Higgs duality as discussed in chapter 2 is anything but self-dual, and the remarkable self-dual holographic result might well be special to the strong 't Hooft coupling limit. One can test this and deform away from this limit by incorporating higher-gradient terms in the bulk gravity. This is further discussed

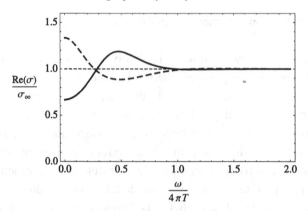

Figure 7.2 The optical conductivity as a function of ω/T for a field theory at zero density in $2+1$ dimensions as computed by holography. In the standard holographic setting one obtains that this scaling function is independent of ω/T (dotted line), a behaviour that originates in the "particle–vortex" self-duality realised in the limit of infinite 't Hooft coupling of the field theory. By including a higher-gradient term with a strength parametrised by γ in the bulk associated with turning the 't Hooft coupling finite in the boundary, this self-duality can be destroyed. On tuning γ to the maximal or minimal admissible value, one finds either a particle-like (dashed line) or a vortex-like (solid line) conductivity [295]. Figure adapted from [295]. (Reprinted with permission from the American Physical Society. © 2011.)

in box 7.3: it turns out that such contributions can indeed spoil the self-duality, and one finds for the optical conductivity the result shown in Fig. 7.2. These results might be already quite realistic, as suggested by very recent quantum Monte Carlo results [244, 296].

We discuss the holographic computation for the conductivity in detail below in box 7.3. It is among the most elementary computations of the kind. As with the viscosity, the conductivity is determined through a Kubo formula in terms of the current–current propagator in the boundary. Via the GKPW prescription for this two-point function, this translates into the ratio of amplitudes of the leading and sub-leading contributions near the boundary of an infinitesimal photon field in the bulk. These are derived from the equations of motion for the $U(1)$ gauge field in the AdS–Schwarzschild background of the Einstein–Maxwell system (7.47). This does indeed reproduce the engineering scaling result of a frequency-independent conductivity in $2+1$ dimensions at zero temperature (pure AdS). Quite exceptionally, the finite-temperature solutions for the "probe" gauge field in the presence of the Schwarzschild black hole can also be obtained analytically and these reveal that the optical conductivity is indeed temperature-independent in the large 't Hooft coupling limit.

Box 7.3 **The optical conductivity of the CFT$_3$ at zero and finite temperature from holography**

In linear response, the optical conductivity can be computed from the two-point retarded current–current correlation function

$$\sigma(\omega) = -\frac{i}{\omega} G_{xx}^{R}(\omega, \mathbf{k} = 0), \tag{7.49}$$

where

$$G_{xx}^{R}(\omega, \mathbf{k}) = -i \int dt \, d\mathbf{x} \, e^{i\omega t - i\mathbf{k}\cdot\mathbf{x}} \theta(t) \langle [J_x(t, \mathbf{x}), J_x(0, 0)] \rangle. \tag{7.50}$$

Here $J_\mu(t, \mathbf{x})$ are the $U(1)$ current operators in the field theory. Such a global $U(1)$ symmetry is the defining component of a superfluid system. Later in the text we will also sometimes (incorrectly) call the superfluid a "superconductor". This follows the traditions of condensed matter physics, where the weak long-range Coulomb interactions of matter are reintroduced perturbatively at the very end by weakly gauging the $U(1)$. It is from this perspective where electromagnetism is external that the currents encode the conductivity of the system.

According to the dictionary, the global currents in the boundary are dual to Maxwell-gauge fields in the bulk, and we are especially interested in the specific case of a $(2+1)$-dimensional quantum critical state. The bulk dual will thus be AdS$_4$. To compute the two-point correlation function of the currents, we need to know the dynamics of the fluctuations of the dual gauge field (photons) in the bulk. For photons, the first step is to choose a gauge in the bulk. Given that the components in the radial direction have no physical meaning in the boundary, a convenient choice is the radial gauge $A_r = 0$. We are interested in the zero-momentum finite-frequency conductivity, and can therefore limit our attention to the spatially uniform fluctuation $A_x(t) = \int d\omega \, a_x(r, \omega) e^{-i\omega t}$. In the AdS–Schwarzschild background it turns out that this fluctuation decouples from the other polarisations A_y and A_t, and the evaluation will be straightforward. For finite momentum $\mathbf{k} \neq 0$, on the other hand, there will be cross-coupling between the fluctuations, and the fluctuations have to be diagonalised first [24].

Let us first recall what happens at zero temperature. This just repeats the classic AdS/CFT computation (5.12) of chapter 5, but for a vector field. At zero temperature the geometry is pure AdS$_4$, and from the Maxwell equation of motion in curved space

$$\frac{1}{\sqrt{-g}} \partial_\mu \left[\sqrt{-g} g^{\mu\nu} g^{\alpha\beta} F_{\nu\alpha} \right] = 0 \tag{7.51}$$

Box 7.3 (Continued)

one obtains straightforwardly the equation for the zero-momentum fluctuation $a_x(r)$ in radial gauge

$$a_x'' + \frac{2}{r}a_x' + \frac{L^4\omega^2}{r^4}a_x = 0. \tag{7.52}$$

Here we denoted radial derivatives with a prime. The in-falling solution is simply deduced in this case; it is

$$a_x = c_0 e^{i\omega L^2/r}, \tag{7.53}$$

where c_0 is a constant. Near the boundary ($r \to \infty$) it follows immediately that

$$a_x = a_x^{(0)} + \frac{a_x^{(1)}}{r} + \cdots, \tag{7.54}$$

with $a_x^{(0)} = c_0$ and $a_x^{(1)} = ic_0\omega L^2$. Using the GKPW prescription, we find that

$$\langle J^x(-\omega)J^x(\omega)\rangle = \frac{1}{L^2 g_F^2}\frac{a_x^{(1)}}{a_x^{(0)}} = i\omega L^2, \tag{7.55}$$

where the normalisation follows from the action Eq. (7.47). This clearly displays the characteristic scaling $\langle JJ\rangle \sim \omega^{2\Delta-d-1}$ of a conserved current with dimension $\Delta = d$ in $d = 2$ dimensions. Thus the conductivity equals

$$\sigma(\omega) = -\frac{i}{\omega}G_{xx}^R(\omega, \mathbf{k} = 0) = \frac{1}{L^2 g_F^2}\frac{-i}{\omega}\frac{a_x^{(1)}}{a_x^{(0)}} = \frac{1}{g_F^2}. \tag{7.56}$$

The fact that it is constant follows directly from the protected scaling dimension of the conserved current.

Another way of understanding this result is as follows. The leading-order contribution of the gauge field $a_x^{(0)}$ near the boundary gives the electric field on the boundary: $E_x = -\dot{a}_x^{(0)}$, and the VEV of the induced current (response) is the sub-leading term near the boundary $J_x = (1/(L^2 g_F^2))a_x^{(1)}$. From Ohm's law, we have

$$\sigma = \frac{J_x}{E_x} = \frac{1}{L^2 g_F^2}\frac{A_x^{(1)}}{-\dot{A}_x^{(0)}} = \frac{1}{L^2 g_F^2}\frac{-ia_x^{(1)}}{\omega a_x^{(0)}}, \tag{7.57}$$

which is coincident with the result of the previous paragraph.

Let us now compute the optical conductivity for the CFT at *finite* temperature: we repeat the above calculation but now the background is the AdS–Schwarzschild black hole. This changes the radial equation of motion Eq. (7.52) to

$$a_x'' + \left(\frac{2}{r} + \frac{f'}{f}\right)a_x' + \frac{L^4\omega^2}{r^4 f^2}a_x = 0, \tag{7.58}$$

where $f(r) = (1 - r_0^3/r^3)$ is the emblackening or redshift factor $f(r)$. As explained in chapter 6, to obtain the retarded Green function in this real-time formalism we need to impose in-falling boundary conditions for the photons at the horizon. From chapter 6 we also know that near the horizon the emblackening factor becomes

$$f = \frac{4\pi T L^2}{r_0^2}(r - r_0) + \mathcal{O}\big((r - r_0)^2\big). \tag{7.59}$$

Upon inserting this into Eq. (7.58) to determine the solution near the black-hole horizon one finds two branches of solutions near the horizon:

$$a_x \sim e^{\pm(i\omega/(4\pi T))\ln(r-r_0)}\big(1 + \mathcal{O}(r - r_0)\big). \tag{7.60}$$

Since $A_x = a_x(r)e^{-i\omega t}$, near the horizon, the "+" and "−" branches correspond to the out-going and in-falling waves. We need the in-falling one, which behaves as $a_x \propto (r - r_0)^{-i\omega/(4\pi T)}$ when $r \to r_0$.

Remarkably a closed solution for the radial differential equation for this boundary condition exists – for anything more complicated one has to rely on numerics. This explicit solution is

$$a_x = c_0\left(\frac{4\pi T L^2 - 3r}{\sqrt{9r^2 + 12\pi T L^2 r + (4\pi T)^2 L^4}}\right)^{-i\omega/(4\pi T)}$$

$$\times \exp\left(-\frac{i\omega}{4\pi T}\right)\left(\sqrt{3}\arctan\left[\frac{1}{\sqrt{3}}\left(1 + \frac{6r}{4\pi T L^2}\right)\right] - \frac{\sqrt{3}}{2}\pi - i\pi\right), \tag{7.61}$$

where c_0 is an integration constant. It is easy to check that this satisfies the in-falling near-horizon boundary condition. The near-boundary ($r \to \infty$) behaviour now becomes

$$a_x^{(0)} = c_0, \quad a_x^{(1)} = ic_0\omega L^2. \tag{7.62}$$

Upon inserting this into (7.57) we find, surprisingly, that the optical conductivity is a strict constant that is completely independent of ω/T

$$\sigma(\omega/T) = \frac{1}{g_F^2}. \tag{7.63}$$

Box 7.3 **(Continued)**

Holographic conductivity at zero density and finite 't Hooft coupling.

The reason why the holographic computation yields a constant optical conductivity can immediately be traced to the special feature of the two-derivative Einstein–Maxwell gravity theory. $U(1)$ Maxwell theory in $3 + 1$ dimensions in the absence of sources is self-dual. In terms of the dual field strength $\tilde{F}_{\mu\nu} = \frac{1}{2}\epsilon_{\mu\nu\alpha\beta}F^{\alpha\beta}$ one has exactly the same action, but with $\tilde{g}_4^2 = 1/g_4^2$ [24]. A more generic action will not possess this manifest self-duality. It is certainly not a feature of most quantum critical condensed matter systems, with the Bose–Hubbard model as prime example. Since the generic ingredients are the same on both sides, however, the most obvious culprit is the large-N limit inherent in two-derivative gravity. Indeed, moving away from the strong 't Hooft coupling limit by adding higher-order gradient terms to the Einstein–Maxwell theory is generically incompatible with self-duality. We will do so here and show that as a result the scaling function for the optical conductivity is no longer a constant. The lowest-order correction in this sector lacking this self-duality is [295]

$$S = S_{\text{EM}} + \int d^4x \sqrt{-g} \frac{\gamma L^2}{g_F^2} C_{\mu\nu\rho\sigma} F^{\mu\nu} F^{\rho\sigma}, \qquad (7.64)$$

where γ is a dimensionless parameter, L is the AdS radius, g_4 is the gauge coupling constant and $C_{\mu\nu\rho\sigma}$ is the Weyl tensor

$$C_{\mu\nu\rho\sigma} = R_{\mu\nu\rho\sigma} - \frac{1}{2}(g_{\mu\rho}R_{\nu\sigma} + g_{\nu\sigma}R_{\mu\rho} - g_{\mu\sigma}R_{\nu\rho} - g_{\nu\rho}R_{\mu\sigma})$$

$$+ \frac{1}{6}(g_{\mu\rho}g_{\nu\sigma} - g_{\mu\sigma}g_{\nu\rho})R. \qquad (7.65)$$

Combinations other than the Weyl tensor renormalise the conductivity at zeroth order, and we can therefore absorb them in a redefinition. For $\gamma \neq 0$ the extra term clearly spoils the self-duality in the bulk. An interesting aspect is that causality constraints enforce the restriction $|\gamma| < 1/12$. This gives a taste of some of the surprising insights into matter that can be gained from the AdS gravity perspective. It turns out that the extremal values $\gamma = \pm 1/12$ precisely correspond to a theory of free particles or vortices [297]. In the presence of this correction the procedure outlined in the first section of this box is no longer analytically tractable. However, it can be solved numerically, and the results are indicated in Fig. 7.2. The conductivity is no longer a constant. Instead one can demonstrate that the relation between the $\sigma_0 = \sigma(\omega = 0)$ and $\sigma_\infty = \sigma(\omega = \infty)$ transport coefficients becomes

$$\frac{\sigma_0}{\sigma_\infty} = 1 + 4\gamma, \quad \sigma_\infty = \frac{1}{g_F^2}. \tag{7.66}$$

The dissipative part σ_0 is now no longer identical with the (unaltered) high-frequency σ_∞.

7.4 Hydrodynamics and quantum anomalies

Let us finish this chapter with an illustration of the power of the fluid/gravity corre-spondence by addressing a recent development in generalising hydrodynamics. The constitutive relations of the macroscopic thermal fluid can include a non-dissipative term that encodes the quantum anomalies of the microscopic quantum field the-ory. Anomalies are symmetries of the classical Lagrangian that are destroyed in the quantum theory. We refer here in particular to *chiral* anomalies of Yang–Mills theories coupling to massless chiral fermions in $d + 1 =$ even-dimensional theories, with of course a special interest for the case of $3 + 1$ dimensions. Clas-sically, for a theory with massless charged fermions we can define two kinds of currents J_μ, a parity-even $U(1)_V$ vector-current counting the total contribu-tion from charged fermions, and a parity-odd $U(1)_A$ axial current counting the difference between the left-handed and right-handed fermions. At the classical level the handedness of massless fermions is conserved ($\partial_\mu J_A^\mu = 0$). However, this symmetry is broken in the quantised theory. One finds in $d + 1 = 3 + 1$ dimensions [298]

$$\partial_\mu J_A^\mu = -\frac{C}{8} \epsilon_{\mu\nu\alpha\beta} F^{\mu\nu a} F^{\alpha\beta a}, \tag{7.67}$$

where

$$F_{\mu\nu}^a = \partial_\mu A_\nu^a - \partial_\nu A_\mu^a + g f_{bc}^a A_\mu^b A_\nu^c \tag{7.68}$$

is the field strength of the external (non-)Abelian gauge field A_μ^a under which the fermions are also charged, while C is the "anomaly constant" which is related to the excess of right-handed over left-handed fermions. The right-hand side of Eq. (7.67) is topological in nature. To understand what the anomaly means, consider the system in an external electromagnetic field. Then one finds that $\partial_\mu J_V^\mu = 0$ and $\partial_\mu J_A^\mu = C E_\mu B^\mu$ with E_μ and B_μ equal to the electrical and magnetic field strengths, respectively. The condensed matter reader will recognise the θ-vacuum terms as found at the surface of topological band insulators [58]. The non-conservation is due to a spectral flow that happens when the gauge vacuum tunnels between configurations that are topologically different. The definition of the fermions is different in the different vacua, and this causes the energy levels to gradually shift upwards for the particles and downwards for the anti-particles, with the effect that a net particle creation takes place.

It has been known for a long time that such anomalies are present in the electroweak sector of the Standard Model. In the context of AdS/CFT they were among the earliest tests of the correspondence [3] – as topological quantities they are independent of the 't Hooft coupling. Recently in AdS/QCD they have been invoked to explain fluctuations of electrical charge separation in non-central heavy-ion collisions via the "chiral magnetic effect" [299, 300, 301]. As a very recent development in condensed matter physics, it has been argued that such $U(1)$ triangle anomalies should also be active in doped Weyl semi-metals: the metallic siblings of the $(3 + 1)$-dimensional topological insulators [302, 303, 304, 305].

With the discovery of the fluid/gravity duality and the manifest presence of anomaly terms on the gravity side, the question of how the anomaly terms in the (microscopic) quantum field theory imprint on the behaviour of the macroscopic classical fluid now arises. Given that the anomalies originate in topology, one does anticipate that they must have macroscopic effects as well, but one would like to have a precise hydrodynamical description. This was recently derived by Son and Surowka [286] (see also [306, 307, 308]). One finds novel anomaly-induced flows that are parallel to the vorticity in the fluid ("chiral vortical effect") as well as parallel to an external magnetic field ("chiral magnetic effect"). These discoveries were directly inspired and guided by the fluid/gravity correspondence [309, 310]. In $(4 + 1)$-dimensional gravity with a Maxwell field, it is very natural to add a Chern–Simons term; it is indeed present in many top-down constructions. Similarly to quantum Hall physics, this Chern–Simons term leads to the chiral anomaly for the dual boundary field theory [286, 309, 310]. Through the fluid/gravity correspondence it is then straightforward to deduce the ramifications for the hydrodynamical theory, as we will elucidate in box 7.4. (See e.g. [311] for a review on this subject.)

The presence of parity-breaking terms in the constitutive relations is not limited to anomalies in $(3 + 1)$ dimensions. Another interesting example of such a use of holography is the recent development revolving around hydrodynamics in systems where parity is broken in a lower dimension. An example is the A-phase of helium-3, where the chiral p-wave condensate breaks parity spontaneously. For instance, for a $(2 + 1)$-dimensional boundary theory the parity-breaking terms can arise in the dual theory from the presence of topological θ terms for the gauge field or the gravitational Riemann tensor in the $(3 + 1)$-dimensional gravitational bulk theory [285, 312] (see also [313]). This then leads to predictions for the Hall viscosity, edge currents as well as the spontaneous generation of angular momentum [313, 314, 315].

Box 7.4   The impact of quantum anomalies on the dynamics of the thermal fluid

We consider here the usual relativistic fluid in $3+1$ dimensions with a $U(1)$ quantum number. If the current corresponding to the charge is anomalous, it is no longer conserved; instead, in the presence of a background field, the divergence of the current equals

$$\partial_\mu J^\mu = -\frac{C}{8}\epsilon^{\mu\nu\alpha\beta} F_{\mu\nu} F_{\alpha\beta}. \tag{7.69}$$

Note that here we considered only one anomalous $U(1)_A$. In realistic models, such as RHIC or Weyl semi-metals, one should consider both $U(1)_V$ and $U(1)_A$, for which we can redefine $J_V = J_R + J_L$, $J_A = J_R - J_L$ and we have Eq. (7.69) for J_R and similarly for J_L with opposite values of C. Here for simplicity we only consider a single $U(1)$ and it is straightforward to generalise to the $U(1)_V \times U(1)_A$ case [306, 316]. In the presence of an external electric and magnetic field, the hydrodynamic equations (7.5) and (7.10) for the charged fluid system change, due to the chiral anomaly, to

$$\partial_\mu T^{\mu\nu} = -F^{\mu\nu} J_\mu, \quad \partial_\mu J^\mu = C E_\mu B^\mu, \tag{7.70}$$

where $E^\mu = F^{\mu\nu}u_\nu$ and $B^\mu = \frac{1}{2}\epsilon^{\mu\nu\alpha\beta}u_\nu F_{\alpha\beta}$ are the background electric and magnetic fields. On writing out the constitutive relations in terms of T and u_μ but also E_μ and B_μ, one finds that, up to the first derivative, the energy–momentum tensor for this anomalous fluid is still given by Eq. (7.7) in the Landau frame where the fluid flow is equal to the energy flow. The expression for the hydrodynamical (covariant) current has novel terms, however [286, 306, 307]:

$$J^\mu = nu^\mu - \sigma T\left(g^{\mu\nu} + u^\mu u^\nu\right)\partial_\nu(\mu/T) + \sigma E^\mu + \xi_V\omega^\mu + \xi_B B^\mu, \tag{7.71}$$

where n is the charge density, σ is the DC conductivity, T is the temperature, and $\omega^\mu = \frac{1}{2}\epsilon^{\mu\nu\lambda\rho}u_\nu \partial_\lambda u_\rho$ is the vorticity in the fluid. The first three terms are standard and parity-conserving, but the last two terms are induced by the anomaly; they violate parity. The term proportional to the kinetic coefficient ξ_V expresses that a current is induced in the direction of the vorticity

Box 7.4 (Continued)

present in the fluid: the "chiral vortical effect" (CVE). The last term is responsible for the "chiral magnetic effect" (CME): for finite ξ_B one finds that a current is induced parallel to an applied magnetic field. Note that both of the novel parity-violating terms are non-dissipative, as follows immediately from the way in which they transform under time reversal. In terms of the anomaly coefficient C, in the Landau frame the existence of an entropy current with non-negative derivative completely fixes the coefficients ξ_V and ξ_B to be [286, 307, 317]

$$\xi_V = C\mu^2 + C_G T^2 - \frac{2n}{\epsilon + P}\left(\frac{C\mu^3}{3} + C_G \mu T^2\right),$$

$$\xi_B = C\mu - \frac{1}{2}\frac{n}{\epsilon + P}\left(C\mu^2 + C_G T^2\right), \tag{7.72}$$

where ϵ and P are the energy density and pressure of the fluid, μ is the chemical potential and C_G is related to the gravitational anomaly [318]. In particular, if the fluid lives in a non-trivial space-time background, we have an extra term in the current conservation equation, namely $\partial_\mu J^\mu = -(C_G/8)\epsilon^{\mu\nu\rho\sigma} R_{\mu\nu\alpha\beta} R_{\rho\sigma}{}^{\alpha\beta}$.

Let us now show how Eqs. (7.71) and (7.72) follow from the gravity computation. Since we are considering a charged fluid the gravity theory will contain Maxwell fields, in addition to the Einstein–Hilbert action. The topological nature of the anomaly is reflected through the dictionary: on the gravity side the AdS$_5$ Maxwell action is extended by a topological Chern–Simons term [286, 309, 310]. In units where the AdS radius is set to $L = 1$, the action is

$$S = \frac{1}{16\pi G}\int d^5 x \sqrt{-g}\left[R + 12 - \frac{1}{4}F_{\mu\nu}F^{\mu\nu} + \frac{\alpha_{CS}}{3}\epsilon^{\beta\mu\nu\rho\lambda} A_\beta F_{\mu\nu} F_{\rho\lambda}\right], \tag{7.73}$$

where $\epsilon^{\mu\nu\rho\sigma\tau} = (1/\sqrt{-g})\varepsilon^{\mu\nu\rho\sigma\tau}$ with $\varepsilon_{0123r} = -\varepsilon_{0123} = 1$. For simplicity let us focus on the chiral anomaly, ignoring the terms related to the gravitational anomaly, which can be addressed by adding the gravitational Chern–Simons term [317].

The equations of motion for (7.73) are

$$R_{\mu\nu} - \frac{1}{2}g_{\mu\nu}(R + 12) - \frac{1}{2}\left(F_{\mu\beta}F_\nu{}^\beta - \frac{1}{4}g_{\mu\nu}F^2\right) = 0, \tag{7.74}$$

$$\nabla_\nu F^{\nu\beta} + \alpha_{CS}\epsilon^{\beta\mu\nu\rho\lambda} F_{\mu\nu} F_{\rho\lambda} = 0. \tag{7.75}$$

The Chern–Simons term does not affect the asymptotic behaviour of the gauge field A_μ near the AdS boundary. Since A is a massless field dual to a marginal operator, the leading term will be a constant, and in $4 + 1$ dimensions the sub-leading term will behave as r^{-2},

$$A_\mu(r, x)|_{r \to \infty} = A_\mu^{(0)}(x) + \frac{A_\mu^{(2)}}{r^2} + \cdots . \tag{7.76}$$

We identify the expectation value of the *covariant* current J^μ as the dual to the gauge field equal to $J_\mu(x) = A_\mu^{(2)}/(8\pi G)$ with $8\pi G$ determined by the particular normalisation of the Maxwell action used in Eq. (7.73). Using the equation of motion Eq. (7.75) and comparing with (7.69), we find the relation

$$C = \frac{1}{2\pi G} \alpha_{CS}. \tag{7.77}$$

To derive the fluid/gravity duality for this theory, we start with the solution dual to an equilibrium fluid at finite temperature T and finite chemical potential μ. This solution is the charged AdS Reissner–Nordström black hole – we will address it in detail in the next chapter. For now we summarise its properties here. The charged black hole is a static solution to these equations of motion with an electrostatic potential and metric that take the following form in Eddington–Finkelstein coordinates:

$$ds^2 = 2\,dv\,dr - r^2 f(r, m, q)dv^2 + r^2\,d\mathbf{x}^2, \quad A = -\frac{\sqrt{3}q}{r^2}\,dv. \tag{7.78}$$

The emblackening factor equals $f(r, m, q) = 1 - m/r^4 + q^2/r^6$ and q is the charge of the black hole. The temperature T and chemical potential μ of the dual fluid are determined by the mass m and charge q of the black hole as

$$T = \frac{2r_0^2 m - 3q^2}{2\pi r_0^5}, \quad \mu = \frac{\sqrt{3}q}{r_0^2}, \quad n = \frac{2\sqrt{3}q}{16\pi G}, \quad P = \frac{\epsilon}{3} = \frac{m}{16\pi G}, \tag{7.79}$$

with r_0 determined by $f(r_0, m, q) = 0$.

This solution can be rewritten in the form of a fluid with boost velocity u^μ and background gauge field A_μ^{bg}:

$$ds^2 = -2u_\mu\,dx^\mu\,dr + r^2(P_{\mu\nu} - f u_\mu u_\nu)dx^\mu\,dx^\nu, \tag{7.80}$$

$$A = \frac{\sqrt{3}q}{r^2} u_\mu\,dx^\mu.$$

Here $P_{\mu\nu}$ is the projector orthogonal to u_μ. Note that in the rest frame $u_\mu\,dx^\mu = -dv$ and the above solution reduces to Eq. (7.78). As in box 7.2, we now solve for the perturbations to this background order by order in the gradient expansion [286, 309, 310]. From the equations of motion for the first-order corrections the anomalous transport coefficients are determined to be

Box 7.4 **(Continued)**

$$\xi_V = \frac{3q^2\alpha_{CS}}{2\pi Gm}, \quad \xi_B = \frac{\sqrt{3}(3r_0^4 + m)q\alpha_{CS}}{8\pi Gmr_0^2}. \tag{7.81}$$

We can see that these two coefficients are exactly the same as (7.72) by substituting Eq. (7.79). This precise agreement between the fluid/gravity calculation and the outcomes of the hydrodynamical analysis (7.72) can also be confirmed by computing the transport coefficients in linear response using the Kubo formula [317].

8

Finite density: the Reissner–Nordström black hole and strange metals

The material systems studied in condensed matter laboratories are formed from a finite density of conserved entities such as the number of electrons or (cold) atoms. This is a priori quite distinct from the vacuum states which have been discussed in the previous two chapters. These purely scaleless critical states can be mimicked in the laboratory, but this involves meticulous fine tuning to the critical point. The famous example is the cold-atom superfluid Bose Mott-insulator system [319]. As we discussed in chapter 2, according to the established wisdom of condensed matter physics the zero-temperature states of matter which are understood, are generically stable or "cohesive" states. These typically break symmetry spontaneously, with a vacuum that is a short-range entangled product state. In addition, the Fermi liquid and the incompressible topologically ordered states are at present understood as stable states that are "enriched" by the long-range entanglement in their ground states. Finally, by fine tuning of parameters one can encounter special unstable states associated with continuous quantum phase transitions, but these are understood within the limitations of "bosonic" field theory, which relies on the statistical physics (in Euclidean space-time) paradigm.

The AdS/CMT pursuit was kickstarted in the period 2008–2009, inspired by this finite-density perspective underlying condensed matter physics. The first shot was aimed at spontaneous symmetry breaking: the holographic superconductor. We shall discuss this at length in chapter 10. We will find out there that much of the physics of this iconic zero-temperature state of matter is impeccably reproduced by holography, with the results acting thereby as a powerful confidence builder. However, in the development that followed it became increasingly clear that holography insists on the existence of an entirely new class of "non-cohesive" finite-density states at zero temperature. These emergent quantum critical *phases* are best called "strange metals" [159, 320]. This quantum criticality is *not* tied to the conformal invariance and supersymmetry of the zero-density CFT inherent in the AdS/CFT name. These symmetries are badly broken by the finite chemical potential. Instead, there emerges in the infrared a new type of scale-invariant quantum dynamics that does not need fine tuning at all. It is in principle realised regardless of the precise values of UV parameters. In addition, these quantum critical states are characterised by scaling properties that are alien to the critical states of statistical physics. On the one hand, one naturally finds an emergent dynamical critical exponent **z** that expresses how the time dimension scales with respect to space in the IR quantum critical regime. This exponent can become very large, even infinite. At the same time these states readily exhibit "hyperscaling-violation" scaling behaviour that does not seem to have a counterpart in the standard Wilsonian renormalisation group.

These holographic strange metals earned their name because they have suggestive traits in common with the laboratory variety discussed at length in chapter 3. This includes the "local quantum criticality", referring to the scaling behaviour characterised by $z \rightarrow \infty$ which was first discovered by the experimentalists. In the remainder of this text, these resemblances will be a recurrent theme. However, it is at present unclear whether the holographic strange metals are just a useful metaphor or whether holography reveals a general principle that is also at work in experiment. On the one hand, the identification of the mechanisms at work in an explicit boundary field-theoretical language is still completely lacking. Are the holographic strange metals "UV-sensitive" and tied to the un-realistic origins of a large-N super-conformal UV, or does this reveal a general emergence principle tied to strongly interacting quantum matter that is behind the "fermion-sign horizon"? On the other hand, in the remainder of this book various experimental tests for holographic strange-metal behaviour will be presented, but they await experimental confirmation. More than anything else, the holographic strange metals presented in this chapter form the heart of AdS/CMT: it is the greatest surprise coming from holography, with the greatest potential to impact on experiment, being at the same time utterly mysterious.

A brief look at the AdS/CFT dictionary reveals right away that in order to describe the case of finite density in the boundary, one has to introduce an electrical monopole charge in the interior of the bulk. The simplest theory that can do so is two-derivative (strong 't Hooft coupling), classical (large-N) AdS–Einstein–Maxwell theory. The equations of motion have a single, unique family of static, rotationally invariant solutions: the AdS version of charged black holes discovered in 1916/1918 by Reissner and Nordström. As in the previous chapter, in Poincaré coordinates, the rotationally invariant black hole translates into the Lorentz-invariant black brane with a planar horizon. In this family of charged black holes there is one special member. When the mass of the black hole is coincident with its electromagnetic energy, it can no longer shed energy due to charge conservation. For this "extremal" black hole the Hawking temperature therefore vanishes, as it cannot radiate. The state dual to this extremal black hole is the zero-temperature finite-density strange metal in the boundary. Given its maximally simple, elegant and universal nature, we will analyse it in detail in this chapter. In section 8.1 we will discuss the solution of the charged black hole in AdS, and derive the thermodynamics of the boundary theory with an emphasis on its mysterious ground-state entropy. In section 8.2 we will analyse in detail the special role played by the special near-horizon "AdS$_2$" geometry. This will be dual to the emergence of local quantum criticality in the boundary as well as the "volume" hyperscaling-violating scaling. We will study its macroscopic properties in section 8.3. The optical conductivity exhibits the generic perfect-metal behaviour, but the strange metal curiously continuous to behave like a hydrodynamical liquid even at zero temperature. It possesses a zero sound mode similar to the Fermi liquid (chapter 2), albeit with a quite peculiar $T = 0$ attenuation.

Although this holographic strange-metal development started with the RN metal, it has become clear that it is just an "extremal" member of a large family of strange metals, all characterised by emerging quantum criticality but exhibiting a spectrum of distinct scaling behaviours. The key is that as long as one can get away with classical gravity the emergent deep-infrared physics should be universally governed by an extension of Einstein–Maxwell gravity with a so-called dilaton field [321, 322, 323, 324]. This field is dual to the dominant relevant operator in the boundary field theory to complement the conserved energy–momentum and charge currents. We have already met the dilaton field in section 6.3, as a field that is ubiquitous in string-theoretical top-down constructions. Its effect is in essence to dynamically screen the electrical charge such that it can vary along the radial direction in the on-shell bulk solutions. Technically this allows for solutions with an electrical flux piercing through the boundary without a black hole in the interior. This describes finite-density states in the boundary with vanishing ground-state entropy, but one still has an emergent scaling geometry in the deep interior. As we will discuss in section 8.4, one can construct in this way a bottom-up "scaling atlas" classifying the emergent scaling behaviours of holographic strange metals, which is admissible by virtue of the universality of the gravity in the bulk.

8.1 The Reissner–Nordström strange metal

Let us depart from the boundary field theory. In the last section of the previous chapter, section 7.4, we explained that to account for matter we introduce a global $U(1)$ symmetry in the boundary that encodes for a conserved quantity like the number of electrons. In the absence of electromagnetic sources in the bulk, the theory describes electromagnetic fluctuations, but the state of the system is still at zero density. The chemical potential sits right at the Dirac node of the relativistic theory. In order to lift this to a finite density one has to expose the field theory to a chemical potential that is conjugate to the density. From the GKPW construction, we immediately see that the source for the density operator is the bulk electrostatic potential A_t. For a spatially uniform time-independent density this can only depend on the radial direction. Therefore, to encode for the chemical potential in the boundary a radial electrical field is required, with field lines "piercing" through the boundary in a spatially uniform way.

How can one accomplish this in the AdS gravity theory? In the case of the compact boundary space (section 6.2), we have to store an electrical point charge "somehow" in the deep interior, while for the non-compact case one needs a translationally invariant condenser plate. At finite temperature (in the deconfined phase) one has to marry this charge in one way or another with the requirement that a black hole (or black brane) has to be present as well. The simplest way out is to combine the two requirements and consider the electrically charged black hole/black brane: this is the Reissner–Nordström (RN) black hole/brane. In fact, for two-derivative Einstein–Maxwell theory in the bulk this is the unique stationary solution.

This black hole has a long history. Directly after Einstein's seminal GR work, and Schwarzschild's discovery of the metric named after him, Reissner and Nordström independently showed that the Einstein–Maxwell system allowed for a new solution: the combined electric field and space-time of a black hole with a net electric charge Q and overall mass M. The Einstein–Maxwell action including the cosmological constant for the general $(d + 2)$-dimensional space-time is

$$S_{\text{EM}} = \int d^{d+2}x \sqrt{-g} \left[\frac{1}{16\pi G} \left(R + \frac{d(d+1)}{L^2} \right) - \frac{1}{4g_{\text{F}}^2} F_{\mu\nu} F^{\mu\nu} \right]. \tag{8.1}$$

The saddle point of Eq. (8.1) in AdS can be written in the form of a metric and gauge potential A_t as

$$ds^2 = \frac{r^2}{L^2}(-f \, dt^2 + d\mathbf{x}^2) + \frac{L^2 \, dr^2}{r^2 \, f},$$

$$f(r) = 1 + \frac{Q^2}{r^{2d}} - \frac{M}{r^{d+1}}, \quad A_t = \mu \left(1 - \frac{r_0^{d-1}}{r^{d-1}} \right), \tag{8.2}$$

where

$$\mu = \frac{g_{\text{F}} Q}{2c_d \sqrt{\pi G} L^2 r_0^{d-1}}, \quad \text{with } c_d = \sqrt{\frac{2(d-1)}{d}}. \tag{8.3}$$

The difference in the geometry compared with the now-familiar AdS–Schwarzschild solution (6.19) is encapsulated by the altered emblackening factor $f(r)$. The horizon is, as always, still determined by the zero of this function, $f(r_0) = 0$. One infers directly that this factor allows for two solutions, representing an *inner* and *outer* horizon. In the holographic context the inner horizon does not play any role. The outer horizon – the larger root of the emblackening factor at the radial position r_0 – takes over the role of the horizon of the Schwarzschild black hole. Note that we have used gauge invariance to conveniently shift the electrostatic potential Eq. (8.2) such that it vanishes at the horizon.

Using the recipe for the black-hole thermodynamics of chapter 6, the Reissner–Nordström black-hole temperature and entropy are

$$T = \frac{(d+1)r_0}{4\pi L^2}\left(1 - \frac{(d-1)Q^2}{(d+1)r_0^{2d}}\right), \quad s = \frac{1}{4G}\left(\frac{r_0}{L}\right)^d.$$

(8.4)

The relation between the outer horizon position r_0 and the mass and charge of the black hole is

$$M = r_0^{d+1} + \frac{Q^2}{r_0^{d-1}}.$$

(8.5)

One clearly sees that when the mass M of the RN black hole is large compared with its charge Q the black hole forgets that it is charged and one recovers the results for the Schwarzschild black hole. This is not particularly interesting, since it just means that the temperature in the boundary is large compared with the chemical potential μ.

The interesting new regime is when T becomes less than μ. There one expects to find the physics associated with finite density to show up. The RN black hole behaves quite interestingly in this regard. Upon lowering the mass M (temperature) for a fixed charge Q (chemical potential), the horizon will recede towards the deep interior. However, although the energy stored in the space-time fabric decreases when the black hole shrinks, the energy stored in the electromagnetic field is fixed, given the fixed charge. The total energy cannot become less than the energy contained in this fixed electromagnetic field. Violating this intuitive restriction would result in a space-time with a naked singularity. The solution Eq. (8.2) is therefore bounded by

$$M \geq \frac{4(d+1)^{(d+1)/(2d)}}{c_d^2(d-1)^{(d+1)/(2d)}} Q^{(d+1)/d}.$$

(8.6)

The interesting case is when this inequality is exhausted. This is the special "extremal black hole" characterised by a mass that is entirely due to its electromagnetic charge.

It was recognised early on that in the context of black-hole thermodynamics such extremal black holes are fascinating objects. Substituting the saturation value into the formula for the Hawking temperature, Eq. (8.4) shows that its temperature is zero. On further thought, this makes sense. The Schwarzschild black hole is unstable since it emits Hawking radiation and as a consequence its mass should decrease with time. However, when its mass is entirely due to its conserved electrical charge it should be impossible to reduce this mass further. Energy conservation forbids the further emission of radiation: its temperature has to be zero.

The extremal RN black hole therefore codes for a *zero-temperature* finite-density state. But now comes the remarkable "strangeness" of this state: when the bound Eq. (8.6) is saturated there is still a black-hole solution with a horizon at

$$r_* = \left(\frac{d-1}{d+1}\right)^{1/(2d)} Q^{1/d}. \tag{8.7}$$

The horizon (and in particular its area) is therefore manifestly of a macroscopic size. This implies in turn that the extremal black hole carries a *finite* Bekenstein–Hawking entropy $S = A/(4G)$. The gravitational structure that is supposedly the most elementary way to encode finite density in the field theory predicts a ground state that is highly degenerate: this is a state characterised by zero-temperature entropy.

Finite ground-state entropy states are abundant in *classical* matter. In condensed matter they are called frustrated systems. An elementary example is the Ising anti-ferromagnet on a triangular lattice, which is characterised by an extensive number of different spin configurations that all have the same minimal energy. However, the ground-state degeneracy of the RN metal state somehow has a very different origin that is at present not understood at all. A system with an extensive ground-state degeneracy violates Nernst's third law of thermodynamics. This pragmatic "law" refers to the simple wisdom that such states are very fragile: any influence will lift the degeneracy. In the case of classical geometrically frustrated systems the "frustration volume" of ground states in configuration space is spanned by spatial configurations, but that does not seem to be the case for the RN metal. The zero-temperature entropy scales like $S_0 = \mu^d N^2$, indicating that it relates to the decon-fined degrees of freedom (N^2), while it is obviously present only at a finite density ($\mu \neq 0$). Referring to the (emergent gauge) fractionalisation ideas discussed in chapter 2, the RN metal is called by some a "fractionalised phase" (contrasting with the "cohesive phases" like the Fermi liquid and the superconductor), but it remains to be seen whether there is any relationship with this "particle-physics fractionalisation", given the mysterious nature of the RN ground-state entropy.

Summary 5   The "Reissner–Nordström" strange metal is characterised by a finite entropy at zero temperature. The ground-state degeneracy is of a new kind: differently from the frustrated systems of condensed matter physics, it arises in the "$\hbar \to \infty$" limit.

As one's intuition prompts, the RN black hole is an incredibly unstable state. Let us take a moment to emphasise how utterly surprising this is from a purely gravitational point of view. Conventional wisdom states that black holes, being the inescapable end state of all matter, should be infinitely stable. In AdS this turns out to be simply not true, and it is no exaggeration to state that it is this surprise that governs a large part of the AdS/CMT programme.

Starting in section 8.4 and continuing in subsequent chapters, we will discuss these aspects in detail. First we shall delve further into the remarkable properties of the RN metal. Using the GKPW rule explained in box 8.1 to derive the free energy of the boundary field theory, we shall find that the first law of thermodynamics in a system with finite density $d\epsilon = T\,ds + \mu\,d\rho$ is neatly reproduced by the "charged" bulk. Intriguingly, the entropy density at finite temperature is of the form $s(T) = s_0 + cT + \mathcal{O}(T^2)$ when $T \ll \mu$. Besides the zero-temperature contribution s_0, one finds for temperatures well below the "degeneracy" scale μ a finite-temperature contribution that is linear in T. This implies that the specific heat is proportional to the temperature whatever the dimensionality of the system. At low temperature the RN metal exhibits the same (Sommerfeld) specific heat as the Fermi liquid. This is a first example of a very strange metal that nevertheless behaves quite normally in some respects. For instance, the cuprate strange metals exhibit a Sommerfeld specific heat and it is a reflex to interpret this as a signal that somehow Fermi-liquid physics is at work.

Box 8.1   Reissner–Nordström thermodynamics

For the RN black hole we use the same recipe to compute the thermodynamics as discussed in chapter 6. In the field theory, turning on a finite chemical potential μ for the global $U(1)$ current J^μ means perturbing the field theory with

$$\delta S_{\mathrm{FT}} = \mu \int d^{d+1}x\; J^t. \tag{8.8}$$

According to the dictionary, the boundary value of A_μ is the source for the corresponding operator J^μ, thus $A_t(r \to \infty) = \mu$. As in chapter 6, we will work in the grand canonical ensemble and the thermal potential for the dual field theory can be computed from the Euclidean on-shell action. We simply quote the result; unlike for the gravitational action, no special boundary term is needed for the Maxwell part and we straightforwardly obtain

$$\Omega = -T \ln \mathcal{Z} = T S_{\mathrm{E}}[g_{\mathrm{E}}]$$
$$= -V_d \left(\frac{2(d-1)Q\mu}{\sqrt{2}\kappa c_d g_{\mathrm{F}} L^{d+1}} - \frac{2\pi}{\kappa^2}\left(\frac{r_0}{L}\right)^d T \right), \tag{8.9}$$

Box 8.1   (Continued)

where $V_d = \int d^d x$ and T is the temperature (8.4). Thus the charge density ρ and entropy density s and the energy density ϵ for the boundary field theory are

$$\rho = -\frac{1}{V_d}\frac{\partial\Omega}{\partial\mu} = \frac{2(d-1)}{c_d}\frac{Q}{\sqrt{2}\kappa L^{d+1}g_F},$$

$$s = -\frac{1}{V_d}\frac{\partial\Omega}{\partial T} = \frac{2\pi}{\kappa^2}\left(\frac{r_0}{L}\right)^d, \qquad (8.10)$$

$$\epsilon = \frac{\Omega}{V_d} - \mu\rho = \frac{d}{2\kappa^2}\frac{M}{L^{d+2}}.$$

The pressure is that of a conformal fluid

$$P = \frac{\epsilon}{d}. \qquad (8.11)$$

Since $L^d/(2\kappa^2) \sim N^2$ and $g_F \sim \kappa/L$, all these thermodynamical quantities are of order N^2. One can check that the thermodynamical relation $\epsilon + P = \mu\rho + Ts$ as well as the first law of thermodynamics for a finite-density system,

$$d\epsilon = T\,ds + \mu\,d\rho, \qquad (8.12)$$

are indeed satisfied.

It is convenient to introduce a length scale r_* to parametrise Q as

$$Q = \sqrt{\frac{d+1}{d-1}}r_*^d. \qquad (8.13)$$

In terms of r_*, all the physical quantities can be expressed as functions of r_0 and r_*, and we have

$$\rho = \frac{1}{\kappa^2}\left(\frac{r_*}{L}\right)^d\frac{1}{e_d},$$

$$\mu = \frac{d(d+1)r_*}{(d-1)L^2}\left(\frac{r_*}{r_0}\right)^{d-1}e_d, \qquad (8.14)$$

$$T = \frac{(d+1)r_0}{4\pi L^2}\left(1 - \frac{r_*^{2d}}{r_0^{2d}}\right),$$

where $e_d = Lg_F/\left(\kappa\sqrt{d(d+1)}\right)$ is a dimensionless number, using natural units $k_B = \hbar = c = 1$.

At zero temperature ($T = 0$), we have $r_0 = r_*$ and $M = (2d/(d-1))r_*^{d+1}$. Note that the horizon area is non-zero. Thus we have non-vanishing ground-state entropy. The entropy density is

$$s = (2\pi e_d)\rho. \tag{8.15}$$

At low temperature, $T/\mu \ll 1$, we have

$$s = \frac{2\pi}{\kappa^2}\left(\frac{r_0}{L}\right)^d = (2\pi e_d)\rho + \frac{4\pi r_*^{d-1}}{(d+1)\kappa^2 L^{d-2}}T + \mathcal{O}(T^2). \tag{8.16}$$

One can easily compute the specific heat by taking $c_v = T\,\partial s/\partial T \propto T$. When the temperature is very high, we know that RN AdS is quite similar to that containing a Schwarzschild black hole and we have $s \propto r_0^d$ and $T \propto r_0$, and thereby $c_v = T\,\partial s/\partial T \propto T^d$ as expected from the $(d+1)$-dimensional CFT at finite temperature. For calculational convenience, one can use dimensionless quantities by scaling

$$r \to r_0 r, \ (t, \mathbf{x}) \to \frac{L}{r_0}(t, \mathbf{x}), \ A_t \to \frac{r_0}{L^2}A_t, \ M \to r_0^{d+1}M, \ Q \to r_0^d Q \tag{8.17}$$

and the RN solution can be easily characterised by the dimensionless quantity T/μ.

8.2 The AdS$_2$ near-horizon geometry and the emergent local quantum criticality

We now present one of the most revealing insights from AdS/CMT: the emergence of the local, purely temporal quantum criticality in the strange (unstable) RN metal [159, 320]. For condensed matter the physics of interest is the emergent physics in the deep IR. Following the dictionary, this physics can be found in the deep interior, near the horizon, of the AdS black hole. Everything that really counts can be deduced from what happens to the geometry in the near-horizon limit of the zero-temperature, extremal black hole. This theme of focussing on the near-horizon geometry to isolate the IR physics will recur a number of times in the remainder of this book. The RN black hole is a most interesting example of this holographic phenomenon, so let us therefore step slowly through the analysis.

The crucial information is carried by the near-horizon behaviour of the emblackening factor $f(r)$ in Eq. (8.2). The horizon r_0 is determined by the vanishing of $f(r_0) = 0$ and we can Taylor expand $f(r)$ near the horizon in terms of a small $r - r_0$,

$$f(r) = f'(r_0)(r - r_0) + f''(r_0)(r - r_0)^2 + \cdots, \tag{8.18}$$

where "$\cdots$" denotes higher-order terms in $r - r_0$. The leading order in this expansion is representative for the IR physics in the field theory. From the discussions around Eq. (6.18) in chapter 6 we learned that $f'(r_0) \propto T$. The zero-temperature solution is therefore the solution where also $f'(r_0)$ vanishes. This is a universal feature: zero-temperature systems have a double zero at the horizon. Consider the RN emblackening factor at $T = 0$ explicitly. In terms of the $T = 0$ horizon coordinate r_* (Eq. (8.13)), this becomes

$$f(r) = 1 - \frac{2d}{d-1}\left(\frac{r_*}{r}\right)^{d+1} + \frac{d+1}{d-1}\left(\frac{r_*}{r}\right)^{2d}$$

$$= d(d+1)\frac{(r-r_*)^2}{r_*^2} + \cdots \quad \text{(when } r \to r_*\text{)}. \tag{8.19}$$

We recognise the double zero at the horizon. What has happened is that the inner and outer horizons of the finite-temperature RN black hole merge into a single double horizon for the extremal zero-temperature case. This analysis immediately shows that the zero-temperature near-horizon geometry, which is governed by the second-order coefficient $f''(r_0)$, is fundamentally different from the finite-temperature near-horizon geometry, which is governed by the first-order coefficient $f'(r_0)$.

Inserting the near-horizon emblackening factor into the full metric yields the near-horizon geometry in the regime $(r - r_*)/r_* \ll 1$,

$$ds^2 = -\frac{d(d+1)(r-r_*)^2\,dt^2}{L^2} + \frac{L^2\,dr^2}{d(d+1)(r-r_*)^2} + \frac{r_*^2}{L^2}\,dx^2 + \cdots, \tag{8.20}$$

while the gauge potential becomes, straightforwardly,

$$A_t = \frac{d(d+1)e_d}{L^2}(r-r_*). \tag{8.21}$$

The manipulations are simple, and a new geometry has emerged, but what is remarkable is the structure of this near-horizon space-time. One infers that, in terms of the near-horizon coordinate $r - r_*$, the metric factors multiplying dt^2 and dr^2 acquire a structure similar to that of the bare AdS_{d+2} metric Eq. (1.5): this is an effective anti-de Sitter geometry, except that the space directions (dx^2) are multiplied by the constant $(r_*)^2$. Therefore, the space directions form just a flat space, while an effective two-dimensional anti-de Sitter geometry is realised in the time–radial-direction plane. To render this more explicitly, let us re-parametrise the near-horizon metric in terms of a radial coordinate ζ, which is the inverse of the distance from the horizon, and a radius L_2:

$$\zeta = \frac{L_2^2}{r-r_*}, \quad L_2 = \frac{L}{\sqrt{d(d+1)}}, \tag{8.22}$$

$$ds^2 = \frac{L_2^2}{\zeta^2}(-dt^2 + d\zeta^2) + \frac{r_*^2}{L^2}\,dx^2, \quad A_t = \frac{e_d}{\zeta}. \tag{8.23}$$

With the knowledge of AdS metrics which we have gained, this is readily recognised as a space-time with an $AdS_2 \times \mathbb{R}^d$ geometry.

Recall now the "central dogma" of holography that the isometries in the bulk code for the global space-time and scaling symmetries in the boundary field theory, as we highlighted in chapters 1 and 5. On comparing with the scaling isometry of the pure AdS_{d+2} geometry Eq. (4.87) (z is the radial direction),

$$t \to \lambda t, \quad z \to \lambda z, \quad \mathbf{x} \to \lambda \mathbf{x}, \tag{8.24}$$

the near-horizon $AdS_2 \times \mathbb{R}^2$ metric Eq. (8.23) has the scaling isometry

$$t \to \lambda t, \quad \zeta \to \lambda \zeta, \quad \mathbf{x} \to \mathbf{x}. \tag{8.25}$$

We directly infer that we are dealing with a (quasi-)local quantum critical state in the boundary field theory: in spatial directions there is no sense of scale invariance, while the dynamics of the fields is scale-invariant merely in time. Such scaling may appear unfamiliar, but it is a natural extension of the anisotropic scaling familiar in non-relativistic condensed matter theories. Emergent continuum theories can exhibit a scaling symmetry where, under a spatial rescaling, time scales with a different dynamical critical exponent $\mathbf{z}$:

$$t \to \lambda^{\mathbf{z}} t, \quad \mathbf{x} \to \lambda \mathbf{x}. \tag{8.26}$$

On redefining $\lambda \to \lambda^{1/\mathbf{z}}$ we recognise the scaling symmetry of the AdS_2 near the horizon in the limit $\mathbf{z} \to \infty$. There is no precedent in the classical/bosonic renormalisation-group theory for such a very large, let alone diverging, dynamical critical exponent. However, as we discussed in chapter 3 at several points, such a behaviour appeared to show up in experiments on the laboratory strange metals, starting with the marginal Fermi liquid of the late 1980s. We also discussed in chapter 3 a similar notion of purely temporal quantum dynamics as revealed by the dynamical mean-field theory for fermionic systems which is controlled by large dimensions. Irrespective of whether or not this is a coincidence, at least the designation "local quantum criticality" for such scaling behaviour has been directly borrowed by the holographists.

Summary 6   The RN strange metals are emergent quantum critical phases, characterised by a dynamics that is only critical in a purely temporal sense: they exhibit *local quantum criticality*, a property they appear to have in common with the empirical strange metals.

This emergent local quantum criticality is almost as powerful as the full $(d+1)$-dimensional quantum criticality in pristine AdS models. It imposes powerful constraints on the behaviour of field-theory correlation functions in the deep IR, given the scaling properties hard-wired into the bulk geometry. Even knowing little about the specifics, the general form of these deep-IR two-point propagators is easy to deduce. The crux is that the near-horizon geometry becomes a direct product of AdS_2 with $\mathbb{R}^2$. The ramification is that the fluctuations indexed by spatial momenta k_i after Fourier transformation along $\mathbb{R}^2$ decouple. For every k_i one has an independent problem associated with the AdS_2 time–radial-direction plane in the bulk. According to the same AdS/CFT dictionary this translates into a CFT_1, a $(0+1)$-dimensional critical "quantum-mechanical" theory. Given the temporal conformal invariance, the local quantum critical propagators for a given k_i have to take the form

$$\mathcal{G}(\omega, k) = c_k e^{i\phi_k} \omega^{2\nu_k}. \tag{8.27}$$

We will confirm this by presenting explicit bulk computations in the next chapter. In this expression the free constants are yet-to-be-determined functions of the independent momentum k_i. The remarkable feature is that the computation does indeed bear out the claim that the scaling exponent ν_k manifestly depends on the momentum k. The reason follows straight from the dictionary: the scaling exponent is given in terms of the mass in the bulk gravity theory, and in terms of the time–radial-direction dynamics in the AdS_2 bulk the Fourier-transformed fluctuations have an "effective mass" $m_{\text{eff}}^2 = m^2 + g^{ij}k_i k_j$ that depends on k. Since the scaling dimension is linearly related to the mass, one invariably finds that $\nu_k \sim \sqrt{k^2 + 1/\xi^2}$, where the length scale ξ is set by the only scale in the problem. This is the chemical potential $\xi \simeq 1/\mu$. This implies that the Euclidean propagator will behave in space-time as $\mathcal{G}_E(\tau, x) \sim 1/\tau^{2\nu_{k=0}}$ at distances $x = |\mathbf{x}| \ll \xi$ while it will decay exponentially in space as $\mathcal{G}_E(\tau, x) \sim \exp(-x/\xi)$ at large distances $x \gg \xi$ [159]. This differing behaviour with reference to the extra spatial length scale ξ has been named "semi-local quantum criticality" by Liu, Iqbal and Mezei. In the deep IR, however, this length scale is set by the chemical potential, which now takes the role of the UV cut-off. In a condensed matter context, ξ would be of the order of the lattice constant and one could just as well drop the prefix "semi" and call it the conventional "local quantum criticality".

Summary 7 In RN strange metals the correlation functions of the field theory at low energy are eventually dominated by local quantum criticality. Temporal scaling insists that they become algebraic functions of energy with exponents that generically depend on the momentum.

Of course, in the RN black hole the AdS$_2$ emerges only in the deep IR, and the exact correlation functions may differ. Nevertheless, we will see that these AdS$_2$ correlators $\mathcal{G}$ play a big role in various spectroscopic properties. More than anything else, these are the real "signals emitted by the finite-density black holes" and perhaps the most interesting challenge for the condensed matter physicists is to find out whether such signals can be isolated from the experimental data. The role of the emergent AdS$_2$ in exact correlation functions will be the sole focus of the next chapter. The GKPW formula together with this AdS$_2$ analysis allows one to carry out an exquisite interrogation of the long-range physics of holographic strange metals. In this chapter we will analyse the macroscopic characteristics of the RN metal and its generalisations.

8.3 The zero sound and conductivity of the RN metal in the Galilean continuum

A most elementary property of a finite-density medium is its capacity to transport its conserved matter. We discussed at length the hydrodynamics and conductivity of the zero-density CFT in the previous chapter, but what happens in the finite-density RN metal? The answer is that the gross qualitative behaviour of the transport properties looks quite conventional, at least in linear response. This involves more than just the finite-temperature properties which are governed by the hydrodynamical principle. An equally meaningful question is to ask for its transport at zero temperature: its "quantum hydrodynamics". In this regard the collective macroscopic transport properties of an RN metal look at first sight similar to those of a Fermi liquid. Here we focus on the transport in the translationally invariant continuum, which is a greatly simplifying and equalising condition. Chapter 12 deals with the rich transport physics which arises in the strange metals when translational symmetry is broken.

Dimensionality is not important in what follows, and we will consider a $(2+1)$-dimensional boundary in the remainder of this chapter. We are interested in the density susceptibility and the conductivity. The former follows from the density–density propagator,

$$G_{tt}^R(\omega, \mathbf{k}) = -i \int dt \, d^2\mathbf{x} \, e^{i\omega t - i\mathbf{k}\cdot\mathbf{x}} \theta(t) \langle [J_t(\mathbf{x}), J_t(\mathbf{0})] \rangle, \qquad (8.28)$$

whereas the conductivity is expressed in terms of the current–current propagator G_{xx}^R (Eq. (7.50)) by $\sigma(\omega) = (-i/\omega)G_{xx}^R(\omega, \mathbf{k} = 0)$. Since the density and current responses are tied together by the continuity equation through $G_{xx}^R(\omega, \mathbf{k}) = (\omega^2/\mathbf{k}^2)G_{tt}^R(\omega, \mathbf{k})$, they in fact contain the same information. In atomic (and molecular) fluids like helium the density susceptibility can be measured in an extended kinematical regime using inelastic neutron scattering supplemented by ultrasound measurements, including the low-temperature Fermi-liquid (^{3}He) and superfluid (^{4}He) regimes. However, in optical experiments on condensed matter electron systems one is limited to zero momentum since the velocity of light is very large compared with the material velocity. In principle one can measure the density susceptibility at large momenta with electron-loss spectroscopy and inelastic X-ray scattering, but at present these experiments lack the required energy resolution. Here we will first discuss in detail the general problem of the density response both at finite momentum and at finite frequency, and conclude this section with the special case of the optical conductivity at zero momentum.

To interpret the RN strange-metal responses at low temperatures, the Fermi liquid serves as a convenient reference point. Although its zero sound at $\mathbf{k} = 0$ is governed by hydrodynamical protection, we also learnt in chapter 2 that it is different from the finite-temperature first sound. It actually does correspond to a coherent breathing of the Fermi surface that remains present at absolute zero. This has the effect that its attenuation mechanism is different from the first sound of the thermal fluid carried by the gas of incoherent, thermally excited quasiparticles. As a consequence there is an attenuation maximum at $\omega \simeq T$ where the system crosses over from its zero-temperature "quantum" protected sound characterised by a damping $\Gamma \simeq \omega^2/\mu$ to the hydrodynamically protected sound of a thermal fluid with a $\Gamma \simeq \mu\omega^2/T^2$. We will see that the RN metal behaves similarly in that it has both a distinguishable zero-temperature "quantum" sound and the high-temperature hydrodynamical sound. However, both these sounds behave very differently from the sounds of the Fermi liquid.

Since at finite density both charge and momentum conservation are involved, the holographic computations become more involved: one has to solve coupled equations of motion in the bulk involving both the photon and the graviton. This reflects the fact that charge and momentum transport are no longer independent [325, 326, 327, 328]. In box 8.2 we will exhibit this for both the longitudinal and the transverse current–current responses. Intriguingly, the "hydrodynamics" of the zero-temperature RN metal is actually understood only on the linear-response level, for infinitesimal external perturbations. The "fluid/gravity" procedure we highlighted in section 7.2 linking the non-equilibrium hydrodynamics to the near-horizon gravity turns out to become singular in the zero-temperature limit [328]! It is urgently necessary to establish what the equivalent of the Navier–Stokes equations for this zero-temperature RN quantum fluid would look like. To the best of our knowledge nobody has yet found a way to reliably analyse the non-stationary gravity of an extremal RN black hole.

The results for the linear-response calculations are remarkably interesting. To appreciate their special nature, let us first return to the zero-density finite-temperature case of the previous chapter. There we learned that the thermal fluid is governed by Navier–Stokes hydrodynamics. It is a classic exercise to determine the current–current susceptibilities for hydrodynamical fluids [255]. There should be three gapless excitations: a transverse diffusion mode associated with the transverse ("shearing") sound channel, with quadratic dispersion relation

$$\omega = -i\frac{\eta}{\epsilon + P}k^2 + \mathcal{O}(k^4); \tag{8.29}$$

a longitudinal charge diffusion mode; and a longitudinal sound mode with a linear dispersion and a diffusive correction

$$\omega = \pm v_s k - i\frac{d}{d-1}\frac{\eta}{\epsilon + P}k^2 + \mathcal{O}(k^3), \tag{8.30}$$

where $v_s = \sqrt{dP/d\epsilon}$ and η is the shear viscosity. Since our starting point in the UV is a conformal theory, the bulk viscosity should vanish automatically and one need only deal with the shear viscosity η. For the same reasons, the equation of state has to be $\epsilon = dP$, and it follows that the sound velocity is fixed as $v_s = c/\sqrt{d}$. Modes with these particular dispersion relations are precisely seen to exist in the holographic description of the zero-density system in the probe limit [329, 330]. Using $\epsilon + P = sT$, where $s \propto T^d$, while $\eta/s = (1/(4\pi))(\hbar/k_B)$, one finds for the transverse diffusivity the specific value $\mathcal{D} = \eta/(\epsilon + P) = (c^2/(4\pi))(\hbar/(k_B T))$ in explicit units, while the longitudinal sound attenuation $\Gamma_L k^2$ also fits the expectation. Similarly to the case of a Fermi liquid, the damping of the sound decreases for increasing temperature, although the dependence on temperature is different [331].

At finite density the holographic computations become much more complicated. Even in the small-ω, k regime of relevance to hydrodynamics, the problem in the longitudinal channel can only be solved numerically [325, 327]; see box 8.2.

In the transverse channel it is a very different story. In a tour de force the transverse diffusivity has been computed analytically at low temperature using the matching technique explained in the next chapter [328]. The outcome is remarkably simple and physically counterintuitive at the same time. The diffusion constant in the small-frequency regime for $T \ll \mu$ is still given by the hydrodynamical expression

$$\mathcal{D} = \frac{\eta}{\epsilon + P}, \tag{8.31}$$

but now

$$\epsilon + P = \mu\rho + Ts, \quad \eta = \frac{1}{4\pi}\frac{\hbar}{k_B}s(T). \tag{8.32}$$

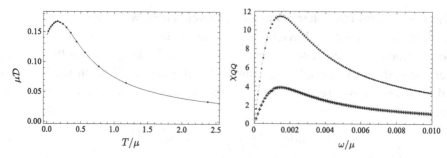

Figure 8.1 The left panel shows the charge diffusion constant $\mu\mathcal{D}$ of the RN metal as a function of temperature. The line shows the hydrodynamical result $\mathcal{D} = \eta/(\epsilon + p)$. This result clearly persists all the way to zero temperature. This can be understood from the viscosity bound $\eta = s/(4\pi)$ together with the non-vanishing ground-state entropy of the black hole. It means that even at zero temperature hydrodynamics continues to apply and energy and momentum can diffuse. The right panel explicitly shows the continued validity of hydrodynamics in the physical response. Here the numerically computed spectral function of the shear perturbation Im $G_{T_{xy}T_{xy}}$ (dots) and charge diffusion Im $G_{J_yJ_y}$ (crosses) for $k/\mu = 0.1$ and $T/\mu = 0.005$ are matched to the elementary hydrodynamics prediction (lines). The match is essentially perfect. Figure source [328]. (Reprinted with kind permission from Springer Science and Business Media, © 2013, SISSA, Trieste.)

The entropy density $s(T) = s_0 + AT + \cdots$ of the RN metal at low temperatures contains both the ground-state entropy s_0 and the leading "Sommerfeld" contribution $\sim T$, Eq. (8.16). At low temperatures, knowing that holographically $\eta = s/(4\pi)$, the diffusion constant thus behaves as $\mathcal{D} \sim s/\mu$. On the other hand, the diffusion at high temperatures behaves universally as $\mathcal{D} \sim 1/T$. Interpolating from low to high temperatures, there is therefore a maximum in the diffusion constant at the crossover temperature $T \sim \mu$; see Fig. 8.1.

Returning to the longitudinal channel, the sound attenuation is given by $\mathcal{D}_{\mathrm{L}}q^2$, where $\mathcal{D}_{\mathrm{L}}$ is the longitudinal diffusion constant. This behaves very similarly to the transverse diffusivity; it appears to be consistent with the numerical outcomes for the longitudinal sound, where one finds that the attenuation constant Γ_{L} is within 10% of the value derived from $\mathcal{D}$ [325]. It follows that there is a sound-attenuation maximum, as in the Fermi liquid. The remarkable conclusion is that the RN fluid continues to behave like a thermal, hydrodynamical liquid dissipating the shear motions all the way down to zero temperature. Similarly, the viscosity continues to be simply proportional to the entropy, while at zero temperature it is determined by the ground-state entropy. As we discussed in chapter 2, in the strongly interacting Fermi liquid there is even a transverse propagating sound. This is clearly missing in the RN liquid.

The punchline is that the RN *zero-temperature* fluid as "probed" by the way in which its sound is damped behaves as if it were a viscous thermal fluid. Adding to this mystery is the finding that the zero-temperature "viscosity" is apparently associated with the ground-state entropy. At present we are not aware of any explanation for this peculiar phenomenon in a field-theoretical language.

Summary 8 The RN strange metals have the very unusual property that they continue to behave like thermal hydrodynamical liquids all the way down to zero temperature. At zero temperature they support a propagating "zero-sound" mode that is damped as if it were a zero-density quantum critical liquid with a viscosity now proportional to the ground-state entropy.

Box 8.2   Computing the sound response of the Reissner–Nordström metal

To compute the density–density propagator in the $(2 + 1)$-dimensional boundary, the fluctuations of the $U(1)$ gauge field in the bulk have to be determined. The novel aspect we encounter at finite density is that these cross-couple to the fluctuations of the metric. Since the fluctuations of the metric correspond to the energy–momentum tensor, this is the bulk expression of the fact that at finite density both charge and momentum conservation are correlated in the boundary field theory. We sketch first the computation of the longitudinal response up to where we must resort to numerics, and next the somewhat simpler transverse sound response.

We decompose the metric and the gauge field as before:

$$g_{\mu\nu} = \bar{g}_{\mu\nu} + \int \frac{d\omega\, dk}{(2\pi)^2} e^{-i\omega t + ikx} h_{\mu\nu}(r),$$

$$A_\mu = \bar{A}_\mu + \int \frac{d\omega\, dk}{(2\pi)^2} e^{-i\omega t + ikx} a_\mu(r). \tag{8.33}$$

Here $\bar{g}_{\mu\nu}, \bar{A}_\mu$ are the RN background fields (8.2), while $h_{\mu\nu}, a_\mu$ are infinitesimal fluctuations in momentum space, where we have exploited rotational invariance to set $k_y = 0$ and $k_x = k$. As usual, we use the *radial gauge* for both the metric and the gauge fields such that $a_r = 0$ and $h_{r\nu} = 0$, with $\nu = \{t, x, y, r\}$.

With respect to the inversion symmetry $y \rightarrow -y$, the fluctuations decouple in linear order into two groups depending on their parity under this transformation: $\{h_{ty}, h_{xy}, a_y\}$ have odd parity corresponding to the *transverse* shear and (charge) diffusion modes of the boundary theory. The even-parity fluctuations

Box 8.2   **(Continued)**

$\{h_{tt}, h_{tx}, h_{xx}, h_{yy}, a_t, a_x\}$ are associated with the *longitudinal* diffusion and sound modes of the boundary theory since these describe the response parallel to the direction of momentum flow [326, 328, 332].

Longitudinal sector

In the *longitudinal* sector the radial gauge does not completely fix the gauge freedom of $h_{\mu\nu}$ and a_μ with no r components. To account for this it is convenient to use gauge-invariant linear combinations of the fluctuations. These can be chosen to be [325, 327]

$$X_1 = \omega a_x + k a_t - \frac{k A_t'}{2r} h_{yy},$$

$$X_2 = \frac{2\omega k}{r^2} h_{xt} + \frac{\omega^2}{r^2} h_{xx} + \frac{k^2}{r^2} h_{tt} + \frac{k^2 f}{r^2} h_{yy} \left(1 + \frac{rf'}{2f} - \frac{\omega^2}{k^2 f} \right). \tag{8.34}$$

One can check explicitly that these combinations are invariant under the gauge transformations $\xi = e^{ikx - i\omega t} \xi(\omega, k)$ and $\Lambda = e^{ikx - i\omega t} \Lambda(\omega, k)$ with

$$h_{\mu\nu} \to h_{\mu\nu} - \bar{\nabla}_\mu \xi_\nu - \bar{\nabla}_\nu \xi_\mu, \quad a_\mu \to a_\mu - \xi^\nu \bar{\nabla}_\nu A_\mu - A_\nu \bar{\nabla}_\mu \xi^\nu, \tag{8.35}$$

$$A_\mu \to A_\mu - \bar{\nabla}_\mu \Lambda, \tag{8.36}$$

where $\bar{\nabla}_\mu$ is the covariant derivative w.r.t. the background metric.

As usual, the expansion (8.33) is inserted into the bulk Einstein–Maxwell action with suitable boundary terms, and this results in an on-shell action for the fluctuations of the form

$$S_{\text{on-shell}} = \int_{r \to \infty} \frac{dw\, dk}{(2\pi)^2} \Bigg[X_i(r, -\omega, -k) K_{ij}\, \partial_r X_j(r, \omega, k)$$

$$+ X_i(r, -\omega, -k) C_{ij} X_j(r, \omega, k) \Bigg], \tag{8.37}$$

with X_i ($i = 1, 2$) satisfying the in-falling boundary condition near the horizon [327].

The retarded Green function can be obtained as follows [333]. With linearly independent in-falling boundary conditions on the horizon for X_i one can obtain two sets of boundary values for the operators X_i, one for each of the independent in-falling conditions. One can choose special initial values at the horizon that can diagonalise the boundary source matrix. The final retarded Green function can be obtained following

$$G_{ij}^{\text{R}}(\omega, k) = 2 \lim_{r \to \infty} \left(K_{ik}\, \partial_r F_{kj} + C_{ij} \right), \tag{8.38}$$

with the gauge-invariant modes expressed as a function of sources $X_i(r) = F_{ij}(\omega, k, r)X_j^{(0)}$, where we can choose $F_{ij}(\omega, k, \infty) = \delta_{ij}$. By substituting the RN AdS$_4$ black-hole solution and fluctuations into the above procedure, one can numerically solve the system to obtain the density–density correlator G_{tt}^{R} (Fig. 8.2). The zero sound follows from the pole of G_{tt}^{R}, with the dispersion relation as has already been discussed in the main text.

An advantage of the choice (8.34) is that it becomes very easy to compute the conductivity as well, since the Ward identity $G_{xx}^{R} = (\omega^2/k^2)\, G_{tt}^{R}$ we referred to in the text is wired in; the optical conductivity is in this way just the special case $k = 0$ in this set-up.

Transverse sector

Let us now turn to the *transverse* sound governed by the shear and charge diffusion, i.e. the fluctuations $\{h_{ty}, h_{xy}, a_y\}$. We can again write the complicated equations of motion in a compact way using the gauge-invariant modes. The gauge-invariant modes that can be constructed from linear combination of the fluctuations can be chosen to be [326, 328, 332]

$$X = h^y_t + \frac{\omega}{k}h^x_y, \quad Y = a_y. \tag{8.39}$$

The Ward identities for the dual field theory can be directly identified from these gauge-invariant expressions (8.39):

$$G_{T^{xy}T^{ty}}^{R} = \frac{\omega}{k}G_{T^{ty}T^{ty}}^{R}, \quad G_{T^{xy}T^{xy}}^{R} = \frac{\omega^2}{k^2}G_{T^{ty}T^{ty}}^{R}, \quad G_{T^{xy}J^y}^{R} = \frac{\omega}{k}G_{T^{ty}J^y}^{R}. \tag{8.40}$$

The equations of motion for X, Y in the RN AdS black-hole background are

$$\left[r^6 f(A_t'X' + f'Y')\right]' + \frac{r^2}{f}(\omega^2 - k^2 f)\left[A_t'X + f'Y\right] = 0, \tag{8.41}$$

$$\left[\frac{r^2 f}{\omega^2 - k^2 f}(r^2 X' + A_t'Y)\right]' + \frac{1}{f}X = 0, \tag{8.42}$$

with $f(r)$ and A_t the emblackening factor and the background electrostatic potential of the RN black hole (8.2). For convenience we have set $2\kappa^2 = g_F = 1$.

The on-shell action is

$$S_{\text{on-shell}} = \int_{r \to \infty} \frac{d\omega\, dk}{(2\pi)^2}\left[-\frac{r^4 f k^2}{2(\omega^2 - k^2 f)}X(r, -\omega, -k)X'(r, \omega, k)\right.$$
$$\left. - \frac{r^2 f}{2}Y(r, -\omega, -k)Y'(r, \omega, k)\right]. \tag{8.43}$$

Box 8.2   (Continued)

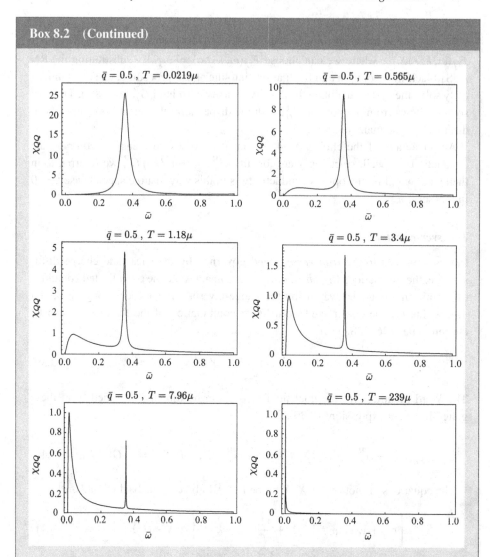

Figure 8.2 The numerical charge-density spectral function Im G_{tt} in units of $2r_0/\kappa^2$ for $q = 0.5\mu$ extracted from the gauge-invariant correlation function for X_1 as the temperature is increased. At low temperatures energy and charge transport are not independent, and both are controlled by the sound mode. The expected maximum in the sound attentuation (the width) discussed in the text is not visible here because the value of $\bar{q} \equiv q/\mu$ is too high for the system to stay in the linear regime; it is present for lower values of $\bar{q}$ [327]. The charge-density spectral function does visualise the crossover between sound-dominated charge transport at low temperature and diffusion domination at high temperature. Figure source [327]. (Reprinted with kind permission from Springer Science and Business Media, © 2011, SISSA, Trieste.)

As for Eq. (8.38), the retarded Green function can be obtained by solving the equations with in-falling boundary conditions.

At finite temperature, in the hydrodynamics limit $\omega \ll T$, this can be done analytically with the matching procedure which will be discussed in detail in chapter 9. The idea is to divide the space-time into two regions: the near-horizon region $(r - r_0)/r_0 \ll 1$ and the far region $\omega/(r_0^2 f'(r_0)) \ll (r - r_0)/r_0$. Solving equations (8.41) in both these two regions and matching the solutions in the overlapping region $\omega/(r_0^2 f'(r_0)) \ll (r - r_0)/r_0 \ll 1$ allows one to determine the integration constant in the far region from the in-falling near-horizon boundary condition of the near region [328]. One then uses the standard GKPW rule to extract the Green functions. One finds that they have a diffusion pole in the low-frequency limit with dispersion relation

$$\omega = -i \frac{\eta}{\epsilon + P} k^2 + \cdots, \tag{8.44}$$

where η is the shear viscosity, ϵ is the energy density, P is the pressure and "..." denotes higher-order terms in k; see Fig. 8.1.

Moreover, (8.44) holds all the way down to zero temperature [326]. As was first found by Edalati, Jottar and Leigh, now the shear viscosity is intriguingly set by the ground-state entropy $\eta = -\lim_{\omega \to 0}(1/\omega)\mathrm{Im}\, G^{\mathrm{R}}_{T_{xy}T_{xy}}(\omega, k = 0) = s/(4\pi)$ [332].

The final transport quantity of interest is the *optical conductivity*. We explained at the beginning of this section that the computation of this quantity [17, 19] is just a special case of the determination of the zero sound spectrum due to the correlated response of charge and matter and the Ward identities. The results for the real and imaginary parts of the conductivity as functions of ω/T are shown in Fig. 8.3. As μ/T is increased and particle/hole symmetry is broken, we see quite a bit more structure than for the constant, ω/T-independent, conductivity at zero density. One finds a depletion of spectral weight at energy scales less than μ. As a fully unitary map between theories, the holographic translation is consistent with the f-sum rule, and this missing weight accumulates in the Drude peak at zero frequency: as usual, when momentum is conserved and density is finite, this is a delta function at zero frequency with a pole strength (Drude weight) that measures the number of "carriers" contributing to the perfect metallic conduction. At small frequency, the conductivity behaves like

$$\sigma(\omega) = \sigma_{\mathrm{E}} + \frac{\rho^2}{\epsilon + P}\left(\delta(\omega) - \frac{1}{i\omega}\right), \tag{8.45}$$

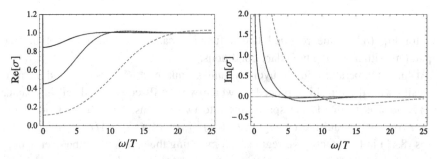

Figure 8.3 The conductivity of the $(2 + 1)$-dimensional RN metal as a function of ω/T computed numerically [17]. The three curves show how the conductivity changes as a function of increasing chemical potential; the solid line is for $\mu \ll T$, the dashed line $\mu \lesssim T$ and the dotted line $\mu \sim T$. For $\omega \to 0$, the imaginary part of the optical conductivity behaves as $1/\omega$ in all cases; this reveals the existence of a delta function $\delta(\omega)$ in Re σ. For $\omega \gg \mu$ the conductivities asymptote to the zero-density result discussed in section 7.3, corresponding in the $(2 + 1)$-dimensional case to a constant as a function of ω/T. For the reasons explained in the main text, the spectral weight is depleted for frequencies below μ, and this missing weight accumulates in the Drude delta function at zero frequency in the real part, showing that the metal has to be a perfect conductor in the Galilean continuum. Figure adapted from [17]. (Reproduced by permission of IOP Publishing. © IOP Publishing. All rights reserved.)

where ρ, ϵ, P are the thermodynamical quantities of the dual field theory, Eqs. (8.10) and (8.11). The first "σ_E" contribution is actually quite surprising, as we will discuss in a moment. The second term is according to expectation: it is completely dictated by the hydrodynamical principle. In the case discussed in the previous chapter at zero density, one can have finite spectral weight at zero frequency since momentum conservation does not constrain the transport at all: the particles and anti-particles moving in opposite directions with zero net momentum can create a finite value, as we explained in chapter 7. At finite density, however, momentum conservation takes control over the excess of particles and forces a finite weight to accumulate in the Drude delta function. The f-sum rule together with the scale set by the chemical potential takes care of the remainder. The form of the optical conductivity $\sigma(\omega)$ for a translationally invariant system is therefore fixed by the general hydrodynamical principle and hence it is not particularly revealing even with regard to the vast differences between the RN strange metal and normal metals. The situation will be acutely different when we break translational symmetry and momentum conservation is lost. This will be the main subject dealt with in chapter 12.

The σ_E term is actually very surprising: it defeats the hydrodynamical wisdoms of the previous paragraph. According to the hydrodynamical principle, the momentum conservation should be felt also by the alternating currents when their frequency becomes less than the chemical potential. The spectral weight should therefore be exponentially suppressed when $\omega < \mu$, as is found in all cases where the conductivity is computed in a controlled fashion (including the condensed matter examples). The value of σ_E is instead rising algebraically at small ω like $\sigma \sim \omega^2$, seemingly defeating the hydrodynamical principle. This is at present understood as a consequence of a deeper phenomenon: these systems appear to scale back in the deep infrared to an *effectively neutral* system, characterised by an emergent charge-conjugation symmetry. This affair is, however, far from settled, and it is perhaps the greatest enigma of the whole holographic portfolio.

8.4 The scaling atlas of emergent holographic quantum critical phases

As we have repeatedly emphasised, the surprise revealed by the RN black hole is the existence of an emergent locally quantum critical *phase* in the infrared characterised by a scaling behaviour that is unfamiliar from the point of view of classic RG. Is this RN metal a singularly unique physical system, or are there more phases of this kind that can exist in principle? If such a family exists, is it possible to "classify" these phases phenomenologically, in the sense of a general Kadanoff–Widom scaling theory of the classical critical state, with the difference that it now relies through AdS/CFT on the uniqueness of GR solutions in the bulk?

A very interesting conjecture answering this question has been proposed [321, 322, 323, 324], although there is no proof that the classification is complete. It is beyond doubt, however, that the RN metal is far from unique. According to generic properties of the bulk gravity, a large family of strange metals is allowed in principle. Within this class the RN metal does remain special. It is the sole member with a finite macroscopically large ground-state entropy; for all other cases it will vanish.

The characterisation of these phases will be according to their scaling behaviour. We will find that there is an interesting variety of scaling properties, and here we will present the "geography" of this scaling "landscape". When one is dealing with a quantum critical *phase*, the classical thermodynamic exponents α, β, γ, δ and the correlation-length exponent ν have no meaning since these are associated with an isolated unstable fixed point. The scaling is quantified firstly in terms of the values of the anomalous dimensions of particular operators (correlation-function exponents). These are fixed in top-down constructions, but they are free parameters in the present bottom-up setting. The remaining scaling dimensions are the dynamical critical exponent z and the "hyperscaling-violating" exponent θ which is not familiar from the theory of the classical critical state.

Let us see what can be learned from the gravity side. We learned that in order to describe a finite charge density one has to include a flux of an electrical field in the bulk piercing through the boundary. In the simplest two-derivative Einstein–Maxwell theory the only way to implement such a flux is the RN black hole. As we discussed at length, one then automatically pays the price of a finite extremal horizon dual to a finite ground-state entropy in the boundary. However, this limitation to minimal Einstein–Maxwell theory is to a degree an over-simplification. In any top-down string-theory-based construction of the duality, the bulk will be littered by many other fields. In fact, there are usually infinitely many such fields, but they typically have very large masses since they follow from the Kaluza–Klein reduction over the string-theory dimensions that do not form AdS. In the bottom-up approach one simply adds additional fields with the desired properties by hand. In the next chapters the roles of these fields will be placed centre stage. Often they are familiar: an addition of a charged "scalar" (a Higgs) field or charged fermion fields. They will be responsible for many interesting phenomena, including spontaneous symmetry breaking, Fermi liquids, etc.

But these familiar fields do not exhaust the possibilities. In string theory one finds that the low-energy spectrum left over after Kaluza–Klein compactification contains generically so-called dilaton fields. These have non-linear and non-minimal couplings to the gravity and Maxwell sector that may be unfamiliar to the non-string-theoretical reader. Nevertheless, these are ubiquitous in top-down holographic constructions. For the present purposes, a dilaton field ϕ is a real scalar field with the defining property that it turns the gauge coupling into a dynamical quantity, by multiplying the Maxwell term by a function $Z(\phi)$ of its field value. In principle it does the same for the gravitational coupling, but for a single dilaton this can be removed by a field redefinition. Conventionally the Einstein–Hilbert action remains the same. In addition the dilaton will have its own potential and the typical form of such an "Einstein–Maxwell dilaton" (EMD) gravity is [321, 322, 323, 324]

$$\mathcal{L} = R - \frac{1}{2}(\partial_\mu \phi)^2 - \frac{Z(\phi)}{4}F^2 - V(\phi). \tag{8.46}$$

The effective gauge coupling in EMD gravity is thus $g_{\text{eff}}^2 = Z^{-1}(\phi)$. Since it is a dynamical function of the scalar field $\phi(r)$, Gauss's law is modified and the Maxwell equation in EMD gravity equals

$$\nabla_\mu \big(Z(\phi) F^{\mu\nu} \big) = 0. \tag{8.47}$$

Equivalently $\partial_\mu \big(\sqrt{-g} Z(\phi) F^{\mu\nu} \big) = 0$. Typically we are most interested in the time-independent (static) translationally and d-spatially rotationally invariant (homogeneous and isotropic) solution with only the A_t component of A_μ non-vanishing. A_t

will therefore depend solely on the radial direction r and the only non-trivial component of $F_{\mu\nu} = F_{rt}$. In this case, Maxwell's equation becomes the bulk Gauss law $\partial_r(\sqrt{-g}Z(\phi)F^{rt}) = 0$.

Let us now present some AdS/CFT gymnastics. From the GKPW rule we know that the charge density in the boundary field theory is given by $\rho_{\text{bdy}} = Z(\phi(r))\sqrt{-g}F^{rt}|_{r=\infty}$. If there is no (additional) charged matter field in EMD, it then follows from Gauss's law that

$$\rho_{\text{bdy}} = Z(\phi(\infty))\left(\sqrt{-g}F^{rt}\right)|_{r=\infty} = Z(\phi(r_0))\left(\sqrt{-g}F^{rt}\right)|_{r=r_0}, \qquad (8.48)$$

where r_0 is the location of the horizon; in the absence of a horizon at zero temperature $r_0 = 0$. In the standard Einstein–Maxwell gravity $Z(\phi(r)) = 1$, and thus $\left(\sqrt{-g}F^{rt}\right)|_{r=r_0} = \rho_{\text{bdy}}$. This signals that in the absence of charged matter in the bulk, the horizon must carry a finite charge density to encode finite-density matter in the boundary: the horizon area has to be finite all the way down to zero temperature. In EMD gravity, on the other hand, the prefactor can change this situation. Borrowing from string theory, we may approximate $Z(\phi)$ with an exponential $Z(\phi) \propto e^{c\phi}$ both at the boundary and in the deep interior of AdS. The solutions to these equations of motion typically exhibit $\phi(r \to \infty) \propto 0$ or $\phi(r \to 0) \propto -\alpha \ln r$ with positive α, where the horizon is at $r = 0$ at zero temperature (see e.g. box 8.3). In the latter case we find from Gauss's law (8.48) that near the horizon $\left(\sqrt{-g}F^{rt}\right) \propto \rho r^{c\alpha}$. The electric flux therefore vanishes in this area and the horizon area is allowed to shrink to zero. This is the distinctive appealing aspect of EMD gravity. It possesses charged solutions without ground-state entropy for quite generic functions $Z(\phi)$ and $V(\phi)$.

The deep-interior geometries of these charged zero-temperature horizonless states turn out to describe the landscape of Lifshitz and hyperscaling-violating quantum critical phases. In full generality, the IR metrics describing the deep interior at zero temperature are of the form [334]

$$ds_{\text{NHL-EMD}}^2 = r^{-2\theta/d}\left(-r^{2z}\,dt^2 + r^2\,dx_i^2 + \frac{dr^2}{r^2}\right) + \cdots, \qquad (8.49)$$

where $r = 0$ is the deep interior: there is in fact no longer a horizon. The parameters z and θ correspond precisely to the dynamical critical exponent and the hyperscaling-violation exponent, respectively.

In standard field theories $\theta = 0$ – they obey hyperscaling – and in this case the metric Eq. (8.49) reduces to the so-called Lifshitz solution, as first studied in [335]:

$$ds_{\text{NHL-EMD}}^2 = -r^{2z}\,dt^2 + r^2\,dx_i^2 + \frac{dr^2}{r^2}. \qquad (8.50)$$

It is easy to check that this solution is invariant under the Lifshitz scaling transformation,

$$t \rightarrow \lambda^{z} t, \quad x_i \rightarrow \lambda x_i, \quad r \rightarrow \lambda^{-1} r, \tag{8.51}$$

which describes a scaling in the field theory characterised by a dynamical critical exponent z. When $z = 1$, this Lorentz-invariant scaling symmetry is precisely coincident with the "pristine" zero-temperature and zero-density AdS_{d+2} space-time. An alternative metric for the same Lifshitz space-time is obtained by changing the coordinates according to $r \rightarrow e^{-y/z}, t \rightarrow t/z, x_i \rightarrow x_i/z$, such that

$$d\tilde{s}^2_{\text{NHL-EMD}} = z^2 \, ds^2 = -e^{-2y} \, dt^2 + e^{-2y/z} \, dx_i^2 + dy^2. \tag{8.52}$$

On taking the limit $z \rightarrow \infty$, the geometry of $d\tilde{s}^2$ in Eq. (8.52) becomes $AdS_2 \times \mathbb{R}^d$. This becomes even more explicit after another coordinate transformation $y = \log \zeta$ such that the metric becomes $d\tilde{s}^2 = (1/\zeta^2)(-dt^2 + d\zeta^2) + dx_i^2$. This metric exactly equals Eq. (8.23). However, as we will discuss in a moment, the simple limit $z \rightarrow \infty$ is not quite right for the purpose of obtaining the near-horizon physics of the RN black hole from a generic Lifshitz space-time.

Let us now turn to the novelty: the hyperscaling violation. Hyperscaling violation is the failure of the free energy (and the entropy density) of the critical boundary theory to scale with its naive engineering dimension. This scaling property is alien to the classical critical state; it points to the fact that strange metals are true quantum matter, possibly governed by long-range quantum entanglement. However, before we discuss this, let us first inspect in more detail how it arises as a property of the deep-interior geometry in the bulk. The metric Eq. (8.49) is manifestly written in a way that exhibits the hyperscaling violation. For any non-zero θ, it is not invariant under the scale transformation Eq. (8.51), but instead transforms as

$$ds^2_{\text{NHL-EMD}} \rightarrow \lambda^{2\theta/d} \, ds^2_{\text{NHL-EMD}}. \tag{8.53}$$

This means that the metric Eq. (8.49) is no longer invariant under a Lifshitz scale transformation, but it does remain *conformal to* Eq. (8.51): when the scaling is combined with an additional explicit transformation of the metric $g_{\mu\nu} \rightarrow \lambda^{-2\theta/d} g_{\mu\nu}$ we do recover the Lifshitz metric.

How does this translate into a hyperscaling-violating boundary field theory? EMD gravity generically allows for black-hole solutions describing the finite-temperature boundary. Given the zero-temperature geometry Eq. (8.49) and the notion that the black-hole horizon ascends along the radial direction as one increases the temperature, a direct geometrical scaling argument suggests that the entropy density (horizon area density) scales with temperature as

$$s \sim \sqrt{\prod_i g_{ii}} \sim r_0^{d-\theta} \sim t^{(\theta-d)/z} \sim T^{(d-\theta)/z}, \tag{8.54}$$

where the third proportionality follows from Eq. (8.51). This demonstrates that the entropy/free energy is violating the simple engineering scaling anticipated in the quantum state for such an extensive quantity. In the Lorentz-invariant case ($z = 1$) time and space scale in the same way, and, since temperature scales reciprocally with time, engineering scaling yields $s \sim (1/L)^d \sim T^d$; Lifshitz engineering where t scales as x^z gives $s \sim T^{d/z}$. The peculiarity revealed by Eq. (8.54) is that this critical state behaves like that of the quantum field theory textbook, which now, however, has decided to live in an effective space dimension $d - \theta$ instead of the manifest dimension d.

Is it possible to impose constraints on the actual values of the exponents θ and z, on basis of a general principle? These near-horizon solutions are in fact natural in EMD gravity. Taking a cue from top-down examples of holographic duals, we approximate the dilaton potentials with exponentials $Z(\phi) = Z_0 e^{\alpha\phi}$ and $V(\phi) = -V_0 e^{\beta\phi}$. With these potentials the equations of motion of EMD gravity can be solved exactly. The solution is a black-hole metric,

$$ds^2 = r^{-2\theta/d}\left(-r^{2z} f(r)dt^2 + r^2\, dx_i^2 + \frac{dr^2}{r^2 f(r)} \right), \tag{8.55}$$

with

$$f(r) = 1 - \left(\frac{r_0}{r}\right)^{d+z-\theta},$$

where θ and z are functions of the coefficients α, β [336]:

$$\theta = \frac{d^2\beta}{\alpha + (d-1)\beta}, \quad z = 1 + \frac{\theta}{d} + \frac{2(d(d-\theta)+\theta)^2}{d^2(d-\theta)\alpha^2}. \tag{8.56}$$

This two-parameter family defines a large class of EMD gravities. It is at present believed that these might even completely enumerate all renormalised infrareds that can form within the limitations of classical gravity in the bulk. This would mean that a complete "scaling atlas" is available for the strange metals, but, as we emphasised at the beginning of this section, this claim is conjectural. Within this family of EMD theories there is a special sub-class of strange metals that we will name, a bit awkwardly, "conformal-to-AdS$_2$" metals, for lack of a better term. We will work out an example of this special class of EMD holography in box 8.3. This class will play a significant role in later chapters. On the one hand, it has an embedding in string theory: it follows from a specific truncation of type IIB supergravity on AdS$_5 \times S^5$ [321] (see chapter 13). On the other hand, it is particularly appealing viewed from the condensed matter side. It shares with the RN metal the attractive properties of the local quantum criticality, but this metal has a *unique* ground state and it does not suffer from the ground-state entropy problem. It turns out also to form the theatre for a top-down implementation for the "Fermi-surface phases" discussed in the next chapter [337].

How is this stable but local quantum critical "conformal-to-AdS$_2$" metal characterised in terms of its scaling behaviour? The local quantum critical scaling property implies that somehow the deep interior should exhibit an AdS$_2$-type geometry, i.e. $z \to \infty$. However, at the same time the zero-temperature entropy has to vanish. These requirements can be simultaneously accomplished by invoking a compensating divergent behaviour of the hyperscaling violation. The exponents θ and z are no longer chosen to be independent. For positive z, θ is bounded from above by $\theta \leq d$ imposed by thermodynamical stability: for any finite z, $\theta > d$ implies that the specific-heat exponent becomes negative. However, it is not bounded from below and the special "conformal-to-AdS$_2$" class is characterised by $z \to \infty$, $\theta \to -\infty$ under the condition that $\eta = -\theta/z$ stays fixed. One infers from Eq. (8.54) that this gives a vanishing ground-state entropy with a specific heat $c_v \sim T^\eta$; for $\eta = 1$ this is the standard Sommerfeld scaling.

The exceptional nature of this class follows directly from the geometry in this double limit. In order to inspect the metric Eq. (8.49) in this limit, we first change coordinates to $r \to \tilde{r} = r^z$. The metric reduces to the following form after taking the limit (we ignored the tilde on top of r),

$$ds^2_{\text{NHL-EMD}} = r^{2\eta/d}\left(-r^2\,dt^2 + r^{2/z}\,dx_i^2 + \frac{1}{z^2}\frac{dr^2}{r^2}\right). \tag{8.57}$$

The AdS$_2$ factor as desired in a direct product with $\mathbb{R}^d$ is directly recognised, but there is also an additional overall conformal factor $r^{2\eta/d}$. For any finite $\eta > 0$ this conformal factor precisely lifts the ground-state entropy while preserving the local quantum critical physics encoded in the AdS$_2$.

There is now an obvious question. Given that this scaling geometry yields a controlled conformal-to-AdS$_2$ geometry, what is actually going on with the scaling geometry of the extremal RN when it is viewed as a special limit of the EMD scaling geometries? At first sight it appears that it is just Lifshitz with $z \to \infty$ and $\theta = 0$. However, this limit is more subtle. In terms of the original parameters of the theory α, β (Eq. (8.56)) there are many scalings that give the right geometry. One needs extra information to determine which is the correct one. We will see in section 14.2.3 that this is provided by the entanglement entropy. It is argued that the correct limit is to first set $\theta = d - 1/z$, and then take $z \to \infty$. According to the thermodynamics of the RN metal addressed in the first section, the entropy should be linear in temperature, but according to Eq. (8.54) it should be constant, given the scaling dimensions. The resolution is that this yields only the leading-order behaviour of the entropy [338, 339], and the contribution linear in T is sub-leading.

As we have learned in the previous paragraphs, violation of hyperscaling is quite a natural property when one is dealing with the deep-interior bulk geometries. However, it reveals a quite unfamiliar behaviour in the boundary field theory. The name "hyperscaling violation" is familiar from the conventional Wilson–Fisher critical state. It is a phenomenon that occurs above the critical dimension. Consider a regular statistical physics isolated unstable fixed point, characterised by an upper critical dimension d_{uc}. One finds that for $d > d_{uc}$ the Kadanoff scaling relations between the thermodynamical exponents $\alpha, \beta, \gamma, \delta$ are violated. This is a consequence of the fact that the self-interaction term is now dangerously irrelevant, vanishing at the critical point but taking over for any finite value of the order parameter, such that mean-field scaling relations are instead satisfied above d_{uc}. The hyper scaling violation in the present context has no relation whatever to this textbook phenomenon – apart from other considerations, we are not dealing here with an isolated critical point, and it is simply impossible to identify a dangerously irrelevant operator associated with an order parameter.

The bottom line is that at present we lack a general understanding in a field-theoretical language of these quantum critical phases characterised by the $\theta, \mathbf{z}$ scaling exponents emerging in finite density. As we will discuss in much more detail in the final chapter, chapter 14, the results for the behaviour of the bipartite von Neumann entropy in the strange metals can be used to arrive at an interpretation of the origin of hyperscaling violation as long as both θ and $\mathbf{z}$ are positive. The entanglement entropy appears to indicate that it is associated with the effective dimensionality spanned by the massless excitations in momentum space. In the case of the "regular" $\theta = 0$ theories one has isolated *points* in momentum space where the massless excitations reside, and this translates into the usual specific heat $\sim T^{d/\mathbf{z}}$, scaling with the space dimension of the system, while $\mathbf{z}$ encodes for the rate at which the number of accessible degrees of freedom increases as a function of energy. The only conventionally understood generic system which behaves differently is the Fermi liquid. This is characterised by a hypersurface in momentum space of dimensionality $d-1$ where massless excitations occur – the Fermi surface. The Fermi liquid therefore has a hyperscaling-violating exponent $\theta = d - 1$, and together with the fact that the quasiparticle excitations are effectively relativistic ($\mathbf{z} = 1$) it follows from Eq. (8.54) that $c_v \sim T$. This is the reason why the Sommerfeld law is independent of dimensionality – its degrees of freedom count always like a $d = 1$, $\theta = 0$ theory. The RN metal appears to follow the same counting logic: the AdS$_2 \times \mathbb{R}^2$ form of the correlation functions at low energies indicates that the system is like a massless $d = 0$ theory for *every point* in momentum space, and this should imply that $\theta = d$.

As we emphasised in chapter 2, the Fermi surface is rooted in the "kindergarten" anti-symmetrisation entanglement of the Fermi gas: in this sense its hyperscaling violation is a strong signal giving away its non-product "quantum matter" nature. The observation that hyperscaling is violated generically in the holographic strange metals might be an indication that they correspond to much more dramatically entangled forms of matter. We will elaborate on this theme in the final chapter, where we will present further support for this interpretation as obtained from the holographic von Neumann entanglement entropy.

In the next chapters, where we turn to holographic descriptions of "cohesive states" (the holographic superconductors and Fermi liquids), we will see the general applicability of this "scaling atlas". The deep-infrared scaling geometry will generically be of the Lifshitz type. The abstract description here in terms of a flow by the most relevant operator will there be replaced by the backreaction of concrete "stellar matter" in the bulk on the background space-time. Let us conclude with a comment on the large-N limit. On its own the Lifshitz geometry with $z > 1$ is not a fully consistent solution, but is characterised by a so-called "null singularity" at the origin $r = 0$. The metric is still smooth in the sense that there is no curvature singularity, but the tidal forces diverge at the origin [340, 341]. The meaning of the singularity is evident from the behaviour of the dilaton. It diverges near $r = 0$ and one therefore expects that quantum corrections encoding $1/N$ and $1/\lambda$ corrections should grow in importance. This implies that at finite N and λ one cannot wholly trust the Lifshitz form of the solutions in the very deep IR. Quite absurdly, the best guess which is available as yet is that eventually a crossover to an emergent AdS$_2$ will follow. Harrison *et al.* [342] attempted to account for quantum corrections by adding finite "gauge-coupling" corrections near the horizon. They found that this quite generally re-establishes an AdS$_2$ geometry in the very deep infrared (see also [343] for an alternative cure). This appears not to be restricted to the Lifshitz geometry per se: also for the hyperscaling-violating cases a singularity with a diverging curvature invariant in the deep IR is generically present [344]), and, as in the Lifshitz case, corrections to the gauge coupling stabilise to an AdS$_2$ geometry in the deep IR [345].

Box 8.3   Einstein–Maxwell-dilaton gravity and the "conformal-to-AdS$_2$" strange metal

A particularly interesting example of $(3 + 1)$-dimensional Einstein–Maxwell-dilaton gravity theory Eq. (8.46) departs from a particular choice for the potentials [321] originating in a top-down procedure that will be further discussed in chapter 13. This exact choice is

$$Z(\phi) = e^{\phi/\sqrt{3}}, \quad V(\phi) = -\frac{6}{L^2} \cosh\left(\frac{\phi}{\sqrt{3}}\right). \tag{8.58}$$

One finds the EMD black-hole solution,

$$ds^2 = e^{2A}(-h \, dt^2 + d\mathbf{x}^2) + \frac{e^{2B}}{h} \, dr^2,$$

$$A_t = \frac{\sqrt{3Q\mu}}{r+Q} - \frac{\sqrt{3Q\mu^{1/3}}}{L^{2/3}}, \tag{8.59}$$

$$\phi = \frac{\sqrt{3}}{2} \ln\left(1 + \frac{Q}{r}\right),$$

where

$$A = \ln\left(\frac{r}{L}\right) + \frac{3}{4}\ln\left(1 + \frac{Q}{r}\right), \quad B = -A, \quad h = 1 - \frac{\mu L^2}{(r+Q)^3}.$$

The black hole becomes extremal when $\mu L^2 = Q^3$, but there is no finite-size horizon left. This step forwards – there is no ground-state entropy – hides a problem. The solution now describes a naked singularity at $r = 0$. The immediate reaction could be that this is bad since naked singularities are not supposed to describe physically viable systems (e.g. the cosmic censorship hypothesis). However, lessons from string theory have taught us that certain "mild" naked singularities exist, namely the so-called "good singularities", which are physically consistent. This happens when the singularity arises from a consistent truncation (see chapter 13) of a higher-dimensional theory, in this case from eleven-dimensional M-theory [321]. One constructs a healthy Reissner–Nordström black hole in eleven-dimensional space-time, but in the compactification to five dimensions the "projection" of this higher-dimensional black hole, which is effectively encoded by the dilaton, turns it into a naked singularity. Clearly, however, there was no problem in the original eleven-dimensional theory. Approaching the singularity simply means that a part of the eleven-dimensional physics that was ignored becomes important. The singularity can thus readily be resolved, and this classifies it as "good". One of the primary indications that a naked singularity is good is that, at finite temperature, this good singularity will be shielded by the finite-temperature horizon.

The near-horizon geometry of the extremal solution turns out to be

$$ds^2 = \left(\frac{L_2^2}{2Q\zeta}\right)\left[\frac{L_2^2}{\zeta^2}(-dt^2 + d\zeta^2) + \frac{Q^2}{L^2} \, d\mathbf{x}^2\right], \quad A_t = -\frac{L_2^3}{2Q\zeta^2}, \tag{8.60}$$

where

$$\zeta = \frac{L_2^2}{2\sqrt{Q}r}, \quad L_2 = \frac{2L}{\sqrt{3}}. \tag{8.61}$$

Box 8.3   (Continued)

One recognises the $AdS_2 \times \mathbb{R}^2$ piece, but there is also an overall factor. It is easy to see that under the scaling transformation Eq. (8.25) $t \rightarrow \lambda t, \zeta \rightarrow \lambda \zeta, \mathbf{x} \rightarrow \mathbf{x}$ we have $ds^2 \rightarrow \lambda^{-1} ds^2$. Thus the near-horizon geometry is conformal to $AdS_2 \times \mathbb{R}^2$, as explained in the main text. The conformal factor precisely ensures that the zero ground-state entropy vanishes, indicating that the geometry Eq. (8.59) is stable, while the AdS_2 part preserves the local quantum critical scaling.

   This near-horizon geometry Eq. (8.60) is a special case of a class of more general geometry with $\mathbf{z} = \infty$ and $-\theta/\mathbf{z} = \eta > 0$ [346],

$$ds^2 = \frac{1}{\zeta^{-2\eta/d}} \left(\frac{-dt^2 + d\zeta^2}{\zeta^2} + d\mathbf{x}^2 \right).$$

(8.62)

This geometry corresponds to a local quantum critical state with an entropy density that vanishes at low temperatures as $s \sim T^\eta$. When $d+1 = 2+1$ and $\eta = -\theta/\mathbf{z} = 1$, Eq. (8.62) becomes precisely Eq. (8.60).

9

Holographic photoemission and the RN metal: the fermions as probes

The general nature of the holographic strange metals as discussed in the previous chapter took some time to be appreciated. Triggered by the studies dating from 2007 by Herzog, Kovtun, Sachdev and Son of quantum critical transport which we described in chapter 7, the AdS/CMT development as focussed on quantum matter started in 2008 with the discovery that the RN metal is quite susceptible to spontaneous symmetry breaking: this is the holographic superconductor [347, 348]. What thoroughly accelerated the interest in the condensed matter applications of AdS/CFT was the discovery in 2009 of the "MIT–Leiden fermions" [349, 350]. This will be the topic of this chapter. In hindsight, this was a highlight of the cross-disciplinary exchange between condensed matter physics and string theory. Driven by experimental progress during the last twenty years or so, angular-resolved photoemission (ARPES) (and in principle also its "unoccupied-states sibling" inverse photoemission) has acquired a very prominent status as a means to "observe" strongly interacting systems in solids. It has been further fortified by the quite recent development of scanning tunnelling spectroscopy (STS), which can be seen as the real-space partner of ARPES. Together these spectroscopies have produced a barrage of serendipitous surprises during the last 15 years, and now play a prominent role in the flourishing of the experimental study of the strongly correlated electron systems.

Both methods probe the single-fermion two-point function. For string theorists with their traditional focus on high-energy experimentation and cosmology, the importance of the single-fermion propagator as an observational tool is less obvious. Its unique powers come into play in *finite-density* systems. Compared with the collective "bosonic" current responses which were in the foreground in the previous two chapters, it yields complementary but yet quite different information regarding the vacuum structure of the interacting system. One advantage is of a rather pragmatic nature. In practice one can measure the bosonic responses often only at small momenta, where general "hydrodynamics" principles tend to equalise the outcomes – see the discussion of zero sound and optical conductivity of the RN metal in section 8.3. On the other hand, photoemission (and its siblings) accesses the full kinematical range associated with the electrons in solids: state-of-the-art spectrometers probe length scales ranging from multiple nanometres to the sub-Ångström scale and energies ranging from sub-kelvin to electron volts.

The other advantage of photoemission and STS is more fundamental. The probe now "carries around a fermion sign" – it knows of quantum statistics – and this is a great benefit when one wants to learn about the quantum entangled nature of the finite-density ground state caused by the fermion signs. The case in point is the Fermi liquid. We emphasised in chapter 2 its "baby" long-range entanglement associated with the anti-symmetrisation requirement. This is responsible for the "non-classical object" controlling its vacuum structure: the Fermi surface. *The only way to directly measure the presence of the Fermi surface is through the single-fermion propagator.* This is encapsulated by infinitely sharp quasiparticle peaks right at the Fermi surface in the single-fermion propagator. The Fermi surface does reverberate in "bosonic" measurements, but one at best integrates over this information (e.g. the momentum distribution) or one needs a "searchlight theory" to interpret the data, which typically has to depart from the assertion that a Fermi liquid exists (all of transport, including the quantum oscillations).

It is not known in general how to use this wisdom in the terra incognita of non-Fermi-liquid "fermionic quantum matter". The appeal of the experimental data is to a great degree that they seem to reveal features of this vacuum entanglement that call for a profound explanation: the nodal–antinodal dichotomy, the Fermi arcs, the incoherent backgrounds and so forth discussed in chapter 3. In fact, the holographic narrative that will unfold in this chapter is another lively illustration of this still-mysterious observational power of the fermionic probe.

The two senior authors of this book started their collaboration shortly after the first contact between AdS/CFT and condensed matter through holographic quantum critical transport in 2007 [24]. In the first discussions the plan was to compute holographic photoemission for the RN metal. A code was quickly written, but our work was delayed by a hidden bug. Other parties had the same idea [351, 352], and in 2009 the numerics was put into working order, at roughly the same moment at MIT [349] and in Leiden [350]. The impact of this work was much helped by the appeal to the familiar: both groups reported the presence of a Fermi surface and Fermi-liquid-like behaviour. In hindsight these results should be treated with extreme care. They are particular to the probe limit and ignore the highly unstable nature of the RN black hole. Nevertheless, this work showed the possible controlled emergence of true non-Fermi liquids, and much of this physics insight carries over to more specialised but stable EMD metals such as the conformal-to-AdS$_2$ metal analysed at the end of the previous chapter.

The subsequent deciphering of the holographic mechanism behind these fermionic responses is fully credited to the MIT group of Faulkner, Iqbal, Liu, McGreevy and Vegh [320]. In the process they also laid the foundation of the macroscopic transport responses of the strange metals which was the focus of the previous chapter. In this chapter we will take the reader through this probe story and its deciphering, more or less following the historical order of the development. Given the supersymmetric roots of AdS/CFT, fermions have their natural place but their status in holography is different from that of the bosonic operators in charge of the collective response. Fermions are inherently quantum-mechanical. Through the magic of the AdS/CFT dictionary that relates quantum corrections in the bulk to $1/N$ corrections in the boundary, this means that fermionic fields on the semi-classical gravity side are dual to fermionic operators in the field theory that have to be $1/N$ suppressed. In the probe limit considered in the present chapter, by definition the fermions in the bulk do not contribute to the overall charge or energy density. The bulk fermions are just described by the Dirac equation as defined in the classical curved space-time in the presence of the classical electromagnetic fields. They do not backreact on the Maxwell field or the metric. With the natural $1/N$ suppression the probe limit is almost automatic, and the fermion propagator in the boundary which is dual to these bulk Dirac wave functions can be considered as the leading-order corrections in the perturbation theory around the large-N limit.

Calibrating our studies at zero density as before, we will first show how holography again produces the correct answer for the fermion propagators of the conformal field theory. It has the same "un-particle" branch-cut form as for the bosons except that now both positive- and negative-energy states are present. Then we turn to finite density in section 9.2 and show that the holographic strange metal fully deserves its name on account of its non-Fermi-liquid-like response to fermionic probes. However, the responses can also show a novel "algebraic pseudogap" behaviour. We will compare these responses with those in the non-Fermi-liquid regime and argue that these "algebraic pseudogaps" are more truly representative of the RN metal. Section 9.3 is devoted to the mathematical machinery discovered by the MIT group which makes it possible to fully decipher the physics and how this dualises to the boundary fermion propagators. Closed semi-analytical expressions for the low-frequency behaviour of the boundary propagators can be obtained from this matching approach between the near-horizon and near-boundary Dirac waves in the bulk. This machinery in the form of the near–far-matching method has proven to be widely applicable in AdS/CMT, such as the transport coefficients in linear response in the previous chapter and the holographic superconductor in the next.

However, to understand the origin of the Fermi-liquid quasiparticle-like features in the boundary, a more revealing analysis is to cast the Dirac equation in terms of an effective Schrödinger potential. This amounts to a clever choice of coordinates that translates the propagation of the Dirac waves in the curved space-time of the bulk into an effective one-dimensional Schrödinger equation associated with the radial coordinate, where the effects of the curvature are swapped for an effective potential living in a flat space-time. One can now rely on the methods of quantum-mechanical potential scattering theory to understand the physics of the fermions in the bulk. What this reveals is the presence of an effective potential well inside the region where the near-boundary AdS geometry changes into the RN black-hole geometry. For a large charge-to-mass ratio of the fermionic operator, i.e. for "relevant" scaling dimensions, this potential well supports bound states. The side barriers act quite like the "walls" described in section 6.3, except that there is a finite probability of tunnelling to the horizon in the deep interior. This reveals that in the finite-density system the field-theory excitations want to confine themselves into particle-like gauge singlets but now of fermionic rather than bosonic nature – "mesinos" rather than "mesons". These are not infinitely stable, but they still decay in the fractionalised deep infrared of the RN metal. These long-lived gauge singlets are responsible for the (non-)Fermi-liquid-like behaviour discovered by MIT and Leiden. However, it turns out that well before these quasi-bound states are formed one finds indications in the form of (Breitenlohner–Freedman-bound-violating) "log-oscillatory responses" that the RN black hole becomes unstable in the presence of the fermions. To study this further one needs to include the backreactions of the fermions in the bulk, which will be the subject of chapter 11.

In hindsight perhaps the most important insight to be gained from the matching/Schrödinger analysis is what happens in the regime where the charge-to-mass ratio is small. Then the fermionic operator is in some sense "irrelevant"; the potential no longer supports bound states and the local quantum criticality of the RN metal takes over. We will show that this local quantum critical essence reveals itself in photoemission through a characteristic "algebraic-pseudogap" behaviour of the fermion spectral functions imposed by the near-horizon scaling geometry. The precise statement is that the photoemission spectral functions behave as $A(\omega, k) \sim \omega^{2\nu_k}$, where the momentum dependence is in the exponent $\nu_k \sim \sqrt{1/\xi^2 + k^2}$, where ξ is a length scale of the order of the lattice constant. Our challenge to the experimentalists is that they should try to see whether such responses can be present in their data.

9.1 The holographic encoding of fermions

Fermions are a natural part of the AdS/CFT correspondence. In Maldacena's canonical $AdS_5 \times S^5$ example of the correspondence the dual is an $\mathcal{N} = 4$ super-symmetric Yang–Mills theory in which bosons and their supersymmetric fermionic partners occur in a perfect symmetric harmony. The subtlety with fermions is that they are intrinsically quantum-mechanical. The defining mutual anti-symmetry of fermion wave functions is meaningful only for finite $\hbar$. The curious aspect of the AdS/CFT correspondence is that the quantum physics in the bulk is dual to *semi-classical* large-N physics in the field theory of interest. This is what the GKPW rule reveals so explicitly. One of the most powerful aspects of AdS/CFT is precisely that it identifies a new matrix large-N (gravitational) saddle point in a fully quantum field theory. Including bulk AdS fermions around classical gravity is thus a re-quantisation of the system around the large-N limit, and the inherent quantum-ness of bulk fermions means that their dual operators are automatically $1/N$ suppressed in the boundary field theory. At first sight this looks bad since a quantised bulk – i.e. quantum gravity – is supposed to be a tall order. However, in the probe limit we consider in this section the fermions in the boundary have a merely linear-response status and the leading order in the $1/N$ expansion amounts to a natural amplitude suppression of the fermion propagators. We are studying a perturbative quantum field in a fixed curved space-time. The true quantum-ness of the fermions will matter once we consider many of them beyond linear response. In chapter 11 we will turn to objects formed from finite-density fermion matter in the bulk, dual to "cohesive" Fermi-liquid-like states in the boundary, and in this context the inherent quantum nature of the bulk physics will come to the fore.

There are some further caveats associated with the precise identity of the fermionic operators in holography that one should be aware of in comparison with conventional fermions in condensed matter physics. It is a typical example where top-down constructions provide additional crucial information. In the canonical Maldacena example of $AdS_5 \times S^5$ on which we rely implicitly for such questions, the microscopic fermions always come in a very particular guise. These fermions are just the super-partners of the force-carrying gauge bosons and therefore occur in the *adjoint* representation of the Yang–Mills gauge group. This is quite different from the fermions of the Standard Model of particle physics, or elementary electrons, which are in the *fundamental* representation of the gauge-group. From these microscopic gauge-charged "deconfined" fermions, one can construct single-trace, gauge-invariant and hence observable, fermionic excitations in terms of a boson–fermion composite. As we described in chapter 4, it is these single-trace gauge-invariant composite operators in the boundary field theory that holographically have large-N duals in terms of a classical gravity theory with now fermionic matter. These fermionic gauge singlet operators are sometimes casually referred to as "mesinos", the supersymmetric partners of the mesons of QCD. QCD mesons are bosonic gauge singlets formed from fermionic quark–antiquark pairs, $\bar{\psi}_a \psi^a$, where the implicit sum is over the gauge colour indices. For fermions in the adjoint representation the analogous meson is $\bar{\psi}_{ab} \psi^{ba} = \text{Tr}[\bar{\psi}\psi]$. In supersymmetric

Yang–Mills theory one can also form fermionic composites $\mathrm{Tr}[\phi\psi]$, where ϕ is a bosonic parton. This is the "mesino"; it bears some similarity to the spinon–holon fractionalisation construction familiar from the slave-fermion theories of condensed matter, except that these are associated with vector large-N, instead of the matrix large-N of the supersymmetric Yang–Mills theories.

Let us note that one can construct top-down models, so-called Dp/Dq-brane intersections, where the microscopic fermions are in the fundamental vector representation of the gauge group. This more complicated gravitational dual where one has a dynamical domain wall in the space-time will be discussed in chapter 13.

9.1.1 Fermion correlation functions from holography

To build confidence, let us first inspect how fermion correlation functions are computed in the bare-bones conformal field theory at zero density. Here the symmetry should already completely determine the form of the fermion propagators. In the next section we will then turn to the surprises revealed by the "holographic photoemission" in the strange metals formed at finite density.

Given the general symmetry considerations underpinning the dictionary, matching quantum numbers on both sides and the requirement that the fermions are encoded by a quantum-mechanical wave equation that supports Fermi–Dirac statistics, one has no other choice than to postulate that the fermionic operators in the boundary field theory are encoded by Dirac fermions in the bulk. This also follows from top-down constructions, since Dirac fields are the natural entities encoding fermions in any quantum field-theoretical setting. Since complex fermions always have a $U(1)$ phase symmetry, and global symmetries become local in the bulk, the minimal set-up is the AdS–Einstein–Maxwell action of the previous section, with a Dirac term added to describe the fermions,

$$
S = \int d^{d+2}x\sqrt{-g}\left[\frac{1}{16\pi G}(R - 2\Lambda) - \frac{1}{4g_{\mathrm{F}}^2}F_{\mu\nu}F^{\mu\nu} \right.
$$
$$
\left. - i\bar{\Psi}\left(e_a^\mu\Gamma^a\left(\partial_\mu + \frac{1}{4}\omega_{\mu bc}\Gamma^{bc} - iqA_\mu\right) - m\right)\Psi\right].
$$
(9.1)

There are two crucial aspects to note. In accordance with the fully relativistic nature of the duality, we have used a relativistic Dirac action for fermions with spin $1/2$. The Dirac spinor in even space-time dimensions has length $2^{d/2+1}$ with $2^{d/2}$ particle polarisations and $2^{d/2}$ anti-particle polarisations, while the Dirac fermions now live in a curved space-time. This is accomplished through the introduction both of a vielbein, the "square root" of the metric defined by $e_\mu^a e_\nu^b\eta_{ab} = g_{\mu\nu}$, and of a spin connection $\omega_{\mu ab}$ defined through the demand that the vielbein should be covariantly constant, $D_\mu e^a_\nu = 0 = \partial_\mu e^a_\nu - \Gamma^\sigma_{\mu\nu} e^a_\sigma + \omega^a_{\mu b} e^b_\nu$. A brief introduction to the vielbein and spin connection can be found in chapter 4 in box 4.4. The vielbein naturally maps between proper vectors and elements in the flat tangent space to the space-time, whereas the spin connection determines how the basis of the tangent space is rotated as one travels along the curved manifold. A spinor, a spin-1/2 representation of the local group of rotations, naturally takes values in the tangent space. As a consequence the gamma matrices also naturally live there, which implies that their defining anticommutation relation reads $\{\Gamma^a, \Gamma^b\} = 2\eta^{ab}$. In particular, this means that the gamma matrices remain constant matrices of numbers, irrespective of the location in space-time.

From a holographic viewpoint the fact that the size of the spinor representation depends on the dimension may immediately sound troublesome. Since the bulk and the boundary do not have the same dimension, how can the dictionary work? The resolution of this problem lies in the famous first-order nature of the Dirac action. Only half of the components of the Dirac field correspond therefore to physical degrees of freedom. The other half correspond to the conjugate momentum. Their values at the boundary naturally translate to the source and the response of the dual operator (box 9.1). A proper projection onto these physical degrees of freedom therefore resolves the conundrum.

Let us also emphasise that the Dirac equation is quite different from classical field equations: in the bulk one studies the *quantum-mechanical* wave functions of these non-interacting relativistic fermions; there is no coherent classical state that can solve the saddle-point equation.

Summary 9 Fermionic operators in the strongly interacting boundary field theory are dual to the quantum mechanics of free Dirac fermions, propagating in the curved space-time of the bulk. Since this involves quantum physics in the bulk, the fermions appear as leading order $1/N$ corrections to the large-N limit of the boundary field theory.

With the dual gravitational formulation of a conformal field theory with fermionic operators in hand, we can now formally apply the GKPW rule to compute its correlation functions. Technically there are some subtleties related to the fermionic nature and the spin structure, which will be discussed in detail in box 9.1. A question that is easy to answer is the following: what should one expect for the behaviour of the two-point fermion propagators in the boundary field theory at zero density, given that the system is highly constrained by conformal and Lorentz invariance? One can identify a physical configuration in condensed matter physics, where this question is very natural. Consider a graphene-like situation where the band structure gives rise to Dirac fermions at low energy. These suffer from fermion doubling, and an even more literal metaphor would be the helical fermions formed at the surface of a three-dimensional topological insulator as discussed in chapter 2 since the fermions at the boundary are truly relativistic. One keeps this system at zero density, with the Dirac node at the chemical potential. One now envisages that the system is driven to a quantum critical point associated with the onset of e.g. an antiferromagnet. One can demonstrate on quite general grounds that the Yukawa coupling with the critical bosonic field is finite at the critical point, with the effect that the fermions land on the critical surface of the strongly interacting critical state [353]. The above question is then the same as asking what the photoemission spectrum would look like in such a circumstance.

The answer is that, as for every other field, the spectral functions will exhibit the "un-particle" branch-cut form as discussed in chapter 2. The form of all two-point functions including the fermion propagator is completely fixed by conformal and Lorentz invariance. The only free parameter is the scaling dimension Δ of the fermion operator $\mathcal{O}_\psi$. The only difference with the propagators of conformal bosons is that the fermion propagator describes both the negative (occupied) and the positive (unoccupied) energy states. For spacelike momenta $k^2 > 0$ we have

$$\langle \mathcal{O}_\psi \mathcal{O}_\psi^\dagger \rangle \sim k^{2\Delta - d - 1}. \tag{9.2}$$

One clearly sees that there is no pole; these have nothing to do with free Dirac particles. Instead, the branch-cut behaviour as a function of energy tells us again that we are dealing with highly collective, strongly coupled fermionic critical fluctuations. Exactly this form of the correlation function is reproduced from the action Eq. (9.1) from the holographic GKPW calculation (box 9.1). As before, this shows that the scaling dimension of the boundary fermions is set by the *mass* of the Dirac field in the bulk through the relation

$$\Delta = \frac{d+1}{2} + mL. \tag{9.3}$$

Although the scaling argument does not reveal it here, the explicit evaluation of the propagator in box 9.1 below shows that the Dirac system in the bulk encodes automatically for the presence of a Dirac sea in the boundary field theory. The exact boundary fermion propagator correctly describes both the occupied and the unoccupied states. Although this is far from being a free-fermion system, the critical fermions surely do remember the notion of occupied and unoccupied states.

In this zero-density situation we know what we are doing since the fermion signs cancel out because of charge-conjugation invariance, and this is a sufficient condition to ensure that the general form of the propagator follows from kinematics. At the same time, the fermions in the bulk form a free system, which is also characterised by positive- and negative-energy states, i.e. a free Dirac sea. We infer a simple one-to-one dictionary relation between the Fermi–Dirac information in the bulk and the way in which fermion statistics governs the strongly interacting zero-density critical state of the boundary. A typical outcome for the zero-density fermion spectral function is illustrated in the inset in the top panel of Fig. 9.1 (in section 9.2).

Box 9.1   Computing fermion correlation functions holographically

Owing to the first-order nature of the Dirac action, there are some technical changes to the prescription of applying the AdS/CFT dictionary. That such changes are necessary is most readily apparent simply on counting components of fermions. In dimensions $d + 1 = 2n$ and $d + 1 = 2n + 1$ a spinor has 2^n components. This means that if one has an even-dimensional bulk with $d + 1 = 2n$ the spinor has double the number of components one would naively expect from the dimensionality $d + 1 = 2n - 1$ of the boundary. Owing to the first-order nature of the Dirac action describing relativistic fermions this counting is wrong. The first-order action simultaneously describes the fluctuation – half of the components – and its conjugate momentum – the other half. Clearly only the former should correspond to a boundary degree of freedom. We can use the extra direction to make this split manifest by projecting onto eigenstates of $\Gamma^L \Psi_\pm = \pm \Psi_\pm$. We can call one the fluctuation, say Ψ_+. The other, Ψ_-, is then the conjugate momentum.

We henceforth specialise to the AdS$_4$/CFT$_3$ situation: this is, however, not critical and in other dimensions it is similar. Under this projection the Dirac action reduces to

$$S = \int d^4x \sqrt{-g} \left(-i\bar{\Psi}_+ \slashed{D} \Psi_+ - i\bar{\Psi}_- \slashed{D} \Psi_- + im\bar{\Psi}_+ \Psi_- + im\bar{\Psi}_- \Psi_+ \right). \quad (9.4)$$

The second issue is that the AdS/CFT correspondence instructs us to derive the CFT correlation functions from the on-shell action. The Dirac action, however, is proportional to its equation of motion. This reflects the inherent quantum nature of fermions, namely that they never influence the saddle point. In a theory with a boundary, this is not quite true, as we shall see. Having chosen Ψ_+ as the fundamental degree of freedom, we will choose a boundary source $\Psi_+^0 = \lim_{r \to \infty} \Psi_+(r)$. Then the boundary value Ψ_-^0 is not independent but related to that of Ψ_+^0 by the Dirac equation.

Box 9.1   (Continued)

We should therefore not include it as an independent degree of freedom when taking functional derivatives with respect to the source Ψ_+^0. Instead it should be varied to minimise the action. To ensure a well-defined variational system for Ψ_-, we add a boundary action,

$$S_{\text{bdy}} = \oint_{r=r_\epsilon} d^3x \sqrt{-h}\, i\bar{\Psi}_+\Psi_-, \tag{9.5}$$

with $h_{\mu\nu}$ the induced metric, which is similar to the Gibbons–Hawking term (6.32) encountered earlier, and $r_\epsilon = 1/\epsilon$ with $\epsilon \to 0$ is the usual cut-off in the radial direction to regulate the asymptotic behaviour near $r = \infty$ [354]. The variation of $\delta\Psi_-$ from the boundary action,

$$\delta S_{\text{bdy}} = \oint_{r=r_\epsilon} d^3x\sqrt{-h}\, i\bar{\Psi}_+\,\delta\Psi_-\Big|_{\Psi_+^0 \text{ fixed}}, \tag{9.6}$$

now cancels out the boundary term from the variation of the bulk Dirac action,

$$\delta S_{\text{bulk}} = \int d^4x\sqrt{-g}\left(-i\,\delta\bar{\Psi}(\slashed{D}-m)\Psi - i(\overline{(\slashed{D}-m)\Psi})\delta\Psi\right)$$
$$+ \oint_{r=r_\epsilon} d^3x\sqrt{-h}\left(-i\bar{\Psi}_+\,\delta\Psi_- - i\bar{\Psi}_-\,\delta\Psi_+\right)\Big|_{\Psi_+^0 \text{ fixed}}. \tag{9.7}$$

This results in a well-defined variational problem, and the explicit boundary term implies that the action no longer vanishes on-shell. This boundary term remains.

The next complication is that the fermionic correlation function is in general a matrix between the various spin components. In a curved background the evolution generically mixes these components. A completely covariant formulation exists [350, 354], but by a clever choice each independent spin component can be separated out [320]. We first undo the projection unto Γ^r, and write down the full covariant first-order Dirac equation,

$$\left(e_a^\mu\Gamma^a\left(\partial_\mu + \frac{1}{4}\omega_{\mu bc}\Gamma^{bc} - iqA_\mu\right) - m\right)\Psi = 0. \tag{9.8}$$

Note that the boundary term does not contribute to the equation of motion.

Metrics that depend on only a single parameter, such as AdS, where the only parameter is r, have the property that, by a redefinition of the Dirac fields,

$$\Psi = (-gg^{rr})^{-1/4}\chi, \tag{9.9}$$

the spin connection $\omega_{\mu bc}$ can be removed. The Dirac equation thereby acquires the form

$$\left[e^\mu_a \Gamma^a (\partial_\mu - iq A_\mu) - m \right] \mathcal{X} = 0. \tag{9.10}$$

We will typically be interested in electrostatic configurations where only $A_t \equiv \Phi$ is non-zero. In a flat space-time we would now Fourier transform and project the four-component Dirac spinor onto four-component spin eigenstates. Since the radial direction of anti-de Sitter space breaks the four-dimensional Lorentz invariance, this cannot be done here. However, there exists a similar projection onto *transverse helicities*, where the spin is always orthogonal both to the direction of the boundary momentum and to the radial direction. We choose the particular basis for Dirac matrices

$$\Gamma^{\underline{r}} = \begin{pmatrix} -\sigma_3 \mathbb{1} & 0 \\ 0 & -\sigma_3 \mathbb{1} \end{pmatrix}, \quad \Gamma^{\underline{t}} = \begin{pmatrix} i\sigma_1 \mathbb{1} & 0 \\ 0 & i\sigma_1 \mathbb{1} \end{pmatrix},$$

$$\Gamma^{\underline{x}} = \begin{pmatrix} -\sigma_2 \mathbb{1} & 0 \\ 0 & \sigma_2 \mathbb{1} \end{pmatrix}, \quad \Gamma^{\underline{y}} = \begin{pmatrix} 0 & \sigma_2 \mathbb{1} \\ \sigma_2 \mathbb{1} & 0 \end{pmatrix}, \quad \dots \tag{9.11}$$

where the underlined index means that they are along the tangent-space direction (the Latin indices in the vielbein) and σ_i ($i = 1, 2, 3$) are the Pauli matrices. Next, using rotational invariance to choose the boundary momentum along the x direction, $\vec{k} = (k, 0)$, we project onto the t-helicities $\mathcal{X}_{1,2}$ that are eigenstates of $\Gamma^5 \Gamma^x$. This reduces the Dirac equation to

$$\sqrt{\frac{g_{ii}}{g_{rr}}} \left(i\sigma_2 \partial_r - m\sqrt{g_{rr}} \sigma_1 \right) \mathcal{X}_\alpha(r; \omega, k)$$

$$= \left((-1)^\alpha k_x \sigma_3 - \sqrt{\frac{g_{ii}}{-g_{tt}}} (\omega + q A_t) \right) \mathcal{X}_\alpha. \tag{9.12}$$

Here we have used the fact that, for a diagonal metric $g_{\mu\nu}|_{\mu \neq \nu} = 0$, the vielbein e^a_μ is literally the square root $e^a_\mu = \sqrt{|g_{\mu\mu}|} \delta^a_\mu$, where μ does not sum.

It suffices to consider only $\mathcal{X}_1$ henceforth, since the results for $\mathcal{X}_2$ simply follow on changing $k \to -k$. As in the scalar case, near the AdS boundary the two-component spinor $\mathcal{X}_1$ has the asymptotic behaviour

$$\mathcal{X}_1(r) = a \begin{pmatrix} 0 \\ 1 \end{pmatrix} r^{mL} + b \begin{pmatrix} 1 \\ 0 \end{pmatrix} r^{-mL} + \cdots, \quad \text{when } r \to \infty. \tag{9.13}$$

For a generic mass the first, leading, component is not normalisable – this will be the source of the dual operator – but the second, sub-leading, one is: this will be the response.

Box 9.1 (Continued)

By judiciously squaring the first-order equation, one can obtain a second-order equation for each of the components of $\mathcal{X}_1$, with two independent solutions. Given these homogeneous solutions to the Dirac equation, the (bulk) Green function for $\mathcal{X}_1$ (still a two-by-two matrix) can be constructed. It is

$$\mathcal{G}(\omega, k, r; \omega', k', r') = \frac{\psi_b(r) \otimes \bar{\psi}_{\text{int}}(r')\theta(r - r') - \psi_{\text{int}}(r) \otimes \bar{\psi}_b(r')\theta(r' - r)}{\frac{1}{2}\left(\bar{\psi}_{\text{int}}(r)\Gamma^{\underline{r}}\psi_b(r) - \bar{\psi}_b(r)\Gamma^{\underline{r}}\psi_{\text{int}}(r)\right)}. \tag{9.14}$$

Here $\psi_b(r)$ is the normalisable solution with leading coefficient $a = 0$, and $\psi_{\text{int}}(r)$ is determined by the appropriate boundary conditions in the interior.

We can now use this bulk Green function to evaluate the on-shell Dirac action from which we can find the boundary correlation function. The on-shell Dirac action comes fully from the boundary term Eq. (9.5). There we postulated that the fundamental field is $\Psi_+ \equiv \frac{1}{2}(1 + \Gamma^{\underline{r}})\Psi$. For the t-helicity $\mathcal{X}_1$ this reduces to the projection with respect to

$$\frac{1}{2}(1 - \sigma_3) = \begin{pmatrix} 0 & 0 \\ 0 & 1 \end{pmatrix}.$$

At the same time one should think of Ψ_- as dependent on the fundamental field Ψ_+. We can indeed see that $\lim_{r\to\infty} \Psi_+(r)$ picks out the leading component and plays the role of the source. The dependence of the conjugate momentum Ψ_- on the source follows directly from the bulk Green function as in the case for scalars in box 5.1

$$\Psi_-(r) = \lim_{\epsilon \to 0} \oint_{r'=\epsilon^{-1}} \frac{d\omega'\, dk'}{(2\pi)^2} \frac{1}{2}(1 - \Gamma^r)\mathcal{G}(\omega, k, r; \omega', k', r')\Psi_+(r'). \tag{9.15}$$

On substituting this Green function into the boundary action (9.5), one obtains, after projection onto the independent t-helicity $\mathcal{X}_1$,

$$S = -\lim_{r \to \infty} \int \frac{d\omega\, dk}{(2\pi)^2} \sqrt{-h}$$

$$\times \left(i\bar{\mathcal{X}}_1^0(r)\frac{1}{2}(1 + \sigma_3)\frac{\psi_{\text{int}}(r) \otimes \bar{\psi}_b(\infty)}{\frac{1}{2}\left(\bar{\psi}_{\text{int}}(r)\sigma^3\psi_b(r) - \bar{\psi}_b(r)\sigma^3\psi_{\text{int}}(r)\right)}\mathcal{X}_1^0 \right). \tag{9.16}$$

In the t-helicity basis the boundary source (the non-normalisable coefficient) has only a lower component

$$\mathcal{X}_1^0 = \begin{pmatrix} 0 \\ J \end{pmatrix}. \tag{9.17}$$

Writing also

$$\psi_{\text{int}}(r) = \begin{pmatrix} b_{\text{int}}r^{-mL} + \cdots \\ a_{\text{int}}r^{mL} + \cdots \end{pmatrix}, \quad \psi_b(r) = \begin{pmatrix} br^{-mL} + \cdots \\ 0 + \cdots \end{pmatrix},$$

one finds

$$S = \lim_{r \to \infty} \int \frac{d\omega\,dk}{(2\pi)^2} \sqrt{-h}\, r^{-2mL} \frac{(J^\dagger b_{\text{int}})(b^\dagger J)}{\frac{1}{2}(b^\dagger a_{\text{int}} + a_{\text{int}}^\dagger b)}. \qquad (9.18)$$

The final step is that an inspection of the Dirac equation reveals that *at zero density in pure AdS* one can always choose b and a_{int} real (but not b_{int}):

$$S = \lim_{r \to \infty} \int \frac{d\omega\,dk}{(2\pi)^2} \sqrt{-h}\, r^{-2mL} J^\dagger \frac{b_{\text{int}}}{a_{\text{int}}} J. \qquad (9.19)$$

Differentiating the on-shell action w.r.t. J and $J^\dagger$, and renormalising the overall r^{-2mL} term away, gives the expression for the fermionic CFT correlation function,

$$G_{\text{fermions}}(\omega, k) = \frac{b(\omega, k)}{a(\omega, k)}, \qquad (9.20)$$

where we dropped the subscript int and a, b are the coefficients in Eq. (9.13) of $\mathcal{X}$ with the appropriate boundary condition near the horizon. It might appear that the denominator a is purely real, with the consequence that there is no interesting emergent physics in this system. As we learned in chapter 6, however, at finite temperature and density one cannot substitute the solution into the manifestly real action and differentiate to obtain the Green function. A very careful proper analysis [243, 320] reveals nevertheless that the analytic continuation of the answer (9.20) is correct.

Holographic fermion propagator at zero density

Let us now apply this to the computation of the conformal fermion propagators for the pure AdS case (zero density and zero temperature). The metric for pure AdS$_{d+2}$ in units where $L = 1$ is given by

$$ds^2 = r^2(-dt^2 + d\mathbf{x}^2) + \frac{dr^2}{r^2}. \qquad (9.21)$$

Equation (9.12) then becomes

$$r^2 \left(i\sigma_2\, \partial_r - \frac{m}{r}\sigma_1 \right) \chi_1 = (-k_x\sigma_3 - \omega)\chi_1. \qquad (9.22)$$

Box 9.1   (Continued)

By squaring this equation one obtains a Bessel equation,

$$-r^2 \, \partial_r (r^2 \, \partial_r \chi_1^u) + (-mr^2 + m^2 r^2 - (k_x + \omega)^2) \chi_1^u = 0, \qquad (9.23)$$

$$-r^2 \, \partial_r (r^2 \partial_r \chi_1^d) + (-mr^2 + m^2 r^2 - (k_x - \omega)^2) \chi_1^d = 0, \qquad (9.24)$$

where $\chi_1 = (\chi_1^u, \chi_1^d)^T$. The solution which is regular in the interior can be obtained, and by applying the procedure explained above the fermion Green function can be derived. This is the Feynman time-ordered Green function. This alternative Green function corresponds to choosing different boundary conditions in the interior. The "in-falling" boundary conditions yield the retarded Green function, which is always complex, and we must use the analytically continued version of (9.20). The Bessel functions with in-falling boundary conditions depend sensitively on the nature of the excitations. Again we have to distinguish spacelike and timelike momenta, and within the timelike family there is now a distinction between positive and negative frequencies. One finds for the fundamental field

$$\chi_+ = \begin{cases} r^{-(d+2)/2} K_m(|k|/r) a_+, & \text{if } k^2 > 0; \\ r^{-(d+2)/2} H_m(|k|/r) a_+, & \text{if } \omega > |\mathbf{k}|; \\ r^{-(d+2)/2} K_m(|k|/r) a_+, & \text{if } \omega < -|\mathbf{k}|. \end{cases} \qquad (9.25)$$

In the standard quantisation case this translates into the retarded Green function [243]

$$\langle \mathcal{O} \mathcal{O}^\dagger \rangle_R = \begin{cases} \dfrac{2 e^{-(m+\frac{1}{2})\pi i}}{k^2} \dfrac{\Gamma(-m+\frac{1}{2})}{\Gamma(m+\frac{1}{2})} \left(\dfrac{k}{2}\right)^{2m+1} (\gamma \cdot k)\gamma^t, & \omega > |\mathbf{k}|, \\[4mm] \dfrac{2 e^{(m+\frac{1}{2})\pi i}}{k^2} \dfrac{\Gamma(-m+\frac{1}{2})}{\Gamma(m+\frac{1}{2})} \left(\dfrac{k}{2}\right)^{2m+1} (\gamma \cdot k)\gamma^t, & \omega < -|\mathbf{k}|, \end{cases} \qquad (9.26)$$

where γ^μ are $(d+1)$-dimensional gamma matrices in the boundary field theory.

9.2 The discovery of the holographic Fermi surfaces

We have now arrived at the exercise that caused the big splash in 2009 [320, 349, 350, 351]. What do the fermion spectral functions look like in the strange metals realised at finite density as introduced in chapter 8? We follow the historical development, considering the elementary Reissner–Nordström metal for simplicity. Strictly speaking, the severe instability of the RN black hole makes even the probe fermion calculation unreliable in many cases. The results are nevertheless still instructive in giving a sense of the gross changes in the fermionic response in finite-density systems, which carry over to a large extent to other backgrounds that we catalogued in chapter 8.

Following the GKPW rule, the spectral function of the fermions follows directly from the imaginary part of the retarded Green function deduced from solutions to the Dirac equation with in-falling boundary conditions at the black-hole horizon. The difference from earlier linear-response computations is that, in addition to the geometry of the RN black hole in the deep interior, the fermions in the bulk will now also interact directly with the electrical potential of the black hole via their bulk charge q. In other words, the dual fermionic operators in the field theory will respond directly to the presence of the finite chemical potential μ. Technically the computations are straightforward. For generic frequency and momentum one must rely on numerics, since closed solutions can no longer be obtained, although for small frequencies approximate analytical solutions can be obtained with a matching method. We shall discuss this in the next section.

But what should one expect? The extreme power of this computation is that it is utterly novel. There is no general controlled mathematical method to address this question on the field-theory side, given the obstruction arising from the fermion-sign problem. An example of what holography predicts for the fermion spectral function in a $(2 + 1)$-dimensional quantum critical boundary field theory at finite density is reproduced in Fig. 9.1 [349, 350]. In the upper panel the fermion spectral functions for very high temperature T are shown in units of μ for a variety of momenta. In this case the effect of the chemical potential is negligible and we do indeed find the expected conformal behaviour of $A(\omega, k) \sim (\sqrt{k^2 - \omega^2})^{2(2\Delta - d - 1)}$, modulated by finite-temperature effects. One notices that the scaling dimension of the fermion operator is deliberately chosen to be small, translating into the "relevant" rise of the spectral weight towards small energy.

The breakthrough result is the result for a large finite density in units of the temperature shown in the lower panel. The spectrum completely rearranges itself, and one finds upon varying the momentum that at a *large* momentum $k_F \simeq \mu$, roughly consistent with the Luttinger-volume Fermi momentum, a very sharp peak develops right at zero energy [349, 350]. At first sight the result appears to reveal a familiar outcome: a well-developed Fermi surface forms with an associated quasiparticle peak. However, depending on the scaling dimension this peak can disperse linearly like a regular Fermi liquid, but it can also, surprisingly, exhibit a non-linear dispersion. The familiarity is deceptive; the holographic computation for the first time revealed the controlled emergence of both normal liquids and non-Fermi liquids from a critical state! Let us emphasise how profound this result is. A state with no discernible signature of the Pauli principle can be deformed to yield the pristine

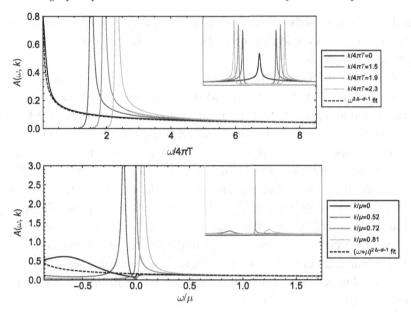

Figure 9.1 The discovery of "holographic photoemission". The top panel shows the energy dependence of the spectral functions for the zero-density conformal system at a small but finite temperature, for various momenta. In the main panel only the positive energy is shown; the inset shows the full range. These plots visually show the branch-cut behaviour of correlation functions in a critical system. In the lower panel the same spectral functions are shown, but at finite density $\mu \gg T$. One immediately sees that a very sharp peak develops at a large Luttinger-volume Fermi momentum k_F, while the peak disperses away from the "Fermi energy", suggesting a linear dispersion and broadening at the same time. In the inset the full height of the peak is given, highlighting the delta-function nature of the quasiparticle peak at the Fermi surface. Figures adapted from [349, 350].

characteristic signature of Fermi statistics: the Fermi surface. Moreover, the associated quasiparticles need not be those of the Landau Fermi liquid. On probing further, two other remarkable features emerge. As one increases the momentum, at low momenta, long before we reach the Luttinger scale $k \sim \mu$, the spectral function displays a curious oscillatory behaviour in $\ln(\omega)$ (Fig. 9.2). In the simplest approximation of the Green function as a power law $\omega^{2\Delta - d - 1}$ this would imply a complex scaling dimension Δ. In chapter 5 we have already met such a complex scaling dimension, as it arises in bosonic systems when the Breitenlohner–Freedman bound is violated. We learned that this is associated with an instability in the system. This is also the case here, even though for fermions the nature of the instability is not simply associated with a field being at the top of a potential. We will discuss this in more detail in chapter 11, where we will confirm the view that the

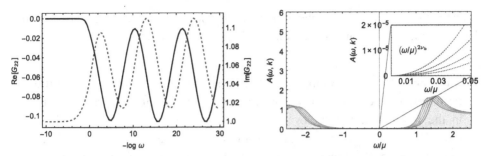

Figure 9.2 The left panel shows the log-oscillatory behaviour of the spectral function of a fermion probing the RN metal with small mass/charge ratio. The log-oscillatory region, signalling an instability of the ground state, is observed for momenta for which the scaling exponent ν_k is imaginary. The Fermi surfaces of Fig. 9.1 appear only for special higher momenta at which ν_k is always positive. For generic momenta, however, the low-energy behaviour of the spectral function is controlled by the local quantum critical or "algebraic pseudogap" behaviour $\omega^{2\nu_k}$. This is the reflection of an emergent AdS$_2$ region in the near-horizon RN geometry (right panel). This pseudogap is in fact the *generic* behaviour: for large mass/charge ratio it is seen for all momenta. Figures adapted from [349] and [355].

log-oscillatory behaviour reflects an instability of the simple Reissner–Nordström black hole.

A second remarkable aspect is associated with what happens when the scaling dimension is increased. We started by choosing deliberately a small mass/low scaling dimension to optimise the sensitivity of the fermionic probes to the new IR of the Reissner–Nordström black hole. When we increase the scaling dimension, however, the peaks characteristic of the Fermi surface disappear. However, we do not recover the branch-cut scaleless spectral functions with no weight at $\omega < |k|$ of zero density. Instead one finds at low ω an emergent power-law scaling $A(\omega, k) \sim \omega^{2\nu_k}$ with a momentum-dependent power. The fluctuations are now directly probing the local quantum critical AdS$_2 \times \mathbb{R}^2$ structure of the near-horizon RN black hole we presaged in the previous chapter. Indeed, by analysing the Dirac equation in the near-horizon regime (box 9.2), one precisely finds that the exponents ν_k are of the form

$$\nu_k = \sqrt{\frac{g_F^2 k^2}{\mu^2} + \frac{m^2 L^2}{6} - \frac{g_F^2 q^2}{6}},$$

which matches the numerical results. Here g_F is the electromagnetic coupling constant defined in Eq. (9.1).

Formally there is one further parameter we can vary: the charge of the fermion. For a fixed scaling dimension, tuning the charge has the same effect as tuning the scaling dimension. It interpolates between the local quantum critical and (non-) Fermi-liquid regimes at small and large charges, respectively.

Summary 10   Depending on the UV scaling dimension of the probe fermions, the spectral functions show a very different behaviour at low energy in the RN metal. For "relevant" UV fermions with a small mass/charge ratio there is a strong tendency to show particle-like behaviour, suggesting the emergence of a (near) Fermi liquid characterised by a large-"Luttinger-volume" Fermi surface. For "irrelevant" UV fermions with large mass/charge ratio one finds instead the "algebraic pseudogap" response $A(\omega, k) \sim \omega^{2\nu_k}$, where $\nu_k \sim \sqrt{1/\xi^2 + k^2}$. This algebraic scaling is a consequence of the local quantum critical near-horizon geometry.

Box 9.2   Fermion correlation functions at finite density

To obtain the spectral function of fermionic operators we need to solve the Dirac equation in AdS–Einstein–Maxwell theory

$$\left[e_A^\mu \Gamma^A \left(\partial_\mu + \frac{1}{4}\omega_{\mu AB}\Gamma^{AB} - iq A_\mu \right) - m \right] \Psi = 0, \qquad (9.27)$$

where the spin connection $\omega_{\mu AB}$ and the gauge field A_μ are now given by their RN background values. For simplicity we will specialise to the case of a $(3 + 1)$-dimensional AdS space dual to a $(2 + 1)$-dimensional field theory. We rescale the fermion $\Psi = (-gg^{rr})^{1/4}\mathcal{X}$ to absorb the spin connection and project onto partially Fourier transformed two-component $\Gamma^{\underline{r}}$ eigenspinors

$$\Psi_\pm = (-gg^{rr})^{-1/4} e^{-i\omega t + ik_i x^i} \begin{pmatrix} -iy_\pm \\ z_\pm \end{pmatrix}. \qquad (9.28)$$

Using isotropy to set $k_y = 0$ and $k_x = k$, one can choose a basis of Dirac matrices, $\Gamma^r = \sigma^3 \otimes \mathbb{1}, \Gamma^{t,x,y} = \sigma^{t,x,y} \otimes \sigma^1$, where one obtains two decoupled sets of two simple coupled equations

$$\sqrt{g_{ii}g^{rr}}(\partial_r \mp m\sqrt{g_{rr}})y_\pm = \pm(k - u)z_\mp,$$
$$\sqrt{g_{ii}g^{rr}}(\partial_r \pm m\sqrt{g_{rr}})z_\mp = \pm(k + u)y_\pm, \qquad (9.29)$$

where $u = \sqrt{-g_{ii}/g_{tt}}(\omega + qA_t)$. In this basis of Dirac matrices the GKPW CFT Green function $G = \langle \bar{\mathcal{O}}_{\psi_+} \gamma^0 \mathcal{O}_{\psi_+} \rangle$ is

$$G = \lim_{\epsilon \to 0} \epsilon^{-2mL} \begin{pmatrix} \xi_+ & 0 \\ 0 & \xi_- \end{pmatrix} \Bigg|_{r=1/\epsilon}, \quad \text{where } \xi_+ = \frac{y_-}{z_+}, \quad \xi_- = \frac{z_-}{y_+}. \qquad (9.30)$$

Rather than solving the coupled equations (9.29) it is convenient to solve for $\xi_\pm$ directly,

$$\sqrt{\frac{g_{ii}}{g_{rr}}} \, \partial_r \xi_\pm = -2m\sqrt{g_{ii}}\xi_\pm \mp (k \mp u) \pm (k \pm u)\xi_\pm^2. \qquad (9.31)$$

To obtain the retarded Green function, the in-falling boundary conditions should be imposed near the horizon. These are $\xi_\pm|_{r=r_0} = i$ for non-zero ω, whereas for $\omega = 0$ we should impose [349]

$$\xi_\pm|_{r=r_0,\omega=0} = \left(m - \sqrt{k^2 + m^2 - \frac{\mu^2 q^2}{6} - i\epsilon} \right) \Bigg/ \left(\frac{\mu q}{\sqrt{6}} \pm k \right). \qquad (9.32)$$

The numerical solutions to this equation are given in Fig. 9.1.

Locally quantum critical scaling from the near-horizon geometry AdS$_2$ × $\mathbb{R}^2$

The low-frequency local quantum critical scaling behaviour of the fermion spectral functions for large mass/charge ratio can be heuristically understood from the emergence of local quantum criticality in the deep IR. In the next section we will prove that this heuristic answer is correct. Taking to heart that the IR physics is given by the near-horizon AdS$_2$ × $\mathbb{R}^2$ geometry of the zero-temperature RN black-hole solution, we simply approximate the full space-time by this geometry introduced in Eq. (8.23),

$$ds^2 = \frac{L^2}{6\zeta^2}(-dt^2 + d\zeta^2) + \frac{r_*^2}{L^2} \, d\mathbf{x}^2, \quad A_t = \frac{1}{\sqrt{3}\zeta}, \qquad (9.33)$$

where $r_* = \mu/(2\sqrt{3})$ and we have set $2\kappa^2 = g_{F=1}$ in Eq. (9.1). In these coordinates $\zeta = \infty$ is the horizon and $\zeta = 0$ corresponds to the boundary of this near-horizon geometry.

In this background the equation of motion Eq. (9.12) of the spinor $\mathcal{X}_\alpha$ becomes

$$\partial_\zeta \mathcal{X}_\alpha + U\mathcal{X}_\alpha = 0, \qquad (9.34)$$

where

$$U = \frac{m}{\sqrt{6}\zeta}\sigma^3 - i\left(\omega + \frac{q}{\sqrt{3}\zeta}\right)\sigma^2 - (-1)^\alpha \frac{\sqrt{2}}{\mu\zeta}k\sigma^1. \qquad (9.35)$$

Here again we assume that the momentum is in the x direction and

$$\mathcal{X}_\alpha = \begin{pmatrix} \mathcal{X}_{\alpha 1} \\ \mathcal{X}_{\alpha 2} \end{pmatrix}.$$

Box 9.2   **(Continued)**

Equation (9.34) is solvable in terms of Whittaker functions. In order to obtain the retarded Green function, we impose in-falling boundary conditions for $\mathcal{X}_\alpha$ at $\zeta = \infty$. The in-falling solutions for the two independent components of the spinor are

$$
\begin{pmatrix} \mathcal{X}_{\alpha 1} \\ \mathcal{X}_{\alpha 2} \end{pmatrix}
$$

$$
= \begin{pmatrix} c_{(\alpha,\text{in})} i \zeta^{-1/2} \left[W_{1/2+iq/\sqrt{3},\,v_k}(-2i\omega\zeta) \right. \\ \left. \qquad + \left(\frac{m}{\sqrt{6}} + \frac{i(-1)^\alpha \sqrt{2}k}{\mu} \right) W_{-1/2+iq/\sqrt{3},\,v_k}(-2i\omega\zeta) \right] \\[2ex] c_{(\alpha,\text{in})} \zeta^{-1/2} \left[W_{1/2+iq/\sqrt{3},\,v_k}(-2i\omega\zeta) \right. \\ \left. \qquad - \left(\frac{m}{\sqrt{6}} + \frac{i(-1)^\alpha \sqrt{2}k}{\mu} \right) W_{-1/2+iq/\sqrt{3},\,v_k}(-2i\omega\zeta) \right] \end{pmatrix},
$$

$$
\tag{9.36}
$$

where

$$
v_k = \sqrt{ \frac{2k^2}{\mu^2} + \frac{m^2}{6} - \frac{q^2}{3} }
$$

while $c_{(\alpha,\text{in})}$ are arbitrary constants.

At the boundary $\zeta \to 0$, the solutions $\mathcal{X}_\alpha$ asymptote as

$$
\mathcal{X}_\alpha = v_{-\alpha} \zeta^{-v_k} + v_{+\alpha} \mathcal{G}_\alpha(\omega) \zeta^{v_k} + \cdots, \tag{9.37}
$$

where the leading coefficients $v_{\pm\alpha}$ are the eigenvectors of U in Eq. (9.35) with eigenvalues $\pm v_k$,

$$
v_{\pm\alpha} = \tilde{c}_0 \begin{pmatrix} i(-1)^\alpha \sqrt{2}k/\mu \pm i v_k - im/\sqrt{6} - iq/\sqrt{3} \\ (-1)^\alpha \sqrt{2}k/\mu \pm v_k + m/\sqrt{6} + q/\sqrt{3} \end{pmatrix}, \tag{9.38}
$$

and the "IR Green function" can be simply calculated from the boundary expansion of the in-falling solutions to be

$$
\mathcal{G}_\alpha(\omega) = e^{-i\pi v_k}
$$

$$
\times \frac{\Gamma(-2v_k)\Gamma(1 + v_k - iq/\sqrt{3})\left((-1)^\alpha \sqrt{2}k/\mu - im/\sqrt{6} - v_k - iq/\sqrt{3} \right)}{\Gamma(2v_k)\Gamma(1 - v_k - iq/\sqrt{3})\left((-1)^\alpha \sqrt{2}k/\mu - im/\sqrt{6} + v_k - iq/\sqrt{3} \right)}
$$

$$
\times (2\omega)^{2v_k}. \tag{9.39}
$$

This is exactly of the form predicted for an effective local quantum critical theory (e.g. Eq. (8.27)),

$$\mathcal{G}_k(\omega) = c_k e^{i\phi_k} \omega^{2\nu_k} \quad \text{with} \quad \nu_k = \sqrt{\frac{2k^2}{\mu^2} + \frac{m^2}{6} - \frac{q^2}{3}}, \tag{9.40}$$

where c_k and ϕ_k are both analytical and real functions of k, which can be read off from the full expression Eq. (9.39).

At low temperature $T \ll \mu$, the near-horizon geometry of the AdS RN black hole becomes a black hole in $AdS_2 \times \mathbb{R}^2$ and the retarded Green function for a probe spinor in this background can also be computed. The procedure of the computation is exactly the same as above, with as the final result [22, 356]

$$\mathcal{G}_\alpha(\omega, T) = \left[\Gamma(-2\nu_k)\Gamma(1 + \nu_k - iq/\sqrt{3})\Gamma\left(\frac{1}{2} + \nu_k - \frac{i\omega}{2\pi T} + \frac{iq}{\sqrt{3}}\right) \right.$$
$$\times \left((-1)^\alpha \sqrt{2}k/\mu - im/\sqrt{6} - \nu_k - iq/\sqrt{3}\right) \right]$$
$$\times \left[\Gamma(2\nu_k)\Gamma(1 - \nu_k - iq/\sqrt{3})\Gamma\left(\frac{1}{2} - \nu_k - \frac{i\omega}{2\pi T} + \frac{iq}{\sqrt{3}}\right) \right.$$
$$\times \left. \left((-1)^\alpha \sqrt{2}k/\mu - im/\sqrt{6} + \nu_k - iq/\sqrt{3}\right) \right]^{-1} (4\pi T)^{2\nu_k}.$$

The local quantum critical scaling is of course universal and holds for any other field as well. We can easily repeat the near-horizon analysis for such other cases. Specifically, for a charged massive boson one finds an IR Green function at zero temperature

$$\mathcal{G}_{\text{boson}}(\omega) = e^{-i\pi\nu_k} \frac{\Gamma(-2\nu_k)\Gamma(\frac{1}{2} + \nu_k - iq/\sqrt{3})}{\Gamma(2\nu_k)\Gamma(\frac{1}{2} - \nu_k - iq/\sqrt{3})} (2\omega)^{2\nu_k}, \tag{9.41}$$

with

$$\nu_k = \sqrt{\frac{1}{4} + \frac{2k^2}{\mu^2} + \frac{m^2}{6} - \frac{q^2}{3}}. \tag{9.42}$$

and at finite temperature

$$\mathcal{G}_{\text{boson}}(\omega, T)$$
$$= \frac{\Gamma(-2\nu_k)\Gamma(\frac{1}{2} + \nu_k - iq\sqrt{3})\Gamma(\frac{1}{2} + \nu_k - i\omega/(2\pi T) + iq/\sqrt{3})}{\Gamma(2\nu_k)\Gamma(\frac{1}{2} - \nu_k - iq/\sqrt{3})\Gamma(\frac{1}{2} - \nu_k - i\omega/(2\pi T) + iq/\sqrt{3})} (4\pi T)^{2\nu_k}. \tag{9.43}$$

9.3 Computing fermion spectral functions: Schrödinger potentials and the matching method

Shortly after their co-discovery of the "MIT–Leiden fermions" [349, 350], the MIT group of Faulkner, Liu, McGreevy and Vegh published a seminal paper, where they showed how to understand the physics of the fermionic responses described above in a precise semi-analytical language [320]. Their key insight is that the AdS–RN geometry is such that one can formally obtain closed solutions for the wave propagation at low frequencies by matching asymptotic expansions from infinity and the interior, with the only unknowns in the form of matching coefficients that can easily be determined numerically. This amazingly powerful and general method will also be of great value when we are dealing with bosonic probes, such as the holographic superconductivity of chapter 10. Let us therefore discuss it here in some detail, in the context where it was first used for AdS/CMT – the fermions.

Travelling along the radial direction, in the natural coordinates Eq. (8.2), the geometry of the AdS–RN space-time changes drastically and rather suddenly from the zero-density AdS_{d+2} to the near-horizon $AdS_2 \times \mathbb{R}^d$ geometry of the extremal Reissner–Nordström black hole. This change is so rapid that one can picture it as a geometrical "domain wall" that acts quite like a tunnelling barrier. One can therefore seek solutions to the wave equation (be it for bosons or fermions) corresponding to AdS_{d+2} geometry on the boundary side and $AdS_2 \times \mathbb{R}^d$ on the black-hole side, which are matched at the domain wall.

Before we address the matching method in detail, let us briefly discuss a most noteworthy new aspect of the finite-density spectrum: the sharp peaks at $\omega = 0$. A peak in the spectral function of course signals the presence of a pole in the Green function. Such a pole has a very natural reflection in the bulk physics. From the GKPW formula that expresses the Green function as the ratio of the asymptotic coefficients of the normalisable solution to those of the non-normalisable solution, we see that a pole arises whenever there is an exact normalisable solution, given a set of fixed interior boundary conditions. Although Dirac fermions in a curved space-time behave in ways that are hard to grasp with flat-space intuition, there is a useful trick to tease out precisely these normalisable solutions. By a coordinate transformation plus field rescaling one can transform the problem to an effective one-dimensional (radial-direction) relativistic Schrödinger equation where the influence of the geometry is translated into a simple potential. Subsequently, one can employ textbook potential scattering theory of one-dimensional quantum mechanics to analyse the bulk physics.

This procedure is outlined in box 9.3. The caveat is that this potential will generically allow normalisable tunnelling solutions with a complex frequency – so-called quasinormal modes [213]. The normalisable boundary condition is a real boundary

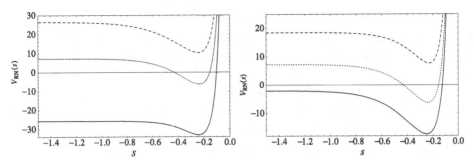

Figure 9.3 The Schrödinger potential $V(s)$ for the fermion component z_+ in the AdS–RN background. In these coordinates $s = 0$ is the AdS boundary and $s = -\infty$ is the near-horizon region. We use as parameters $r_+ = 1$, $\mu = \sqrt{3}$, $g_F = 1$, $mL = 0.4$, $c_0 = 0.1$. On the left the dependence on the momentum $k = 3, 2, 1$ (top to bottom) for charge $q = 2.5$ is shown. On the right the dependence on the charge q for the values $q = 2, 2.5, 3$ (top to bottom) for a fixed momentum $k = 2$ is shown. In both figures the lowest solid-line potentials correspond to the oscillatory region $v_k^2 < 0$, the middle dotted potentials show the generic shape that can support an $\omega = 0$ bound state, and the top dashed potentials are strictly positive and no zero-energy bound state is present. Figure source [357].

condition. Together with the fact that the equations are real, these modes imply that the only possible normalisable solution for real frequencies with in-falling boundary conditions at the horizon occurs at $\omega = 0$. All peaks are therefore sharpest at $\omega = 0$. We can therefore limit our attention to this value. One finds in that case a simple, one-dimensional Schrödinger-like wave equation for the fundamental fermion field z_+ in Fourier space which is dual to the source of the fermion operator in the boundary,

$$\partial_s^2 z_+ - V(s)z_+ = 0, \qquad (9.44)$$

where one has to look for solutions with zero energy eigenvalue. Here s is a reparametrisation of the radial coordinate and the potential $V(s)$ (Eq. (9.49) below) is predictably a function of the fermion mass, charge and chemical potential. But, because we are approximating a higher-dimensional evolution equation with an effective one-dimensional equation, it also depends strongly on the momentum of the fermions. In Fig. 9.3 we show representative cases for this potential and it now becomes quite easy to read off the physics of the Dirac fermions in the AdS/RN space-time.

One can discern three typical situations, as indicated in Fig. 9.3. In all cases the potential decreases from the boundary (on the right) towards the interior with a potential well in the middle and a constant asymptote approaching the horizon (on the far left). The barrier at the boundary $s = 0$ is the effect of the curved AdS background. It shows why one can think of AdS as a box in which a (massive) particle cannot escape to infinity. The most interesting aspect, however, is the potential well. This corresponds to the "geometrical domain wall" alluded to above, dividing the near-boundary AdS geometry from the near-horizon geometry of the RN black hole. We can now analyse what happens as we vary the various parameters k, m, q, μ on which the potential depends. Direct inspection already shows that the potential depends only on the combination μq. The second observation is that to a first approximation the effect of increasing k is to move up the overall level of the potential, with a slight steepening of the well. For large mass m and small q, the potential typically remains strictly positive for all momenta and all zero-energy solutions are over-damped. This means that waves emanating from the boundary just fall towards the black-hole horizon. Since there is nothing that "distracts" this bulk propagation from "events" that happen half-way along the radial direction, the propagators are expected to directly represent the deep-IR physics encoded in the near-horizon geometry. As we will further elucidate underneath, this is the regime where one finds the local quantum critical "algebraic pseudogap" fermion spectral functions of Fig. 9.2.

Upon lowering the mass and/or increasing the charge q one finds that the near-horizon Schrödinger potential turns negative for some range of small momenta (the lowest curves in Fig. 9.3). This signals that something novel happens in the RN geometry. From the Schrödinger-potential point of view one can now have a zero-energy oscillating state emanating from the horizon while being reflected back at the boundary. To understand how this connects back to the spectral function, we notice that the explicit form of the near-horizon potential is directly proportional to the square of the scaling exponent of the local quantum critical correlation function $V_{\text{near-hor}} \sim v_k^2$ (box 9.3). The negative value of the potential near the horizon thus means that the scaling exponent $v_k = \sqrt{v_k^2}$ has become imaginary. We thus see that the existence of an undamped zero-energy state is responsible for the mysterious log-oscillatory behaviour in the spectral function, signalling an instability.

This insight can be confirmed in other ways as well. One can show that in purely gravitational language the onset of a negative near-horizon potential corresponds to a spontaneous discharge of the black hole by spontaneous pair creation due to the strong curvature near the black hole [358]. For bosons this signals the "tachyonic waterfall down the potential". For fermions it is more subtle, since the Pauli principle forbids them from building up coherently and simply condensing. Nevertheless, the discharge will create a cloud of matter in the bulk. This cloud of matter will be pulled together gravitationally and a material "star-like" object should take over in the deep interior. This phenomenon, representing the gravitational dual description of the formation of a cohesive macroscopic state for bosons and fermions, respectively, will be the topic of chapters 10 and 11.

In the probe limit we will ignore this instability. Instead we can explore the Schrödinger potential further. The third typical outcome is that the near-horizon potential is still positive but the domain-wall potential well dips below zero. This means that a *quasi-bound state* can form at zero energy. Generically, it is not infinitely long-lived since there is a finite probability of tunnelling to the horizon and when one waits long enough the fermion will eventually be absorbed by the black hole. The fermion falling in from the boundary will now first get trapped in this bound state, representing a "halo" surrounding the black hole, to tunnel at a much later time to the horizon where it gets absorbed by the black hole. These zero-energy bound states need a finite momentum, corresponding to the large-"Luttinger-volume" momenta $k_F \simeq \mu$, to render the potential well steep enough. It is these "domain-wall quasi-bound states" which are responsible for the sharp quasiparticle peaks seen in Fig. 9.1.

Summary 11 Probe waves in an asymptotically AdS space-time can be rewritten in an effective one-dimensional Schrödinger equation in the radial direction where the effects of the curvature are collected in an effective potential. The qualitative features of the spectral function can be understood in terms of quantum-mechanical potential theory with peaks corresponding to bound states, instabilities to oscillatory solutions and scaling behaviour to over-damped wave functions.

Box 9.3   The geometrical domain wall and the Schrödinger potential

To derive the Schrödinger equation related to the zero-frequency Dirac equation, we start with the $\omega = 0$ equation (9.27) for one set of components Eqs. (9.29),

$$\sqrt{g_{ii}g^{rr}}\partial_r y_- + m\sqrt{g_{ii}}\,y_- = -(k - \hat{\mu})z_+,$$
$$\sqrt{g_{ii}g^{rr}}\partial_r z_+ - m\sqrt{g_{ii}}\,z_+ = -(k + \hat{\mu})y_-, \tag{9.45}$$

where $\hat{\mu} = \sqrt{-g_{ii}/g_{tt}}\,qA_t$ and we will drop the subscript x on k. Here g_{ii} and $g_{rr} = 1/g^{rr}$ are the components of the RN metric, or any other metric that is static, spatially rotationally invariant and depends only on the radial direction r, e.g. any Lifshitz space-time. In our conventions z_+ (and y_+) is

Box 9.3   (Continued)

the fundamental component dual to the source of the fermionic operator in the CFT. Rewriting the coupled first-order Dirac equations as a single second-order equation for z_+ gives

$$\partial_r^2 z_+ + \mathcal{P}\,\partial_r z_+ + \mathcal{Q}z_+ = 0,$$

$$\mathcal{P} = \frac{\partial_r(g_{ii}g^{rr})}{2g_{ii}g^{rr}} - \frac{\partial_r\hat{\mu}}{k+\hat{\mu}}, \tag{9.46}$$

$$\mathcal{Q} = -\frac{m\,\partial_r\sqrt{g_{ii}}}{\sqrt{g_{ii}g^{rr}}} + \frac{m\sqrt{g_{rr}}\,\partial_r\hat{\mu}}{k+\hat{\mu}} - m^2 g_{rr} - \frac{k^2-\hat{\mu}^2}{g_{ii}g^{rr}}.$$

The first thing one notices is that both $\mathcal{P}$ and $\mathcal{Q}$ diverge at some $r = r_*$ where $\hat{\mu}+k = 0$. Since $\hat{\mu}$ is (chosen to be) a positive semidefinite function that increases from $\hat{\mu} = 0$ at the horizon, this implies that for negative k (with $-k < \hat{\mu}_\infty$) the wave function is qualitatively different from the wave function with positive k which experiences no singularity. The analysis is straightforward if we transform the first derivative away and recast it in the form of a Schrödinger equation by redefining the radial coordinate:

$$\frac{ds}{dr} = \exp\left(-\int^r dr'\,\mathcal{P}\right) \quad\Rightarrow\quad s = c_0\int_{r_\infty}^r dr'\frac{|k+\hat{\mu}|}{\sqrt{g_{ii}g^{rr}}}, \tag{9.47}$$

where c_0 is an integration constant whose natural scale is of order $c_0 \sim q^{-1}$. In the new coordinates the equation (9.46) is of the standard form

$$\partial_s^2 z_+ - V(s)z_+ = 0, \tag{9.48}$$

with potential

$$V(s) = -\frac{g_{ii}g^{rr}}{c_0^2|k+\hat{\mu}|^2}\mathcal{Q} \tag{9.49}$$

$$= \frac{1}{c_0^2(k+\hat{\mu})^2}\left[(k^2 + m^2 g_{ii} - \hat{\mu}^2) + mg_{ii}\sqrt{g^{rr}}\,\partial_r\ln\left(\frac{\sqrt{g_{ii}}}{k+\hat{\mu}}\right)\right]. \tag{9.50}$$

For g_{ii} and g_{rr}, the components of the RN metric, this is the potential presented in Fig. 9.3.

One can repeat exactly the same exercise for a massive charged boson. It obeys the following equation of motion:

$$(\nabla^\mu - iqA^\mu)(\nabla_\mu - iqA_\mu)\phi - m^2\phi = 0. \tag{9.51}$$

By Fourier expanding the scalar field $\phi = \phi(r)e^{-i\omega t + ikx}$ for a single momentum $k_x \neq 0$, and assuming that the background metric is static, rotationally invariant and only depends on the radial direction r, we obtain

$$\partial_r^2 \phi + \frac{\partial_r(g_{xx}g^{rr})}{g_{xx}g^{rr}}\partial_r\phi + g_{rr}\left[\left(-m^2 - \frac{k^2}{g_{xx}}\right) + g_{rr}(\omega + qA_t)^2\right]\phi = 0. \quad (9.52)$$

Similarly to Eq. (9.47), we introduce a new "tortoise" coordinate

$$s = c_0 \int_{r_\infty}^{r} dr' \frac{1}{g_{xx}g^{rr}}. \quad (9.53)$$

In this new coordinate, we have the equation of motion

$$\partial_s^2 \tilde{\phi} - V(s)_{\text{boson}}\tilde{\phi} = 0, \quad (9.54)$$

with the effective potential

$$V(s)_{\text{boson}} = -\frac{g_{xx}^2 g^{rr}}{c_0^2}\left[\left(-m^2 - \frac{k^2}{g_{xx}}\right) + g_{rr}(\omega + qA_t)^2\right]. \quad (9.55)$$

The near-horizon potential and local quantum criticality

In the simplifying special case of the extremal near-horizon $\text{AdS}_2 \times \mathbb{R}^2$ metric we can solve for the near-horizon potential V explicitly in terms of the new coordinate s. For the fermionic case, in units $r_0 = 1$ (i.e. $\mu = 2\sqrt{3}$) we have $s = (c_0/\sqrt{6})(k + q/\sqrt{2})\ln(r - 1) + \cdots$ when $r \to 1$. As noted in [320], one obtains that the near-horizon potential for $s \to -\infty$ is proportional to the self-energy exponent:

$$V(s) = \frac{6}{c_0^2(k + q/\sqrt{2})^2}v_k^2 + \cdots. \quad (9.56)$$

This shows that a negative potential at the horizon corresponds to an imaginary scaling exponent v_k. This is the signature of the instability of the RN black hole.

Note that the potential appears to have a singularity in the potential for $k < 0$. A detailed analysis reveals that no qualitatively new features appear for this case [359].

For the scalar case we have from equations (8.20) and (9.53)

$$s = \frac{-c_0}{6(r - 1)} + \cdots \quad (9.57)$$

when $r \to 1$ and the potential $V(s)_{\text{boson}}$ behaves as

$$V(s)_{\text{boson}} = -\frac{1}{6s^2}(2q^2 - m^2 - k^2) + \cdots$$

$$= \frac{1}{s^2}\left(-\frac{1}{4} + v_k^2\right) + \cdots, \tag{9.58}$$

where v_k is defined as in Eq. (9.42). Thus the oscillatory regime for the scalar field is $v_k^2 < 0$.

Analytic low-frequency correlators from the matching method

Armed with these qualitative insights into the fermion spectral function, we will now place it on a solid footing using the explicit GKPW rule. In box 9.2 it was explained how to compute the fermionic Green function in general. At high frequencies the only option is to use brute-force numerics, but in the low-frequency regime $\omega \ll \mu$ one can obtain very useful "universal" expressions for the general form of the Green function [320] by matching across the domain wall. The derivation is explained in detail in box 9.4. Conceptually it is straightforward: it is just an implementation of the matching method which is well known from quantum-mechanical scattering theory. Consider a smooth potential that behaves quite differently at small and large distances. Solutions can be found both in the small and in the large-distance regime, and one subsequently matches these solutions in the intermediate-distance regime where the potential changes. In the present problem of evolution in the holographic bulk, one has, as a function of the radial coordinate, a far (from the horizon) regime that can be approximated by a perturbative solution (in frequency) in the regime $r - r_* \gg \omega$ and a near-horizon regime $r - r_* \ll \mu$ with its familiar $\text{AdS}_2 \times \mathbb{R}^2$ geometry where we solved for the fermions in box 9.2. As long as $\omega \ll \mu$ these two regions overlap and the solutions can be matched in the intermediate region where these geometries meet (the geometrical domain wall).

The result (see box 9.4) is an analytical expression for the full boundary propagator in the low-frequency $\omega \ll \mu$ regime,

$$G_R(\omega, k) = \frac{b_+^{(0)} + \omega b_+^{(1)} + \mathcal{O}(\omega^2) + \mathcal{G}_k(\omega)\left(b_-^{(0)} + \omega b_-^{(1)} + \mathcal{O}(\omega^2)\right)}{a_+^{(0)} + \omega a_+^{(1)} + \mathcal{O}(\omega^2) + \mathcal{G}_k(\omega)\left(a_-^{(0)} + \omega a_-^{(1)} + \mathcal{O}(\omega^2)\right)}. \tag{9.59}$$

Here $a_\pm^{(i)}$ are the matching coefficients of the non-normalisable near-boundary solution a with the leading (+) and the sub-leading (−) AdS_2 solution, subsequently Taylor expanded in ω to order i; $b_\pm^{(i)}$ is the similar matching coefficient for the normalisable solution. Each coefficient is real but still a function of the spatial momentum k. Except for the special case $m = 0$ [360], these can only be computed numerically.

Since the coefficients $a_{\pm}^{(i)}$, $b_{\pm}^{(i)}$ are real, this result immediately explains the quantum critical scaling behaviour in the large-mass/small-charge regime. The matching result shows that the exact spectral function $\operatorname{Im} G_R(\omega, k)$ is directly proportional to the spectral function of the IR AdS$_2$ correlation function

$$\operatorname{Im} G_R(\omega, k) \propto \operatorname{Im} \mathcal{G}_k(\omega). \tag{9.60}$$

This proportionality of spectral functions also explains right away the log-oscillatory behaviour when the AdS$_2$ scaling exponent ν_k turns imaginary.

The real power of the matching procedure is the explanation it provides for the origin and details of the quasiparticle peaks of Fig. 9.1. What is different in this regime is that inspection of the numerical results at low mass/charge ratio reveals that the UV coefficient $a_+^{(0)}(k)$ in Eq. (9.69) can *vanish* at a specific momentum $k = k_F$. Upon expanding $a_+^{(0)}(k) = v_F(k - k_F) + \cdots$ around $k = k_F$ Eq. (9.69) reduces to a deceptively attractive form, given that all other UV coefficients are finite near $k = k_F$,

$$G_R(\omega, k) \simeq \frac{Z}{\omega - v_F k_\perp - \Sigma(k, \omega)} + \cdots, \tag{9.61}$$

with

$$Z = \frac{b_+^{(0)}(k_F)}{a^{(1)}(k_F)}, \tag{9.62}$$

$$v_F = -\frac{\partial_k a^{(0)}(k_F)}{a^{(1)}(k_F)}, \tag{9.63}$$

$$\Sigma(k, \omega) = -\frac{c_{k_F}}{a^{(1)}(k_F)} e^{i\phi_{k_F}} \omega^{2\nu_{k_F}}. \tag{9.64}$$

If we recall that the distance in the radial direction corresponds to the scale, we can literally interpret this as the formation of a bound state at the domain wall at $r \sim \mu$. This is how the bound state in the Schrödinger potential is reflected in the form of a pole in the boundary propagator. The fact that these bound states eventually decay by tunnelling through to the horizon translates into a complex self-energy $\Sigma(\omega, k)$ that is directly governed by the local quantum criticality in the near-horizon limit.

Such "critical self-energies" $\Sigma \sim \omega^\alpha$ have quite some history in condensed matter physics, and accordingly this discovery caused a stir. Bound states are governed by such poles for which the imaginary part of $\Sigma(\omega, k)$ vanishes at $\omega = 0$. These states therefore span a well-defined Fermi surface at the location of the poles, $k = k_F$. However, they are quite unlike the usual Fermi liquids. There the damping of the quasiparticles is governed by the weak interaction of the quasiparticles

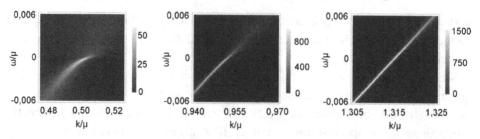

Figure 9.4 The fermion spectral functions in the momentum–energy plane presented in false-greyscale ARPES style. These are the results numerically obtained in the "Fermi-surface" regime of the probe fermions. The three panels show the qualitative change in the spectral response as a function of the scaling dimension ν_{k_F} which sets the nature of the "quantum critical" self-energy. Left: the "singular" Fermi liquid, characterised by $\nu_{k_F} < 1/2$, showing a Fermi surface but no quasiparticles. Middle: the marginal Fermi liquid with $\nu = 1/2$, identical to the form deduced in the late 1980s from experiment. Right: the "non-Landau" Fermi liquid with $\nu_{k_F} > 1/2$, showing well-defined quasiparticles but an unconventional energy dependence of their lifetimes. To give the peaks a visible width, the spectral functions are computed with a fixed small negative imaginary component for the frequency ω. Figure adapted from [361]. (Reprinted with permission from AAAS.)

with themselves, leading to the universal Im $\Sigma(\omega, k) \sim \omega^2$. But in these holographic metals the damping is governed by the local quantum critical deep infrared instead. The precise nature of these "non-Landau Fermi-surface liquids" is governed by the value of the exponent $\nu_{k=k_F}$. Qualitatively they fall into three classes, with the respective spectral functions given in Fig. 9.4. When $\nu_{k_F} > 1/2$, the inverse lifetime $\Gamma(\omega, k) = \text{Im}\,\Sigma(\omega, k)$ of the quasiparticles is smaller than their energy ω as we approach the pole at $k = k_F$ and therefore these are well-defined quasiparticle excitations. Their pole strengths and masses are all finite, and in the scaling limit a genuine Fermi gas of non-interacting quasiparticles is realised. This is therefore a Fermi liquid, but not of the Landau kind, due to its anomalous lifetime.

The truly novel case is the pole with $\nu_{k_F} < 1/2$. This is a state with a Fermi surface, but no true quasiparticles. The putative quasiparticles are over-damped ($\Gamma(\omega) > \omega$ as $\omega \to 0$). The pole strength vanishes and this is no longer a free IR fixed point, but it is still characterised by a Fermi surface as a locus in momentum space where massless excitations occur. By definition this is a true non-Fermi liquid.

At the boundary between the two where ν_{k_F} is exactly $\nu_{k_F} = 1/2$ one finds for the real and imaginary parts of the self-energy Re$\Sigma \sim \omega \ln \omega$ and Im $\Sigma \sim \omega$, respectively. Remarkably, this is precisely the same form as postulated on the basis of experimental information in 1988 by Varma, Littlewood, Schmitt-Rink, Abrahams and Ruckenstein [152]. It was christened the marginal Fermi liquid, and we discussed this deeply inspired guess in chapter 3: more than anything else, its

Table 9.1 *The zoo of "Fermi-surface liquids" present in holographic locally quantum critical metals. These are essentially never of the Landau Fermi-liquid type. They have the structure of the marginal Fermi-liquid theory deduced in the 1980s on the basis of experimental information in the cuprate strange metals [152], but the notion of a "marginal" local quantum critical deep-infrared damping of the quasiparticles is generalised to a variety of deep-infrared scaling behaviours possible in an RN strange metal. These are at present understood as artefacts of the large-N and probe limits.*

$\Sigma \sim \omega^{2\nu_{k_F}}$	Fermi-system phase	Quasiparticle properties
$2\nu_{k_F} > 1$	"Non-Landau" FL	$\omega_*(k) = v_F k_\perp,$ $\Gamma(k)/\omega_*(k) \propto k_\perp^{2\nu_{k_F}-1} \to 0,$ $Z = \text{constant}.$
$2\nu_{k_F} = 1$	Marginal FL	$G_R = \dfrac{h_1}{c_2\omega - v_F k_\perp + c_R \omega \ln \omega},$ $Z \sim \dfrac{1}{\ln \omega_*} \to 0.$
$2\nu_{k_F} < 1$	Singular FL	$\omega_*(k) \sim k_\perp^{1/2\nu_k},$ $\Gamma(k)/\omega_*(k) \to \text{constant},$ $Z \propto k_\perp^{(1-2\nu_{k_F})/(2\nu_{k_F})} \to 0.$

profundity lies in the local quantum criticality of the self-energy, which here arises quite naturally from holography.

In this bottom-up context, the marginal Fermi liquid requires fine tuning of the exponent ν_{k_F} and a priori it has no special status. It could simply be that no exact top-down dual holographic pair that has fermions with the required mass and charges can be found within string theory, or even that the charge/mass ratio in all string-theoretic constructions is never large enough to exhibit a Fermi surface. In fact, the bulk fermion masses are generically rather large in top-down constructions, with the implication that these systems will behave like the "local quantum critical" AdS$_2$ metal phase, and exhibit no sign of non-Fermi-liquid behaviour. To establish the validity of the result, at least one such top-down construction that does show Fermi surface behaviour in the probe limit needs to be found. One such model has been identified [362, 363]. It corresponds to a sub-sector of the canonical Maldacena example. Rather than an exact RN black hole, in this sub-sector a "conformal-to-AdS$_2$" deep-interior scaling geometry is realised as introduced in section 8.4. As we discussed there, this is characterised by a vanishing ground-state entropy, while the local quantum critical infrared is preserved. In this top-down model the parameters characterising the fermions are fully determined, and these give rise to Fermi surfaces of the novel non-Fermi-liquid type which we just discussed. There is even one Fermi surface characterised by the marginal exponent $\nu_{k_F} = 1/2$ [362, 363]. This provides a proof of principle that these probe results are physically consistent. We discuss this in detail in chapter 13.

We have discussed here the application of the matching method to the simple RN black hole. The method, however, extends to the conformal-to-AdS$_2$ metals with no major alterations. The crucial necessary ingredient is only that the deep-interior/near-horizon geometry is locally quantum critical.

Summary 12   At low frequencies the retarded Green function of a holographic theory with emergent local quantum critical behaviour is of the form

$$G_R(\omega, k) = \frac{b_+^{(0)} + \omega b_+^{(1)} + \mathcal{O}(\omega^2) + \mathcal{G}_k(\omega)\left(b_-^{(0)} + \omega b_-^{(1)} + \mathcal{O}(\omega^2)\right)}{a_+^{(0)} + \omega a_+^{(1)} + \mathcal{O}(\omega^2) + \mathcal{G}_k(\omega)\left(a_-^{(0)} + \omega a_-^{(1)} + \mathcal{O}(\omega^2)\right)},$$

(9.65)

with real coefficients $a_\pm^{(i)}$ and $b_\pm^{(i)}$, and $\mathcal{G}(\omega) = c_k e^{i\phi_k} \omega^{2\nu_k}$ equal to the purely local quantum critical IR Green function. In the absence of poles it completely determines the spectral function: Im $G_R \sim$ Im $\mathcal{G}$. Poles are determined by the vanishing points of the leading coefficient $a_+^{(0)}$ at $\omega = 0$.

Box 9.4   Boundary propagators in the deep infrared: the matching method in the bulk

The analytical properties of the retarded Green function ensure that any pole must arise at $\omega = 0$. The full retarded Green function for fermions in the extremal RN black-hole background is given by the solution of the Dirac equation

$$\sqrt{\frac{g_{ii}}{g_{rr}}} \left(i\sigma_2 \partial_r - \sqrt{g_{rr}}\sigma_1 m\right) \mathcal{X}_\alpha(r; \omega, k) = \left((-1)^\alpha k_x \sigma_3 - \sqrt{\frac{g_{ii}}{-g_{tt}}}(\omega + qA_t)\right)\mathcal{X}_\alpha,$$

(9.66)

with the metric (8.2). In general this equation can only be solved numerically, but one can try to exploit the smallness of ω/μ to obtain analytical expressions in order to obtain deeper insight into the structure of any pole. A direct attempt to solve Eq. (9.66) perturbatively order by order in ω/μ fails. The double pole for g_{tt} of the extremal RN black hole limits the range of applicability to $r - r_0 \gg \omega$.

This problem is cured by the near–far matching method. The space-time is divided into two regions: the near region and the far region. The far region is defined as the region above $r - r_0 \gg \omega$. The near region is the near-horizon region where $r - r_0 \ll \mu$. Here the geometry becomes the now-familiar $AdS_2 \times \mathbb{R}^2$. We already solved the equations in the near region and found that the linearly independent solutions $\eta_\pm(r)$ for in-falling boundary conditions behave near the outer edge of the near region

$$\mathcal{X}^{\text{near}} = \eta_{\alpha+}^{\text{near}}(r) + \mathcal{G}(\omega)\eta_{\alpha-}^{\text{near}}(r)$$

$$= (r - r_0)^{\frac{1}{2}+\nu_k} + \mathcal{G}(\omega)(r - r_0)^{\frac{1}{2}-\nu_k} + \cdots \quad \text{when } r \to r_0. \quad (9.67)$$

In the far region, we Taylor expand both the equation and the solution in ω. Each higher-order term satisfies an inhomogeneous equation that is uniquely determined, once we know the lowest-order $\omega = 0$ homogeneous solution. For this lowest order we can now use the completeness and independence of the two independent solutions to the linear Dirac equation to match. This is trivial at this order, since for $\omega = 0$ the far region fully overlaps with the near region. We can call the extension of the solution $\eta_\pm^{\text{near}}(r)$, $\eta^{(0)}(r)$. Thus the full solution will be schematically of the type $\mathcal{X}^{(0)}(r) = \eta_+^{(0)}(r) + \mathcal{G}(\omega)\eta_-^{(0)}(r)$. Since each higher-order term is uniquely determined, we can now immediately conclude that $\mathcal{X}(r) = \eta_+^{(0)}(r) + \omega\eta_+^{(1)}(r) + \cdots + \mathcal{G}(\omega)(\eta_-^{(0)}(r) + \eta_-^{(1)}(r)) + \cdots$.

To extract now the Green function at small ω, we recall that, at the asymptotic AdS_4 boundary of the space-time $r \to \infty$, each solution must take the form

$$\eta_\pm^{(n)} = a_\pm^{(n)} r^m \begin{pmatrix} 0 \\ 1 \end{pmatrix} + b_\pm^{(n)} r^{-m} \begin{pmatrix} 1 \\ 0 \end{pmatrix}. \quad (9.68)$$

On dividing the response by the source, we immediately see that the full retarded Green function in the boundary is of the form

$$G_R(\omega, k) = \frac{\left[b_+^{(0)} + \omega b_+^{(1)} + \mathcal{O}(\omega^2)\right] + \mathcal{G}_k(\omega)\left[b_-^{(0)} + \omega b_-^{(1)} + \mathcal{O}(\omega^2)\right]}{\left[a_+^{(0)} + \omega a_+^{(1)} + \mathcal{O}(\omega^2)\right] + \mathcal{G}_k(\omega)\left[a_-^{(0)} + \omega a_-^{(1)} + \mathcal{O}(\omega^2)\right]}. \quad (9.69)$$

One still has to determine the a, b matching coefficients, and in practice they can only be determined numerically.

9.4 The physics of the holographic fermions: confinement, semi-holography and black-hole stability

With the insights we have obtained in previous chapters – emergent local quantum criticality at the horizon, the matching method and the $1/N$ nature of the fermionic probe approximation – we can in fact construct a rather complete understanding of these "holographic Fermi surfaces" in the language of the boundary field theory.

Consider the form of the fermion propagator as follows from the matching procedure in the Fermi-surface regime, Eq. (9.64). This has an unambiguous interpretation in the field theory: it describes a system of free fermions forming a Fermi gas (the $\omega = v_F(k - k_F)$ poles) coupled to a second system dominating the deep infrared characterised by the RN-metal local quantum criticality (the $\mathcal{G}_k(\omega) \sim \omega^{2\nu_k}$ propagator). The form of the propagator implies that the interaction between these two subsystems is governed by the second-order perturbation-theory diagram indicated in Fig. 9.5, "naively" re-summed in a Dyson series [364]. The "un-particles" of the AdS$_2$ metal just act as a simple "heat bath" damping the free fermions. One notices that this is precisely coincident with the central assumption of the original marginal-Fermi-liquid phenomenology. In the original marginal-Fermi-liquid context, the puzzle has all along been why there are these two subsystems. But, even if one takes this for granted, the more pressing question is why can one get away with a naive re-summation of second-order perturbation theory. In a single system of interacting electrons on the microscopic scale, diagrammatics would insist that one has to include vertex corrections as well as "boson" self-energies, and it has been argued many times that these also should become singular [365]. An old proverb states that Fermi liquids are like pregnancy, in that it is impossible to be marginally pregnant.

How can we understand the emergence of nearly free fermions and the validity of re-summed second-order perturbation theory in the holographic result? The first key is from section 6.3. We learned that by capping off the deep interior geometry one finds a confining state in the boundary. In the radial direction the bulk turns into an effective box and the standing waves formed in this box dualise in *non-interacting* gauge-singlet mesons in the boundary. They are truly free because the interactions through pion exchange are $1/N$ suppressed. Here the deep potential well manifest in the Schrödinger formulation acts in a very similar way to this confining box. The only difference is that the bulk Dirac fermion has a finite probability of tunnelling into the RN horizon. This dualises in the "local quantum critical" self-energy that provides a finite width to the otherwise free fermions.

Holography teaches us here a valuable lesson regarding the physics of strongly interacting Yang–Mills fields when the density becomes finite. Fermionic probes in the "deconfining" or "fractionalised" vacuum of the local quantum critical AdS$_2$ metal – in chapter 8 we found out that e.g. the ground-state entropy scales with N^2 – can stabilise into long-lived gauge-singlet "mesinos" if these are sufficiently relevant at zero density. One approaches closely a confining state of a nearly free

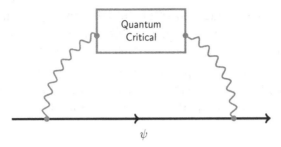

Figure 9.5 A cartoon illustrating the structure of the self-energy diagram behind the "Fermi-surface" holographic phases highlighted in Fig. 9.4.

Fermi gas, but at sufficiently long times these mesinos decay in the continuum of "fractionalised un-particles": the proper excitations of the AdS_2 vacuum. Given that their density of states vanishes at zero energy, the mesinos acquire an infinite lifetime right at the Fermi surface of the mesino gas.

This insight that at the domain wall one essentially has a new long-lived confined state is the content of what has been coined "semi-holography" by Faulkner and Polchinski [366]. If one pictures the fermionic operator as built out of partons, the long-lived excitation is a bound state. For time scales comparable to the lifetime of this new confined bound state the physics will be most appropriately given in terms of the independent dynamics of this new state. This is exactly analogous to Cooper-pair dynamics in a superconductor below the pairing transition. The natural effective theory therefore has a free fermion χ corresponding to the bound state, which interacts linearly with the composite operator of the strongly coupled theory,

$$S = \int dt\, d^d x \left(\chi^\dagger (i\, \partial_t - \varepsilon(k) + \mu)\chi + g\chi^\dagger \Psi + g\Psi^\dagger \chi \right) + S(\Psi). \qquad (9.70)$$

Supplying this with the additional information that the strongly coupled theory has a matrix large-N limit with its implied factorisation, i.e. to leading order in large-N only the two-point function survives, the full correlation function for χ follows trivially from the Dyson summation

$$\langle \chi^\dagger \chi \rangle = \sum_n g^n G^{\text{free}}_{XX} \left(\mathcal{G}_{\Psi\Psi} G^{\text{free}}_{XX} \right)^n. \qquad (9.71)$$

This theory is semi-holographic in that one still needs holography to compute the two-point function of the strongly coupled operator Ψ. Taking the locally quantum critical theory as the strongly coupled theory with $\mathcal{G}_{\Psi\Psi} = c_k e^{i\phi_k} \omega^{2\nu_k}$, we obtain the effective Fermi propagator of a non-Landau Fermi liquid. The confinement and large-N scaling explain fully why the holographic results are of the form they are. Phenomenologically the controlled emergence of non-Fermi liquids is interesting, but it raises the poignant question of the meaning of large-N matrix theory for e.g. the electron systems of the cuprates. Similarly to the criticism of the original marginal Fermi liquid, what is the other subsystem in this context, and, in addition, why can the large-N approximation be used to "harness" the low-order perturbation theory? This is quite obscure, and the similarity between the marginal Fermi liquid of the cuprates and the holographic fermion propagators might well be coincidental.

Summary 13   The physics of holographic (non-)Fermi liquids can be understood in terms of the formation of a confined long-lived bound state coupled to a deconfined locally quantum critical deep infrared. The large-N limit in the local quantum critical theory ensures that it is effectively Gaussian, and this validates a second-order perturbative treatment of the self-energy.

In the case of the simple Reissner–Nordström black hole there is an additional issue associated with the self-consistency within holography. We have noted already that the fermion spectral function in the RN background might display "log-oscillatory" behaviour whenever ν_k becomes imaginary, signalling a lingering instability of the vacuum. In $2 + 1$ dimensions this happens whenever the scaling dimension of the fermionic probe obeys

$$\Delta < \sqrt{2}q + \frac{3}{2}. \tag{9.72}$$

A WKB analysis of the Schrödinger potential, however, reveals that poles occur only for [320]

$$\Delta < 2\frac{q}{\sqrt{3}} + \frac{3}{2}. \tag{9.73}$$

If we consider the theory as a function of both Δ and q, the poles defining the Fermi surface at "large" k_F occur only when there is already a log-oscillatory region present at small momenta: this situation is summarised in Fig. 9.6. This log-oscillatory behaviour signals that we are dealing with an instability. The regime where the Fermi surfaces are formed is thus a false vacuum, and to fully understand the theory the ramifications of this fermionic instability for the bulk physics have to be found out. Note that this instability is not immediately tied to the ground-state entropy of the RN black hole. It is more fundamental, insofar as it persists in the conformal-to-AdS$_2$ top-down solutions in [362, 363]. We will show in chapter 11

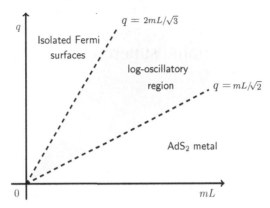

Figure 9.6 The "diagram of holographic finite-density RN theories" deduced from the fermion spectral functions in the probe limit. At large bulk fermion mass m and small fermion charge q one finds the AdS_2 metal behaviour with fermion spectral functions exhibiting algebraic pseudogaps as in Fig. 9.2. At large q and small m the "Fermi-surface" phases with "marginal-Fermi-liquid-like" behaviour of Fig. 9.4 associated with the formation of the quasi-bound states at the geometrical domain wall in the bulk are realised. However, well before these domain-wall bound states start to form, the Breitenlohner–Freedman bound of the fermions in the bulk is violated at small momenta. This gives rise to the "log-oscillatory" behaviour of the fermion propagators, which indicates a possible instability of the RN black hole towards the backreaction of the bulk fermions. The criterion for the occurrence of the domain-wall quasi-bound states is not known analytically: the quoted value $q = 2m/\sqrt{3}$ is obtained from a WKB analysis [320].

that the resolution of this complicated problem is the formation of fermion matter in the bulk.

A different matter is the "AdS_2" metal found for large mass and small q in the bulk (Fig. 9.6), which can be in principle a stable strange metal phase. As we have repeatedly emphasised, its observational fingerprints are the "algebraic pseudogap" fermion spectral functions. To seek any relationship with experiment, the incoherent backgrounds which are routinely observed in the ARPES of cuprates and so forth would be prime suspects. However, as we discussed in chapter 3, these go hand in hand with the "nodal–antinodal dichotomy" in momentum space. Such momentum-space structure cannot possibly be realised in a translationally invariant background, and to arrive at a meaningful comparison the holographic description will need to include strong periodic potentials that break spatial translations. It is a tall order to combine this with the GR in the bulk, and we will discuss the first attempts in this direction in chapter 12.

10

Holographic superconductivity

The AdS/CMT pursuit aimed at addressing the physics of finite-density quantum matter started seriously in 2008 with the discovery of holographic superconductivity, as first suggested by Gubser [347] and subsequently implemented in an explicit minimal bottom-up construction by Hartnoll, Herzog and Horowitz [348, 367]. This triggered a large research effort in the string-theory community. The underlying physics of spontaneous symmetry breaking means that this aspect of AdS/CMT is now quite well understood theoretically, much more so than, for instance, the fermion physics of chapters 9 and 11.

Holographic superconductivity is quite an achievement, however, since it is also far more than the straightforward physics of symmetry breaking. From the condensed matter perspective it should be viewed as the first truly mathematical theory for the mechanism of superconductivity that goes beyond the Bardeen–Cooper–Schrieffer (BCS) theory. As we emphasised in chapter 2, the Cooper mechanism, the central wheel of the BCS theory, critically depends on the normal state being a Fermi liquid. We continued by arguing in chapter 3 that the BCS vacuum structure does not need a Fermi-liquid "mother". Starting from the RVB wave function Ansatz, we illuminated the case in which the BCS vacuum structure should be viewed in full generality as a long-range entangled state formed from a charge-$2e$ Bose condensate living together with a Z_2 spin liquid that is responsible for the "spinon" Bogoliubov excitations.

Holography takes this a step further by demonstrating that a *generalisation of the Cooper mechanism is at work in the holographic strange metals* which were the focus of the previous chapters. As in the Fermi-liquid case, the fermion pair/order parameter channel is singled out as the source of the instability. The phenomenology is very similar, up to the point that one can construct holographic Josephson junctions. One can contemplate s-wave superconductors but also p- and d-wave pairs and so forth. Crucially, a gap opens up at T_c showing a BCS (mean-field) temperature evolution, while at low temperature one might find long-lived Bogoliubov fermions. The differences from BCS are that this gap opens in the incoherent "unparticle" excitations of the strange metal – the sharp Bogoliubov particle poles develop only deep in the superconducting state. Also the rules determining the transition temperature drastically change: a "high" T_c becomes easy to accomplish. In fact, the holographic superconductors mimic the BCS variety so closely that on the basis of standard measurements it is quite hard to tell the difference. The exception is the dynamical pair susceptibility of the order parameter. Inspired by the holographic lessons, we will present in this chapter how this unconventional experiment can act in this regard as the proverbial smoking gun.

Holographic superconductivity is simultaneously a marvel when it is viewed from the string-theory side. It is a direct counterpoint to the long-standing relativist's wisdom, first formulated by Wheeler, that "black holes do not have hair" [368]. This refers to the "no-hair theorems" stating that in an asymptotically flat space-time the Einstein equations allow only black-hole solutions that are uniquely characterised by an overall mass, charge and (angular) momentum but nothing else. It formalises the intuitive idea that black holes swallow everything around them and no probe of their internal structure can possibly exist. Black holes behave in this regard like elementary particles without any structure (the "hair"). The surprise is that the no-hair theorems do not apply to asymptotically AdS space-times. This subtle difference is essential to describe spontaneous symmetry breaking in the boundary. As we will discuss, the dictionary insists that a finite order parameter in the boundary translates into an AdS black hole with hair. More precisely, we will see that the dual of the order parameter corresponds to a Higgs field in the AdS bulk action. Upon lowering the temperature, the AdS–Reissner–Nordström black hole becomes *unstable* and develops an "atmosphere" formed from a finite-amplitude Higgs field. This atmosphere will dualise into the superfluid order parameter of the boundary. In the context of relativity this represents a stunningly novel class of black-hole solutions, discovered by asking a natural question of the boundary field theory.

In section 10.1 we will introduce the basics, construct the "hairy black hole" and discuss its consequences for the boundary in terms of the holographic order. In section 10.2 we will present a selection of highlights from the large literature dealing with phenomenological properties of the holographic superconductors, closely mimicking in many regards the properties of conventional superconductors. Using the "fluid/gravity correspondence" one derives precisely the Tisza–Landau two-component fluid dynamics – this works so well that holography has been used to shed light on the problem of superfluid turbulence. In the microscopic realm, one finds that holographic SNS Josephson junctions behave exactly like junctions involving conventional superconductors and metals. Just as in the Landau–Ginzburg–BCS superconductor, the onset of the order goes hand in hand with the opening of a gap in the optical conductivity, and the highlight is the quite literal BCS behaviour as revealed by the fermion spectral functions. Finally, it turns out that "unconventional" (p- and d-wave) superconductors are also part of the holographic portfolio.

Given that conventional experimental signatures do not sharply discriminate between the BCS and holographic superconductors, where should we look for the differences in the underlying physics? In section 10.3 we zoom in on how the order-parameter propagator (dynamical pair susceptibility) develops in the normal state, as we approach the superconducting instability. This sheds light on the mean-field nature of the holographic mechanism, but it also shows where to look for the signals in experiments that discriminate qualitatively between conventional and "strange-metal" superconductivity. In section 10.4 we take this a step further. We will show how one can tune holographic superconductivity using the parameters of the UV theory (the scaling dimensions of the order parameter, the double-trace deformation). We shall find that one can achieve zero-temperature quantum phase transitions from the superconducting state to the normal metal that are of more than one unconventional ("holographic BKT", "hybridised") kind.

The final section 10.5 is dedicated to a great surprise. Approaching zero temperature, the energetics of the Higgs field becomes a dominating ingredient in the deep interior of the bulk. According to Einstein's equations this should affect the nature of the space-time geometry. Treating its gravitational "backreaction" effects seriously, the geometry in the deep interior of the bulk completely changes. The extremal black hole fully disappears; instead one finds a "Higgs star", a quasi-localised condensate of a self-gravitating scalar field. Upon inspecting its geometry, the IR usually displays a remarkable scaling behaviour that translates into a Lifshitz critical point with an emergent dynamical critical exponent z. Remarkably, this quantum critical phase is stabilised by the order itself.

10.1 The black hole in AdS with scalar hair

Let us delve into the specifics of the mechanism of holographic superconductivity. Elegantly, the symmetry rules underlying the correspondence are so tight that they do not leave room for any ambiguity regarding the nature of the bulk physics. According to the Ginzburg–Landau wisdom, the fundamental physics of superconductivity is the spontaneous symmetry breaking of a $U(1)$ symmetry. The broken state that is established is strictly speaking a superfluid, when the $U(1)$ symmetry is a global one. However, one can use a procedure that is standard in condensed matter physics: the superfluid can be promoted to a superconductor by weakly coupling in a $U(1)$ gauge field in the boundary afterwards. We thus consider a boundary theory with a $U(1)$ conserved current, and we now seek to break this global symmetry.

The dictionary, however, insists that its current is dual to a $U(1)$ gauge field in the bulk. The gauge symmetry in the bulk therefore has to be "broken" as well. We know how to do so consistently in only one way: in the bulk a Higgs condensate has to form. The dictionary insists that a *superfluid in the boundary* is dual to a form of *superconductivity in the bulk*. To allow this to happen, one has to add a charged complex scalar (relativistic Ginzburg–Landau, Higgs) field Φ to the bulk action. On coupling this field minimally to gravity and the Maxwell gauge fields, one has the AdS bulk action

$$S = \int d^{d+2}x \sqrt{-g} \left[\frac{1}{16\pi G} \left(R + \frac{d(d+1)}{L^2} \right) - \frac{1}{4g_F^2} F^{\mu\nu} F_{\mu\nu} \right.$$
$$\left. - |\partial_\mu \Phi - iq A_\mu \Phi|^2 - V(\Phi) \right], \tag{10.1}$$

where q is the charge of the scalar field. The only quantity regarding which one needs more information is the potential of the scalar field $V(\Phi)$. In the following we will consider $d = 2$, corresponding to a $(3+1)$-dimensional AdS description of the $(2+1)$-dimensional boundary. In the Higgsed vacuum this scalar field develops a profile and the translation to the boundary is immediate. It has to be the dual to the order parameter whose expectation value parametrises the broken global symmetry in the boundary. The leading and sub-leading components of the near-boundary asymptote of Φ are as usual associated with the source and the expectation value of the response of this scalar operator. We have arrived at a key point: spontaneous symmetry breaking in the boundary corresponds to a vacuum expectation value (VEV) of an operator that stays non-zero even in the absence of an external source (which would break the symmetry explicitly). This requires a classical scalar field configuration in the AdS bulk with the property that near the boundary its leading piece vanishes (the source of the explicit symmetry breaking) while its sub-leading piece (the response/VEV) stays finite. Thanks to the gravitational potential well provided by the asymptotically AdS solution this sub-leading solution is normalisable and therefore corresponds to an allowed finite-total-energy contribution. This signals the discovery of the hairy black hole: a solution with a scalar hair profile in the bulk that exhibits precisely this normalisable near-boundary asymptote.

Given the surprising nature of a solution where matter does not automatically fall into the black hole but can remain stable, the obvious question is what physics allows this solution to exist. The essence of the mechanism turns out to be already contained in the "minimal holographic superconductor" where one ignores any self-interactions of the Higgs scalar, and the potential is just quadratic

$$V(\Phi) = m^2 |\Phi|^2. \tag{10.2}$$

This truncation is physically justified for temperatures close to the transition temperature. There the amplitude of the Higgs field will be small and the quadratic action should suffice. Let us now consider what happens to this massive charged scalar field in the presence of the RN black hole. To highlight the physics let us consider the extremal case. The crucial trigger will be the non-vanishing electrostatic potential in this background. The fluctuations of the charged scalar field around a general Einstein–Maxwell background are governed by the equation

$$\frac{1}{\sqrt{-g}} \, \partial_\mu \left[\sqrt{-g} \big(g^{\mu\nu} \, \partial_\nu \phi - iq A_\nu \, g^{\mu\nu} \phi \big) \right] - iq \frac{1}{\sqrt{-g}} g^{\mu\nu} A_\nu \partial_\mu (\sqrt{-g} \phi)$$
$$- m^2 \phi - g^{\mu\nu} q^2 A_\mu A_\nu \phi = 0. \quad (10.3)$$

If just an electrostatic potential $A_t \neq 0$ is present the last two terms become

$$- m^2 \phi + |g^{tt}| q^2 A_t A_t \phi \quad (10.4)$$

due to the opposite sign of the time–time component of the metric. This indicates that the electrostatic potential acts as a *negative* mass squared for the charged scalar field. If this term becomes large enough it can trigger an instability. In chapter 5 we learned, however, that an instability sets in not when the effective mass squared changes sign, but at a somewhat negative mass squared where the Breitenlohner–Freedman (BF) bound is violated. Recall that we can think of the RN black-hole geometry as characterised by a domain wall between pure AdS$_4$ in the UV and the near-horizon AdS$_2 \times \mathbb{R}^2$ solution in the IR. In the pure AdS$_4$ UV the additional contribution $g^{tt} A_t A_t \sim \mu^2/r^2$ is sub-leading to the explicit mass and one has the standard BF bound

$$(mL)^2 \geq -\frac{(d+1)^2}{4} = -\frac{9}{4} \quad \text{for } d = 2. \quad (10.5)$$

When this bound is obeyed the vacuum is stable. However, in the AdS$_2 \times \mathbb{R}^2$ IR we read off from Eq. (8.20) that the combination $g^{tt} A_t A_t$ is no longer sub-leading, but rather is constant. In addition, in that region the dimensions of the AdS space have also changed. On correcting for this, one finds a second stability bound applicable to the extremal RN deep IR geometry,

$$(mL)^2 - \frac{q^2 g_F^2 L^2}{8\pi G} \geq -\frac{3}{2}. \quad (10.6)$$

This bound is more stringent than the bound associated with the zero-density UV. Therefore, a window exists as a function of masses and/or charges,

$$-\frac{(d+1)^2}{4} \leq (mL)^2 \leq -\frac{3}{2} + \frac{q^2 g_F^2 L^2}{8\pi G}, \quad (10.7)$$

where a charged scalar field that is stable in the UV becomes *unstable* in the IR of an RN black hole.

What happens when this instability sets in? We have to rely on the numerical procedure explained in box 10.1 to address this quantitatively. The qualitative physics, however, is easy to explain. From the probe calculations in the previous chapter, chapter 9, we know that the violation of the IR BF bound corresponds to the onset of spontaneous pair production. For scalar fields these quanta can condense in a macroscopic (classical) condensate. This capacity is reflected in the properties of the retarded probe Green function of the scalar field. Violating the IR BF bound now corresponds literally to its pole moving into the upper half plane [320]. Near the charged black-hole horizon the relative strength of the electric field is such that the local vacuum starts to discharge: in the black-hole literature this phenomenon is known as super-radiance. The negatively charged quanta will be doubly attracted by the charged black hole: they will fall into it and simultaneously neutralise the black-hole charge. The positively charged quanta in turn will tend to escape to infinity. At this point the unique aspect of anti-de Sitter space comes into play, namely that it allows the existence of black-hole hair. The AdS background provides a gravitational potential barrier near infinity that can balance the repulsive electromagnetic force. This has the consequence that a charged scalar atmosphere starts to develop around the black hole. At each stage some charge of the black hole is carried to this atmosphere, weakening the electric field near the horizon, and this process continues until the production of quanta ceases. This understanding also explains why at higher temperatures the self-interaction of the Higgs field can be ignored: the black-hole discharging effect is in this regime quantitatively much more important than the self-interactions. When the size of the order parameter (the energy density in the scalar atmosphere) becomes comparable to the temperature, this is no longer the case and, especially at zero temperature, self-interactions become very important, as we will find out in section 10.5.

The exact final equilibrium configuration with both a scalar hair profile and a black hole can be numerically determined [348], and a typical outcome is illustrated in Fig. 10.1. Confirming the qualitative insights of the previous paragraph and consistent with the interpretation of the radial direction as representing the energy scale, most of the scalar field energy is concentrated in the deep interior. It is maximal at the horizon, and it monotonically decreases towards the boundary as the boundary system increasingly forgets the presence of a broken symmetry in the ground state when it is probed at higher energies. Mathematically this decay is precisely governed by the sub-leading normalisable solution, and this allows us to use the dictionary to read off the spontaneous symmetry-breaking VEV of the order parameter dual to the scalar field. The "lump of Higgs field" clearly fills all of space, and we will find out that at zero temperature it will replace the black

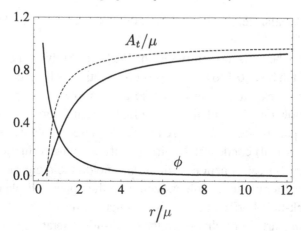

Figure 10.1 The profiles of the gauge field (A_t/μ) and the scalar hair ϕ along the radial direction for $T/T_c = 0.324$ in a holographic superconducting state (solid lines). The horizon is at $r/\mu = 0$ on the left. Here $m^2L^2 = -2$, $q = 3$ and we focus on the case of standard quantisation (see box 10.1). The gauge field in the RN metal state is shown by a dashed line at the same temperature $T/\mu = 0.051$.

hole completely. Although this "lump" has no edge, it is concentrated in the deep interior and, because of its similarity with the even more concentrated solutions we will encounter in chapter 11, one might want to call this object a "Higgs star".

Summary 14 Holographic superconductivity is gravitationally encoded in the form of a "discharging" instability of the AdS–Reissner–Nordström black hole. For temperatures below a critical point, the black hole develops an "atmosphere" in the form of a Higgs field that acquires a finite amplitude with a maximum near the black-hole horizon ("scalar hair"). The non-vanishing Higgs field in the bulk is dual to the non-vanishing order parameter in the boundary field theory.

To remove any doubt regarding the correctness of the black-hole hair encoding of the spontaneous symmetry-broken state in the boundary field theory, let us turn to the analysis of its thermodynamics. The free energy can be straightforwardly computed numerically using the machinery explained in chapter 6, and typical results for the finite-temperature physics are shown in Fig. 10.2. The holographic superconductor has indeed a lower free energy than the RN metal below T_c, demonstrating its thermodynamical stability. One also infers that the system exhibits a second-order thermal phase transition, as expected for a simple symmetry breaking.

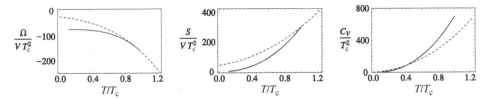

Figure 10.2 The thermodynamics of the holographic superconductor. The Gibbs energy (Ω), entropy (S) and specific heat (C_V) as functions of the temperature T (all in units of T_c), for the normal state (dashed line) compared with the holographic superconductor (solid line), as encoded by a bulk RN and "hairy" black hole, respectively. These results are for standard quantisation with $q = 3$, $m^2 L^2 = -2$ and $\mu = 1$. One infers immediately that the thermodynamics behaves as a standard second-order phase transition to a low-temperature symmetry-broken phase.

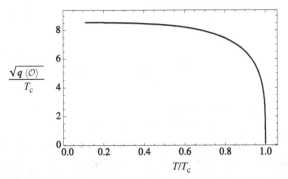

Figure 10.3 A typical outcome for the expectation value of the order parameter of the holographic superconductor $\langle \mathcal{O} \rangle$ as a function of temperature for $q = 3$, $m^2 L^2 = -2$ in the grand canonical ensemble (in standard quantisation; see box 10.1). Near the critical temperature, $\langle \mathcal{O} \rangle$ behaves as $\langle \mathcal{O} \rangle \propto (T_c - T)^{1/2}$, indicating that the transition is precisely of the Landau mean-field kind, in fact occurring independently of the dimensionality of the boundary.

On closer inspection, it turns out to correspond with a precise mean-field transition. This can be seen from the thermal evolution of the superconducting order parameter. At least close to T_c the VEV of the order parameter $\langle \mathcal{O} \rangle$ exhibits a precise $(T_c - T)^{1/2}$ temperature dependence, characteristic of a Landau mean-field second-order phase transition (Fig. 10.3). The surprising aspect is that it displays this behaviour in all dimensions $d \geq 2$. For the breaking of a $U(1)$ symmetry one would expect mean-field behaviour only at and above the upper critical space dimension $d_{uc} = 4$, while below d_{uc} one expects the anomalous scaling dimensions of the XY universality class associated with the strongly interacting thermal critical state. In fact, in $d = 2$ space dimensions true long-range order due to a continuous symmetry breaking at finite temperature cannot even exist according to the Coleman–Mermin–Wagner theorem.

There is, however, a good reason why the holographic superconductor shows such mean-field behaviour in all dimensions. In the 1970s it was discovered that in matrix theories the thermal fluctuations in the global symmetry sector are suppressed in the large-N limit. This has much the same effect as large dimensionality in vector theories. Notice that we encounter here a concrete circumstance where an artefact of the matrix large-N limit is manifest. This shortcoming is, however, not too serious, and in principle it is fairly well understood how to restore the order-parameter fluctuations. The case in point is the demonstration by Anninos, Hartnoll and Iqbal [369] that by semi-classically re-quantising the bulk hair one picks up the order-$1/N$ thermal fluctuations. These have also been shown to turn the long-range order of the large-N limit into the algebraic long-range order expected for finite-temperature spontaneous symmetry breaking in $d = 2$ dimensions.

Summary 15   Second-order phase transitions in holography universally display mean-field behaviour when computed with classical gravity. This is caused by the implicit matrix large-N limit in the boundary field theory. For the same large-N reason, the system can evade the implications of the Coleman–Mermin–Wagner theorem and exhibit spontaneous symmetry breaking of global symmetries at finite temperatures in $d \leq 2$ dimensions.

In real-life superconductors in condensed matter physics the entities that condense are pairs of electrons. We learned from the two previous chapters that the RN black hole describes a state that concerns fermions as well, but do these fermionic degrees of freedom play any role in holographic superconductivity? In the bottom-up approach one has no clue. In bottom-up phenomenological contructions, the explicit form of the UV theory is simply unknown. To answer this question one has to rely on top-down constructions where this information is available. Most top-down constructions, such as the ones that will be discussed in chapter 13, include a global $U(1)$ "flavour" symmetry in the boundary, and have scalar fields that are charged under this symmetry in the bulk. Particularly illuminating are so-called Dp/Dq-brane constructions where the explicit form of the symmetry-breaking field has been identified as a pair of fundamental fermions ψ in the UV theory [370],

$$\mathcal{O} = \mathrm{Tr}[\psi\psi] + \cdots . \tag{10.8}$$

These fundamental fermions form a gauge-invariant Cooper pair (the trace is over the colour degrees of freedom). This demonstrates that the holographic order parameter can quite literally represent the VEV of a fermion-pair operator. As we have stressed repeatedly, the normal state is not a Fermi liquid, and the pair-binding mechanism is therefore different from the Cooper mechanism, which depends critically on the presence of a sharp Fermi surface. Holographic superconductivity appears to signal the existence of a generalised form of the Cooper mechanism at work in the quantum critical strange metals. As a caveat, the "$\cdots$" in this equation signals that in a strongly interacting IR there is no a-priori reason to expect that only pair correlations exist, and in principle one has to include also multi-point correlators in Eq. (10.8) with the same quantum numbers. However, these all need to have the same symmetry as the lowest-order pairs. This is just the appropriate way to think about "Cooper pairs" in such a strongly interacting quantum soup.

Box 10.1   Computing the minimal holographic s-wave superconductor

To be concrete, we consider the holographic superconductor in $2 + 1$ dimensions corresponding to a $(3+1)$-dimensional gravity with the action (10.1) and the potential (10.2). The equations of motion are

$$R_{\mu\nu} - \frac{1}{2}g_{\mu\nu}R - \frac{3}{L^2}g_{\mu\nu} = T_{\mu\nu}^{\text{Maxwell}} + T_{\mu\nu}^{\text{Charged Scalar}},$$

$$(\nabla^\nu - iqA^\nu)(\nabla_\nu - iqA_\nu)\Phi - m^2\Phi = 0, \qquad (10.9)$$

$$\nabla_\mu F^{\mu\nu} = iqg_F^2[\Phi^*(\partial^\nu - iqA^\nu)\Phi - \Phi(\partial^\nu + iqA^\nu)\Phi^*],$$

where

$$T_{\mu\nu}^{\text{Maxwell}} = \frac{\kappa^2}{g_F^2}\left(F_{\mu\rho}F_\nu{}^\rho - \frac{1}{4}F^2 g_{\mu\nu}\right),$$

$$T_{\mu\nu}^{\text{Charged Scalar}}$$

$$= \kappa^2\bigg[(\partial_\mu + iqA_\mu)\Phi^*(\partial_\nu - iqA_\nu)\Phi + (\partial_\mu - iqA_\mu)\Phi(\partial_\nu + iqA_\nu)\Phi^*$$

$$- g_{\mu\nu}\big(|(\partial_\alpha - iqA_\alpha)\Phi|^2 + m^2|\Phi|^2\big)\bigg].$$

Probe limit

At the simplest level of approximation we will ignore the backreaction on the metric. In the limit $\kappa L/qg_F \ll 1$ the matter terms on the right-hand side of Einstein's equation Eq. (10.9) become negligible. This means that the Maxwell–scalar sector decouples from the gravity, and the gravity background therefore is the regular AdS–Schwarzschild solution

Box 10.1   (Continued)

$$ds^2 = \frac{r^2}{L^2}\left(-f(r)dt^2 + dx^2 + dy^2\right) + \frac{dr^2}{f(r)r^2},$$

$$f(r) = 1 - \frac{M}{r^3},$$
(10.10)

with temperature $T = 3M^{1/3}/4\pi L^{4/3}$. We will set $L = 1$ as usual. We next take the Ansatz that the solution is homogeneous in the field-theory directions parallel to the boundary and only has a radial evolution,

$$A_t = h(r), \quad A_i = 0, \ \Phi = \phi(r).$$
(10.11)

Moreover, from the r component of the equation of motion for the gauge field, we know that the phase of the complex scalar field should be constant. Here we choose the phase to be zero, i.e. $\phi(r)$ real without loss of generality. On substituting this into the equations of motion (10.9) we obtain

$$\phi'' + \left(\frac{f'}{f} + \frac{4}{r}\right)\phi' + \frac{h^2}{r^4 f^2}\phi - \frac{m^2}{r^2 f}\phi = 0,$$

$$h'' + \frac{2}{r}h' - \frac{2\phi^2}{fr^2}h = 0.$$
(10.12)

The holographic superconductor is the non-trivial solution to these equations with the appropriate boundary conditions. It is straightforward to see that near the AdS$_4$ boundary the two equations decouple and one has the known asymptotics

$$h(r) = \mu - \frac{\rho}{r} + \cdots,$$

$$\phi(r) = J_{\mathcal{O}} r^{\Delta-3} + \langle\mathcal{O}\rangle r^{-\Delta},$$
(10.13)

with $\Delta = \frac{3}{2} + \sqrt{\frac{9}{4} + m^2 L^2}$. For the holographic superconductor encoding spontaneous symmetry breaking we seek normalisable solutions where we have a finite VEV $\langle\mathcal{O}\rangle \neq 0$ without a source $J_{\mathcal{O}} = 0$. Near the horizon one readily deduces that $h(r_0) = 0$ and $\phi' = (m^2/f'(r_0))\phi$.

The resulting system of equations can now be solved using a numerical shooting method. For high temperatures the only solution is $\phi = 0$ and $h = \mu - \rho/r$. Below a critical temperature T_c, on the other hand, one finds non-trivial hairy-black-hole solutions with $\phi \neq 0$. Usually there are multiple solutions characterised by the same asymptotic behaviour, but with a different number of nodes in the scalar profile. The solution with a monotonic scalar profile is the correct one minimising the free energy.

Including backreaction

The scalar hair that encodes the non-vanishing order parameter in the boundary carries gradient energy. For generic values for which $\kappa L \sim q g_{\mathrm{F}}$, this energy is no longer negligible and will backreact on the geometry through Einstein's equations. Here we focus on the finite-temperature backreacted solution; the zero-temperature solution will be discussed in section 10.5.2. The hairy-black-hole Ansatz for the metric field, gauge field and scalar field is

$$ds^2 = -f(r)e^{-\chi(r)}dt^2 + \frac{dr^2}{f(r)} + r^2(dx^2 + dy^2),$$
$$A_t = h(r), \tag{10.14}$$
$$\Phi = \phi(r).$$

One can substitute this Ansatz into the equations of motion (10.9) and obtain two second-order equations for $h(r), \phi(r)$ and two first-order equations for $f(r), \chi(r)$.

One has now to look for solutions with an asymptotic AdS$_4$ geometry at the boundary. The system becomes more involved than the probe limit to be solved. Let us present here finally a sketch of the strategy used in the numerics. The full set of equations can be solved by integrating the fields numerically from the horizon r_+ as determined by $f(r_+) = 0$ up to infinity (the conformal boundary). In total there are four physical fields that need to be solved: $f(r), \chi(r), h(r), \psi(r)$. We demand that $h(r)$ vanishes at the horizon ($h(r_+) = 0$) in order for the gauge one-form to be well defined at the horizon. We have therefore only four independent boundary values,

$$r_+, \ \chi(r_+), \ h'(r_+), \ \phi(r_+), \tag{10.15}$$

since $f(r_+) = h(r_+) = 0$ and $\phi'(r_+)$ can be determined from these four parameters by using Einstein's equation. We can get solutions of this system by integrating the equations of motion, given the initial values at the horizon of the four parameters we just specified.

In the numerical calculations one can exploit the three scaling symmetries of this system to simplify the calculations,

$$\mathrm{I}: \ r \to br, t \to bt, L \to bL, q \to q/b;$$
$$\mathrm{II}: \ r \to br, (t, x, y) \to (t, x, y)/b, f \to b^2 f, h \to bh;$$
$$\mathrm{III}: \ e^\chi \to b^2 e^\chi, t \to bt, h \to h/b;$$

having as consequence that one can set $L = 1$ (from I), $r_+ = 1$ (from II) and $\chi(\infty) = 0$ (from III). Near the AdS$_4$ boundary, the metric fields behave as

Box 10.1 **(Continued)**

$$f(r) = r^2 + \cdots - \frac{M}{r} + \cdots,$$

$$e^{-\chi} f = r^2 - \frac{M}{2r} + \cdots \tag{10.16}$$

and the matter fields as

$$h = \mu - \frac{\rho}{r}, \quad \psi = \frac{\mathcal{O}_1}{r^{\Delta_-}} + \frac{\mathcal{O}_2}{r^{\Delta_+}}, \tag{10.17}$$

where $\Delta_\pm$ corresponds to the conformal dimension of the dual operator. By imposing the normalisable boundary condition one can identify two regimes. For $-9/4 < m^2 L^2 < -5/4$, one can choose either $\mathcal{O}_1 = 0$ (standard quantisation) or $\mathcal{O}_2 = 0$ (alternative quantisation), while for $-5/4 \leq m^2 L^2$ we can choose only the standard quantisation condition.

Since it follows from scaling property III that $\chi(r_+)$ can be fixed by putting $\chi(\infty) = 0$, there remain only two independent parameters $h'(r_+)$, $\phi(r_+)$ at the horizon that can be used as initial values. At the boundary there are in total five parameters encoding for the properties of the dual field theory: $\mu, \rho, \mathcal{O}_1, \mathcal{O}_2$ and M. The result is a map for the integration from the horizon to infinity:

$$(h'(r_+), \phi(r_+)) \mapsto (\mu, \rho, \mathcal{O}_1, \mathcal{O}_2, M). \tag{10.18}$$

With the constraints coming from the normalisable boundary condition for the order parameter, the map reduces to a one-parameter family of solutions for each choice of m^2 and q. We can think of this parameter as being the temperature of the theory at a fixed charge density (canonical ensemble) or fixed chemical potential (grand canonical ensemble).

The resulting system of equations can now be solved using the numerical shooting method. Similarly to the probe limit case, the solution with a monotonic scalar profile is the correct one minimising the free energy, and a typical example is shown in Fig. 10.1.

10.2 The phenomenology of holographic superconductivity

The discovery of holographic superconductivity triggered a concerted effort aimed at studying the physical properties of the holographic superconductors. It is by now very well established that the holographic superconducting state is a close sibling of the standard BCS superconductor. The phenomenology of the holographic superconductor as established through the observables measured in standard laboratory experiments is to quite a degree indistinguishable from run-of-the-mill superconductivity. This literature is too large to review here in any detail. We present instead a selection of highlights, emphasising the ability of the hairy-black-hole gravity to reproduce a strikingly normal boundary superconductor. There are some subtle differences: these will be highlighted in the next sections.

In the macroscopic realms of the superfluid hydrodynamics the physics is completely governed by the principles of symmetry breaking and it is therefore not surprising that the holographic version gets it all right. We will first discuss this highlight, then "ascend" increasingly into the microscopic realms where one finds sensitivity to the specialities of the BCS theory. Particularly with regard to these spectroscopic responses, the similar appearance of the gross experimental features in holographic and BCS superconductivity is striking.

10.2.1 Superfluid hydrodynamics and turbulence

A superfluid such as that realised in ultracold helium does flow under the influence of external forces. Even at zero temperature it is subjected to the basic symmetry principles underpinning hydrodynamics, but the actual "quantum hydrodynamics" is quite different from the hydrodynamics of a classical fluid. The key reason is the additional role of a spontaneously broken global $U(1)$. The associated Goldstone boson is protected and always survives as a long-range dynamical excitation. This phase mode or "superfluid second sound" is responsible for the super-current: $\Psi(\vec{x}) = |\Psi| \exp(i\phi(\vec{x}))$, and when $|\Psi| \neq 0$ the super-current $J_\mu(\vec{x}) \sim \partial_\mu \phi(\vec{x})$ flows forever because of the Goldstone protection. Since the super-current is proportional to the gradient of a scalar function this corresponds to a circulation-free potential flow, which is therefore also dissipationless. The vorticity is massive and quantised. The effect is that when a vessel with superfluid is rotated the fluid stays at rest until a critical angular frequency is reached, whereupon vortices start to precipitate, forming a regular lattice.

As we learned in chapter 7, hydrodynamics is a stronghold of holography, and this wisdom surely extends to holographic "superconductivity" (in fact, superfluidity). This is just controlled by the generic consequences of spontaneous symmetry breaking on the macroscopic scale. Given the tight symmetry relations between bulk and boundary, one anticipates that the quantisation of the circulation in the boundary will be impeccably reproduced in the bulk since the Higgs field in the bulk shares the periodicity of the superfluid order parameter in the boundary. The difference is that the bulk is gauged and the holographic dual of the superfluid vortex in the boundary is therefore like an Abrikosov fluxoid in the bulk. For ease of visualisation, let us consider a static "particle" (or "pancake") vortex

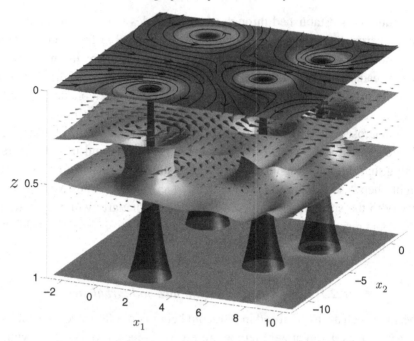

Figure 10.4 Holographic description of a two-dimensional superfluid with vortices [375]. The vertical axis labelled by $z = r_0/r$ corresponds to the radial direction of AdS$_4$, with the AdS boundary at the top ($z = 0$) and the horizon ($z = 1$) at the bottom. The lighter-grey surface in the middle is a surface of constant bulk charge density, with the region between the two slices defining a "slab" of condensate where most of the bulk charges reside. The vector field of energy flux (its length represents its amplitude) vanishes very quickly below the slab. The vortices, with energy flux circulating around them, punch holes through this screening slab, providing avenues for excitations to fall into the black hole. The surface at $z = 0$ also shows the condensate on the boundary (with the darkest colour representing zero condensate), with flow lines of the superfluid velocity superposed. The flux-tubes show a surface of constant $|\Phi|^4/z^4$, which coincides with the boundary condensate while it also shows the flux of energy through the horizon. Figure source [375]. (Reprinted with permission from the AAAS.)

in the $d = (2 + 1)$-dimensional boundary. This just extends in the bulk as an Abrikosov flux line along the radial direction terminating in the deep infrared, disappearing at the black-hole horizon when the temperature is finite. Numerical solutions for such explicit holographic vortices can be found in the "early" literature [371, 372, 373, 374]. In Fig. 10.4 we show a particularly appealing graphical representation of what such a holographic vortex lattice looks like.

There is, however, more to the hydrodynamics of superfluids. At any finite temperature, some of the "bosons" are "leaving the condensate" because of thermal excitations, forming a classical dilute fluid coexisting with the "rigid" superfluid.

One must now account for the velocity both of the superfluid and of the normal fluid in the constitutive equations, and the generalised Navier–Stokes theory for the non-relativistic superfluid is known as the Tisza–Landau two-component fluid [376, 377]. This is described in terms of a normal fluid with density n_n and flow velocity u_μ, as well as a superfluid with density n_s and flow velocity v_μ, which is the unit vector of $-\partial_\mu \phi$, with a total density $n = n_n + n_s$.

As for the normal fluid of chapter 7, the hydrodynamical equations for the relativistic two-component fluid are more elegant. These were derived first in the early 1980s by Israel, Khalatnikov and Lebedev [378, 379, 380] and very recently again [283, 381]. The conservation laws are, as usual,

$$\partial_\mu T^{\mu\nu} = 0, \quad \partial_\mu J^\mu = 0. \tag{10.19}$$

But now the "rigid" response of the superfluid is governed by the Josephson equation: $u^\mu \partial_\mu \phi + \mu = 0$ coming from the dynamics of the Goldstone mode. The constitutive equation at leading order becomes

$$T_{\mu\nu} = (\epsilon + P)u_\mu u_\nu + P\eta_{\mu\nu} + \mu n_s v_\mu v_\nu, \quad J_\mu = n_n u_\mu + n_s v_\mu, \tag{10.20}$$

with the constraint imposed by the Josephson equation $u^\mu v_\mu = -1$.

This relativistic two-fluid hydrodynamics can be reconstructed and generalised to higher gradients using the "fluid/gravity correspondence" explained for the classical fluid in box 7.2 in chapter 7 [282, 283, 284, 382], except that one considers now as background the hairy black hole including its scalar field atmosphere. By boosting the static solution to include leading-order gradients one obtains, for the metric, gauge and scalar field expressions of the form

$$ds^2 = -2hu_\mu \, dx^\mu \, dr - r^2 f u_\mu u_\nu \, dx^\mu \, dx^\nu + r^2(\eta_{\mu\nu} + u_\mu u_\nu)dx^\mu \, dx^\nu$$
$$+ r^2(2Cu_{(\mu}n_{\nu)} - Bn_\mu n_\nu)dx^\mu \, dx^\nu + \frac{2Ch}{f}n_\mu \, dx^\mu \, dr,$$
$$A = (-\phi u_\mu + \varphi n_\mu)dx^\mu - \frac{\phi h}{r^2 f} \, dr, \tag{10.21}$$
$$\Phi = \xi e^{iq\alpha},$$

where $f, h, B, C, \phi, \varphi, \xi, \alpha$ are functions of r, u_μ is the boosting parameter with $u^\mu u_\mu = -1$ and n_μ is a normal vector of u_μ with $n^\mu u_\mu = 0$. One inserts this Ansatz into the equations of motion of the Einstein–Maxwell–Higgs system Eq. (10.1) to solve the system order by order in the gradients. The solution is finally dualised to the boundary in the same way as in box 7.2. To leading order one recovers precisely the form Eq. (10.20) for the relativistic two-fluid hydrodynamics [382]. When it comes to bearing consequences for empirical physics, this superfluid hydrodynamics appears at this moment in time to be near the top of the list of holographic accomplishments. The problem of the precise nature of turbulence in normal fluids is among the great unsolved problems of classical physics. But the nature of *superfluid turbulence* is even less understood. This came into focus in the helium community about ten years ago, and has been revived by the cold-atom community studying atomic Bose–Einstein clouds. A deep insight into the nature of classical turbulence is provided by the *Kolmogorov scaling law*. One stirs a fluid hard with the effect that it gets littered with eddies. In the classical fluid this injected energy is passed from one scale to another without substantial loss. This energy can now cascade upwards in the form of smaller and smaller vortices or downwards, whereby smaller vortices turn into larger ones with slower currents. Kolmogorov conjectured that, when the energy distribution in the fluid is self-similar, it should, on the basis of dimensional analysis, scale with the wave vector with a 5/3 exponent but a priori it is unclear whether the energy will be transported to smaller or larger distances (the "direct" and "inverse" cascades, respectively). This is particularly well understood for classical fluids in two dimensions, where one can exploit the conservation of the "enstrophy" (the square of the vorticity) to demonstrate that an inverse cascade is realised.

Turning now to the superfluid, by "stirring this hard" one can now create in this case a highly non-equilibrium distribution of quantised vortices. It turns out that the arguments for the Kolmogorov scaling of the classical fluid do not apply straightforwardly to the quantised vorticity. There was an attempt to address this problem using conventional (Gross–Pitaevskii) means, but this led to conflicting claims with regard to whether the cascade is direct or inverse. The recent striking result is that the solution to this problem is offered by holography.

We briefly discussed the progress made in addressing the turbulent regime of classical hydrodynamics using holography in section 7.2. In this strong non-equilibrium regime of the boundary, one has to deal with serious non-stationary gravity in the bulk but this can now be addressed using numerical general relativity in the bulk. In a landmark effort, Adams, Chesler and Liu [288, 375] studied the turbulent superfluid in $d = 2 + 1$ dimensions by computing the time evolutions in the bulk using numerical GR [23], departing from highly non-equilibrium configurations of vortices in the boundary. The conclusion is that this superfluid turbulence exhibits Kolmogorov scaling, which is, however, associated with the direct cascade. This highlights the difference from the classical fluid. It appears that the improvement compared with the Gross–Pitaevskii simulations is in the proper treatment of the dissipation caused by the moving vortices. This is hard to include correctly in the "direct" simulations, while it is automatically included in holography, where this dissipation is described in terms of the flux moving over the black-hole horizon.

Summary 16 The circulation is quantised in holographic superconductors in the same way as in conventional superconductors. Using the fluid/gravity correspondence, the Tisza–Landau two-component fluid dynamics can be directly derived from the "hairy-black-hole" bulk. The flow properties of the superfluid are so well described by holography that it sheds light on fundamental questions regarding the nature of superfluid turbulence.

10.2.2 Spectroscopy, tunnelling and holographic superconductivity

The macroscopic properties such as the superfluid hydrodynamics are very generic, and it is not surprising that a holographic superconductor in this regard behaves like those in the laboratories. It is less evident that the holographic variety will behave in a way similar specifically to a BCS superconductor when its *microscopic* physics is probed by spectroscopic means. As we will discuss here, although there are differences, some gross behaviours appear to be surprisingly similar. We will here present some of these results, in order of increasing "sensitivity" to the particulars of the microscopic physics.

The holographic Josephson junctions

The Josephson effects are a first iconic example involving such microscopic aspects. A Josephson junction corresponds to two superconductors that are separated by a thin barrier that is either metallic or insulating, forming an SNS or SIS junction, respectively. Horowitz, Santos and Way [383] devised an SNS junction holographically by constructing a black hole where the electrical flux emanating from the horizon, encoding the chemical potential in the boundary, makes a brief drop in a middle barrier region. The system is then tuned such that in this region the BF bound is not violated. This corresponds to two holographic superconductors in the boundary separated by a barrier where the RN metal is still stable due to the smaller chemical potential (see [384] for an alternative construction). On computing the current through the barrier induced by a phase difference $\delta\phi$ between the two superconductors, the standard DC Josephson relation $J = J_{\max}\sin(\delta\phi)$ is found to be satisfied. In a metallic weak link the Josephson coupling originates (at least at higher temperatures) from the proximity effect, the fact that superconductivity is induced in the barrier metal sensing the gap function of the nearby superconductors. It follows for such a proximity-effect-mediated Josephson coupling that $J_{\max}/T_c^2 \sim \exp(-l/\xi)$, where l is the width of the junction and ξ is the coherence length of the superconductor. Similarly, the order parameter in the middle of the barrier as induced by the proximity effect should behave as $\langle \mathcal{O} \rangle_{J=0} / T_c^2 \sim \exp{(-l/(2\xi))}$. Horowitz, Santos and Way studied these quantities by varying the width of the barrier and found that this applies also to their holographic SNS junction. The conclusion is that the proximity coupling between two holographic superconductors separated by a barrier formed from the RN strange metal behaves identically to the coupling of standard BCS superconductors through a Fermi liquid.

Summary 17 Holographic SNS junctions that behave in the same way as conventional superconductors have been constructed. Even the coupling of the superconducting phases via the proximity effect in the metal junction is quantitatively similar.

The optical conductivity: gapping due to order

The next question to ask is what happens to the optical response when superconductivity sets in? We discussed the "normal-state" optical conductivity of the RN metal in detail in section 8.3. Let us again focus on $d = 2 + 1$ dimensions. We found that as a function of decreasing frequency the constant spectral weight of the optical conductivity of the zero-density $(2 + 1)$-dimensional CFT started to deplete at a frequency set by the chemical potential. The f-sum rule states that this missing spectral weight has to go somewhere. Because of the momentum conservation – an important aspect for transport at finite density – it has to accumulate in the delta function ("Drude peak") at zero frequency in the optical conductivity, signalling that the metal is a perfect conductor. When the system turns into a superconductor, not that much more can actually happen: the metal is in some ways already a "super" conductor thanks to momentum conservation. In a conventional superconductor in fact nothing would happen if it were to live in the Galilean continuum. Only when translational symmetry is broken is there spectral weight at finite frequencies in σ (inter-band transitions, broadening of the Drude peak due to quenched disorder). Generically, this spectral weight at energies less than the BCS gap will be transferred to the diamagnetic peak when the superconducting order develops. In the holographic version a similar phenomenon happens. There is now finite frequency-spectral weight in the metal, even in the continuum. This is a "remnant" of the zero-density system, see Eq. (8.45). As we will discuss in more detail in section 10.4, it is easy to construct very "high-T_c" superconductors where the energy scale associated with the superconductivity is of the order of the chemical potential itself. When this is the case, one finds that

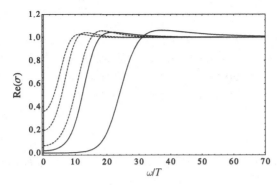

Figure 10.5 Electrical conductivity for holographic superconductors of $\mathcal{O}_2$ with $q = 3, m^2 L^2 = -2$ (solid line) and RN black holes (dashed line) in the grand canonical ensemble ($\mu = 1$) for decreasing temperatures. The lowest of the dashed curves is at $T = T_c$; it precisely coincides with conductivity in the superconducting phase at T infinitesimally close to T_c. There is a delta function at the origin in all cases.

the incoherent spectral weight at energies of order μ is transferred to the diamagnetic peak when the holographic superconducting order develops. In this sense, the optical conductivity of the continuum holographic superconductor mimics the way in which the BCS gap imprints on the optical conductivity of a conventional superconductor. In chapter 12 we will discuss the effects of broken translational symmetry.

The holographic optical conductivity can be straightforwardly computed using the prescription in box 7.3 in chapter 7 but now in the background of the hairy black hole. Typical results are shown in Fig. 10.5, where the same chemical potential is used such that the optical conductivity of the RN-metal "normal state" is the same. One then varies the charge of the scalar field with the effect that T_c varies from $T_c \ll \mu$ to a "strongly coupled" superconductor with a T_c of the order of μ. The conductivity is then computed at a "low" temperature $T \simeq T_c/4$. For the weakly coupled case ($T_c \ll \mu$) there is barely a discernible difference between the metal and the superconductor. However, in the strongly coupled case one finds that at high energies the spectral weight gets further depleted in the presence of the condensate, and is transferred to the diamagnetic peak at $\omega = 0$. However, even at the lowest temperatures there is still some spectral weight at low frequencies. A simple computation based on the matching method reveals that it is not a hard gap, but an algebraic pseudogap, behaving as $\text{Re}[\sigma] \sim \omega^2$ at low frequencies. The reason for this behaviour will be discussed in section 10.5. Surprisingly, we will see that some weight will remain at low frequencies.

Summary 18 The optical conductivity of the holographic superconductor behaves in a way that is quite similar to what is found in a BCS superconductor: a gap opens up in the optical response, and the spectral weight is transferred to the diamagnetic peak at zero frequency.

Photoemission and the Bogoliubov fermions of the holographic superconductors

Optical conductivity is surely not the preferred experimental observable to find out whether one is precisely dealing with a structure of the excitation spectrum associated with a BCS ground state. Instead, one would like to directly interrogate the fermion spectral function. In chapter 3 we emphasised that the crucial difference between a BCS superconductor and the condensate of charge-$2e$ bosons lies in the fact that the former has a vacuum structure that supports the Bogoliubov fermionic quasiparticle excitations. On the basis of the RVB Ansatz, we showed that, even when the normal state is not a Fermi liquid, Bogoliubov fermions can still arise in the superconducting state. Upon focussing on the fermionic response of the holographic superconductor, a remarkably explicit realisation of such physics is found. One departs from a non-Fermi-liquid normal state devoid of fermionic quasiparticles, but upon entering the superconducting state one encounters a parameter regime where the spectrum reconstructs completely into a response that looks quite conventional, being dominated by very sharp Bogoliubov excitations that closely mimic the BCS results.

The appearance of sharp Bogoliubov excitations deep in the superconducting state is not generic: it requires some fine tuning. In the fermion-probe limit, where the backreaction of the fermions on the geometry is ignored, one has to be in the regime where the "geometric-domain-wall" fermionic bound states discussed in the previous chapter are formed in the normal state. As we emphasised, in this regime fermionic backreaction should actually be included: this is a difficult problem that becomes even harder when the superconductivity too has to be taken into account. Although there is a sense of a Fermi surface in this regime, one can now tune the parameters such that these "quasiparticles" are strongly overdamped at the superconducting transition temperature, due to their rapid decay into the RN quantum critical continuum. Upon entering the superconducting state, these states are rapidly cohering, turning into very sharp quasiparticle peaks at low temperature, and exhibit the typical dispersion relation of Bogoliubov fermions as determined by the presence of the s-wave gap function: see Figs. 10.6 and 10.7 [385, 386].

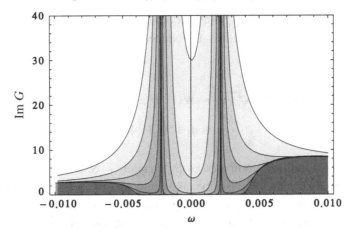

Figure 10.6 The fermionic spectral function $A(\omega, k)$ in the holographic superconductor background as a function of ω for different temperatures. The momentum
k is chosen near to k_{F}. One clearly sees the gap characteristic of the Bogoliubov fermions. The superconducting background is for the order parameter values
$q = 1, m^2 L^2 = -1$. The probe fermion parameters are $q_{\mathrm{f}} = 1/2, m_{\mathrm{f}} = 0$ and
$\eta_5 = 0.025$. Figure source [385].

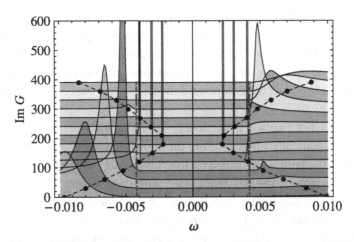

Figure 10.7 The fermionic spectral function $A(\omega, k)$ in the holographic superconductor background as a function of ω for different momenta at a fixed low
temperature $T \ll T_{\mathrm{c}}$. The dotted line traces the location of the peak. Beyond the
dashed line the incoherent part of the spectral density is completely suppressed,
and the lifetime of the quasiparticle is infinite. The superconducting background is
for the order-parameter values $q = 1, m^2 L^2 = -1$. The probe fermion parameters
are $q_{\mathrm{f}} = 1/2, m_{\mathrm{f}} = 0$ and $\eta_5 = -0.025$. Figure source [385].

To exhibit this, an explicit Yukawa coupling between the fermions and the superconducting order parameter is introduced into the bulk gravitational action [385],

$$S_{\text{probe}} = \int d^4 x \sqrt{-g} \left[-i\bar{\Psi}(\Gamma^\mu \mathcal{D}_\mu - m_f)\Psi + \eta \Phi^* \bar{\Psi}_c \Psi \right.$$
$$\left. + \eta_5 \Phi^* \bar{\Psi}_c \Gamma^5 \Psi + \text{c.c.} \right]. \tag{10.22}$$

This is a new ingredient compared with chapter 9. Since these couplings are allowed by symmetry, they have to be taken into account. (This is crucial because the η_5 coupling is responsible for the opening of a gap in the fermion spectral function of the boundary.) In the bulk these couplings act much like the usual Bogoliubov–de Gennes (BdG) couplings, with the scalar field taking the role of the pair potential in this fermion-probe limit. The fermion "hybridises" with its charge-conjugate partner and, since these become degenerate at $\omega = 0$, a gap opens up by level repulsion.

We have already learned to appreciate that the "mesinos" described by bound states in the bulk away from the deep interior are in one-to-one dual correspondence to free fermions in the boundary. This rule is also in effect when the bulk fermions interact with potentials. We will discuss this in detail for the particle–hole channel in section 12.3. Here this wisdom is at work in the BdG particle–particle channel. Accordingly, the Bogoliubov excitations which are formed in the bulk by the action of the Yukawa coupling to the scalar hair dualise in Bogoliubov fermions in the boundary. This is the mechanism behind the BCS appearance of the quasiparticle dispersion in Fig. 10.7.

But this does not explain immediately why the lifetime of these excitations increases so dramatically upon entering the superconducting state. As we learned, the damping in the bulk is described by the "tunnelling" of the domain-wall bound states to the horizon of the RN black hole. The issue is now that this deep interior geometry is strongly altered in the presence of the scalar hair. In fact, as we will discuss in section 10.5, the backreaction of the scalar hair on the geometry will have the effect that the black hole completely disappears, and the deep interior geometry is replaced by a different geometry associated with the deep interior of the "Higgs star". Using the rewriting in terms of the Schrödinger potential, one can now show that, when the scalar hair develops, the effect is that the potential well associated with the black-hole horizon disappears, getting replaced at low temperatures by a potential *maximum* [385]. The effect of this in the boundary is that the quasiparticle can no longer decay in the quantum critical continuum and becomes long-lived, much like the BCS Bogoliubov fermion which is protected by the gap.

10.2.3 Unconventional superconductivity: the case of the triplet

Up to this point we have been dealing with a singlet s-wave superconductor. As discussed in chapter 3, the superconductivity in strongly correlated electron systems appears to be invariably unconventional in the sense that the superconductor breaks in addition spatial and internal spin symmetry: thus we are dealing with p-wave, d-wave, etc. superconductors.

The question of principle is whether such unconventional superconducting order parameters can be described holographically. The answer is not fully settled. On the one hand, one can construct holographic order parameters that can faithfully be called p-wave [387, 388], having even a sound top-down embedding in brane intersections [370]. Similarly, there are also proposals for d-wave order parameters (e.g. Refs. [389, 390, 391]), although there are still debates regarding whether these are fully consistent in the bulk. These invoke dynamical massive spin-2 tensors, which interfere with the massless spin-2 graviton, and are notoriously inconsistent as a quantum theory. On the other hand, these phenomenologically motivated holographic superconductors are entirely different from the unconventional superconductors of condensed matter, where the spin and orbital momentum relates to the wave function of the Cooper pair. It appears that the encoding of such "simple" non-s-wave superconductors in the gravitational bulk still remains to be found out.

The phenomenological theory for the holographic p-wave superconductor has a vector order parameter as a fundamental building block [387, 388]. Its dual field will be a vector field in the bulk. In analogy with the scalar order parameter, one now has to make this vector field charged so that it can respond to a finite density and break the symmetry in the ground state. As we discuss in detail in box 10.2, the only way to do so consistently in the bulk is to turn the vector field into a non-Abelian gauge field. Instead of Einstein–Maxwell–Higgs, we must consider Einstein–Yang–Mills theory. The simplest Yang–Mills theory has $SU(2)$ as the gauge group, corresponding to a triplet of vector currents in the boundary field theory. The holographic implementation of p-wave superconductivity is therefore rather different from the condensate of a spin-1 pair operator. The $U(1)_3$ subgroup of this $SU(2)$ gauge group is dual to the usual global $U(1)$ in the boundary, and this leaves the "1, 2" components of the gauge field in the bulk to be dual to order-parameter fields in the boundary carrying spin-1. The oddity is now that the spin-1 condensate in the boundary goes hand in hand with a *current* operator in a particular spatial direction (the spatial "p-wave") acquiring a VEV: $\langle J_x^1 \rangle \neq 0$, where the upper label refers to spin. This is a state carrying some form of "spontaneous spin current". Insofar as they are understood, the hydrodynamical properties of this "superfluid" are quite different from those of normal superfluids, including the continuum triplet variety realised in ^{3}He.

Summary 19 Unconventional (p-wave and d-wave) holographic superconductors can be constructed, but this involves non-Abelian gauge fields or
charged spin-2 fields in the bulk. The condensates are spin-1 currents and
spin-2 tensors, respectively.

Box 10.2   The holographic triplet superconductor

The first bottom-up p-wave superconductor in the $(2 + 1)$-dimensional boundary was
constructed in Refs. [387, 388]. The p-wave here refers to the condensate of a current
operator $\langle J_\mu^1 \rangle \neq 0$. To achieve this through the same instability mechanism as for the
standard s-wave holographic superconductor, the vector field A_μ^1 dual to the current
has to be charged. This uniquely leads to the consideration of $(3 + 1)$-dimensional
AdS–Einstein–Yang–Mills theory,

$$S = \frac{1}{2\kappa^2} \int d^4x \sqrt{-g} \left[R + \frac{6}{L^2} - \frac{1}{4g_F^2}(F_{\mu\nu}^a)^2 \right], \tag{10.23}$$

where g_F is now the Yang–Mills coupling constant. Here the field strength $F_{\mu\nu}^a$ of the
$SU(2)$ gauge field A_μ^a,

$$F_{\mu\nu}^a = \partial_\mu A_\nu^a - \partial_\nu A_\mu^a + f^{abc} A_\mu^b A_\nu^c, \tag{10.24}$$

with $f^{abc} = \epsilon^{abc}$, manifestly exposes the charged nature of the gauge fields with
respect to each other.

We now use this to let a $U(1)$ subgroup of the $SU(2)$ gauge group provide a dual
chemical potential and a finite charge density in the boundary. We take the gauge field
to be of the form

$$A = \Phi(r)\tau^3 \, dt + w(r)\tau^1 \, dx, \tag{10.25}$$

where τ^a are the generators of the $SU(2)$ gauge group satisfying $[\tau^b, \tau^c] = \tau^a f^{abc}$.
The subgroup generated by τ^3 is indicated by $U(1)_3$ and this subgroup is dual to the
$U(1)$ conserved charge of the boundary. The component $w(r)\tau^1 \, dx$ will be the dual of
the order parameter. Since this order parameter corresponds to a spin-1 operator, and
it breaks the symmetry of $U(1)_3$ spontaneously, the dual field theory is a holographic
p-wave superconductor.

Let us again consider the simplest bulk construction with the gauge field in the
probe limit, corresponding to the large charge-to-Newton's-constant ratio. The fully
backreacted solution for the holographic p-wave superconductor can be found in Refs.

[392, 393]. In the probe limit the background metric is again the AdS$_4$ Schwarzschild black hole Eq. (10.10). By redefining $\tilde{\Phi} = L^2\Phi$, $\tilde{w} = L^2 w$ and setting $r_0 = 1$, the equations of motion for the gauge field become

$$\tilde{\Phi}'' + \frac{2}{r}\tilde{\Phi}' - \frac{1}{r(r^3 - 1)}\tilde{w}^2\tilde{\Phi} = 0,$$

$$\tilde{w}'' + \frac{1 + 2r^3}{r(r^3 - 1)}\tilde{w}' + \frac{r^2}{(r^3 - 1)^2}\tilde{\Phi}^2\tilde{w} = 0. \tag{10.26}$$

Note the similarity to the s-wave holographic superconductor equations (10.12). At the asymptotic AdS boundary $r \to \infty$, we impose the boundary conditions

$$\tilde{\Phi} = a_0 + \frac{a_1}{r} + \cdots, \quad \tilde{w} = \frac{w_1}{r} + \cdots, \tag{10.27}$$

where according to the dictionary the boundary chemical potential $\mu = a_0$, while the boundary charge density $\rho = -a_1$. We demand that there is no constant term in $\tilde{w}$, since we insist that the symmetry breaking is spontaneous. Its leading constant term corresponding to an explicit source must therefore vanish. w_1 then encodes the symmetry-breaking expectation value of the current J_i^a, i.e.

$$\langle J_x^1 \rangle \sim w_1. \tag{10.28}$$

The near-horizon behaviours of these two fields are

$$\tilde{\Phi} = \Phi_{\text{hor}}(r - 1) + \cdots, \quad \tilde{w} = w_{\text{hor}} + \cdots, \tag{10.29}$$

where we require that the gauge potential vanishes at the horizon. There are two independent initial parameters, Φ_{hor} and w_{hor}, while there is one normalisable boundary condition at the boundary. Thus there is a one-parameter family of solutions. This system is then solved numerically by a shooting method as in the holographic s-wave set-up. Numerics shows that solutions with p-wave hair can exist only below a certain temperature T_c [387] with characteristic holographic mean-field behaviour

$$w_1 \sim \sqrt{T_c - T}, \quad \text{when } T \to T_c. \tag{10.30}$$

Unlike the holographic s-wave superconductor case where we have $\sigma_{xx} = \sigma_{yy}$ and $\sigma_{xy} = 0$, owing to the vector nature of the order parameter the conductivity is no longer isotropic, and it will differ in different directions. Transverse to the order parameter, the σ_{yy} conductivity just follows from the solution for the perturbation

---

Box 10.2   (Continued)

$$\delta A = e^{-i\omega t} a_y^3(r) \tau^3 \, dy. \tag{10.31}$$

It turns out the equation of motion for $a_y^3(r)$ is similar to the s-wave case and thus the behaviour of σ_{yy} is qualitatively the same as s-wave results.

For the σ_{xx} conductivity parallel to the order parameter, one has to take into account that the perturbation of $\delta A_x = e^{-i\omega t} a_x^3(r) \tau^3$ couples to other modes. The general form of the perturbation of A_μ with modes that couple to δA_x should be

$$\delta A = e^{-i\omega t} \left[\left(a_t^1(r)\tau^1 + a_t^2(r)\tau^2 \right) dt + a_x^3(r)\tau^3 \, dx \right]. \tag{10.32}$$

On substituting into the bulk Maxwell equation, one now has to solve the linearised equations of motion for these modes. Near the boundary the expansion of these fluctuations becomes

$$a_x^3 = a_x^{3(0)} + \frac{a_x^{3(1)}}{r} + \cdots,$$

$$a_t^1 = a_t^{1(0)} + \frac{a_t^{1(1)}}{r} + \cdots, \tag{10.33}$$

$$a_t^2 = a_t^{2(0)} + \frac{a_t^{2(1)}}{r} + \cdots,$$

and the dual conductivity of σ_{xx} can be obtained from

$$\sigma_{xx} = -\frac{i}{\omega L^2 a_x^{3(0)}} \left(a_x^{3(1)} + w_1 \frac{i w L^2 a_t^{2(0)} + a_0 a_t^{1(0)}}{a_0^2 - w^2 L^4} \right). \tag{10.34}$$

By choosing in-falling boundary conditions for all these modes the conductivities can be computed numerically [387, 394] (Fig. 10.8). The longitudinal conductivity in the direction of the condensate shows the expected gap. At very low frequencies the result actually matches a Drude model very well. This is somewhat curious, since, as we shall discuss at length in chapter 12, there is no obvious momentum dissipation, even though rotational invariance is clearly broken.

Finally, as a next step, one can also turn on the A_y^2 mode, together with the A_x^1 mode, yielding

$$A = \Phi(r)\tau^3 \, dt + w_x(r)\tau^1 \, dx + w_y(r)\tau^2 \, dy. \tag{10.35}$$

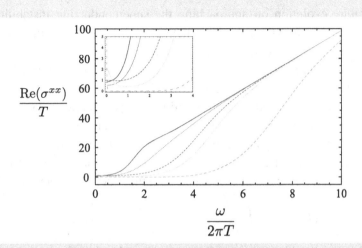

$$\frac{\mathrm{Re}(\sigma^{xx})}{T}$$

$$\frac{\omega}{2\pi T}$$

Figure 10.8 The conductivity of a $(4+1)$-dimensional holographic p-wave superconductor dual to a $(3+1)$-dimensional field theory at various temperatures. In all cases the behaviour at large ω recovers the critical scaling of a $(3+1)$-dimensional CFT where $\sigma \sim \omega$. At lower ω one sees that a gap develops at successively lower T. The inset zooms in near the orgin to show the spectral weight transfer into the superconducting pole at $\omega = 0$. Figure source [394]. (Reprinted with kind permission from Springer Science and Business Media, © 2013, Springer.)

This corresponds to a $(p + ip)$-wave superconductor at the boundary [388]. It can be shown numerically that a $(p + ip)$-wave superconductor is not as stable as a pure p-wave superconductor.

10.3 Observing the origin of T_c: the pair susceptibility of the strange metal

The exposition in the above shows that holographic superconductivity is quite like conventional superconductivity. This appears to be the case to such a degree that on the basis of available experimental information it is impossible to identify a precise, qualitative criterion on the basis of which to prove or disprove whether anything approaching holographic superconductivity is at work in known exotic superconductors. The fundamental physics at work is, however, completely different: a normal state of un-particles as opposed to electronic quasiparticles at a Fermi surface that undergo Cooper pairing. What quantity that is sensitive to this gross difference in physics could and should be measured in the laboratory?

We have actually already answered this question in section 2.4 in chapter 2: one should ask this question directly of the main actor itself, namely the dynamical susceptibility of the order parameter. By measuring it directly, and by inspecting its temperature evolution on approaching the superconducting instability in the normal state, one can extract the information regarding the origin of the superconductor. It is a considerable challenge to realise this in the laboratory. It can be done using the proximity effect in a heterogeneous Josephson junction between two different superconductors [395, 396, 397, 398] (see section 2.4). In the 1970s this was successfully carried out for an ordinary superconductor, but it has never been performed on the cuprates or other exotic superconductors. Holographic superconductivity provides an excellent motivation to take up this challenge [72].

Quite generally, the dynamical susceptibility $\chi(\omega, \mathbf{k})$ is just another word for the (retarded) two-point propagator associated with the order parameter field $\mathcal{O}(x, t)$,

$$\chi(\omega, \mathbf{k}) = \int \frac{d^4x}{(2\pi)^4} e^{-i\omega t + ikx} \langle \mathcal{O}(x, t)\mathcal{O}(0, 0)\rangle. \tag{10.36}$$

This is to be evaluated in real time. One normally takes $k = 0$, since this is where the instability resides. When one is dealing with a superconducting instability in the condensed matter context, $\mathcal{O}(x, t)$ should be interpreted as the fermion pair operator, and the order-parameter susceptibility can be directly associated with the dynamical pair susceptibility $\chi_p(k, \omega)$ (for $k = 0$, where the instability resides) as defined by Eq. (2.29) in section 2.4. Given that the (conventional) BCS thermal phase transition is governed by mean-field theory because of the large coherence length, the full susceptibility in the normal state can be reliably approximated by time-dependent Hartree–Fock/RPA,

$$\chi_p(\omega, T) = \frac{\chi_p^0(\omega, T)}{1 - V \chi_p^0(\omega, T)}, \tag{10.37}$$

where $\chi_p^0(\omega, T)$ is the pair susceptibility of the Fermi liquid in the absence of the attractive interactions, while V is the pairing interaction. Upon approaching the instability from the metallic side one finds that the low-frequency regime is governed by the relaxation "Ornstein–Zernike" mean-field dynamics, which gives rise to a relaxational peak in the imaginary part Im $\chi_p(\omega, T)$ (Eq. (2.36)).

Although it is a large-N pathology, the mean-field nature of the order parameter dynamics of the holographic superconductor becomes quite convenient at this stage. It makes it possible to compare apples with apples, and thereby arrive at a maximally sharp contrast between mean-field BCS superconductivity (and its perturbative extensions) and the holographic variety. As we shall see, with regard to its susceptibility the holographic superconductor behaves precisely according to the "quantum critical BCS" phenomenology we explained in chapter 2, section 2.4. We focus here on the minimal holographic superconductor. The metallic normal state is here the holographic RN metal with its characteristic AdS$_2 \times \mathbb{R}^2$

near-horizon IR. The key to rediscovering the RPA behaviour is in the matching technique between this local quantum critical IR and the AdS$_4$ UV which we introduced for the fermions in section 9.3. This carries through exactly the same for the scalar field in the bulk, dual to $\mathcal{O}(x, t)$. In the small-frequency regime the order-parameter susceptibility can therefore be written as

$$
\chi_p^H(\omega, T) \sim \frac{b_+^{(0)} + b_+^{(1)}\omega + \mathcal{O}(\omega^2) + \mathcal{G}(\omega, T)\left(b_-^{(0)} + b_-^{(1)}\omega + \mathcal{O}(\omega^2)\right)}{a_+^{(0)} + a_+^{(1)}\omega + \mathcal{O}(\omega^2) + \mathcal{G}(\omega, T)\left(a_-^{(0)} + a_-^{(1)}\omega + \mathcal{O}(\omega^2)\right)}, \quad (10.38)
$$

where $\mathcal{G}(\omega, T)$ is the near-horizon propagator of the scalar field. At temperatures $T \gg T_c$ deep in the metallic state the scalar field just "falls freely" from the boundary to the horizon and Im $\chi_p^H(\omega, T) \sim$ Im $\mathcal{G}(\omega, T)$. Since the system is far away from the instability, the pair susceptibility is just equal to that of the background metal. When this metal is quantum critical, either because $T > \mu$ or because the transition happens at $T \ll \mu$ deep in the AdS$_2$ emergent local quantum critical regime, the pair susceptibility will reflect this quantum criticality through the IR Green function $\mathcal{G}(\omega, T)$ at finite temperature $T > T_c$. This exhibits a scaling that is related to the zero-temperature local quantum critical Green function $\mathcal{G}(\omega) \sim \omega^{2\nu}$, but it is not the same. We will compute it in the next paragraph.

What happens when we approach the transition at $T = T_c$? We now have to inspect the matching coefficients. Just like for the fermions, they can only be determined numerically. However, differently from the fermion case, for scalars we know that it is the $k = 0$ scalar fields which will start to condense. We can therefore restrict our attention to $\chi_p(\omega, T)|_{k=0}$. The occurrence of the pole at $\omega = k = 0$ translates into a behaviour of the matching coefficients such that at low frequency in the approach to T_c only $a_-^{(0)}, a_+^{(1)}$ and $b_+^{(0)}$ contribute (i.e. $a_+^{(0)}(T_c) = 0$). These can be parametrised as $\gamma_0 = b_+^{(0)}(T_c)$, $\beta_0 = \partial_T a_+^{(0)}(T_c)$, $\beta_1 = \lim_{\omega \to 0}(1/(i\omega))\mathcal{G}(\omega, T_c)a_-^{(0)}(T_c)$ and $\beta_2 = a_+^{(1)}(T_c)$. It follows that the pair susceptibility at low frequency can be written as

$$
\chi_p^H(\omega, T) \sim \frac{\gamma_0}{\beta_0(T - T_c) + i\omega\beta_1 + \omega\beta_2}. \quad (10.39)
$$

One infers that the susceptibility acquires precisely the Ornstein–Zernike form Eq. (2.36),

$$
\chi_p^H(\omega, T) = \frac{\chi'^H_p(\omega = 0, T)}{1 - i\omega\tau_r - \omega\tau_\mu}, \quad (10.40)
$$

where one recognises the Curie–Weiss-like "bulk pair susceptibility"

$$
\chi'^H_p(\omega = 0, T) = \frac{\gamma_0}{\beta_0(T - T_c)}. \quad (10.41)
$$

The order-parameter relaxation time is given by,

$$\tau_r = \lim_{\omega \to 0} \frac{i}{\omega \beta_0 (T - T_c)} \mathcal{G}(\omega, T_c) a_-^{(0)}(T_c) \tag{10.42}$$

in terms of the IR Green function $\mathcal{G}(\omega, T)$. The final step is to obtain the detailed form of the IR propagator at finite T. In the case of a strongly coupled holographic superconductor with $T_c \simeq \mu$, the AdS_2 region is shielded behind the finite-temperature horizon; we may consider the geometry as essentially indistinguishable from AdS Schwarzschild. Near the horizon the Schwarzschild geometry is to leading order a Rindler space-time. For $\omega \ll T$, the Rindler near-horizon IR Green function at $k = 0$ takes in this case the universal form $\mathcal{G}(\omega, T) = -i\omega/(4\pi T)$ as shown in box 10.3. Therefore $\tau_r = \alpha_0/(T - T_c)$. Alternatively, when the transition happens at very low temperature deep in the RN-metal regime, we should use instead the AdS_2-type deep-infrared propagator with a small temperature modification. This has the universal finite-temperature general scaling form $\mathcal{G}(\omega, T) \sim T^{2\nu} F(\omega/T)$ and it is easy to find out that $F(\omega/T) \sim F_0 + \cdots$, when $\omega \ll T \ll \mu$ (box 10.3). It follows that also in this regime $\tau_r = \alpha_0/(T - T_c)$.

Notice that we find in addition the factor $\omega \tau_\mu$ controlled by the particle–hole asymmetry parameter $\tau_\mu = -\beta_2/(\beta_0(T - T_c))$, which is absent from the elementary expression (2.36). This time scale becomes significant only when the breaking of charge conjugation in the metal due to the finite chemical potential influences the superconducting order, i.e. when T_c becomes of order μ. It is easy to check that this is the ubiquitous way for particle–hole asymmetry to enter this relaxational regime.

Box 10.3 The universality of the finite-temperature IR Green function

In this box we will consider the IR Green function for a chargeless scalar field with mass m in the AdS RN black-hole background at finite temperature and show that the IR Green function is linear in ω, irrespective of T/μ. This result then readily carries over to a charged field, since it is easy to see that the finite-temperature horizon determines the IR Green function irrespective of the presence or absence of an electrostatic potential.

Any finite-temperature black hole has a single pole in g_{tt}, and in the deep IR we may therefore approximate the geometry by the Rindler space-time

$$ds^2 = -\alpha(r - r_+)dt^2 + \frac{dr^2}{\alpha(r - r_+)} + r_+^2(dx^2 + dy^2), \tag{10.43}$$

with $\alpha = 4\pi T$. In this region the equation of motion for a neutral scalar field can be simplified to

$$\phi'' + \frac{1}{r - r_+}\phi' + \left(\frac{\omega^2}{\alpha^2(r - r_+)^2} - \frac{m^2}{\alpha(r - r_+)}\right)\phi = 0. \qquad (10.44)$$

The analytic solution of this equation is the Bessel function of the first kind

$$\phi_{\text{Rindler}} = A_0 J_{-2i\omega/\alpha}\left[2m\sqrt{\frac{r - r_+}{\alpha}}\right] + B_0 J_{2i\omega/\alpha}\left[2m\sqrt{\frac{r - r_+}{\alpha}}\right]. \qquad (10.45)$$

The in-falling boundary condition imposes that $B_0 = 0$. By expanding this in-falling solution (10.45) near the matching region $\omega \ll r - r_+ \ll r_+$, i.e. $(r - r_+)/\alpha \to 0$ and $\omega/\alpha \to 0$, we obtain $\phi \simeq A_1 + B_1 \ln(r - r_+)$, with $B_1/A_1 \simeq -i\omega/(4\pi T) + \mathcal{O}(\omega^2/T^2)$.

The conclusion is that one finds both in BCS theory and in the holographic superconductor that the order-parameter susceptibility shows a ubiquitous mean-field relaxational behaviour upon approaching the transition from the metallic side. Of course, this should be the case since the thermal transition is of mean-field nature itself. The universal symmetry-breaking dynamics fully controls this regime. Any difference must therefore arise at higher frequencies. Although the comparison in this regime can only be done numerically, it will be quite informative to compare the result also with the simple "quantum critical BCS" (QCBCS) phenomenological model [71] which we discussed in section 2.4. QCBCS just asserts that one can still use the RPA expression for the susceptibility, but instead of using the χ_p^0 of the Fermi liquid one assumes that this has the conformal form $\chi_p^0(\omega) \sim 1/(i\omega)^{\Delta_p}$ at zero temperature, and the "energy–temperature" scaling form $\chi_p^0(\omega, T) = (1/T^{\Delta_p}) \mathcal{F}(\omega/T)$ at finite temperature. The only free parameter characterising the metal is then the anomalous dimension Δ_p. In the actual QCBCS computations the scaling function $\mathcal{F}$ is modelled by $(1 + 1)$-dimensional CFT expressions.

Figure 10.9 shows the imaginary part of the susceptibility obtained from QCBCS (panel (c)), side by side with the holographic results (panels (d) and (e)) as a function both of ω and of the reduced temperature, such that the origin coincides with the superconducting instability. One finds in both cases a large peak developing near the origin, corresponding to the relaxational peak that turns into a δ function at $\omega = 0$ right at T_c. Upon raising the temperature this relaxational peak moves to higher energy, broadening proportionally, signalling that the transient order-parameter correlations in the metal gradually disappear.

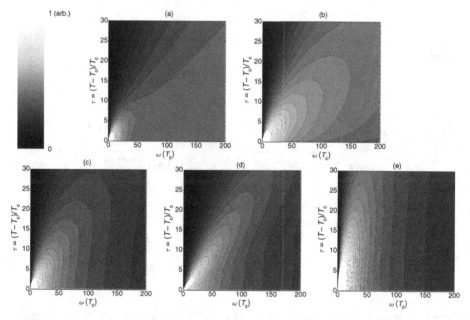

Figure 10.9 False-colour plots of the imaginary part of the pair susceptibility Im $\chi(\omega, T)$ in arbitrary units as a function of ω (in units of T_c) and reduced temperature $\tau = (T - T_c)/T_c$, for five different cases: case (a) represents the traditional Fermi-liquid BCS theory, case (b) is the Hertz–Millis-type model with a critical glue, case (c) is the phenomenological "quantum critical BCS" theory, case (d) corresponds to the "large-charge" holographic superconductor with AdS$_4$-type scaling, and case (e) is the "small-charge" holographic super-conductor with an emergent AdS$_2$-type scaling. Im $\chi(\omega, T)$ should be directly proportional to the measured second-order Josephson current, which can be measured by a Ferrell–Scalapino experiment (section 2.4). In the bottom left of each plot is the relaxational peak that diverges (white coloured regions are off-scale) as T approaches T_c. This relaxational peak looks qualitatively quite similar for all five cases. Only at larger temperatures and frequencies do qualitative differences between the five cases become manifest. Figure source [72].

In fact, the parameters have deliberately been chosen in such a way that matters are as similar as possible in panels (c)–(e) [72]; in particular, the same anomalous dimension $\Delta_p = 0.5$ is used in all cases. This is not at all obvious in the raw data of Fig. 10.9. This similarity becomes manifest on using the hidden quantum criticality in both models. This predicts that the susceptibility should be a function of ω/T only. Figure 10.10 shows T^{Δ_p} Im χ replotted versus ω/T. One now clearly sees that, at temperatures equal to a couple of times T_c, the scale associated with the superconducting instability is forgotten, while the susceptibility exhibits the scaling behaviour associated with the pristine quantum critical metal.

Let us now compare this with conventional condensed matter physics models. Panel (a) in Figs. 10.9 and 10.10 shows the dynamical pair susceptibility for

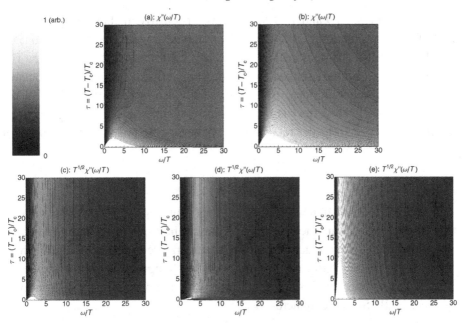

Figure 10.10 False-colour plots of the imaginary part of the pair susceptibility, as in Fig. 10.9, but now the horizontal axis is rescaled by temperature while the magnitude is rescaled by temperature to a certain power: here we plot $T^{\Delta_p} \operatorname{Im} \chi (\omega/T, \tau)$, in order to show energy–temperature scaling at high temperatures. For the quantum critical BCS (case (c)), AdS_4 (case (d)) and AdS_2 (case (e)), with a suitable choice of the exponent $\Delta_p > 0$, the contour lines run vertically at high temperatures, meaning that the imaginary part of the pair susceptibility acquires a universal form $\operatorname{Im} \chi(\omega, T) = T^{\Delta_p} \mathcal{F}(\omega/T)$, with $\mathcal{F}$ a generic scaling function, the exact form of which depends on the choice of different models. Here we choose in cases (c)–(e) $\Delta_p = 1/2$, by construction. The weak-coupling Fermi-liquid BCS case (a) also shows scaling collapse at high temperatures, but with a marginal exponent $\Delta = 0$. In the quantum critical glue model (case (b)) energy–temperature scaling fails: for any choice of Δ, at most a small fraction of the contour lines can be made vertical at high temperatures (here $\delta = 0$ is displayed). Figure source [72].

standard weakly coupled BCS theory, whereas panel (b) shows the result for Hertz–Millis theory, one of the prime candidates for the strange metal as we discussed in section 2.5. Interestingly, the weakly coupled BCS superconductor of panel (a) also yields to the scaling collapse. In fact this is no wonder, since the pair susceptibility of the Fermi gas is a conformal propagator characterised by the marginal scaling dimension $\Delta_p = 0$ (e.g. Eq. (2.30)). This is, however, coincidental, since the Fermi liquid is a stable form of matter where the Fermi energy sets the scale. As a consequence, as soon as perturbative corrections are included the normal-state pair susceptibility will violate the energy–temperature scaling. The Hertz–Millis result in panel (b) of Figs. 10.9 and 10.10 illustrates this clearly. It shows the result of a tour-de-force computation [72] for the pair susceptibility of the metal formed right at the Hertz–Millis quantum critical point [74], as discussed in section 2.5. The Fermi-gas is here "brutally shaken" by the quantum critical fluctuations associated with a magnetic quantum phase transition, which also mediate the pairing of the fermions [78, 399]. The effect is that the pair susceptibility is no longer conformal, as can be seen from the total failure of the scaling collapse in panel (b) of Fig. 10.10.

Summary 20   The dynamical pair susceptibility is the observable quantity that reveals the unique signature of the mechanisms of quantum critical (holographic) superconductivity: at high temperature the pair susceptibility is governed by the conformal symmetry of the strange metal, which implies an "energy–temperature" scaling collapse. This is not present in a strongly interacting BCS superconductor.

As a prequel to the next section, we still have to explain the difference between the systems illustrated in panels (d) and (e) of Figs. 10.9 and 10.10. The parameters in the case of panel (d) are chosen such that the superconducting T_c is of the order of the chemical potential μ (as in Fig. 10.5). The scaling behaviour one sees at high temperatures is actually associated with the zero-density CFT since the energy and temperature scales here exceed the chemical potential. The situation in panel (e) is such that T_c has been strongly reduced so that it is now small compared with μ, and accordingly the scaling at higher temperatures and frequencies now reflects the RN strange metal. This is actually derived by taking a "high"-T_c "bare" superconductor (like that in panel (d)), while subsequently a double-trace deformation is used as a "repulsive interaction" to strongly reduce T_c. As we will see next, this amounts to an explicit realisation of the RPA formula in this context of holographic superconductivity.

10.4 The phase diagram of holographic superconductivity

We have seen that the holographic strange metal resembles the strange-metal normal state in high-T_c cuprates most notably in the non-Fermi-liquid response and that it can be distinguished by energy–temperature scaling in the susceptibility. An obvious question is, however, what sets T_c? Is it typically high in a holographic superconductor, and which ingredients determine the onset of the superconducting instability? These quantitative matters are predictably determined entirely by the UV of holography.

From the development up to this point a rule of thumb for the "height of T_c" can be extracted: *a relevant scaling behaviour of the pair susceptibility in the normal state is beneficial for a high T_c*. Using the conventions of the previous section, the pair susceptibility in a quantum critical metallic state will behave as $\chi_p^H \sim 1/(i\omega)^{\Delta_p}$ and a "large" positive value of Δ_p will go hand in hand with a high T_c. This rule is hard to circumvent departing from an "RPA-like" mechanism, as we highlighted in the discussion of the quantum critical BCS phenomenology of section 2.4. It is obvious that this should be so as soon as one generalises the pair susceptibility beyond the marginal BCS logarithmic scaling.

In the holographic setting one can read off this rule directly from the bulk. Recall the BF bound for the near-horizon geometry of the charged black hole, Eq. (10.6): $(mL)^2 - q^2 g_F^2 L^2/(8\pi G) \leq -\frac{3}{2}$, where $q g_F$ is the charge of the scalar field. The mass mL of the scalar field is associated with its scaling dimension at zero density. This becomes maximally relevant on approaching the BF (unitarity) bound of the zero-density CFT: $m^2 \to -9/4$ in $3+1$ dimensions. Taking a value for the mass corresponding to a scaling dimension close the unitarity bound $\Delta = 1/2 + \cdots$, one finds the behaviour shown in panel (d) of Figs. 10.9 and 10.10. In this case superconductivity sets in directly at the scale where the finite density becomes discernible, i.e. it is very strongly coupled with a $T_c \sim q\mu$. One can parametrically suppress the superconducting instability by increasing the mass of the scalar field in the bulk, and there is a critical value at which the pair operator becomes "sufficiently irrelevant", with the effect that no transition happens at any temperature. Considering the mass (scaling dimension) as a tunable parameter in the strange metal, this critical value denotes a zero-temperature phase transition of the holographic system.

There is yet another way to tune the stability of the holographic superconductor: the double-trace deformation. In chapter 5 we saw that in the large-N limit this literally reproduces the RPA formula. Though one should always be suspicious of the large-N limit for condensed matter applications, here we can again appeal to the fact that a large-N implies a mean-field response to have some faith in the holographic predictions. To briefly recall the set-up, we add a double-trace deformation to the strongly interacting zero-density CFT in the UV of the form

$$S_{FT} \to S_{FT} - \int d^3x \, \tilde{\kappa} \mathcal{O}^\dagger \mathcal{O}, \qquad (10.46)$$

where $\tilde{\kappa} = 2(3 - 2\Delta)\kappa$, and $\mathcal{O}$ is the operator of the scalar field dual. In the large-N limit this double-trace deformation will factorise according to Eq. (4.29) as $\langle \mathcal{O}^\dagger \mathcal{O} \rangle \to \langle \mathcal{O}^\dagger \rangle \langle \mathcal{O} \rangle + \cdots$. This is the same factorisation property as that which

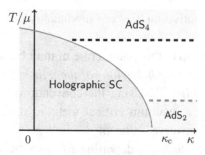

Figure 10.11 A phase diagram of a holographic superconductor including a double-trace deformation with strength κ. For $\kappa = 0$ one has the minimal holographic superconductor, case (d) of Figs. 10.9 and 10.10, where $T_c \sim \mu$. Increasing the value of κ can decrease the critical temperature all the way to $T_c = 0$, if one includes a non-minimal coupling to the AdS-gauge field (see the text). The shaded regions indicate which region of the geometry primarily determines the susceptibility. It is clear that one must turn on a double-trace coupling to describe superconductors whose susceptibility is determined by the local quantum critical AdS$_2$-type physics. This is of interest since AdS$_2$-type physics contains fermion spectral functions that are similar to what is found experimentally. Figure adapted from [72].

controls the conventional Hartree–Fock mean-field theory, and this has the implication that the propagator of the $\mathcal{O}$ field will acquire the RPA form, in a notation appropriate for the holographic superconductor,

$$\chi_p^\kappa(\omega) = \frac{\chi_p^H(\omega)}{1 + \kappa \chi_p^H(\omega)} \tag{10.47}$$

where $\chi_p^H(\omega)$ is the propagator computed in the absence of the double-trace deformation.

Referring to the standard wisdoms, the RPA expression Eq. (10.47) corresponds precisely to the way one accommodates (extra) repulsive or attractive interactions in BCS theory. Given that the single-trace order parameter $\mathcal{O}$ is the equivalent of the fermion-pair operator, the double-trace operator is associated with additional "attractive" ($\kappa < 0$) or "repulsive" ($\kappa > 0$) interactions. One can now at will enhance or diminish T_c by tuning the double-trace deformation strength κ. In particular, one can reduce T_c by a large repulsive κ from the "large-charge" strongly coupled "panel (d)" superconductor until it is a small fraction of μ. This implies that the instability occurs only at a temperature at which the system has turned into a local quantum critical AdS$_2$ strange metal, as illustrated in Fig. 10.11. The panel (e) results in Figs. 10.9 and 10.10 are representative for this regime.

These two-parameter holographic models in terms of the scaling dimension governing the bare susceptibility and the double-trace deformation governing the RPA

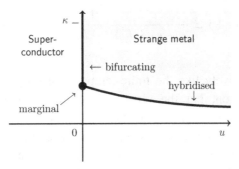

Figure 10.12 The zero-temperature phase diagram of the holographic superconductor as a function of the scaling dimension u in the extremal AdS$_2$ near-horizon
geometry, $u \equiv m^2 L_2^2 + 1/4 - q^2 e_d^2$, and the strength of the double-trace deformation κ in alternative quantisation [160]. The surprise is in the nature of the
quantum phase transitions. The "bifurcating" transition driven by u turns out to be
a "holographic Berezinskii–Kosterlitz–Thouless" (BKT) transition. This shows
the same "infinite-order" behaviour as the BKT transition of $d = 2$ statistical
physics, but this has nothing to do with the unbinding of vortices: it is a large-
N mean-field transition. The "hybridised" transition driven by the double-trace
deformation is qualitatively a more regular transition in terms of a condensing
order parameter, but quantitatively its behaviour is more exotic due to its coexistence with the locally quantum critical AdS$_2$. It has a natural interpretation in
terms of semi-holography. The two types of phase transition meet at a marginal
point. Figure adapted from [160].

correction are the natural minimal extension of BCS theory to strange metals.
They control the onset of superconductivity and the value of T_c. The interesting
part is that there are values of the mass and the double-trace coupling for which
T_c becomes zero, beyond which no instability happens. This is reflected in the
zero-temperature phase diagram in Fig. 10.12 as a function of the (square of the)
AdS$_2$ scaling dimension $\nu^2 = (m^2 L_2^2 + 1/4 - q^2 e_d^2)$ and the double-trace formation κ [160]. Although the finite-temperature phase transition are invariably of the
second-order mean-field kind, the nature of the zero-temperature transitions are
quite unconventional. These are (large-N mean-field) quantum phase transitions of
a kind that has not been identified elsewhere.

The way to understand these novel phase transitions is from the behaviour of the
order-parameter susceptibility in the normal state

$$\chi = \frac{\chi_0}{1 + \kappa \chi_0} = \frac{1}{\kappa + \chi_0^{-1}}. \tag{10.48}$$

For a relativistic boson the inverse bare susceptibility corresponds to the dispersion
relation. At low frequencies this can be determined with the matching method of
the previous chapter. By expanding around $\omega = 0$, $\mathbf{k} = 0$ the susceptibility is found
to have the RPA form [160, 400]

$$\chi(\omega, \mathbf{k}) \simeq \frac{1}{\kappa - \kappa_c + h_{\mathbf{k}}\mathbf{k}^2 - h_\omega \omega^2 + h\mathcal{G}_k(\omega)}, \qquad (10.49)$$

with $\kappa_c = -\chi_0^{-1}$.

This looks qualitatively like the finite-temperature result which we discussed in the previous section, but the physics is quite different from that of the Ornstein–Zernike thermal relaxational fluctuations. This describes a propagating order-parameter field governed by a Gaussian fixed point, imposed by the large-N mean field as at finite temperature. The order parameter has a mass set by the double-trace coupling $m^2 \sim \kappa$. In addition, however, there is the self-energy term $h\mathcal{G}_k(\omega)$, which is the crucial result of the matching method. This indicates that the order parameter acquires an additional damping due to the presence of the massless degrees of freedom associated with the AdS$_2$ deep infrared of the strange metal. We learned in section 8.2 that here propagators have the unusual form $\mathcal{G}_k(\omega) = c_k e^{i\phi_k}\omega^{\nu_k}$, where the exponent equals $\nu_k \sim \sqrt{u + k^2}$, and u can be obtained from Eq. (9.42) by virtue of the specifics of the AdS$_2$ geometry.

The low-energy dynamics of the order parameter therefore submits to the same semi-holographic wisdom as we discussed at length in section 9.4 for the fermions. This simple form rests on the large-N limit [366]: to a first approximation the order-parameter dynamics is that of a Gaussian free field, but it still communicates with a "heat bath" (the AdS$_2$ sector) through a simple linear coupling. It allows one to posit a similar effective semi-holographic field theory,

$$S_{\mathrm{eff}} = S_{\mathrm{AdS}_2}[\Phi] + \int \lambda(k, \omega)\Phi_{-\vec{k}}\Psi_{\vec{k}} + S_{\mathrm{LG}}[\Psi], \qquad (10.50)$$

where $S_{\mathrm{AdS}_2}[\Phi]$ is the action for a strongly coupled locally quantum critical sector in terms of unknown fields Φ, while S_{LG} now equals the usual Ginzburg–Landau–Wilson theory for the order-parameter field, to be literally interpreted in the mean-field sense,

$$S_{\mathrm{LG}} = -\frac{1}{2}\int \Psi_{-\vec{k}}(\kappa_c - \kappa + h_k k^2)\Psi_{\vec{k}} + h_t \int (\partial_t \Psi)^2 + \cdots . \qquad (10.51)$$

This semi-holographic effective GLW theory includes a standard phase transition when κ becomes negative. Nevertheless, due to the interaction with the local quantum critical sector, the physical characteristics differ from a conventional GLW second-order transition. It has therefore been christened a "hybridised" phase transition. In this case, the static susceptibility $\chi(\omega = 0, \mathbf{k})$ near the critical point does exhibit a conventional mean-field behaviour. The dynamical susceptibility, however, at low frequencies will now include the extra self-energy contribution $\mathcal{G}(\omega) \sim \omega^{2\nu}$. For $\nu > 1$ its effect is mild: the order parameter will slowly decay into the quantum critical bath, with a rate $\Gamma \sim \mathcal{G}(\omega)$. Normally this will be enhanced by higher-order self-interactions that are suppressed in the mean-field limit. For $\nu < 1$, however, the dynamical susceptibility at low frequencies will in fact be

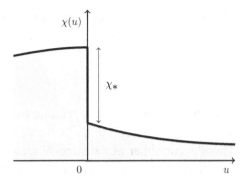

Figure 10.13 A jump in the static susceptibility across the critical point (here located at $u = 0$) is a typical example of the physical properties which the zero-temperature "bifurcation" transition between the RN metal and the holographic superconductor shares with the BKT transition of $d = 2$ statistical physics. Figure source [160].

governed by the self-energy $\chi(\omega, \mathbf{k} = 0) \sim (\kappa + h_\omega \omega^{2\nu})^{-1}$. This implies that the correlation time $\xi_t \sim \kappa^{1/2\nu}$ will now scale differently from the correlation length $\xi \sim \kappa^{1/2}$, with an emergent dynamical critical exponent $\mathbf{z} = 1/\nu$.

There is, however, a second path to a quantum phase transition in this parameter space. This is when the mass or charge of the order parameter is tuned such that the Breitenlohner–Freedman bound in the AdS$_2$ near-horizon region is violated at zero temperature and ν becomes complex. This "bifurcating" transition is charac-terised by scaling properties that are similar to the Berezinskii–Kosterlitz–Thouless (BKT) transition [160, 400], as familiar from the thermal two-dimensional XY sys-tem. However, the underlying physics of the holographic BKT behaviour is rather completely detached from the vortex unbinding underlying the BKT transition in the context of two-dimensional statistical physics. Instead, it is tied to the merg-ing of an IR fixed point with a UV fixed point of the system. This can give rise to the characteristic exponential behaviour of the free-energy difference. Another characteristic feature is that the susceptibility does not diverge when approaching the transition, but that it instead develops a branch-cut singularity ("bifurcation") at the critical coupling [401] (see Fig. 10.13). This correlates with a quite distinct behaviour of the order parameter. Unlike what happens at conventional quantum phase transitions, there is no zero-energy pole developing in the susceptibility: the order-parameter (quantum) fluctuations retain their gap. Instead, the novel physics that drives this quantum phase transition can be deduced from our analysis of the probe fermions in the previous chapter. The complex AdS$_2$ scaling dimension signals that pair production sets in, associated with the fluctuations of the local quantum critical state, which subsequently undergoes Bose condensation.

Finally, the two types of transitions meet at the "marginal" quantum critical point (see Fig. 10.12). Here the susceptibility diverges and bifurcates at the same time,

while it acquires a similar form [160] to the spectrum of the quantum critical fluctuations postulated by the marginal Fermi-liquid theory discussed in sections 2.2
and 9.3.

Summary 21 Differently from the finite-temperature transitions, the quantum phase transitions from the RN metal to the holographic superconductor
are of a new kind. These occur either as the large-N mean-field versions of
the Berezinskii–Kosterlitz–Thouless transitions, or as the "hybridised" transitions where the mean-field order parameter acquires extra damping by its
decay into the deep-infrared AdS_2 quasi-local quantum critical degrees of
freedom associated with the RN extremal horizon.

10.5 The zero-temperature states of holographic superconductors

Let us first comment on the role of the scalar-field profile in the bulk – the scalar
hair – in the geometry. Up to reasonably low temperatures in the superconducting
state these backreactions are at least *qualitatively* unimportant because the finite
horizon of the black hole controls most of the physics. However, at very low and
zero temperature on the superconducting condensed side this gravitational backreaction becomes fundamental. The energy density in the scalar and Maxwell fields
will now dominate over the black hole. We will see that in response to this energy
density in the infrared the nature of the geometry in the deep interior will change
drastically.

The great surprise is that this deep infrared of the zero-temperature holographic
s-wave superconductor is not characterised by a gap in the conductivity or spectral
function of the BCS kind, below which all degrees of freedom disappear except for
the phase mode. Such a gap was also suggested by the phenomenology presented in
section 10.2, but on closer inspection it turns out not to be a real gap. Instead, one
finds that the deep-interior geometry generically reconstructs in a *Lifshitz scaling
geometry*, as introduced in section 8.4. For the boundary this means that a new
quantum critical phase characterised by an unusual dynamical critical exponent **z**
is realised in the deep infrared. This is quite surprising, showing that holographic
superconductors are quite different from the BCS variety.

10.5.1 Confinement and holographic superconductivity: rediscovering BCS

To set the stage, let us first introduce a context that is devoid of such infrared
subtleties. The guess from Goldstone's theorem and BCS theory is that the broken

superconducting state ought to have this gap. On the gravity side this should mean that the geometry gets "capped off". Fundamentally, from the boundary field-theory perspective capped-off geometries describe a confining regime together with a mass gap, as we discussed in section 6.3. What we can do, is first impose this confinement with a gap by hand, and then slowly remove it until the true IR ground state emerges. The scalar field of holographic superconductivity in a confining geometry is the bosonic version of the "mesino" gauge-singlet fermionic quasiparticles of chapter 9. Owing to confinement, it interacts only weakly at low energies, with the difference that this bosonic field will condense at finite density. Such a holographic confining finite-density boson set-up was studied in Refs. [402, 403]. As we will see, it has the comforting outcome that it behaves precisely according to conventional condensed matter expectations. A fully gapped s-wave superconductor that can undergo a phase transition into an insulating state is found. This is quite literally a holographic incarnation of the "dilute-boson" system in condensed matter physics, where as a function of the chemical potential the system turns from a zero-density insulator into a finite-density fully gapped s-wave superconductor [25].

Rather than the hard-wall geometry, the capped-off geometry is provided by the the AdS soliton discussed in section 6.3. This is identical to the Schwarzschild geometry after a double Wick rotation, where the Euclidean compact time circle now becomes one of the spatial directions. In $4 + 1$ dimensions one has

$$ds^2 = \frac{L^2\,dr^2}{r^2 f(r)} + r^2(dx^2 + dy^2 - dt^2) + r^2 f(r)d\phi^2 \qquad (10.52)$$

with $f(r) = 1 - r_0^4/r^4$. Regularity at $r = r_0$ demands that ϕ has to have a period of $\pi L/r_0$. This geometry is capped off at r_0; the periodic boundary conditions in the ϕ direction give all excitations a mass, and the dual boundary field theory at low energy corresponds to a gauge theory characterised by a confining vacuum. The dual optical conductivity $\mathrm{Re}\,\sigma(\omega)$ on this AdS soliton background shows a series of delta functions at increasing energies, indicative of the "mesons", while the vanishing DC conductivity confirms that the state is insulating.

In this background one now studies the Einstein–Maxwell-scalar theory Eq. (10.1). Compared with the minimal holographic superconductor the single difference is in the boundary conditions in the interior – at the AdS boundary one still demands a normalisable solution corresponding to a spontaneous VEV without a source. Physically this translates into there being for small enough μ/r_0 no solution – this is the gap due to the compact geometry. At high temperature $T \gg r_0$ and small chemical potential $\mu \ll r_0$ there is therefore no scalar profile, and neither is the soliton-confinement scale felt. The unique solution is the charged black hole with the horizon hiding the location where the geometry caps off. The dual state

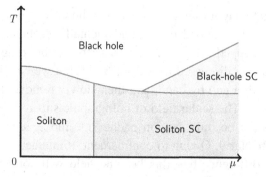

Figure 10.14 The typical phase diagram for a five-dimensional Einstein–Maxwell-scalar system with one compact direction [403]. The "black hole" is the usual RN metal, while the "black-hole SC" corresponds to the usual holographic superconductor. At lower temperatures, however, the confinement scale set by the size of the compact direction takes over. The effect is that as a function of the chemical potential one finds a zero-density confining state that can be interpreted as a Mott insulator that undergoes a phase transition to a fully conventional s-wave superconductor characterised by a collective phase mode and a gap to all other excitations. The precise phase boundary to the superconducting phases depends on the specific charge q and mass m of the scalar field dual to the order parameter. Figure source [403].

is the usual RN metal at finite temperature, albeit with one compact dimension. Upon lowering the temperature at small chemical potential, a transition similar to the Hawking–Page transition of chapter 7 to a confining finite-temperature "soliton" state occurs for $T \lesssim r_0$ (Fig. 10.14). Despite the finite chemical potential, there will be no charge density because it is too small to excite the gapped states ($\mu \ll r_0$).

Returning to high temperatures, but now increasing the chemical potential, the soliton confinement scale is again not felt, and at $\mu \gtrsim T$ the system undergoes a transition to a finite-temperature holographic superconductor of the type discussed in section 10.1. However, when the temperature is now lowered to the point where the confinement scale becomes noticeable, a new type of holographic superconductor forms: the "soliton superconductor". The solution to the full set of Einstein–Maxwell-scalar equations for this state, including the effect on the geometry, is of the form [403]

$$ds^2 = r^2\left(e^{A(r)}B(r)d\phi^2 + dx^2 + dy^2 - e^{C(r)}dt^2\right) + \frac{dr^2}{r^2 B(r)} \qquad (10.53)$$

for the metric, and

$$A_t = \varphi(r), \quad \psi = \psi(r), \qquad (10.54)$$

for the gauge and scalar fields. $B(r)$ vanishes smoothly for the soliton solution at some radius r_0, which is the tip of the soliton. Smoothness at the tip requires that $\phi \simeq \phi + \phi_0$ has to be periodic with a periodicity of

$$\phi_0 = \frac{2\pi e^{-A(r_0)}}{r_0^2 B'(r_0)}. \tag{10.55}$$

The precise form of the radial profiles of the various fields can be computed numerically. The characteristic feature is of course a finite normalisable scalar-field profile, with the correct boundary asymptotics encoding a spontaneous order-parameter VEV in the boundary. Note that there is no charged horizon left. The boundary is at a finite density, but all the bulk charge is now completely absorbed by the scalar hair. Instead, the deep interior geometry (including the scalar field) is capped off, signalling that an absolute energy gap is present in the boundary. This is confirmed by the optical conductivity: this now acquires a delta function at zero frequency, confirming the superconducting nature of the state. In addition, at energies less than the confinement scale one finds isolated delta functions. These are the massive mesons of AdS/QCD we encountered in section 6.3. Except for these additional "narrow bands", this soliton superconductor is quite like the conventional BCS state.

10.5.2 The deconfined holographic superconductor at zero temperature

What happens at zero temperature when the geometry is not capped off "by hand"? Upon lowering the temperature the stress-energy associated with the scalar field continuous to increase near the extremal RN horizon, and the geometry eventually has to react in the (very) deep interior. Unlike the modification in the soliton super-conductor, this backreaction will completely change the physics, and one has to find a new solution to this highly non-linear problem. We will give a rather intuitive derivation of the main results; further details can be found in [160, 404, 405, 406].

First note that even for a *chargeless* scalar field there is a window of masses where the AdS$_2$ BF bound is violated. This occurs when

$$-\frac{9}{4} < (mL)^2 < -\frac{3}{2}. \tag{10.56}$$

A neutral scalar field with a mass in this range will also condense in the presence of a finite chemical potential. This is readily interpreted as the spontaneous symmetry breaking of an Ising $\mathbb{Z}_2$ symmetry instead of a continuous global symmetry. Realising this, however, gives an immediate insight into what the form of the fully backreacted zero-temperature superconductor geometries should be. The neutral Einstein–Maxwell–Higgs theory is just an extremely simple version of the more complicated Einstein–Maxwell-dilaton theories we studied in section 8.4. The zero-temperature ground states of EMD theories generated the atlas of scaling geometries generically of the Lifshitz type characterised by an emergent dynamical critical exponent **z** and hyperscaling-violating parameter θ.

These are the types of backreacted geometries we should therefore expect for the holographic superconductor. Generically they will always obey hyperscaling ($\theta = 0$), but with an emergent Lifshitz scaling in the IR. The black-hole horizon has completely disappeared, and the charge is fully carried by the scalar field that extends all the way to the deep interior, with a profile and an associated electric field forming a consistent solution to the Einstein–Maxwell–Higgs system in the bulk. The RN black hole has "un-collapsed" by the self-gravitation of the Higgs condensate into an object that one can call a "Higgs star" or, perhaps better, a "Higgs lump", since it is lacking an edge. In full accordance with expectation, the Lifshitz deep infrared is associated with a unique ground state: since the extremal RN horizon has disappeared, this zero-temperature holographic superconductor has cured the ground-state entropy problem.

Summary 22 For a holographic superconductor at near zero temperatures, the magnitude of the scalar field in the deep interior of the bulk becomes so large that its gravitational effects have to be fully taken into account. The consequence is that the Reissner–Nordström black hole "un-collapses" into a "Higgs star" characterised by a Lifshitz-type deep-interior geometry with no ground-state entropy.

The semi-local quantum liquid

Something very interesting happens along the way, however. This Lifshitz regime only arises at *exponentially* low energies. This is directly visible in the geometry. In the Lifshitz scaling region the metric will be of the form Eq. (8.50) after a radial coordinate redefinition $r \rightarrow r^{1/z}$,

$$ds^2 = -\frac{r^2}{L^2} dt^2 + \frac{r^{2/z}}{L^2} dx^2 + L^2 \frac{dr^2}{r^2}. \tag{10.57}$$

Switching on a small temperature, i.e. generalising this metric to a black hole, amounts, to a first approximation, to the addition of a warp factor just like for a Schwarzschild solution,

$$ds^2 = -\frac{r^2}{L^2} f(r) dt^2 + \frac{r^{2/z}}{L^2} dx^2 + L^2 \frac{dr^2}{f(r)r^2}. \tag{10.58}$$

The near-horizon metric will look almost identical to the near-horizon metric of the RN black hole: this is the solution with $z = \infty$. The difference is only in the spatial directions along the boundary,

$$ds^2_{\text{RN-BH}} - ds^2_{\text{Lifshitz-BH}} = \frac{1}{L^2}\left(1 - r_0^{2/z}\right)dx^2 . \tag{10.59}$$

For $z \gg 2$ this rapidly vanishes as the temperatures $T \sim r_0$ increases from zero. Only for very small temperatures does the Lifshitz scaling become physically important. This happens when $T/\mu \ll c^{-z/2} = e^{-z\ln|c|/2}$, where c is an arbitrary threshold larger than one. Physically this means that, upon descending along the radial direction from the boundary, one first enters a regime where the finite density becomes relevant. Initially the amplitude of the scalar field is in this regime still too small, though, to make any difference compared with the non-backreacted solution. The system behaves as if it were evolving towards the extremal RN horizon. This implies that the system is effectively still governed by an AdS$_2$-like geometry, and the boundary system behaves like the local quantum critical metal at these intermediate energies, even in the presence of the superconducting order. Only at a scale given by [160],

$$r_{\text{Lif}} = r_* - L_2 e^{-z}, \tag{10.60}$$

does the backreaction by the scalar hair take over, changing drastically the overall solution. In this expression r_* and L_2 are the horizon coordinate and AdS$_2$ radius (8.22) of the RN black hole in the absence of the hair, respectively.

This phenomenon that the system first flows to an apparent locally quantum critical fixed point, only to turn away at the last moment, connects with an old idea of Anderson of *intermediate fixed points* [407]. Mathematically characterised by saddle points in the RG flow (Fig 10.15), they physically will reflect the scenario outlined above. At intermediate scales one may forget for all practical purposes that the fixed point is unstable: it fully controls the behaviour of the physics.

To obtain the full zero-temperature solution including the Higgs and Maxwell field profiles, it is necessary to specify a specific potential $V(\Phi)$ for the scalar field. We already explained that close to T_c the self-interactions are just overwhelmed by the effects of the geometry such that we could get away with ignoring these altogether, but this is no longer the case when we approach zero temperature. At zero temperature the Higgs field needs to be stabilised inside a bounded potential, and a "minimal" $\lambda|\Phi|^4$ self-interaction term in the potential $V(\Phi)$ suffices for this purpose. The hard work is then to find fully consistent non-linear solutions. For generic values of the parameters in the action (m, q, λ, the mass, charge and self-interaction, respectively) one obtains in the deep interior $r \ll 1$ the Lifshitz solution [404]

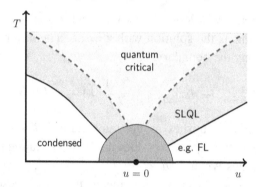

Figure 10.15 Intermediate-fixed-point flows of semi-local quantum liquids. Figure adapted from [160].

$$ds^2 = -r^{2z} dt^2 + g_0 \frac{dr^2}{r^2} + r^2(dx^2 + dy^2) \tag{10.61}$$

in coordinates $r \to r^z$, compared with Eq. (10.57), with

$$g_0 = \frac{2m^2\mathbf{z} + q^2(3 + 2\mathbf{z} + \mathbf{z}^2)}{6q^2}, \tag{10.62}$$

while both the gauge potential and the scalar field become independent of the radial coordinate

$$A_t = \sqrt{\frac{\mathbf{z} - 1}{\mathbf{z}}}, \quad \Psi = \sqrt{\frac{12\mathbf{z}}{2m^2\mathbf{z} + q^2(\mathbf{z}^2 + 2\mathbf{z} + 3)}}. \tag{10.63}$$

The equations of motion relate the dynamical critical exponent $\mathbf{z}$ to the parameters of the action through the equality

$$2q^2(m^2 - 2q^2)\mathbf{z}^3 + (m^4 + 2m^2q^2 - 4q^4 + 12\lambda_0)\mathbf{z}^2 + 4q^2(2m^2 - q^2)\mathbf{z} + 12q^4 = 0. \tag{10.64}$$

For large self-interaction λ_0 this can be solved with the outcome,

$$\mathbf{z} = \frac{6}{q^2(-m^2 + 2q^2)}\lambda_0 + \left(-1 + \frac{m^4}{2q^2(-m^2 + 2q^2)}\right) + \mathcal{O}\left(\frac{1}{\lambda_0}\right). \tag{10.65}$$

To arrive at the complete solution which connects this deep-IR Lifshitz solution to the original AdS$_{d+2}$ UV one needs to resort to numerics [404, 405]. These techniques will be explained in the next chapter in box 11.2.

A Lifshitz geometry is the generic solution. For special values of the parameters one can find alternative IRs. For very small and vanishing λ, the potential becomes very flat, with all the difficulties associated with stability one anticipates on the basis of common sense. Nevertheless, the backreacted solutions have been found [405]. When the charge q of the scalar field vanishes, one finds that the extremal

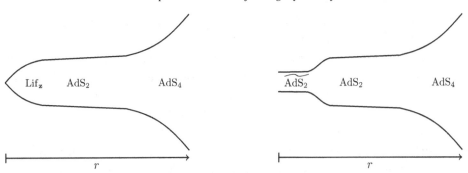

Figure 10.16 The zero-temperature geometry after considering the full gravitational backreaction of the scalar field [160]. The left panel is the generic solution. Upon approaching the interior, first an AdS_2 geometry is realised as if there were still an RN black hole. In this regime the magnitude of the scalar field is just too small to cause appreciable changes to the geometry through backreaction. However, on penetrating further into the deep interior a crossover occurs at the radial scale $r \simeq r_* - L_2 e^{-z}$. Here the backreaction by the scalar field takes over, causing the geometry to turn into a Lifshitz scaling geometry that persists all the way to the origin of AdS. In special cases, e.g. when the charge q of the scalar field vanishes, the local quantum critical nature of the RN black hole can survive at zero temperature. This is shown in the right panel. The deep-infrared geometry is again AdS_2, albeit with a radius that is modified by the scalar field. Figure adapted from [160].

RN horizon survives [406], albeit with a modified AdS_2 radius, as indicated in the left panel of Fig. 10.16. Alternatively, for a sufficiently positive mass squared the emergent IR in the zero temperature superconductor can even "rediscover" a $z = 1$ AdS_4, albeit with a smaller AdS radius.

The physical characteristics of Lifshitz solutions were discussed in section 8.4. For the boundary field theory the deep infrared in the presence of the superconducting order behaves like a quantum critical phase with time scaling as a power of space, in terms of the bulk geometry

$$r \to b^{-1}r, \quad t \to b^z t, \quad \{x, y\} \to b\{x, y\}. \tag{10.66}$$

One particular consequence is that the specific heat will behave as $C_V \sim T^{d/z}$. Although we have already encountered emergent Lifshitz geometries in the scaling atlas of chapter 8, let us emphasise how utterly surprising this result is from the perspective of forming an ordered, symmetry-broken state in the boundary. The scaling atlas itself was already rather counterintuitive: the UV scale invariance is badly broken by the chemical potential, yet an emergent scale invariance is dynamically generated in the infrared, describing quantum critical *phases* with non-classical scaling properties. In the holographic superconductors this is taken a step further. The scale invariance is broken yet again by the cohesive superconducting order, but this fails to generate a gap. We can learn some lessons from the discussion of the probe fermions of the previous chapter. Semi-holographically the single-trace order parameter confines and condenses. Associated with it is a massless Goldstone mode [333, 369]. However, in addition to the Goldstone sector there is a massless critical sector that survives. In the backreacted solution, this sector now exhibits Lifshitz scaling as opposed to the AdS_2 scaling in the probe limit, but it stays massless nevertheless. Why this is so remains a puzzle. It could be a very interesting intrinsic feature of the field theories describable by holography. To a degree the Lifshitz geometry itself might be an artefact of the large-N limit in holography. Lifshitz geometries have a mild singularity, as we saw in chapter 8, and this geometry is expected to be resolved by higher-derivative corrections on the gravity side encoding $1/N$ corrections in the boundary field theory.

We will take up this theme in the final chapter. The Hartree–Fock wisdom that the order parameter goes hand in hand with a gap function is rooted in the product-state structure of the ordered ground state. The violation of this principle as encountered above might be a symptom of long-range entanglement that is preserved in the presence of spontaneous symmetry breaking. A helpful analogy might be found in the physics of "conventional" ordered states formed from fermions. For unconventional (non-s-wave) superconductors, but also weakly coupled charge- and spin-density waves, one finds that massless fermionic excitations survive the order, getting reconstructed by the order in the form of nodal fermions, small Fermi-surface pockets and so forth. As we explained in section 2.2 using the RVB vacuum structure as an example, such states are actually long-range entangled, albeit through the "baby" anti-symmetrisation entanglement of free fermions. It could be that a long-range entanglement physics of a yet very different kind is revealed by the quantum critical phases coexisting with order, as predicted by holography.

Summary 23 A rather mysterious feature of the zero-temperature holographic superconductors is the persistence of gapless degrees of freedom deep inside the superconducting gap. These are governed by a "Lifshitz geometry". This is the gravitational encoding for a scaling behaviour governed by a dynamical critical exponent z that can vary between $z = 1$ (Lorentz invariance) and $z = \infty$ (AdS_2), depending on the details of the system.

11

Holographic Fermi liquids: the stable Fermi liquid and the electron star as holographic dual

In chapter 9 we introduced the holographic description of single-fermion propagators in the finite-density Reissner–Nordström metals. We discussed in particular how sharp Fermi surfaces can arise due to the approximate confinement of fermionic excitations in the potential well created by the geometrical domain wall. The interaction of these fermionic quasi-bound states with the strongly coupled AdS$_2$ IR gives these a finite lifetime. Eventually it decays into the quantum critical horizon, and this imbues the state with its non-Fermi-liquid properties. For particular parameter choices these fermionic responses can closely resemble the "marginal-Fermi-liquid" spectral functions suggested by photoemission experiments in the cuprate strange metals. However, these computations involved an approximation in the form of the probe limit, which assumed that the bulk fermions do not influence the bulk physics.

But the reader is now familiar with holographic superconductivity, a context where the limitations of the probe limit become very explicit: the violation of the BF bound by the fluctuations of a scalar field in the bulk signals an instability of the vacuum, and one has to recompute the response after the backreaction of the scalar field has been fully accounted for. Ignoring this amounts to computing the physics of an unstable, false vacuum. We emphasised in chapter 9 that a similar issue arises in the parameter regime of the fermion scaling dimension and charge where the holographic Fermi surface is formed. Invariably the fermion propagators show in this regime a log-oscillatory behaviour at small momenta, which is caused by a BF-bound violation of the bulk fermions. The physics in the bulk behind this log-oscillatory behaviour is Schwinger pair production in the background of the charged black hole. Although the physics is less straightforward than for the bosonic fields, it does indicate that the system is unstable – the Fermi surfaces of chapter 9 are properties of a false vacuum. To stabilise the bulk it has to be that the fermionic states in the bulk get occupied. In the true ground state the bulk fermions have to have a macroscopic effect themselves, and the only way this can be accomplished is by forming finite-density fermionic matter. The bulk fermions are non-interacting to leading order in $1/N$, but they are still subject to Fermi–Dirac statistics. Thus a finite-density Fermi gas has to form in the bulk with a charge and energy density that will modify the gauge fields and the geometry.

As in the holographic superconductor, this change in geometry and electric field due to the new matter and charge distributions in the gravitational bulk will translate into a change in the physical nature of the state of matter realised in the boundary. As we will explain in this chapter, the outcome is that in analogy with the holographic superconductor, a stable cohesive state is formed. This state is a *real Fermi liquid*. Holography appears to deliver here a remarkable insight into a classic mystery of fermionic quantum matter physics. As we discussed at length in section 2.3, it has long been understood that the Fermi liquid represents some kind of cohesive "orderly" state characterised by its own form of rigidity. The revelation is that in the gravitational dual this rigidity is governed by a principle that is qualitatively similar to bosonic cohesive matter: instead of the "scalar hair" in front of the horizon of the holographic superconductor, the RN black hole acquires "fermionic" hair that is dual to the Fermi liquid in the boundary. An implication is that holography gives for the first time access to the physics that governs the emergence of the Fermi liquid from a non-rigid fermionic state. In this way, one can even attempt to address the general characteristics of quantum phase transitions where the Fermi-liquid "order" forms.

The bulk physics is much more involved in the case of fermions, however, to the degree that at the time of writing it has not been fully brought under control. This can be traced directly to the different nature of the inherently quantum Dirac equation compared with classical bosonic field equations. The appearance of a finite bosonic order parameter is driven by the violation of the BF bound in the deep IR. For a boson this is nothing other than a classic roll-down from an unstable potential through a tachyonic mode. The Schwinger pair production is just the coherent build-up of this mode. In the black-hole literature this stimulated emission is known as super-radiance [408]. Owing to the Pauli principle, fermions cannot do this, and the precise physical mechanism of the formation of the bulk fermion system beyond the onset of Schwinger pair production is not yet fully understood. Nevertheless, we can compare thermodynamically the initial AdS_2 metal with the most likely end state: the backreacted Fermi gas. As we will see, this does show that holographically the Fermi-liquid "order" forms in a way that is remarkably similar to bosonic order. There are still distinct differences, of course, especially at finite temperature. The bosonic super-radiance mechanism clearly translates naturally into continuous *thermal* phase transitions in the field theory, which is consistent with the expectation for the breaking of a global symmetry. However, given its effectively one-dimensional nature, Fermi-liquid "order" will be destroyed at any finite temperature. The conclusion is that the only phase transition that can occur is the van der Waals-type liquid–gas transition driven by a singular change in the density. The holographic behaviour of the bulk Fermi gas when the temperature is increased is consistent with this fundamental property: part of the confined Fermi gas suddenly collapses into a deconfined Schwarzschild black hole when the temperature gets too high, corresponding to a first-order density-driven transition [409]. This true nature of the phase transition is revealed only at first order in $1/N$. In a fluid approximation for the Fermi gas in the large-N limit it is artificially of third order [410, 411]; see section 11.3.1.

In principle the Fermi liquid can also emerge from the AdS_2 metal at zero temperature through a quantum phase transition by tuning the scaling dimension of the fermionic operator from the stable region down to a value that exhibits a log-oscillatory spectral function. This is technically an extremely delicate procedure insofar as it requires a very accurate treatment of the inherent quantum-mechanical nature of the fermions [412, 413]. This has not yet been accomplished. Instead, all approaches to the finite density of fermions in the bulk rely on some approximation to circumvent these difficulties. In section 11.1 we discuss an approach we have seen before. One can introduce an explicit hard wall to manifestly confine the fermions. In this way one removes the strongly coupled charged IR from the Fermi gas. The bulk problem turns into a simple Fermi gas in a box, dualising into a free Fermi gas of composite operators in the boundary. This hard wall clearly illuminates a number of details of correspondence for charged fermions. In particular, it shows how Luttinger's theorem governing the Fermi-surface volume of the Fermi liquid in the boundary is related to the charge density "in front of the horizon" in the bulk.

This hard-wall construction has the drawback that one has removed by hand the potentially important physics associated with the deep interior in the presence of the bulk fermion matter. In section 11.2 we will discuss a different approximation that makes it possible to compute this deep-infrared physics. This is based on a highly intuitive gravitational construction. What is the physical system of degenerate Fermi matter held together by gravity? The answer is a star, such as an astrophysical neutron star in the case of neutral fermions. Since the density is high, one can approximate the bulk Fermi gas by a fluid. This is the well-known semi-classical Thomas–Fermi approximation that takes the level spacing to zero at a fixed total energy and charge. In this fluid limit the gravity system reduces to a well-known exercise: the equations for a self-gravitating fermionic fluid were first written down and solved by Tolman, Oppenheimer and Volkoff in the 1930s to describe neutron stars. Here we have charged fermions and a more appropriate name is an "electron star". The gravitational background of this solution reveals the indiscriminate way in which gravity reacts to matter, whether fermionic or bosonic. As in the holographic superconductor, the geometry in the deep interior of this star is again of the Lifshitz type, with a finite dynamical critical exponent **z**.

The price that is paid by using the fluid limit is revealed when we inspect the fermion propagators in the boundary. This reveals a near infinity of regularly spaced concentric Fermi surfaces, as if it were a system characterised by a near infinity of nearly degenerate bands. Although such Fermi liquids might be realised in special limits of finite-density large-N Yang–Mills theories, it is clear that this is not a very satisfactory situation when one wants to address the Fermi liquids realised in the "single-flavour" electron systems of condensed matter physics. This infinity of Fermi surfaces is surely not a fundamental property of holography. It is just a consequence of the desire to keep the bulk physics tractable. To describe systems with a few Fermi surfaces, or even a single Fermi surface, one has to handle the quantised nature of the fermions forming the matter in the bulk. This is technically a highly complicated affair, which has not quite completely been accomplished yet. In the final section of this chapter, section 11.3, we will first discuss some recent progress in this direction. One can at the very least address leading-order quantum corrections, relying on a semi-classical WKB-style re-quantisation of the electron star in the fluid limit. This shows that the fluid limit is mildly singular: the main result is that the third-order density-driven thermal transition turns into a healthy first-order van der Waals transition as soon as the quantum corrections become finite. We then address the nature of the zero-temperature transition from the strange metal to the holographic Fermi liquid. We do so in an interesting construction that still rests on the fluid limit, but by employing a dilaton in the bulk one can nevertheless tune between the Fermi liquid and the strange metal. It shows that there is a coexistence regime where the charge gradually transfers from the Fermi liquid to the strange metal in a Lifshitz-like transition.

We end this section with a short discussion of other potential instabilities of the "electron-star" Fermi liquid. The standard BCS superconductivity mechanism can be naturally included in holography. The simplest way is to switch on an attractive fermion–fermion interaction in the bulk, with the effect that the fermions in the star will pair as in a neutron star. Given the one-to-one nature of holographic duality for confined Fermi liquids, this describes a literal BCS superconducting instability of the Fermi liquid in the boundary. There is also a more novel instability, although it is natural from the gravity side. If the boundary exists in a finite volume, one finds a zero-temperature/finite-density analogue of the Hawking–Page transition of chapter 6. This is the collapse of an electron star back to a black hole beyond a critical density. It describes a transition from a low-density confined phase back to a high-density deconfined phase in the boundary theory.

11.1 The cohesive Landau Fermi liquid from hard-wall holography

We saw in chapter 6 that a holographic space-time with a hard-wall cut-off is an extreme but simple way to encode confinement with a gap in the dual theory. In the confined phase of the strongly coupled boundary theory single-trace operators that are neutral under similarity transformations of the large-N matrices become nearly free and particle-like. The spectrum no longer exhibits un-particle branch-cuts but has distinct particle peaks above the gap. The radial position r_c of the wall sets the gap in the dual theory. From the perspective of the charge dynamics, this gap gives away that the system is a trivial insulator.

Qualitatively this physics is readily understood from the gravity theory. At the boundary the asymptotically AdS geometry always supplies a potential barrier. With a hard wall in the interior the radial direction becomes an effective box. The normalisable bulk fluctuations are therefore quantised in sharp radial harmonics, while they are still free to disperse in the extended directions parallel to the boundary. As the GKPW rule clearly reveals, each radially normalisable mode corresponds to a peak in the boundary correlation function: these radial harmonics are in a one-to-one correspondence to the spectrum of particles. Moreover, the minimal energy of the lowest radial mode characterises the gap.

Except for the details of the potential shape, this is also the physics behind the quasiparticle-like excitations around the holographic Fermi surfaces of chapter 9. These correspond now to the normalisable bound states in the Schrödinger potential, instead of a hard-wall potential well. Sachdev realised that this means that there is an easy way to understand how to construct a bulk solution containing finite-density fermion matter [414]. By hand we can insert a potential of the hard-wall type into the bulk, and in this box on the gravity side it is straightforward to populate the bulk free-fermion states. This will of course miss precisely the subtle way in which the deep AdS interior might respond to the finite density of ordinary matter. By construction the potential ramifications for a strongly coupled IR in the boundary will be cut off by the gap due to the wall. At the same time, this is the simplifying circumstance which makes it technically straightforward to account for the effects of the backreaction due to the finite-density Fermi gas in the bulk. Thanks to the hard-wall gap, the gravitational backreaction can be in the first instance ignored altogether, and we need to take into account only the effects of the bulk Fermi gas on the electrostatic potential matter. This is simple screening physics that can easily be described by a Hartree mean-field calculation. A typical outcome for the bulk Fermi gas formed from the mean-field-adjusted normalisable fermion modes is shown in Fig. 11.1. One starts with an Ansatz for the gauge potential profile, whose leading term gives the chemical potential in the boundary. By occupying normalisable states in this particular bulk background up

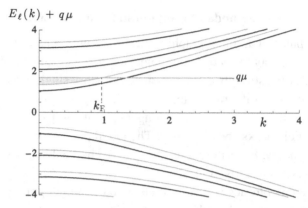

Figure 11.1 Spectrum of normalisable fermionic modes in AdS with a hard wall. This corresponds to a field theory in the confining phase. The darker lines reflect the pure AdS spectrum for $m = 1$, $r_c = 1/3$. Increasing the chemical potential μ there is a critical value at which the energies become negative. Occupying these states changes the electrostatic potential, which in turn modifies the fermion spectrum. The lighter lines show this adjusted spectrum for $\mu = 1$ and $q = \sqrt{3}$. The horizontal line marked with $q\mu$ shows the zero of energy at the effective chemical potential. These results are obtained with a value for $q\mu = \sqrt{3}$. The shaded region shows the occupied states. Figure source [414]. (Reprinted figure with permission from the American Physical Society, © 2011.)

to the chemical potential, one constructs a Fermi gas. Subsequently one computes its Hartree mean-field potential and energy; this corrects the background potential, and one iterates until the solution is self-consistent (box 11.1).

This exercise explicitly reveals how the probe-fermion results of chapter 9 should be interpreted. As Fig. 11.1 shows, there is not a single normalisable state in this hard-wall model, but instead each radial harmonic extends into a continuum. By occupying states in just one such radial harmonic sector, an effectively free Fermi gas is formed, where each state is infinitely sharp and in one-to-one correspondence to a specific Dirac wave in the bulk. The system formed in the boundary is just this bulk Fermi gas "projected on the holographic screen". Moreover, because the bulk system is effectively free, the set of normalisable states is exactly reflected in the fermion spectral function in the boundary. In other words, the peaks are infinitely sharp for this hard-wall star, and *the self-energy completely vanishes*. The reason is easy to understand either from the Schrödinger potential or from the semi-holographic point of view. The hard wall has removed the horizon, and the fermions can no longer tunnel to a deep interior: accordingly, there is in the dual boundary no longer a coupling to a strongly coupled locally quantum critical IR. One can subsequently argue that there are weak interactions between the fermions in the bulk at one order higher in the $1/N$ expansion. Including these corrections will result in the conventional $\Sigma \sim \omega^2$ damping [414]. This should again be in a one-to-one relationship with the fermions in the boundary. The conclusion is striking: *the quantum-corrected holographic hard-wall fermions describe a precise Landau Fermi liquid in the boundary.*

A regular weakly interacting Landau Fermi liquid in the bulk translates into a regular weakly interacting Landau Fermi liquid in the boundary field theory. How does this square with the strong–weak duality inherent in AdS/CFT? The explanation follows directly from the confinement due to the hard wall, together with the large-N factorisation. Confinement precisely means that the underlying forces in the boundary gauge theory are so strong that "neutral" gauge-invariant objects separate out of the collective whole. These gauge-neutral excitations can be cluster decomposed, and at large enough distances they appear as individual excitations. The notorious examples are the mesons, protons, neutrons and other hadrons in QCD. In chapter 6 we showed how hard-wall holography precisely encodes this for meson-like operators; baryons are infinitely heavy in the large-N limit and not easily incorporated into holography [415, 416]. Moreover, in the large-N limit the weak remaining "di- and multi-polar" interactions between the mesons – pion exchange in QCD – are completely suppressed and the gauge-invariant operators become exactly free. What we see here is the same physics at work, but now for the fermionic analogues of the mesons: the "mesino" bound states of a gauge-charged boson and a gauge-charged fermion. These mesinos carry an additional quantum number with respect to which we can then build a finite-density system. In the large-N limit this system is non-interacting, and at finite density these mesinos form a regular Fermi gas.

Now that we know precisely the correct physics in the boundary, we conclude that once again the holographic machinery delivers impeccably. The zero-temperature Fermi-liquid state which is ubiquitous in the "simple" metals as realised by nature can be encoded in terms of the gravitational bulk theory. The only fundamental requirement is that the system is in a confining regime. We also learn that, thanks to the Pauli principle, the low-energy quasiparticles in the bulk below the confinement scale are in one-to-one correspondence to the quasiparticles in the boundary. In holography a (confined) Fermi liquid maps to a (confined) Fermi liquid.

11.1.1 Luttinger's theorem in holography

To elaborate further on the one-to-one holographic relation between radial excitations in the bulk and a weakly coupled Fermi liquid in the boundary, let us consider the fate of Luttinger's theorem in the bulk. This fundamental Fermi-liquid theorem prescribes that the area enclosed by the Fermi surfaces precisely adds up to the microscopic charge density. The connection is again through the identification of the bulk fermion excitations with the boundary ones. By occupying the fermion states associated with the lowest radial harmonic, one constructs the bulk Fermi gas associated with a single Fermi surface. This manifestly satisfies Luttinger's theorem: the full bulk charge density is carried by the free bulk fermions and automatically equals the area of its Fermi surface. Since each state associated with the lowest radial harmonic corresponds to a state in the boundary theory, the Fermi surfaces do match.

The conclusion seems rather trivial, but on careful inspection (box 11.1) it reveals an important insight into the holographic dictionary. To compute the charge density in the bulk, one has to pay special attention to the boundary conditions in the interior. There could be electric flux emanating from the horizon, as in the Reissner–Nordström black hole. However, the natural boundary condition is that the radially oriented electric field vanishes right at the wall. As a consequence, the full charge density in the system arises from the occupied modes in the bulk, and therefore Luttinger's theorem is satisfied. The immediate corollary is that *any holographic system with a non-vanishing electric field on the horizon cannot satisfy Luttinger's theorem.* In the remainder we will encounter this lesson a number of times.

Luttinger's theorem truly confirms that the hard-wall star describes the regular Landau Fermi liquid. There are two aspects of the hard-wall star, however, that are special. From the first chapter onwards we have emphasised the interpretation of the radial direction as the scale in the dual theory. However, in the hard-wall star it is obvious that there are zero-energy states – the quasiparticles at k_F – while the geometry is cut off along the radial direction. The latter should imply the absence of gapless states if the GR = RG rule is taken literally. The hard-wall "star" is an explicit example showing that the relation between the radial direction and the scale of the dual theory is more refined. It still holds in a qualitative sense if one considers momentum scales rather than energy scales. A similar effect can happen for Goldstone modes in holographic superconductors, however. If one considers a charged scalar at finite density in the capped AdS-soliton background of chapter 6, one can drive the system from the gapped insulator to a superconducting state [402]. Irrespective of the Goldstone bosons, the geometry remains capped off on both sides of the transition. What is happening is that, since both the fermions and the bulk collective modes are of order $1/N$, these do not contribute to the semi-classical geometry described by GR. The order-N^2 gauge-charged degrees of freedom are actually gapped by the hard wall. But the $1/N$ sub-leading operators are not, and these span the normal full energy range. The hard-wall star is a prototypical illustration of this exception to the radial-direction/energy-scale rule of thumb.

A second noteworthy aspect is that the boundary Landau Fermi liquid corresponds to occupying only the *lowest* radial harmonic in the bulk. By raising the chemical potential one can start to populate the next radial harmonic as well. One can think of this as a second electronic band. Strictly speaking, the hard wall therefore describes a set of multiple Fermi liquids, with a regular hierarchy in Fermi momenta. It is clear why such a hierarchy emerges from the mathematics behind the dictionary, and this will become important in the discussion below. The physical reason why deforming a conformal field theory with a finite chemical potential gives rise to these multiple Fermi surfaces, instead of just a single one, is presumably related to the rearrangement of the states in representations of the UV conformal group, but this is not fully understood.

Box 11.1   **Computing the backreacted hard-wall electron star**

We follow Sachdev's original construction of backreacting fermions in the hard-wall background [414]. As in chapter 9, the equations of motion follow from the AdS$_4$ Einstein–Maxwell Lagrangian with a minimally coupled massive fermion of charge q dual to a fermionic operator with conformal dimension $\Delta = d/2 + mL$,

$$S = \frac{1}{2\kappa^2} \int d^4x \sqrt{-g} \left[R - 2\Lambda - \frac{\kappa^2}{4e^2} F_{\mu\nu} F^{\mu\nu} \right.$$

$$\left. - \bar{\Psi} \left[e_a^\mu \Gamma^a \left(\partial_\mu + \frac{1}{4}\omega_{\mu ab}\Gamma^{ab} - iqA_\mu \right) - m \right] \Psi \right]. \quad (11.1)$$

Similarly to the holographic superconductor, the gravitational backreaction is controlled by the ratio of the charge to the gravitational coupling constant $\kappa/(qL)$. In the limit that the fermion charge is large we can ignore the gravitational backreaction, and the background geometry is pure AdS:

$$ds^d = \frac{r^2}{L^2}(-dt^2 + dx_i^2) + L^2\frac{dr^2}{r^2}. \quad (11.2)$$

The difference is that in the hard wall we cut off this geometry by hand at $r = r_c$. The dynamical equations are thus the coupled Maxwell–Dirac system in this background:

$$D_\mu F^{\mu\nu} = qi\bar{\Psi}\gamma^\mu\Psi,$$

$$\left(e_a^\mu \left(\partial_\mu + \frac{1}{4}\omega_{\mu ab}\Gamma^{ab} - iqA_\mu \right) - m \right) \Psi = 0. \quad (11.3)$$

Box 11.1 (Continued)

We again remove the spin connection $\omega_{\mu ab}$ by the redefinition

$$\Psi = (-gg^{rr})^{-1/4} \begin{pmatrix} \chi_+ \\ \chi_- \end{pmatrix}. \tag{11.4}$$

Then we Fourier transform, and choose without loss of generality the boundary momentum along the x direction, $\vec{k} = (k_x, 0)$. Projecting onto the t-helicity $\chi_\pm$ eigenstates of $\Gamma^5\Gamma^x$, the Dirac equation reduces to

$$\sqrt{\frac{g_{ii}}{g_{rr}}} \left(i\sigma_2 \partial_r - \sqrt{g_{rr}}\sigma_1 mL \right) \chi_i = \left((-1)^i k_x \sigma_3 - \sqrt{\frac{g_{ii}}{-g_{tt}}} (\omega + q\Phi) \right) \chi_i. \tag{11.5}$$

It suffices to consider only χ_+ from here on, since the results for χ_- simply follow on changing $k_x \to -k_x$.

Occupying states to build a finite-density Fermi gas means that we populate the normalisable solutions to the Dirac equation. Near the AdS boundary the asymptotic behaviour is the same as before in chapter 9,

$$\chi(r) = a \begin{pmatrix} 0 \\ 1 \end{pmatrix} r^{mL} + b \begin{pmatrix} 1 \\ 0 \end{pmatrix} r^{-mL} + \cdots, \tag{11.6}$$

where the normalisable solutions are those with $a = 0$. The full normalisation condition of the Dirac field is affected by the redefinition that removes the spin connection. In terms of the t-helicities it is

$$\int_{r_c}^{\infty} dr \left(\chi_+^\dagger \chi_+ + \chi_-^\dagger \chi_- \right) = 2. \tag{11.7}$$

The value 2 is due to the fact that the relativistic spinor in $3 + 1$ dimensions has two spin degrees of freedom. For simplicity we shall first ignore the spin degeneracy and set $\chi_- = 0$ by hand. We will comment on this choice at the end. Thus we have the simple normalisation condition for the two-component spinor χ_+

$$\int_{r_c}^{\infty} dr \, \chi_+^\dagger \chi_+ = 1. \tag{11.8}$$

We finally need the boundary conditions at the hard wall. To deduce these we insert the Dirac operator in the normalisation condition, take the complex conjugate and integrate by parts. This shows that we must demand

$$\chi^\dagger(r_c)_+ \sigma^y \chi_+(r_c) = 0 \tag{11.9}$$

to ensure real eigenvalues, i.e. on such solutions the Dirac operator is self-adjoint. This condition equals $\bar{\chi}\sigma^x\chi = 0$. Physically it means that there is no flux across the wall.

We now combine these equations with the Maxwell equation. Since the system is isotropic in the $\{x, y\}$ direction, only $A_0 = \Phi$ and A_r are relevant degrees of freedom, and the latter can be set to vanish by a gauge choice. Thus Maxwell's equation reduces to

$$\partial_r^2 \Phi = -q \int \frac{d^2k}{4\pi^2} \langle \bar{\Psi}^\dagger(k)\Psi(k)\rangle. \tag{11.10}$$

The angle brackets on the right-hand side denote the expectation value of the number operator. Formally it is the convolution of the bulk spectrum with the Fermi–Dirac distribution

$$\partial_r^2 \Phi = -q \int \frac{d^2k}{4\pi^2}\frac{d\omega}{2\pi} \operatorname{Im} G_R(\omega, k, r, r)n_F(\omega, k; T). \tag{11.11}$$

For fermions at zero temperature this is just the sum over all "negative"-energy states with normalisable wave functions:

$$\partial_r^2 \Phi = -q \int \frac{d\omega\, d^2k}{(2\pi)^3}\theta(-\omega)\chi_+^\dagger\chi_+. \tag{11.12}$$

Following the by-now-familiar dictionary rules, the AdS boundary values of the electrostatic potential

$$\Phi = \mu - \frac{\rho}{r} + \cdots \tag{11.13}$$

encode the chemical potential and charge density of the dual CFT. In the interior at the hard wall we will demand that the electric field vanishes, $\partial_r\Phi(r_c) = 0$, i.e. there are no sources of charge emanating from behind the wall. This is as it should be if the system is gapped, because then there are no low-energy charge carriers remaining.

We can now initiate a Hartree procedure. We compute the spectrum of normalisable solutions to the Dirac equation Eq. (11.5) for a fixed given background potential $\Phi(r)$ dual to a CFT chemical potential μ, sum over all negative-energy wave functions, find the corrected potential using the Maxwell equation (11.18) and iterate by substituting this potential back into Eq. (11.5).

Starting at $\mu = 0$, the Dirac equation can be solved exactly by Bessel functions [414]

$$\chi(r; \omega, k) = \frac{1}{\sqrt{r}}\begin{pmatrix} -(M_\ell/(k + \omega(\ell, k)))J_{m+1/2}(M_\ell/r) \\ J_{m-1/2}(M_\ell/r) \end{pmatrix}. \tag{11.14}$$

Box 11.1   (Continued)

Here $M_\ell = r_c j_{m-1/2,\ell}$ with $j_{m-1/2,\ell}$ the ℓth zero of the Bessel function $J_{m-1/2}(x)$. This ensures the correct boundary condition Eq. (11.9) at the hard wall. The normalisability condition at the AdS boundary, $a = 0$, picks out distinct energy eigenvalues

$$\omega(k, \ell) = \pm\sqrt{k^2 + M_\ell^2}. \tag{11.15}$$

Figure 11.1 illustrates the spectrum. We now assume that we can ignore all the negative-energy solutions in the Dirac sea, and consider only the solutions $\omega(k, \ell) > 0$. In the absence of occupied fermion states a constant value of $\Phi = \mu$ is a solution to the Maxwell equations of motion with the right boundary conditions. Increasing the chemical potential in this way shifts the eigenvalues $\omega(k, \ell)$ to $\omega(k, \ell; \mu) = \omega(k, \ell) - q\mu$. Nothing happens until one reaches the critical value where $\omega(k, \ell; \mu) = 0$. At this moment one can occupy fermionic states and, through Maxwell's equations, these will subtly alter the AdS electrostatic potential. This effect increases as one increases μ. Using the iterative Hartree algorithm outlined above, this converges rapidly to the corrected spectrum displayed in Fig. 11.1.

We can now check the claim that this state is precisely the holographic dual of the regular Landau Fermi liquid. Recall from Eq. (11.14) that the solutions to the Dirac equation we construct to occupy the states in the bulk can also be used to construct the Green function in the CFT. Precisely for a normalisable solution $a = 0$, one finds a pole in the Green function. In the hard wall this pole becomes a branch-cut along the full dispersion curve in Fig. 11.1. It is immediately clear that this is due to the fact that we are approximating the system by free fermions in the bulk. As a result the self-energy at low frequency in the boundary also vanishes, and we have also a free-boundary-fermion spectrum. Going beyond the bulk free-fermion approximation by including loop effects, the conventional Landau Fermi-liquid argument for interacting fermions in the bulk will give rise to the characteristic self-energy $\Sigma \sim i\omega^2$. One true pole will remain: the defining quasiparticle pole at $\omega = 0$ corresponding to a distinct momentum value k_F.

A holographic version of Luttinger's theorem

From the relation between the boundary Green function and the bulk wave functions, we directly see that the Fermi momentum k_F in the strongly coupled boundary is the same k_F as that corresponding to the bulk weakly coupled Fermi gas. We can immediately draw an important conclusion [414]. For holographic fermionic systems a version of Luttinger's theorem is satisfied if all the charge is carried by the dual bulk fermions. In particular, there should be no charge contained within the horizon.

The argument is straightforward. The macroscopic charge density in the boundary field theory is by definition the differential of the free energy with respect to the chemical potential,

$$\langle Q \rangle = -\frac{\partial F}{\partial \mu}. \tag{11.16}$$

In AdS/CFT the chemical potential is encoded in the asymptotic behaviour of the electrostatic potential,

$$A_t(r) = \mu - \frac{\rho}{r} + \cdots, \tag{11.17}$$

with ρ the charge density, and the free energy is equal to the on-shell Euclidean AdS action. In an isotropic system the relevant part of the action is (we have used gauge freedom to set $A_r = 0$)

$$S_{\text{Eucl}} = \int dr \sqrt{g} \big(g^{rr} g^{tt} (\partial_r A_t)^2 + A_t J \big)$$

$$= \int dr \sqrt{g} A_t (-\partial_r (g^{rr} g^{tt} \partial_r A_t) + J) + \oint_{r=\infty} \sqrt{g} g^{rr} g^{tt} \mu \, \partial_r A_t. \tag{11.18}$$

The first term, the equation of motion, vanishes on-shell and (in isotropic systems) the free energy and charge density in the strongly coupled field theory are thus given by

$$F = S_{\text{Eucl}}^{\text{on-shell}} = \oint_{r=\infty} \sqrt{g} g^{rr} g^{tt} \mu \, \partial_r A_t |_{\text{on-shell}},$$

$$\tag{11.19}$$

$$\langle Q \rangle = - \oint_{r=\infty} \sqrt{g} g^{rr} g^{tt} \partial_r A_t |_{\text{on-shell}}.$$

In other words, the total charge density in the system is the appropriately normalised electric field in the radial direction, evaluated at the boundary. This agrees with Eq. (11.17), as it should.

The electric field in the radial direction also counts the total charge contained within the bulk theory. From Maxwell's equation,

$$\partial_r \big(\sqrt{g} g^{rr} g^{tt} \partial_r A_t \big) = J, \tag{11.20}$$

we can determine that the bulk charge density is given by

$$\sqrt{g} g^{rr} g^{tt} \partial_r A_t |_{r=\infty} - \sqrt{g} g^{rr} g^{tt} \partial_r A_t |_{r=r_{\text{hor}}} = Q_{\text{bulk}}$$

$$= \int_{r_{\text{hor}}}^{\infty} dr \big(\sqrt{g} g^{rr} g^{tt} \partial_r A_t \big). \tag{11.21}$$

Box 11.1 (Continued)

Thus we obtain the relation

$$\langle Q \rangle = Q_{\text{bulk}} - \underbrace{\sqrt{g}\, g^{rr} g^{tt}\, \partial_r A_t\big|_{r_0}}_{\text{flux from horizon}}. \tag{11.22}$$

If the bulk charge Q_{bulk} is carried solely by weakly coupled fermions, its Luttinger theorem states that $Q_{\text{bulk}} \sim k_F^2$. If, in addition, there is no flux from the horizon, then also $\langle Q \rangle = k_F^2$, and from the equality of the bulk Fermi momentum with the boundary Fermi momentum, a boundary Luttinger theorem follows.

Lifting of spin degeneracy from spin–orbit coupling

We ignored the spin degeneracy in the construction of the hard-wall star for simplicity. In truth, except for $\mu = 0$, there is no true spin degeneracy. As is evident from Eq. (11.5), the Dirac equation for the negative- and positive-spin components χ_- and χ_+ is different. Their wave functions are therefore different, even though the dispersion relation is the same for constant μ. The difference in their wave functions will, however, subtly change the backreaction-corrected potential for each spin component. This manifestly lifts the spin degeneracy. On the boundary side this is due to a spin–orbit coupling effect [417]. In the absence of interactions this has no further consequences.

11.2 The electron star as the dual of holographic fermions

In the hard-wall construction one excises by hand the contribution from the quantum critical physics in the deep IR. In the probe limit of chapter 9 we saw that the normalisable fermionic excitations – the bound states in the potential well – are not truly confined. They decay at large times with a locally quantum critical self-energy $\Sigma \sim \omega^{2\nu_k}$. For $\nu_k < 1/2$ arguably there is not even a regime where they appear as particle-like excitations. The hard-wall construction brutally ignores this, and makes the normalisable excitations infinitely long-lived. Confinement is thus exact. As a result, conventional condensed matter wisdom takes over, and the system is a regular Fermi liquid.

Our primary interest, however, is in *novel* states of matter. Given the remarkable non-Fermi-liquid responses revealed by the fermionic probes, the real question is how the system behaves when the bulk geometry does respond to the presence of a finite density of bulk fermion matter. Moreover, as we emphasised repeatedly, this question has to be asked, since the same probes revealed that the pure Reissner–Nordström background is unstable with respect to spontaneous pair production. We cannot track this "black-hole un-collapse" process dynamically, but the final equilibrium state can be addressed in principle by studying the finite-density gravity system beyond the probe limit.

The technical puzzle is how to accurately account for the effect of the finite density of charged fermions on the gauge field and gravity profile in the bulk. In particular, the IR of the geometry should and will be drastically altered in the absence of a hard wall. How will it change? The first lesson we should take to heart is the one from the previous subsection. There will be a Fermi gas in the bulk dual to the "mesinos" and confined to the potential well. The second insight is that gravity responds indiscriminately to matter, regardless of whether it is bosonic or fermionic. Drawing then the lesson from the holographic superconductor, one anticipates that the resultant geometry in the IR is of the Lifshitz type, with an emergent dynamical critical exponent **z**. This Lifshitz sector encodes for yet another emergent quantum critical deep IR, coexisting with the cohesive Fermi liquid, that is somehow stabilised by the presence of the latter.

At finite temperature this two-fold split in the degrees of freedom at low energies becomes even more manifest. The Lifshitz remnants of the conformal degrees of freedom are now encoded by a black hole. This black hole might swallow up some of the fermionic excitations, and it will therefore be generically charged. However, most of the boundary fermion matter continues to be encoded in the bulk by the Fermi gas which is concentrated in the geometrical domain-wall potential, far away from the black-hole horizon [410, 411]. The geometry represents in a quite "visible" way the existence of the two subsystems characterising the boundary matter. As we increase the temperature and the black-hole horizon encroaches on the geometrical domain wall, more and more fermions will fall through the horizon. Eventually, at high enough temperature, there is a density-driven phase transition after which only the thermal RN black hole remains.

We are dealing with fermions, and the Pauli principle is a distinct challenge for the computations that deal with this physics. In the hard-wall case we were able to sum each of the wave functions one by one to obtain a macroscopic charge density that backreacts. However, this exploited the great simplification in the computation for the bulk wave functions. Owing to the hard-wall boundary condition in the deep interior the spectrum discretised and became tractable [412, 413]. In the full geometry a consistent set of boundary conditions in the deep interior is a very challenging task, since one has to fully account for the backreaction of the metric and the electrostatic potential. In particular, since the space-time is now semi-infinite there is a priori no straightforward radial quantisation. To address this deep-IR boundary problem, one could attempt to regulate the theory with a minute hard wall at $r = r_c$, to subsequently remove the regulator r_c. This is very hard work [412, 413], but it hints at yet another approximation. As we send $r_c \to 0$, the level spacing between the radial harmonics will decrease. At fixed chemical potential that should mean that more and more "radial bands" are occupied. In equilibrium, the charge density should stay constant. From the dispersion relations as computed in the previous section it follows that the effect of the backreaction on the wave functions has to become quite severe. We can ameliorate this by reducing the charge of each microscopic fermion, making it less sensitive to the background gauge field. This points to a limit where we take $q \to 0$, while keeping the total charge Q fixed. These statements together should ring a bell; this is nothing other than the well-known "fluid" limit of a finite density of fermions. It is a rephrasing in terms of charge of the Thomas–Fermi approximation where the energy-level splitting has been taken to zero while the density is kept fixed. We shall use this approximation to substantiate the insights that we previewed in the above.

One can already immediately deduce the general form of the boundary fermion spectrum. In this fluid limit one has a dense multitude of occupied "radial channels", with an energy splitting that disappears in the $q/Q \to 0$ limit. The hard-wall star taught us that each radial mode corresponds to a Fermi surface in the boundary. Therefore, the boundary will be characterised by a near infinity of concentric Fermi surfaces that are "nested like a Russian matrioshka doll". This will be confirmed by explicit calculations.

It also implies that there is some pathology at work in the limit where the level spacing becomes exactly zero. Recall the "phase diagram" that was deduced in the probe-fermion chapter, Fig. 9.6: the $q/Q = 0$ "fluid-limit" electron star is found right at the origin, which is a singular point. To comprehend the singular nature of the fluid limit, we have to recall how the formation of the bulk Fermi gas ought to signal itself on approaching it from the AdS_2 metal side by tuning the UV mass down. In the fluid limit, on the other hand, rather than occupying the bulk radial fermion modes one by one, the system jumps from zero to infinitely many Fermi surfaces instantaneously. It is actually rather obscure how one should address this quantum phase transition from the non-Fermi-liquid RN metal to this singular "Russian-doll Fermi liquid". Although this issue is less obvious, the thermal physics associated with the fluid limits has similar pathologies. One finds a third-order continuous transition [410]. This is absurd: the thermodynamical quantity that changes is the density, and a general statistical-physics principle insists that this should be a first-order transition. It turns out that this is one of those famous $1/N$ pathologies. Recently it was demonstrated that for any finite radial quantisation one recovers a healthy first-order van der Waals transition [409].

11.2.1 The semi-classical electron fluids and the "Russian-doll Fermi liquid"

The bright side of the fluid limit is that the gravitational computations become relatively easy. The self-gravitating Fermi gas in the fluid limit is a classic theoretical construction known to any relativist or astrophysicist. For a neutral Fermi fluid this was solved by Tolman, Oppenheimer and Volkoff in the 1930s. The charged electron star in AdS we will describe here is a very close sibling of this neutron star. Notice that, differently from neutron stars, such electron stars do not exist in our universe. The relative strength of the electromagnetic force is so much larger than the gravitational one that cosmologically all matter is neutral. In the virtual cosmos of holographic duality aimed at describing strongly coupled field theories this limitation does not apply, however.

The secret to our understanding of neutron stars is that the typical neutron density is extremely high. The Fermi energy therefore overwhelms all effects of interactions. In condensed matter language, r_s (the ratio of the interaction energy to the Fermi energy) is nearly vanishing. In addition, by postulating that the stars formed from neutrons have astrophysical sizes of many kilometres, one infers directly that the Fermi wavelength of the neutrons at the Fermi surface is expressed in femtometres, and is tiny compared with the gravitational (and, for a charged star, electromagnetic) gradient lengths. The effect is that the Fermi sea is described perfectly in a local density approximation, and, given the largeness of the Fermi energy compared with the interaction scale, one can just treat this in the form of a simple macroscopic equation of state language. This is the Thomas–Fermi-fluid approximation to the Fermi gas.

In the box at the end of this section we will present a detailed derivation of the fluid limit. Since, on astrophysical scales, almost all mean free paths are tiny, this fluid language is the relativist's favourite way to deal with matter in GR. In the fluid approximation all one needs to know regarding this matter is its stress-energy and charge density, as sources to the Einstein and Maxwell equations,

$$R_{\mu\nu} - \frac{1}{2}g_{\mu\nu}(R + 6) = F_{\mu\rho}F_{\nu}{}^{\rho} - \frac{1}{4}g_{\mu\nu}F_{\rho\sigma}F^{\rho\sigma} + T_{\mu\nu}^{\text{matter}},$$

$$D_{\mu}F^{\mu\nu} = -\frac{qL}{\kappa}J_{\text{matter}}^{\nu}. \tag{11.23}$$

Here $\kappa^2 = 8\pi G$ encodes Newton's constant, and we have rescaled $A_\mu \to (eL/\kappa)A_\mu$ to make it easy to compare the effects of gravitation and electromagnetism. This surely makes sense only for macroscopic matter, which behaves like a classical fluid. On comparing this with Einstein–Maxwell–Dirac theory describing the microscopic fermions carrying a charge q, the charge conservation in the gravity system immediately implies that the total number of constituents in this macroscopic fluid is simply set by $Q/q \to \infty$, where Q is the total charge of the star-to-be. We infer immediately that we are aiming at the bottom left corner of the phase diagram Fig. 9.6.

For a relativistic fluid,

$$
\begin{aligned}
T_{\mu\nu}^{\text{matter}} &= T_{\mu\nu}^{\text{fluid}} = (\rho + p)u_\mu u_\nu + pg_{\mu\nu}, \\
J_\mu^{\text{matter}} &= J_\mu^{\text{fluid}} = nu_\mu,
\end{aligned}
\tag{11.24}
$$

where ρ, n and p are the energy density, number density and pressure determined by the equation of state of the charged fluid. As in the standard local density approximation, one exploits the smallness of the Fermi wavelength and computes these quantities assuming a locally constant background/chemical potential $\mu_{\text{loc}}(r)$ and a locally flat space-time: one computes ρ, n and p, assuming that one is dealing with a non-interacting Fermi gas that varies slowly in space. Compared with the classic Tolman–Oppenheimer–Volkoff equations for neutron stars, there are now two differences: our "electron" star is charged, and it lives in an asymptotic AdS universe with radius L. Expressing the fluid parameters in terms of dimensionless integrals, we have

$$
\begin{aligned}
\rho &= \frac{1}{\pi^2} \frac{\kappa^2}{L^2} \int_{mL}^{\mu_{\text{loc}}} dE\, E^2 \sqrt{E^2 - (mL)^2}, \\
n &= \frac{1}{\pi^2} \frac{\kappa^2}{L^2} \int_{mL}^{\mu_{\text{loc}}} dE\, E \sqrt{E^2 - (mL)^2}, \\
-p &= \rho - \mu_{\text{loc}} n,
\end{aligned}
\tag{11.25}
$$

where the local chemical potential is given by

$$
\mu_{\text{loc}}(r) = \frac{qL}{\kappa} e_0^t(r) A_t(r).
\tag{11.26}
$$

The inclusion of the vielbein $e_0^t(r)$ accounts for the change between the coordinates we are using and the natural coordinate system of a local observer who observes a free Fermi gas at his or her position r.

These expressions are now fed as sources to the Einstein–Maxwell equations as in Eq. (11.23). For a homogeneous and isotropic fluid, the background metric and electrostatic potential will respect these symmetries, and will therefore be of the form

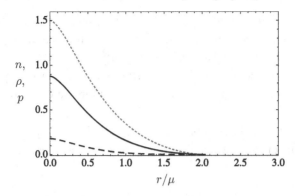

Figure 11.2 The profiles of the number density n (dotted), the energy density ρ (solid) and the pressure p (dashed) as a function of the radial coordinate r of the fluid-limit electron star with $\mathbf{z} = 2$, $\hat{m} = 0.36$. The location of the edge of the star is $r_s/\mu = 2.04$.

$$ds^2 = L^2 \left(-f(r)dt^2 + g(r)dr^2 + r^2(dx^2 + dy^2) \right), \quad A_t = h(r). \qquad (11.27)$$

Here the action is just along the radial direction. One must now solve the equations of motions Eq. (11.23) in a self-consistent fashion, together with the fermionic fluid Eq. (11.26). With the details provided in box 11.2, this can readily be achieved numerically [418].

As in the case of the famous Tolman–Oppenheimer–Volkoff solution, one finds that these zero-temperature solutions look like real stars: the density profile is shown in Fig. 11.2. Differently from the scalar-hair profiles of the holographic superconductors, these electron stars have a real edge where the fermion density drops abruptly to zero. For an astrophysicist this is a satisfying result. In our holographic context it appears at first sight rather paradoxical. In order to read off the properties of the dual field theory, field profiles that extend all the way to the boundary are needed. In chapter 9 this worked fine, since the probe fermions there were described by quantum-mechanical states that are delocalised along the whole of the radial direction. The resolution is that this is an inherent defect of the fluid limit: for any finite level spacing of the fermions, their quantum-mechanical nature will be resurrected, with the effect that they will tunnel through the potential formed by the edge of the star. Eventually, they will extend all the way to the boundary. As a matter of principle, the edge of the quantised holographic star will in fact be fuzzy, and its classical appearance is just an unphysical artefact of the fluid limit.

Last but not least, one would like to address the deep-interior geometry associated with the star solution. It can be demonstrated [418] that this turns into a Lifshitz geometry Eq. (10.61), just as in the case of the holographic superconductor (box 11.2). The Lifshitz horizon has zero entropy and clearly the electron star "has stabilised" the metal in the boundary. In the remainder of this section we will see that the tunnelling between the quasiparticles associated with the electron-star Fermi liquid and the emergent deep-infrared quantum critical phase exhibiting the emergent Lifshitz scaling is exponentially suppressed. As in the holographic superconductor, we find again that according to holography the deep infrared is characterised by two subsystems: the "cohesive" Fermi liquid and an emergent Lifshitz quantum critical phase that forms only in the presence of the cohesive phase.

Box 11.2   An AdS electron star

To construct the electron star [418], we substitute the homogeneous and isotropic Ansatz Eq. (11.27) into the field equations Eq. (11.23),

$$\frac{1}{r}\left(\frac{f'}{f}+\frac{g'}{g}\right)-\frac{gh\hat{n}}{\sqrt{f}}=0,$$

$$\frac{f'}{rf}+\frac{h'^2}{2f}-g(3+\hat{p})+\frac{1}{r^2}=0, \tag{11.28}$$

$$h''+\frac{2}{r}h'-\frac{g\hat{n}}{\sqrt{f}}\left(\frac{rhh'}{2}+f\right)=0,$$

with

$$p=\frac{1}{\kappa^2 L^2}\hat{p},\quad \rho=\frac{1}{\kappa^2 L^2}\hat{\rho},\quad n=\frac{1}{e\kappa L^2}\hat{n}. \tag{11.29}$$

We have

$$\hat{\rho}=\beta\int_{\hat{m}}^{h/\sqrt{f}}d\epsilon\,\epsilon^2\sqrt{\epsilon^2-\hat{m}^2},$$

$$\hat{\sigma}=\beta\int_{\hat{m}}^{h/\sqrt{f}}d\epsilon\,\epsilon\sqrt{\epsilon^2-\hat{m}^2},$$

$$\hat{p}=-\hat{\rho}+\frac{h}{\sqrt{f}}, \tag{11.30}$$

with

$$\beta=\frac{e^4 L^2}{\pi^2\kappa^2},\quad \hat{m}^2=\frac{\kappa^2}{e^2}m^2. \tag{11.31}$$

The non-trivial part of the solution relates to what happens in the deep interior of AdS. On making a scaling Ansatz for the leading behaviour,

$$f=r^{2z},\quad g=g_0 r^2,\quad h=h_0 r^z. \tag{11.32}$$

More surprising is the fact that one can find an exact solution when

$$h_0^2 = \frac{z-1}{z}, \quad g_0^2 = \frac{36(z-1)z^4}{\left((1-\hat{m}^2)z-1\right)^3 \hat{\beta}^2} \tag{11.33}$$

and z as a function of β is defined by inverting the relation

$$\beta = \left(72z^2\sqrt{z-1}\sqrt{z-1-\hat{m}^2 z}\right.$$
$$\times \left(6 + (15\hat{m}^2 - 8)z + \hat{m}^2(9\hat{m}^2 - 10)z^2\right.$$
$$+ \hat{m}^2(\hat{m}^2 - 1)z^3 + 2(\hat{m}^2 - 1)^2 z^4 + 3\hat{m}^4 z^2\sqrt{z-1}\sqrt{z-1-\hat{m}^2 z}$$
$$\left.\left.\times \ln\left(\frac{\hat{m}}{\sqrt{1-\hat{m}^2-1/z}+\sqrt{1-1/z}}\right)\right)\right)^{-1}. \tag{11.34}$$

This solution on its own has no good holographic interpretation because the asymptotic $r \to \infty$ is not AdS. The interpretation of r as the RG direction, however, indicates how we can promote this to a proper holographic space-time. We should interpret this solution as a non-trivial IR to which the boundary conformal field theory flows after a relevant deformation. Or, equivalently, this non-trivial IR should possess an irrelevant deformation that allows one to integrate the RG flow back to an asymptotic AdS space-time. To do so, we add a small perturbation to all the fields, while maintaining homogeneity and isotropy,

$$f = r^{2z}\left(1 + f_1 r^{-\alpha} + \cdots\right),$$
$$g = \frac{g_0}{r^2}\left(1 + g_1 r^{-\alpha} + \cdots\right), \tag{11.35}$$
$$h = h_0 r^z\left(1 + h_1 r^{-\alpha} + \cdots\right).$$

On substituting this into the equations of motion, one finds that there is a solution where f_1 is undetermined (up to a sign), while g_1 and h_1 are proportional to f_1 if the exponent α equals $\alpha = \{\alpha_0, \alpha_\pm\}$, with

$$\alpha_0 = 2 + z, \quad \alpha_\pm = \frac{2+z}{2} \pm \frac{\sqrt{9z^3 - 21z^2 + 40z - 28 - \hat{m}^2 z(4-3z)^2}}{2\sqrt{(1-\hat{m}^2)z - 1}}. \tag{11.36}$$

The value α_0 is "universal"; it is always found in such space-times. Moreover, it is "relevant" from the point of view of the IR theory in that it is positive: the correction therefore grows when $r \to 0$. The understanding of this universal relevant perturbation is that it generates the finite-temperature IR [404, 418]. Of the other two possible exponents, α_+ is also relevant and therefore not of interest to us. We need the deformation α_-, which is irrelevant from the point of view of the IR theory.

One can now use this α_- deformed solution as an initial condition to numerically integrate towards $r \to \infty$. Famously, as Tolman, Oppenheimer and

---

Box 11.2   (Continued)

Volkoff found, one cannot do so all the way. There is a finite value r_* at which the numerics breaks down. By inspection this is precisely the point at which $\mu_{\mathrm{loc}} = mL$: it is the edge of the star beyond which there is no longer any fermionic matter present. Beyond this point, the appropriate equations to solve are thus Eq. (11.28) with $\rho, p, \sigma = 0$. The homogeneous solution to these equations is just the Reissner–Nordström solution,

$$f = c^2 r^2 - \frac{M}{r} + \frac{Q^2}{2r^2}, \; g = c^2/f, \; h = \mu - \frac{Q}{r}, \tag{11.37}$$

with Q the total charge inside the star and M the total mass. The one subtle point is that we have already used the freedom to rescale $r \to \epsilon r$ to set the coefficient of $f(r)$ in Eq. (11.32) to unity. As a consequence the coefficient c in the Reissner–Nordström solution cannot be set to unity in the present case. Its value follows by matching the Reissner–Nordström solution onto the star solution at $r = r_*$. Note that c is the effective speed of light in the UV; i.e. the dual field theory lives in the Minkowski space $ds^2 = -c^2 dt^2 + dx_i^2$.

11.2.2 *Thermodynamics, conductivity and photoemission of the fluid-limit electron star*

The thermodynamics

Although we have argued from the log-oscillatory behaviour of the spectral function that the AdS–RN metal is unstable with respect to the formation of this electron star, the Pauli exclusion of fermions prevents us from tracing the instability dynamically to its end point. We can, however, compare the thermodynamic properties of the electron star with the Reissner–Nordström solution. It is straightforward to show that, due to the Lifshitz scaling, the horizon does not contribute to the total bulk charge density. This quantity is instead simply given by the total charge carried by the bulk Fermi gas forming the star:

$$Q_{\mathrm{bulk}} = c \int_0^{r_*} \sqrt{g} \, n \, ds. \tag{11.38}$$

Similarly the energy density in the system follows from integrating the T^{00} component of the stress tensor over the radial direction,

$$E_{\mathrm{bulk}} = c \int_0^{r_*} \left(\rho + \frac{h'^2}{2fg} \right) s^2 \, ds + \frac{Q^2}{2r_*}. \tag{11.39}$$

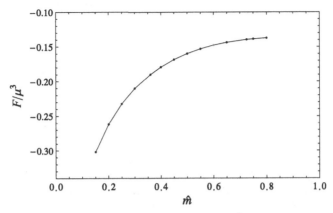

Figure 11.3 The free energy of an electron star F/μ^3 vs. $\hat{m}$, with $\beta = 19.951$. The free energy of the RN metal does not depend on $\hat{m}$, which is defined in Eq. (11.31). For low $\hat{m}$ the electron star is the stabler phase.

Since in the static set-up the energy and charge are conserved quantum numbers, the bulk values also yield the boundary values. We can now compute the (zero-temperature) free energy $F = E - \mu Q$. Notice that, through the (zero-temperature) first law $E + P = \mu Q$, this equals $F = -P$. Moreover, since the UV theory is a $(2+1)$-dimensional CFT, it also means that $E = 2P$. Therefore $F = -\frac{1}{3}\mu Q$, which is now easy to compute. Figure 11.3 shows the behaviour of the free energy for various values of the dynamical critical exponent **z** and mass $\hat{m}$. The Reissner–Nordström black hole is formally recovered in the limit $\mathbf{z} \to \infty$ or $\hat{m} \to 1$. Otherwise it is clear that the electron star is always the preferred solution.

Along similar lines one can obtain finite-temperature solutions, and these are characterised by a thermal horizon in the deep interior. Stationary solutions exist where a thermal RN black hole develops in the deep interior. The star is repelled by the black-hole horizon – this is just the deepening of the domain-wall potential well – and now obtains an inner edge as well. One has in fact a "halo star" with a black hole at its centre, see Fig. 11.4. The total charge density in the system will now be shared between this hollow star and the black hole. Upon raising the temperature further, the fermion-fluid "halo star" will be further sucked up by the black hole, to disappear at a third-order continuous phase transition [410, 411]. Since only the density changes, this is nominally a thermal van der Waals transition that is expected to be of first order. In the next section it will become clear that the third-order transition is a pathology of the fluid limit. Finally, in this large-N limit this Lifshitz quantum critical fluid completely dominates the finite-temperature

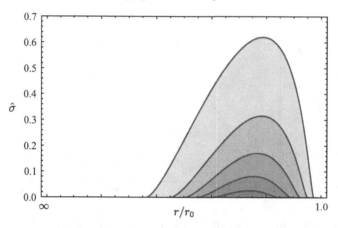

Figure 11.4 The charge density σ of the electron-star solution for various temperatures. The electric field of this black hole repels the bulk Fermi gas and in the fluid limit creates an inner edge to the star near the horizon r_0. As the temperature increases, the amount of charge in the Fermi gas is seen to decrease: more and more charge is freed from the Fermi gas and gets sucked into the black hole. Figure source [410]. (Reprinted with permission from the American Physical Society, © 2011.)

thermodynamics. From the Fermi liquid one would expect a Sommerfeld-law specific heat $C_v \sim T$, but the Lifshitz geometry implies via the rules of holographic thermodynamics that $C_v \sim T^{d/z}$. The reason is of course that the Lifshitz quantum critical sector contributes in the large-N limit, while the fermionic contributions are $1/N$ suppressed.

The optical conductivity

The next ubiquitous physical property of interest is the optical conductivity. As before, this is computed from the linear response to an electric-field perturbation $\delta A_x(\omega)$ around the full electron-star background. We provide the details in box 11.3 below. The typical result is shown in Fig. 11.5 and could have been anticipated on general grounds. As in the other finite-density metals (including the RN metal), the electrical current now carries momentum, which is conserved in the Galilean continuum. As a consequence the system is a perfect metal characterised by the delta-function peak at $\omega = 0$ in the real part of the optical conductivity. To satisfy the f-sum rule this goes hand in hand with a depletion of incoherent spectral weight at $\omega < \mu$. Through the matching method, the low-frequency spectrum is controlled by emergent Lifshitz geometry. This is supported by the electron star, but the bulk Fermi gas influences the conductivity only at second order.

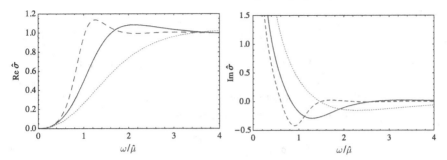

Figure 11.5 The optical conductivity of the $T = 0$ electron star. The conductivity is shown for three different values of the dynamical critical exponent in the emergent IR: the dashed line is for $z = 3$ with $\hat{m} = 0.7$, the solid line is for $z = 2$ with $\hat{m} = 0.36$ and the dotted line is for $z = 1.5$ with $\hat{m} = 0.15$. The $1/\omega$ rise in the imaginary part shows that the real part also contains a delta function at $\omega = 0$. Figure source [418]. (Reprinted with permission from the American Physical Society, © 2011.)

Box 11.3   The optical conductivity dual to the fluid-limit electron star

To compute the optical conductivity in linear response the fluctuations that need to be considered are

$$A_x = \frac{eL}{\kappa}\, \delta A_x(r)e^{-i\omega t}, \quad g_{tx} = L^2\, \delta g_{tx}\, e^{-i\omega t}, \quad u_x = L\, \delta u_x\, e^{-i\omega t}, \qquad (11.40)$$

where u_x is the fluid velocity field in the x direction appearing in the stress tensor and current (Eq. (11.24)) on the right-hand side of Einstein's equations Eq. (11.23). Following the usual procedure, these fluctuations are added to the electron-star background solution, by substituting them in the Einstein–Maxwell equations. Retaining only the first-order terms in the fluctuations, one obtains the fluctuation equations

$$n\,\delta A_x + (p+\rho)\delta u_x = 0,$$

$$\delta g'_{tx} - \frac{2}{r}\,\delta g_{tx} + 2h'\,\delta A_x = 0, \qquad\qquad\qquad (11.41)$$

$$\delta A''_x + \frac{1}{2}\left(\frac{f'}{f} - \frac{g'}{g}\right)\delta A'_x + \frac{h'}{f}\left(\delta g'_{tx} - \frac{2}{r}\delta g_{tx}\right) + gn\,\delta u_x + \omega^2\frac{g}{f}\,\delta A_x = 0.$$

One can readily solve for δg_{tx} and δu_x to obtain the single equation

$$\delta A''_x + \frac{1}{2}\left(\frac{f'}{f} - \frac{g'}{g}\right)\delta A'_x + \left(\omega^2\frac{g}{f} - \frac{gn^2}{p+\rho} - \frac{2h'^2}{f}\right)\delta A_x = 0. \qquad (11.42)$$

Box 11.3   (Continued)

It is immediately inferred how the presence of the charged-fermion fluid with $n \neq 0$ suppresses the response at low frequencies compared with the pure Reissner–Nordström solution. The conductivity then follows as usual from the ratio of the sub-leading $\delta A^{(1)}$ to the leading term in the solution of

$$\sigma = \frac{c}{i\omega} \lim_{r \to \infty} \frac{-r^2 \delta A'_x}{\delta A_x}, \tag{11.43}$$

with the only novel aspect being the additional factor of the speed of light c in the numerator.

Fermion spectral functions

The ultimate test for the existence of a Fermi liquid is of course the form of the single-fermion propagator, as measured by photoemission. We have already alluded to the problem that in the literal fluid limit $q/Q \to 0$ there is no room to probe the system with a charged fermion. The formal fermion charge vanishes in the bulk. This is just an artefact of the singular-fluid limit. A partial remedy is to re-quantise the system in a semi-classical, WKB spirit. These leading-order quantum corrections remove the pathologies of the strict fluid limit. It is already obvious what the probe spectral function should reveal. In order for the fluid limit to exist, one has by construction abandoned the notion of the radial quantisation. The "mesino masses" thus form a continuum. This implies the presence in the boundary of an infinity of infinitesimally close densely packed Fermi-surfaces. Upon re-quantising the charge of the probe the discreteness of the radial quantisation should be restored. From the rescaling below Eq. (11.23) and the expression for the local chemical potential Eq. (11.26), it can be seen that the natural size of the re-quantised charge is set by $q \sim \kappa/L \equiv 8\pi G/L$ in terms of the parameters of the action Eq. (9.1). This matches with the insight that the gravitational coupling constant κ should play the role of Planck's constant (dual to $1/N$ corrections). One can now proceed in a semi-classical perturbative computation of the probe around the fixed electron-star background. A typical outcome for the momentum distribution function at very small energy (i.e. probing the Fermi surfaces) is shown in Fig. 11.6, as a function of the quantisation parameter κ [357]. One does indeed find the expected large number of Fermi surfaces, tending to infinity in the "true" fluid limit. The Fermi momenta k_F are very small compared with the expectation from Luttinger's volume theorem, but this makes sense since the total number of fermionic degrees of freedom has to be shared by the large number of "mesino" species. In fact, one can demonstrate that Luttinger's theorem is exactly obeyed in

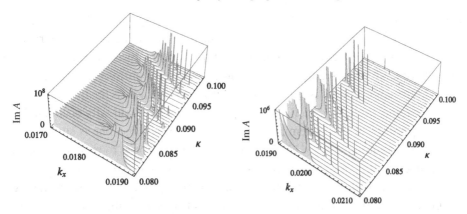

Figure 11.6 Numerically computed electron-star spectral functions showing the dependence both on k (in units of μ) and on κ for $z = 2$, $\hat{m} = 0.36$, $\omega = 10^{-5}$. Because the peak height and weights decrease exponentially, we present the adjacent ranges $k \in [0.017, 0.019]$ and $k \in [0.019, 0.021]$ in two different plots with different vertical scales. For a fixed value of κ there are multiple peaks. As κ decreases they move and become exponentially weaker and narrower. Analytically it can be shown that there are many very narrow peaks below the scale of resolution. They persist all the way up to $k \sim \mu$. Figure source [357].

the same sense as for the hard-wall star [159, 419, 420, 421]. The Lifshitz quantum critical sector in the deep infrared is effectively charge-neutral: all the charge resides in the electron-star/boundary Fermi liquid.

Last but not least, the computation of the spectral functions reveals an exponentially small imaginary part for the self-energy for these quasiparticles [159, 419, 420]: $\text{Im } \Sigma(k, \omega) \sim \exp\left(-(k^{z}/\omega)^{1/(z-1)}\right)$. As can be shown analytically [366], this is a property associated with the tunnelling of the fermion quasinormal modes into the Lifshitz deep infrared, leading to an exponential suppression of the coupling with this "critical bath". This substantiates the claim that the Fermi liquid and the Lifshitz quantum critical phase become decoupled in the deep infrared, with the ramification that the Fermi liquid becomes a real Landau Fermi liquid.

11.3 The landscape of holographic Fermi liquids: radial re-quantisation and instabilities

As we emphasised, the fluid limit follows from computational convenience and from holographic fundamentals. It amounts to completely suppressing the finite "$\hbar$" spacing (in fact, the charge of the AdS Dirac field) associated with the radial quantisation. One anticipates that also a faithfully quantised star-like solution exists, for which the deep interior is determined by backreaction. The quasiparticles in the boundary should then be associated with quantised radial modes in the bulk. This solution should easily give rise to a boundary system with a single Fermi surface, or only a few Fermi surfaces, as in the case of the hard-wall construction. These are obviously the holographic Fermi-liquids of interest to condensed matter physics. To study finite-charge density holographic fermions directly as a function of the microscopic charge is very hard. One is forced to address the quantum nature of the fermions directly – each wave function must be individually accounted for, and the spectrum of the system is not easy to control because one is sitting on the verge of a continuum [409, 412, 413]. The natural solution is to regularise the theory by incorporating a wall as in section 11.1, and then try to remove this regulator reliably. This is technically extremely cumbersome [409, 412, 413]. Attaining a complete description of the fully quantised electron star in the bulk is at present still work in progress.

11.3.1 Perturbative re-quantisation: the "WKB" electron star

An alternative solution builds on the small-charge WKB limit employed fruitfully in the computation of the probe spectral functions, combined with the Hartree summation used in the hard wall. One posits an Ansatz for the potential well that supports the discretised radial modes. If the charge is small and the occupation number is high, the majority of these modes will be faithfully approximated by a WKB solution. The small number of low-lying modes that cannot be approximated in this way should be irrelevant in the limit of a large total charge. The WKB solutions are readily computed and summed to recompute the potential well self-consistently in a Hartree sum. Because this self-consistent sum now also includes the gravitational backreaction, these computations are technically quite demanding, and beyond the scope of this book, but the results are easy to explain [409].

Focussing on the gravitational description in the AdS bulk, we have already mentioned that upon re-quantisation the edge of the star will become fuzzy, since the fermion wave function should extend all the way to the boundary. In this WKB approximation this is manifestly so, because all of the wave functions are accounted for and the potential in the bulk remains finite everywhere. There is therefore always a tail that reaches beyond the edges all the way to the AdS boundary.

There is also a change in the deep interior. The quantum corrections do not affect the healthy properties of the fluid limit [409]: the deep interior still has a Lifshitz geometry, while the Fermi liquid with its finite number of Fermi surfaces which is dual to this "WKB star" obeys Luttinger's theorem. The exponential decoupling of the quasiparticles from the Lifshitz quantum critical infrared is also inherited from the fluid limit. However, the WKB star cures the third striking pathology of the fluid limit. Now the thermal transition is no longer an inexplicable third-order transition but rather turns into a regular density-driven first-order van der Waals transition for any finite re-quantisation [409].

This shows that the fluid limit is only mildly singular, such that its gross features are likely to survive even when the radial quantisation becomes serious. However, there is still a long way to go to the case of most interest to condensed matter physics, characterised by a single Fermi surface. In this context, a most interesting question is the nature of the quantum phase transition from the non-Fermi-liquid strange metal, say of the RN kind, to the holographic single Fermi liquid. Let us finish this chapter with an interesting construction that addresses this question.

11.3.2 The dilaton and the strange-metal–Fermi-liquid quantum phase transition

There is a way to engineer the system so that it drives to a confined/cohesive star state from a deconfined/fractionalised black hole, while maintaining the convenience of the fluid limit. This is achieved by letting a dilaton coupling control the parameters in the Einstein–Maxwell-fluid action [422]. Consider the action

$$S = \frac{1}{16\pi G_N} \int d^4x \sqrt{g} \left[R - \frac{1}{4q^2} Z(\Phi) F_{\mu\nu} F^{\mu\nu} - \frac{1}{2} |\partial \Phi|^2 - \frac{1}{L^2} V(\Phi) \right.$$
$$\left. + \mathcal{L}_{\text{matter}} \right]. \tag{11.44}$$

The effective size of the charge is now set by the value of $q/\sqrt{Z(\Phi)}$. Let us elaborate the problem of a Fermi gas in the fluid limit, coupled to this Einstein–Maxwell-dilaton system.

We are specifically interested in the IR, and in this region the functions $Z(\Phi)$ and $V(\Phi)$ will be controlled by their leading exponential,

$$Z(\Phi) \sim e^{\alpha\Phi}, \quad V(\Phi) \sim e^{\beta\Phi}. \tag{11.45}$$

We therefore simply set $Z(\Phi) = e^{\alpha\Phi}$. For $V(\Phi)$ we must ensure that, in the absence of matter and flux, standard AdS is a solution for $\Phi = 0$. This is achieved by complementing $V(\Phi)$ to the form $V(\Phi) = -6\cosh(\beta\Phi)$. The parameters α, β can then be chosen to be positive without loss of generality.

This dilaton is to be interpreted as the dual of the leading irrelevant operator in the field theory. Notice that the potential is indeed unstable towards both $\Phi \rightarrow \pm\infty$. We can thus drive the system to a different IR by deforming the theory with this operator. This means in concrete terms that the equations of motion should be solved with an explicit boundary source Φ_0 for the dilaton. Depending on whether this drives $Z \rightarrow \infty$ ($\Phi \rightarrow \infty$) or $Z \rightarrow 0$ ($\Phi \rightarrow -\infty$), there will be qualitatively different solutions in the IR. For $Z \rightarrow 0$ the microscopic charge in the IR is very large. Compared with the prototypical electron star of the previous section this merely enhances the tendency of any local chemical potential at the horizon to discharge through pair production into an atmosphere of charged fermionic states forming the usual electron star. This solution is therefore dual to a confined Fermi liquid with nearly identical properties to the normal electron star. On the other hand, for $Z \to \infty$ the effective charge vanishes in the IR. The low-energy fermions will therefore not respond as effectively to a local chemical potential present at the horizon. The system may discharge to some degree, but it suggests that the charged horizon stays mostly stable. These solutions are therefore dual to a mostly deconfined/fractionalised non-Fermi-liquid-like state, with perhaps a small admixture of a regular confined Fermi liquid.

There is a third solution, the special case where $Z = $ constant, or equivalently ($\Phi = $ constant). Since Φ does not evolve along the radial direction in this case – since it does not flow with the RG – the solution exhibits scaling. In general this will be of the Lifshitz type. Making the same Ansatz as before,

$$ds^2 = L^2 \left(-r^{2z} dt^2 + g_0 \frac{dr^2}{r^2} + r^2 (dx^2 + dy^2) \right), \quad A_0 = h_0 r^z, \quad \Phi = \phi_c,$$

(11.46)

one readily finds **z** from the combination of Maxwell's equation,

$$\frac{1}{r^2} \frac{d}{dr} \left(r^2 \frac{Z(\Phi)h'}{\sqrt{fg}} \right) - \sqrt{g}\hat{n} = 0,$$

(11.47)

the dilaton equation of motion for $\Phi = $ constant,

$$\frac{g_0 V'(\Phi)}{4} - \frac{Z'(\Phi)h'^2}{4f} = 0,$$

(11.48)

and a conservation law hidden in the equations of motion. An explicit example is worked out in box 11.4.

After realising that the fixed value ϕ_c for which this third solution exists is determined by the fixed chemical potential μ, a rather beautiful picture emerges. Choosing the boundary value of Φ_0 to be slightly less than ϕ_c, the monotonicity of the radial RG flow directly implies that this state flows in the IR to $\Phi = -\infty$ and one has the confined star solution. For $\Phi_{bdy} > \phi_c$, however, Φ must flow to $\Phi = +\infty$ in the IR and one has a partially or fully deconfined/fractionalised solution in the IR. This explicitly shows how, by tuning the strength of the relevant operator to which the dilaton is dual, one can achieve very different emergent states in the IR: see Fig. 11.7.

How is this strange-metal deconfined ground state related to the confined electron star in this dilaton model, as a function of the charge? The most convenient way to understand this is to start in the confined regime where the microscopic charge is large. Here the deformation parameter is $\phi_0 \ll \phi_c$. As one tunes the

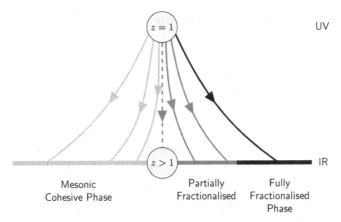

UV

z = 1

z > 1

IR

Mesonic
Cohesive Phase

Partially
Fractionalised

Fully
Fractionalised
Phase

Figure 11.7 Phases of the dilaton electron star. The pristine electron star is a non-trivial IR Lifshitz fixed point of the theory with $z > 1$ obtained through a critical RG flow from the UV Lorentz-invariant ($z = 1$) conformal field theory. By deforming the UV theory, one can flow to different IR theories. Systems where the effective $U(1)$ electrical charge becomes large in the IR easily destabilise the pristine electron star to a fully cohesive phase, which is similar to the pristine star, but with no scaling regime in the IR (light-grey-flows). Systems where the effective $U(1)$ charge becomes weaker in the IR are more resistant to spontaneous discharge. They retain a charged horizon accounting for deconfined/fractionalised degrees of freedom. This horizon can either account for part of the total charge density (grey flows) or the full charge density (dark-grey flows). Figure adapted from [422]. (© IOP Publishing. Reproduced by permission of IOP Publishing. All rights reserved.)

charge down by tuning ϕ_0 towards ϕ_c, one first enters a mixed state at $0 < \phi_0 - \phi_c \ll 1$ after crossing the Lifshitz regime at $\phi_0 = \phi_c$. It does indeed prove to be true that *partially deconfined* solutions containing both a charged horizon and a finite-density fermion gas in the bulk exist. This gas forms a star-like halo around the extremal black hole in the deep interior, similarly to the finite-temperature solution for the electron star. In the fluid limit this is a true halo, with outer and inner edges defined by the locations where the difference between the local chemical potential and the mass gap $\mu_{\text{loc}} - m$ vanishes. As one now tunes further, more and more charge is smoothly transferred from the Fermi gas to the horizon. More and more of the fermions fractionalise, to deconfine and disappear into the strange metal dual to the extremal black hole. Eventually there is only a tiny droplet of Fermi fluid left, which in this model disappears in a third-order transition to the exact strange metal (see Fig. 11.9 later). The transition between the holographic confined Fermi liquid and the strange metal is thus not like a single quantum phase transition. Instead, it is generically a succession of two-phase transitions with an intermediate mixed phase. This lesson turns out not to be an artefact of the fluid limit: it appears to be confirmed in a fully quantised microscopic construction [423].

There is a subtlety that needs to be addressed. As we change the deformation ϕ_0 of the theory through ϕ_c, the IR solutions need not interpolate continuously between a confined and a partially deconfined phase. It can happen that the third Lifshitz scaling solution is in fact dynamically unstable (see box 11.4). In this case the system experiences a first-order transition which avoids the scaling solution, rather than a second-order transition. This happens in particular when the UV charge (where $Z = 1$)-to-mass ratio is rather low. Intuitively this makes sense, since the charged particles are driving the dynamics. For low UV charge the deconfined and confined states may be separated by an effective potential barrier, but for large charge this barrier should be absent and one should have a smooth transition instead. Curiously, this argument fails for extremely low charge, irrespective of the mass, where the transition becomes again continuous [422] for reasons that are at present unclear.

Similarly, the nature of the second phase transition between the partially deconfined and the fully deconfined phase depends on the details of the system. In this dilaton-fluid model it is always of third order. Third-order transitions are well known to occur in large-N matrix models [424], but these become standard transitions at finite N. Insofar as the thermal transition discussed in [409] is concerned, it is likely that upon re-quantisation it will turn into a zero-temperature version of the standard first-order van der Waals transition.

Box 11.4　From confinement to fractionalisation in EMD electron stars

With the tools used to construct the elementary electron star, we can readily construct the electron stars in EMD gravity. For the Einstein–Maxwell-dilaton theory in Eq. (11.44) coupled to a fluid of fermions, the equations of motion become

$$\frac{1}{r}\left(\frac{f'}{f} + \frac{g'}{g}\right) - \frac{gh\hat{n}}{\sqrt{f}} - 2\Phi'^2 = 0,$$

$$\frac{f'}{rf} + \frac{Z(\Phi)h'^2}{2f} - g\left(\hat{p} - \frac{1}{2}V(\Phi)\right) + \frac{1}{r^2} - \Phi'^2 = 0,$$

$$\frac{1}{r^2}\frac{d}{dr}\left(r^2\frac{Z(\Phi)h'}{\sqrt{fg}}\right) - \sqrt{g}\hat{n} = 0,$$

$$\Phi'' + \frac{1}{2}\left(\frac{f'}{f} - \frac{g'}{g} - \frac{4}{r}\right)\Phi' - \frac{gV'(\Phi)}{4} + \frac{Z'(\Phi)h'^2}{4f} = 0.$$

(11.49)

An important role in the analysis is played by an "emergent" conservation law hidden in these equations of motion. This conservation equation is related to the first law of thermodynamics. In the presence of a scalar operator $\mathcal{O}$ of dimension Δ with a possible VEV this states that $E + P = \mu Q - ((\Delta - 1)/2)\Phi_0\langle\mathcal{O}_\Phi\rangle$. Because our holographic theory is, moreover, conformal we know that $E = 2P$. Thus the first law reduces to $\frac{3}{2}E = \mu Q - ((\Delta - 1)/2)\Phi_0\langle\mathcal{O}_\Phi\rangle$. Consider first the situation where $\langle\mathcal{O}_\Phi\rangle = 0$ [422], i.e. the standard electron-star case. By comparing the expression for the energy density,

$$E = c \int_0^{r_*} \left(\rho + \frac{Z(\Phi)h'^2}{2fg}\right) s^2 \, ds + \frac{1}{2r_*} Q^2 \tag{11.50}$$

with the charge density,

$$Q = \int_0^{r_*} \left(r^2 \sqrt{g}n\right) dr, \tag{11.51}$$

one can integrate the second term in Eq. (11.50) by parts, in order then to use Maxwell's equation to relate h'' to the charge density and the equation of state $\rho - (h/\sqrt{f})n = -p$ to find an expression that is nearly coincidental with the first law $E = \frac{2}{3}\mu Q$. Insisting that the first law holds is equivalent to demanding that the following quantity is constant along r:

$$\frac{d}{dr}\left(\frac{2r^2 Z(\Phi)hh'}{\sqrt{fg}} - \frac{r^4(r^{-2}f)'}{\sqrt{fg}}\right) = 0. \tag{11.52}$$

Using the equations of motion, it is straightforward to check that this conservation law is obeyed, even in the presence of a non-vanishing $\Phi(r)$.

A second subtlety is associated with the demand that an explicit source for the dilaton is present. We focus on the case $V(\Phi) = -6 - 4\Phi^2 + \cdots$ when $\Phi \to 0$. This changes the behaviour of the fields near the boundary, giving

$$f(r) = c^2 r^2 \left(1 - \left(E + \frac{1}{3}\phi_0\langle\mathcal{O}\rangle\right)\frac{1}{r^3} + \cdots\right),$$

$$g(r) = r^2 \left(1 - \frac{\phi_0^2}{r^2} + \left(E - \phi_0\langle\mathcal{O}\rangle\right)\frac{1}{r^3} + \cdots\right),$$

$$h(r) = c\left(\mu - \frac{Q}{r} + \cdots\right),$$

$$\Phi(r) = \frac{\phi_0}{r} + \frac{\langle\mathcal{O}\rangle}{2}\frac{1}{r^2} + \cdots, \tag{11.53}$$

where E is the total energy defined in the above.

Box 11.4   (Continued)

The IR one obtains in this way is rather insensitive to the detailed shape of the potential, and we fix the parameters determining the functions $Z(\Phi)$ and $V(\Phi)$ in the following way:

$$Z = e^{2\Phi/\sqrt{3}}, \quad V(\Phi) = -6\cosh(2\Phi/\sqrt{3}). \tag{11.54}$$

The Lifshitz solution

For constant $\Phi = \phi_0$ the theory is essentially an electron star with a Lifshitz metric in the IR,

$$ds^2 = L^2\left(-r^{2z}\,dt^2 + g_0\frac{dr^2}{r^2} + r^2(dx^2 + dy^2)\right), \quad A_0 = h_0 r^z, \quad \Phi = \phi_c. \tag{11.55}$$

Since the dilaton has to be constant, it must satisfy

$$g_0 V'(\Phi) - \frac{Z'(\Phi)h'^2}{f} = 0 \tag{11.56}$$

throughout the radial evolution. In the Lifshitz deep interior this becomes

$$-\frac{6}{4}g_0 \sinh(2\phi_0/\sqrt{3}) - z^2 e^{2\phi_0/\sqrt{3}}h_0^2 = 0. \tag{11.57}$$

Substituting the Ansatz Eq. (11.55) into Maxwell's equation yields

$$2z h_0 e^{2\phi_0/\sqrt{3}} = g_0 n, \tag{11.58}$$

whereas the emergent conservation law reduces to

$$2z e^{2\phi_0/\sqrt{3}} h_0^2 - (2z + 2) = C_1. \tag{11.59}$$

On setting the integration constant to $C_1 = 0$, the last equation can be inverted such that

$$z = \frac{1}{1 - h_0^2 e^{2\phi_0/\sqrt{3}}}. \tag{11.60}$$

Finally, the second equation of motion in Eq. (11.49) can be simplified to become

$$(1 + z)(2 + z) = g_0\Big(p + 3\cosh(2\phi_0/3)\Big). \tag{11.61}$$

We have obtained in this way four equations with four unknowns (h_0, ϕ_0, $\mathbf{z}$, g_0). This system is readily solved numerically.

The Lifshitz IR can now be integrated to the UV using the same procedure as before. We deform the Lifshitz solution infinitesimally:

$$f = r^{2\mathbf{z}}(1 + \delta f\, r^{-\alpha}), \quad g = \frac{g_0}{r^2}(1 + \delta g\, r^{-\alpha}),$$

$$h = h_0 r^{\mathbf{z}}(1 + \delta h\, r^{-\alpha}), \quad \Phi = \phi_0(1 + \delta\phi\, r^{-\alpha}). \tag{11.62}$$

Differently from the Einstein–Maxwell electron star, one finds in this case five exponents. Combined with the trivial exponent $\alpha = 0$, these are associated with the dimensions of three pairs of "sources" J_L and dual operators $\mathcal{O}_L$. Their dimensions always sum to the total scaling dimension $2 + \mathbf{z}$ of the Lifshitz IR. The source with exponent $\alpha = 0$ drives the universal finite-temperature deformation with scaling exponent $\alpha = 2 + \mathbf{z}$. On closer inspection it turns out that, of the remaining two operators, one is always irrelevant while the other is always relevant [422]. We have already encountered the irrelevant operator in the construction of the Einstein–Maxwell electron star, and it can be used to numerically integrate the solution to the UV in order to obtain the full geometry. The relevant operator is the novelty. Its physics is easy to understand: the dilaton potential is unstable in every direction, and the resulting flow is characterised by precisely this relevant operator. The dimension of this relevant operator can in fact be complex (see Fig. 11.8). We learned what this means in the context of the holographic superconductor: this signals that the solution is dynamically unstable. In terms of the dual field theory this means that the Lifshitz solution is a local maximum in the free energy, rather than a minimum. One therefore concludes that upon tuning of the control parameter ϕ_0 through ϕ_c one encounters a first-order phase transition.

The confining solution

Tuning ϕ_0 away from ϕ_c forces ϕ to flow through its equation of motion to either $\Phi = \pm\infty$ in the IR. The solution where $\Phi \rightarrow -\infty$ is straightforwardly understood. In this case the effective charge becomes infinite, which means that the horizon discharges rapidly and all charge flows into the Fermi gas. This solution is therefore qualitatively the same as the simple EM electron star – the dilaton just renders the analysis more involved. In particular, there is no scaling solution in the IR and the solution must be asymptotically constructed instead. One writes down a leading scaling Ansatz that is subsequently solved order by order. One finds that at first subleading order there are deformations δf with a size that is not fixed by the equations of motion

Box 11.4   (Continued)

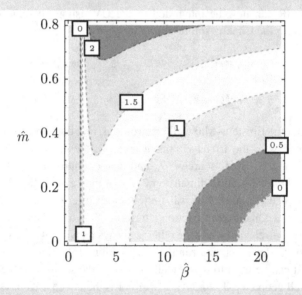

Figure 11.8 The imaginary part of the scaling dimension of the relevant operator in the IR Lifshitz solution of the Einstein–Maxwell-dilaton Fermi fluid star as a function of $\hat{\beta} = q^2\kappa^2/L^2$ and $\hat{m} = m/q$. Figure source [422]. (© IOP Publishing. Reproduced by permission of IOP Publishing. All rights reserved.)

$$f = r^2\left(1 + \sum_{n=1}^{\infty} f_n r^{2n/3} + \delta f\, r^P\right),$$

$$g = \frac{16}{9}r^{-4/3}\left(1 + \sum_{n=1}^{\infty} g_n r^{2n/3} + \delta g\, r^P\right),$$

$$h = h_0 r\left(1 + \sum_{n=1}^{\infty} h_n r^{2n/3} + \delta h\, r^{P-2/3}\right), \tag{11.63}$$

$$\Phi = -\frac{\ln r}{\sqrt{3}} + \sum_{n=1}^{\infty}\left(p_n r^{2n/3} + \delta\phi\, r^P\right).$$

These correspond holographically to the deformation operators in the IR. In this case there is only one such pair, characterised by an exponent for the source,

$$P = -1 + \frac{2}{3}\sqrt{1 + \frac{63h_0^2}{4(h_0^2 - \hat{m}^2)}}. \tag{11.64}$$

It follows from the equations of motion that the local chemical potential for the fermions $\mu_{\text{hor}} = h/\sqrt{f}$ is constant near the horizon. If $\mu_{\text{hor}} > \hat{m}$, the fermion fluid exists all the way to the deep interior. The solutions to the equation of motion show this to be the case: it actually has to be so, since the flux through the horizon $\lim_{r \to 0} \sqrt{-\det[g]}Z(\Phi)F^{tr} \sim r^{1/3}r^{2/3} = r$ vanishes. Luttinger's theorem is therefore obeyed and all the charge must be carried by the bulk Fermi gas. This corresponds in the boundary to the same Russian-doll Fermi liquid as found in the Einstein–Maxwell electron star.

The partially and fully fractionalised solutions

Let us now turn to the novelty found in EMD gravity, which is associated with the case that the effective charge decreases in the IR. This is the solution where Φ flows to $\Phi \to +\infty$. Formally the solution is constructed in the same way as for the fractionalised solution. There is again no exact scaling solution in the IR, but one departs as before from a scaling Ansatz that is corrected order by order using the equations of motion,

$$f = r^6 \left(1 + \sum_{n=1}^{\infty} f_n r^{4n} + \delta f \, r^{-N} \right),$$

$$g = \frac{16}{3} \left(1 + \sum_{n=1}^{\infty} g_n r^{4n} + \delta g \, r^{-N} \right), \tag{11.65}$$

$$h = \frac{1}{\sqrt{2}} r^4 \left(1 + \sum_{n=1}^{\infty} h_n r^{4n} + \delta h r^{-N} \right),$$

$$\Phi = -\sqrt{3} \ln r + \sum_{n=1}^{\infty} p_n r^{4n} + \delta p \, r^{-N}.$$

The scaling dimension of the irrelevant deformation is

$$N = 2 - 2\sqrt{19/3}. \tag{11.66}$$

As before, the solution can be integrated to an asymptotically AdS solution in the UV with this deformation turned on.

The novelty is that the local chemical potential $\mu_{\text{loc}} = h/\sqrt{f} = r/\sqrt{2} \to 0$ in the IR. It is now no longer possible to support a Fermi gas in the deep interior. It could still be the case that the local chemical potential dips down somewhere along the radial direction, such that a bulk Fermi gas can still form in the spirit of the WKB "halo" star. This would then be in turn dual to a Fermi liquid in the boundary. The flux near the horizon $(r^2/\sqrt{fg})Z(\Phi)h'$ is constant, however, and the dual state can never

Box 11.4   (Continued)

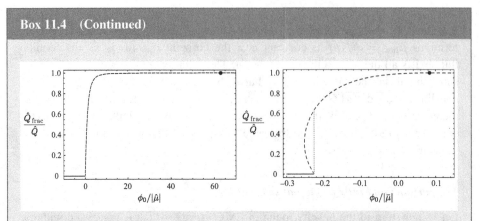

Figure 11.9 The ratio of the deconfined/fractionalised horizon charge to the total charge as a function of the deformation ϕ_0. Negative ϕ_0 gives rise to the IR cohesive phase with $\hat{Q}_{frac} = 0$ (solid line). Positive ϕ_0 gives rise to a partially or fully fractionalised phase with the amount of fractionalised charge denoted by the dashed line. The left plot is for $\hat{\beta} = 20$, $\hat{m} = 0.1$. In this case the relevant parameter in the partially fractionalised/fully fractionalised solution is always real, signalling a continuous transition between the partially fractionalised solution and the confined solution. The right plot is for $\hat{\beta} = 20$, $\hat{m} = 0.5$. Now the relevant parameter in the partially fractionalised phase is complex. The dashed line shows the first-order transition which occurs at $\phi_0 = \phi_c \simeq -0.224$. The dot denotes a third-order transition between the partially fractionalised phase where there is still a finite fermion fluid in the bulk and the fully fractionalised strange-metal phase with no fermion fluid. Figure source [422]. (© IOP Publishing. Reproduced by permission of IOP Publishing. All rights reserved.)

be merely a Fermi liquid. It therefore has to coexist with the fractionalised strange metal. This system can only be solved numerically, and the results are summarised in Fig. 11.9. This confirms the expectation that the Fermi liquid and strange metal coexist over a certain parameter range: on tuning away from ϕ_c an extremal RN black brane forms in the deep interior, which increasingly absorbs the charge. The electron star acquires a halo form with a hole in the deep interior, and gradually loses its charge to the extremal black brane until it has completely disappeared. Then the boundary turns entirely into the strange metal.

11.3.3 Conventional and unconventional instabilities of holographic Fermi liquids

Holographic Fermi liquids – Fermi liquids whose UV is a conformal field theory – therefore resemble (a system of) regular Fermi liquids in many ways. The Fermi liquid itself, however, possesses a famous instability with respect to superconductivity. How should we think about the BCS instability of the Fermi liquid in this holographic context? This requires an attractive four-fermion interaction that we have ignored in the minimalistic phenomenological models which we have considered up to this point. In principle it should be included, but the effects of such an interaction can be confidently predicted without any further calculation. Since the low-energy dynamics of the confined Fermi liquid both in the bulk and in the boundary is controlled by the excitations around the Fermi surface, whatever happens in one will also happen in the other. Adding a four-fermion interaction in the bulk will trigger the usual BCS instability in the electron star, and this will be directly mirrored in a similar instability of the Fermi liquid in the boundary. Technically, one must now solve on the gravity side not a free self-gravitating Fermi gas, but instead an interacting self-gravitating Fermi gas. In the electron star we already learned to handle interactions between fermions due to Coulomb interaction. The bulk BCS state can be handled in a similar way, after a Hubbard–Stratanovich transformation that introduces the usual charge-$2e$ BCS mean field in the particle–particle channel. One can readily construct the associated "BCS star" [425, 426]: according to the expectations it consists of a bosonic scalar condensate core that controls the deep IR, surrounded by a Fermi-fluid halo. It is in fact the exact charged anti-de Sitter analogy of a neutron star with a colour-superconducting core.

The holographic Fermi liquid also appears to possess an instability, which is rather exotic viewed from the perspective of conventional condensed matter. It is well known that astrophysical neutron stars have a Chandrasekhar limit. If the total size of the star is too large, the Fermi pressure can no longer balance the gravitational pressure and the star collapses to a black hole. The dual of a holographic Fermi liquid is qualitatively the *same* object as a neutron star. It is a macroscopic object supported by Fermi pressure. The fermions in the electron star are charged, with the effect that there is some additional electrostatic repulsion, but there should be a size of the object beyond which the gravitational pressure simply wins. The electron star in AdS turns out to be subject to this instability [427, 428], but only if the system is in a finite volume. This is similar to the Hawking–Page example in chapter 6. The boundary is taken to be a sphere, S^3, instead of a flat space-time. One can then show that there exists a critical energy E, or equivalently a critical mass, of the star beyond which the system becomes dynamically unstable towards the formation of an RN black hole. Since we are considering the grand canonical ensemble, the mass/energy is a dependent variable, and it is rather the chemical potential μ that has a critical value beyond which the system destabilises.

The generic expectation is that beyond this energy the system collapses to a black hole. What could this mean for the boundary field theory? The analogy of this transition is known in QCD. Since QCD is an asymptotically free theory, we know that not only at high temperatures, but also at high baryon densities compared with the QCD scale, quarks deconfine. This confinement/deconfinement transition as a function of T and μ is precisely what is studied in the relativistic heavy-ion collisions mentioned in the pursuit of the quark–gluon plasma. In the present holographic context the conformal UV theory is not at all asymptotically free, but the radius of the sphere R plays the role of the intrinsic scale in the theory. We now recognise this re-collapse to a black hole as the finite-density extension of the zero-density Hawking–Page transition between the thermal star solution and the charged black hole, where now the chemical potential plays a role similar to that of temperature. For small μ/R the system is in the confined phase, and correspondingly the mass of the star is less than the critical mass. There is a critical chemical potential $\mu_c \sim R$ at which the system undergoes a phase transition to the deconfined state described by the RN black hole. This might have some relevance for the (de)confinement transitions envisaged in the "slave" theories of condensed matter physics discussed in chapter 3.

12

Breaking translational invariance

Symmetry comes first in physics and we have seen its prominent role in holography. When one wants to compare the results and insights gained in holography up to this point with the physics of electron systems in solids, there is one crucial form of symmetry that does not match. The holographic systems described so far all live in the Galilean continuum while electron systems live in crystals. In crystals, the background lattice formed by the ions breaks the Galilean invariance of space.

This chapter is dedicated to the effects of translational symmetry breaking on holographic matter. At the time of writing, this issue is still being intensely pursued as a research subject. Although rapid progress is being made, it is far from completely settled. The reasons are of a "technical" nature. The breaking of translations in the boundary implies that this symmetry is also broken in the spatial directions in the bulk. Finding solutions of Einstein's equations when translational symmetry is broken is a very challenging exercise – in the absence of this symmetry the non-linear nature of general relativity comes out in force. However, this hard work pays off for the boundary physics: it adds quite a bit of condensed matter realism to the holographic computations, especially when it comes to transport properties.

As we will review in section 12.2, the optical conductivity of the holographic strange-metal/superconductor system in the presence of a lattice potential looks surprisingly similar to the experimental conductivities measured in cuprates, in fact much more so than any other theoretical result that has appeared in the 25-year history of the subject. In section 12.5 two equally credible holographic mechanisms that offer potential explanations for the famous linear-in-temperature resistivity of the cuprate strange metals as discussed in section 3.6.2 will be introduced. Although less is known to date regarding the effects of a lattice on the fermion spectral functions, some first results shed a quite surprising light on the effects of Umklapp on the strange-metal "un-particles" (section 12.3). The discovery of a new form of "algebraic" insulating state will be revealed in section 12.4. Under the influence of a unidirectional periodic potential, these un-particles form states that are in one direction insulating, while they stay metallic in the other directions. Such "quantum smectics" are impossible to realise within the confines of conventional finite-"particle"-density condensed matter physics [148]. Yet these results are tantalisingly suggestive with regard to phenomena seen in nature: high-T_c superconductors exhibit such an "unreasonable" anisotropy of the conductivities parallel and perpendicular to the Cu–O planes. These results all pertain to constructions where the translational symmetry is explicitly broken, but there is also plenty of space in holography to describe the spontaneous breaking of translations, as will be discussed in section 12.6. It is quite natural that such "holographic crystallisation" goes hand in hand with the occurrence of spontaneous current loops similar to the ones that are believed to occur in the pseudogap regime of the cuprates (section 3.6.2).

As we have already emphasised, combining GR with translational symmetry breaking is challenging. Its implication in the boundary is that momentum is no longer conserved. There is yet another way in holography to deal with loss of momentum conservation, by constructing several subsystems that share an overall momentum. When one only studies a "probe" subsystem, it can shed its momentum in the bath formed by the other subsystem(s). This is the general situation realised in the so-called probe-brane constructions which we will discuss in the next chapter. In this chapter we will focus entirely on momentum relaxation in a single system due to the absence of Galilean invariance.

We will start with a general discussion of DC transport properties in the case in which translational symmetry is broken by relatively weak potentials. In the absence of a potential, the fluid/gravity correspondence of chapter 7 controls the macroscopic realms, insisting that at finite temperatures a hydrodynamical fluid is realised in the continuum. A central claim will be that this hydrodynamical behaviour continuous to be relevant as long as the potentials are not too strong. Owing to the very fast Planckian dissipation characterising the quantum critical fluid, local equilibrium will be realised first, such that a hydrodynamical description becomes meaningful, while only at later times will the momentum relaxation take over. Under such circumstances the transport can be described using the powerful phenomenological memory-matrix approach which will be introduced as a general interpretational framework in section 12.1. By applying this to the Reissner–Nordström strange metal in the presence of a weak periodic background lattice, one arrives at a rather surprising temperature scaling of the DC conductivity due to the effects of Umklapp scattering.

The memory-matrix method is restricted to the regime of low frequencies and relatively small potentials. For stronger potentials one has to fully account for the gravitational backreaction, which complicates matters greatly. Such explicit translational symmetry breakings in the bulk imply that one has to deal with *inhomogeneous* bulk geometries. As a consequence, one has to solve coupled partial differential equations (PDEs), which can be accomplished only by employing heavy numerics. As we will elucidate in section 12.2, the first results of this kind for the RN metal and holographic superconductor have become available. They explicitly confirm the results of the memory matrix in the appropriate limit.

Section 12.3 is devoted to a first exploration of the effects of a periodic potential on holographic photoemission spectra. Fermions in inhomogeneous spaces are even more hazardous than the classical fields in the bulk, and only some results associated with the weak-potential limit are available. It already shimmers through, however, that strange-metal "band structure" is a highly counterintuitive and interesting affair with potential ramifications for photoemission experiments in the experimental systems.

It turns out that there is in fact a way to break the translational symmetry in the boundary such that the bulk equations of motion reduce to ordinary differential equations (ODEs) in the radial direction only, similar to the homogeneous bulk geometries we have encountered so far. As we will discuss in section 12.4, an example of this corresponds to a vector field in the boundary that rotates as a helix (Fig. 12.6) in one spatial direction. For a particular helix configuration such systems have an enhanced so-called "Bianchi VII" symmetry that preserves the helix structure uniformly in the radial direction. The bulk equations therefore reduce to ODEs in the radial direction only, which can also be solved in the case of strong potentials. For increasing strength of this periodic potential, there occurs a transition to an "algebraic insulator" formed in the direction of the potential, while in the perpendicular directions the system continues to be a perfect metal: the quantum smectic.

In section 12.5 we abstract this matter of translational symmetry breaking further. According to the symmetry rules of the holographic dictionary, global symmetries translate into local symmetries in the bulk. The breaking of the global translations in the boundary therefore has to be dual to a "Higgsing" of the local coordinate invariance governing the general relativity in the bulk. One of the ramifications is that the corresponding gauge field – the graviton – has to become massive. This is called "massive gravity". One can now employ constructions developed in the relativity community to deal with this massive gravity in the bulk, which are comparatively easy to compute. The results show that this describes a maximally featureless form of quenched disorder active in the boundary. This symmetry-based approach is quite illuminating. For instance, one can extract quite general considerations from the holographic results to argue that a local quantum critical fluid with a Sommerfeld specific heat will generically exhibit a linear-in-temperature resistivity in the presence of a weak random potential.

We turn in the final section to the subject of *spontaneous* translational symmetry breaking. Topological terms in the bulk of the kind we encountered in chapter 7 in the context of anomaly-corrected hydrodynamics also give rise quite naturally to *spontaneous* translational symmetry breaking. The black-hole "hair instabilities" underlying the superconducting transition occur at zero momentum in the absence of the topological terms. However, the effect of the topological terms is to cause a mode coupling that shifts the instability to a finite momentum, and this results among other things, in striped phases that break translations, involving spontaneous currents. This is a very new development, and much more is expected to follow in the near future.

12.1 Transport and un-particle physics: the memory-matrix formalism

The conventional theory of transport by electrons in metals departs from the quasi-particle paradigm of Fermi-liquid theory. One asserts that at zero temperature one has free fermions diffracting against the periodic lattice according to the rule of quantum mechanics. In addition, in the presence of quenched disorder one finds also the portfolio of mesoscopic-physics effects such as Anderson localisation, universal conductance fluctuations and so forth, due to the quantum-mechanical wave interferences in random potentials. At finite temperature the effects of the (irrelevant) quasiparticle interactions become noticeable, but these can be handled by the Boltzmann equation [429]. It implicitly takes as input that at finite temperatures one has a dilute gas of thermally excited particles characterised by a mean free path between collisions that is large compared with $1/k_F$. This mean free path is of the order of $\lambda \sim \hbar E_F v_F/(k_B T)^2$, corresponding to a quasiparticle collision rate $\hbar/\tau \simeq (k_B T)^2/E_F$ as imposed by the Fermi statistics.

In this holographic context we are interested in the effects of a *static* lattice – this art has not yet evolved to a state where the effects of dynamical electron–phonon interactions can be addressed. However, also in the Fermi liquid the leading-order effects of the presence of an ionic lattice on transport are governed by the static part. In the presence of a lattice, the continuum single-particle momentum eigenstates are no longer eigenstates, but have instead admixtures of "Umklapp copies". This mixing is controlled by the ratio of the excitation momentum (close to k_F) and the lattice momentum $k_L \sim 1/a$, and these are of the same order in typical metals. It is now easy to show that at every quasiparticle collision a fraction of the two-particle momentum C, called the "Umklapp efficiency", is absorbed by the lattice. This translates into a momentum relaxation rate $1/\tau_K$ for the macroscopic current, which is just proportional to the microscopic collision rate through the Umklapp efficiency [430]: $1/\tau_K = C/\tau$. In many conventional metals C turns out to be of order unity.

Since $1/\tau_K$ is the momentum relaxation rate governing the macroscopic currents, the Drude formula for the optical conductivity follows immediately:

$$\sigma(\omega) = D \frac{\tau_K}{1 - i\omega\tau_K}. \tag{12.1}$$

Here the Drude weight D – equal to $D = ne^2/m$ for a non-relativistic system – is the weight of the diamagnetic delta function at zero frequency which determines the "strength" of the perfect conductor in the Galilean continuum as in sections 8.3 and 10.2.2. The effect of translational symmetry breaking is that this broadens into a relaxation peak with a width set by the inverse momentum relaxation rate $1/\tau_K$.

Finally, while at small temperatures the dominant contribution is from electron–electron collisions, at zero temperature one finds in the presence of quenched disorder a residual momentum relaxation. This is the story behind the DC resistivity of the Fermi liquid:

$$\rho_{DC}(T) = \rho_0 + AT^2 + \cdots, \tag{12.2}$$

where "$\cdots$" denotes the momentum relaxation due to the coupling with the phonons which become dominant at temperatures of the order of the Debye and/or Einstein temperatures of the phonon system. These will be ignored altogether henceforth.

How, on the other hand, should we think in general about transport in "un-particle" strange metals when the Galilean invariance is broken? We will show that, as long as the potential strength associated with the breaking of Galilean invariance is weak, the macroscopic transport is still governed by collective hydrodynamics, but modified by the lack of momentum conservation at long times. It should once again be emphasised that this is very different from the situation in the Fermi liquid. Its collision time sets the scale of equilibration processes that will lead at much longer times to a hydrodynamical behaviour. However, in a Fermi liquid, momentum conservation is already destroyed before this microscopic time and therefore the hydrodynamical description is irrelevant for the electron system in conventional metals.

Focussing on the linear-response regime, and self-evidently very low frequencies, the effects of translational symmetry breaking in the hydrodynamical setting can be dealt with elegantly by using the "memory-matrix" formalism. The origin of this dates back to the "memory functions" written down in the 1970s [431], which focussed on just a single nearly conserved (charge) current. More recently the formalism was generalised to more than one conserved charge, allowing for a "matrix" of linear responses [432]. This relates directly to the generic situation governing the transport phenomena in finite-density systems. We have already touched upon this theme in the discussion of the conductivity of the RN metal in section 8.3: at finite density one has to account for both the conservation of momentum and the conservation of charge in order to assess the conductivities of the system.

The small quantities that allow this formalism to be useful are the long relaxation rates associated with the weakly broken conserved quantities in the reference system. This is especially so if one rate is parametrically smaller than all the others. When translational symmetry is the most weakly broken symmetry, the quantity of central interest is the momentum relaxation rate. One sees precisely this moral at work in the textbook Drude story in the above: starting from the perfect metal in the continuum, the momentum relaxation enters as a perturbative correction setting the narrow width of the Drude peak. For the un-particle RN metal one therefore anticipates that this will show the same Drude behaviour when the translational symmetry is weakly broken: this will be confirmed in the next section. The characteristic *difference* from the Fermi liquid will be the temperature evolution of the relaxation rate. The power of the universal memory-matrix approach is that one can arrive at surprisingly far-reaching statements regarding the behaviour of such macroscopic quantities by relying solely on generalities of the kinematics and the scaling properties of the strange metals. To give the reader some idea of the power of the memory-matrix approach per se, let us quote here a result that is supposedly completely general, and of quite some relevance for the experimental community. The Lorenz ratio of the (DC) thermal conductivity κ to the electrical conductivity σ at low temperature T obeys the Wiedemann–Franz law in the Fermi liquid; $\kappa/(\sigma T) = (\pi^2/3)(k_B/e)^2$. This is conventionally understood as the signal that quasiparticle-like excitations are solely responsible for both the heat transport and the charge transport. But what should one expect for this ratio in an un-particle system? Hartnoll and Hofman [433] use the memory-matrix formalism to demonstrate on general grounds that this ratio should obey $\kappa/(\sigma T) = (k_B/e)^2(1/T^2)(\chi_{QP}^2/\chi_{JP}^2)$, where χ_{QP} and χ_{JP} are static susceptibilities involving the operators for the total momentum $\vec{P}$, electrical current $\vec{J}$ and heat current $\vec{Q}$ [434]. In the case of the Wiedemann–Franz law, the information on specifics of the momentum relaxation has cancelled out completely, but in full generality there is no reason for the combination of the susceptibilities on the right-hand side of this result to reduce to the Lorentz number of the quasiparticle gas.

Turning to holography, the reader is now of course used to the notion that anything related to symmetry in the boundary is flawlessly encoded in the bulk gravity. This wisdom surely also pertains to the cases where the symmetry is weakly broken and the ramifications for transport in a particular strange metal should therefore be fully trustworthy. The power of the memory matrix on the other hand is that the effects of the weak symmetry breaking can be reconstructed directly in the boundary on the basis of the general properties of the strange metal in the clean limit. We therefore start this chapter by first elucidating this interpretational framework before we turn to the specifics of the holographic constructions.

The focus will be on the linear-response functions, which are associated with two-point correlation functions of the operators of the theory via the Kubo formula. It is convenient to reformulate the correlation functions in terms of "inner products in the space of operators", with time evolution expressed through Liouville operators (see box 12.1 for details). One now projects onto the subspace of "slow" operators. Conventionally these are (nearly) conserved currents such as the total momentum $\vec{P}$ and the total charge-current $\vec{J}$. For the charge-current response, one generically finds that the total momentum and current are coupled, with the immediate consequence that for the conductivity one has to compute a 2×2 matrix problem in this space of operators. Subsequently one switches on a term in the Hamiltonian that mildly destroys the symmetry associated with the momentum current. A typical example in the present context is

$$H = H_0 - g\mathcal{O}(k_L), \tag{12.3}$$

where H_0 is the Hamiltonian of the field theory in the Galilean continuum and $\mathcal{O}(k_L)$ describes a weak periodic potential characterised by a wave vector k_L. The coupling g has to be small since it controls the perturbative considerations that follow. The coupling g runs in general, and, in order for it to remain small in the deep IR at very low frequency and small momenta, it is natural to choose $\mathcal{O}(k_L)$ to be an irrelevant operator. In section 12.4 we will address a situation where this running of the "translational symmetry-breaking coupling" comes to the fore.

This symmetry-breaking operator is generally a "fast" operator compared with the momentum and current. This can be "integrated out" to give rise to a self-energy-like correction to the currents: the "memory matrix" proper. Strictly speaking, the split between "slow" and "fast" operators is somewhat loose, and $\mathcal{O}(k_L)$ can be a slow operator as well (see box 12.1).

It is then easy to derive the general expression for the DC conductivity [433],

$$\sigma_{jj}(\omega = 0, k = 0) = \frac{\chi_{j\vec{P}}^2}{\chi_{\vec{P}\vec{P}}} \tau_K, \tag{12.4}$$

where $\chi_{\mathcal{O}_1\mathcal{O}_2} = \lim_{\omega\to 0}(1/(i\omega))\langle\mathcal{O}_1\mathcal{O}_2\rangle_R$ is the static susceptibility and τ_K is the momentum relaxation time induced by $\mathcal{O}(k_L)$. In a relativistic, finite-density theory the susceptibilities are, to lowest (unperturbed) order in g, determined to equal $\chi_{j\vec{P}} = \rho$, the charge density, while $\chi_{\vec{P}\vec{P}} = \epsilon + P$ equals the sum of the energy density and pressure; in the non-relativistic limit the latter reduces to $\chi_{\vec{P}\vec{P}} = m\rho$. The momentum relaxation rate is set by the expression

$$\Gamma_K = \frac{1}{\tau_K} = \frac{g^2 k_L^2}{\chi_{\vec{P}\vec{P}}} \lim_{\omega \to 0} \frac{\operatorname{Im} G_{OO}^R(\omega, k_L)}{\omega}, \tag{12.5}$$

and one recovers the Drude expression for the DC conductivity, where the combination $\chi_{\vec{J}\vec{P}}^2 / \chi_{\vec{P}\vec{P}} = \rho^2/(\epsilon + P)$ corresponds to the Drude weight. This is the general result for the DC conductivity due to perturbatively weak momentum relaxation associated with a single Umklapp wave vector k_L. One can view the main alternative method of momentum loss – quenched disorder – as the result of an averaged potential over random "lattice" vectors. The expression then immediately generalises to (see section 12.5)

$$\Gamma_K^{\text{disorder}} = \frac{1}{\tau_K^{\text{disorder}}} = \int d^d k \, \frac{g^2 k^2}{\chi_{\vec{P}\vec{P}}} \lim_{\omega \to 0} \frac{\operatorname{Im} G_{OO}^R(\omega, k)}{\omega}. \tag{12.6}$$

The momentum relaxation rate is thus determined by the low-frequency asymptote of the spectral function associated with the symmetry-breaking operator $\mathcal{O}(k_L)$ at the Umklapp momentum. The universality of this formula means that it also applies directly to the Fermi liquid, as we mentioned. The operator $\mathcal{O}(k_L)$ of relevance there is the four-quasiparticle interaction term, where one has to respect the fact that in collisions momentum is conserved only modulo the Umklapp [433],

$$\mathcal{O}(k_L) = \int \prod_{i=1}^{4} d^{d+1} p_i \, \psi^\dagger(p_1) \psi(p_2) \psi^\dagger(p_3) \psi(p_4) \delta(p_1 - p_2 + p_3 - p_4 - k_L). \tag{12.7}$$

Upon evaluating this at the Fermi surface one finds precisely the right scaling dimension such that $\Gamma_K \sim T^2$.

One infers immediately from Eq. (12.6) that a rapidly increasing density of zero-energy excitations is required at the large Umklapp momentum k_L in order to find a sizeable momentum relaxation rate. In Fermi liquids with densities achieved in normal metals, k_F is of order k_L, which is the reason why this condition is fulfilled. To avoid this "collisional" momentum relaxation one has to go to very-low-density systems [435]. Generically, however, it is very hard to find systems with appreciable low-energy spectral weight at finite momentum. The only other known natural systems aside from the Fermi liquid are locally quantum critical systems such as the RN strange metal.

Our probe calculations in chapter 9 revealed that RN metals or conformal-to-AdS metals have characteristic spectral densities

$$\operatorname{Im} G_{OO} \sim \omega^{2\nu_k} F(\omega/T), \tag{12.8}$$

with $\lim_{\omega \to 0} F(\omega/T) \sim (T/\omega)^{2\nu_k - 1} + \cdots$ (box 12.2). The magical power of the memory-matrix method is that nothing more is needed in order to determine the temperature dependence of the transport relaxation time. It follows that

$$\Gamma \sim \lim_{\omega \to 0} \frac{\operatorname{Im} G^R_{J^t J^t}(\omega)}{\omega} \sim T^{2\nu(k_L) - 1}, \tag{12.9}$$

where ν_{k_L} is the IR scaling dimension of the lattice operator. To determine this precisely, one can use a holographic model, but, once it is known, the scaling behaviour of all DC transport is universally determined.

Box 12.1   The memory matrix approach to the DC conductivity

The goal is to compute the DC conductivity as given by the Kubo formula

$$\sigma = \lim_{\omega \to 0} \frac{\operatorname{Im} G^R_{\vec{J}\vec{J}}(\omega)}{\omega} \tag{12.10}$$

in the presence of the small translational-symmetry-breaking perturbation Eq. (12.3) as governed by the operator $\mathcal{O}(k_L)$.

We now rely on a particular way of formulating linear-response theory. We refer the reader to the literature for a detailed account [255]. We divide the space of *operators* into a "slow" sector and a "fast" sector that we will ignore. The "slow" sector always contains the operators which one wants to measure ($\vec{J}$) as well as (nearly) conserved currents such as the total momentum $\vec{P}$. We now integrate out the "fast" sector by projecting the full theory onto the set of "slow" modes. To do so, we need an inner product on the "space of operators". A natural inner product is provided by the dynamical susceptibilities χ_{AB}, or the two-point correlation functions themselves. The conveniently normalised inner product is

$$C_{AB}(t) \equiv (A|e^{-iLt}|B) = (A(t)|B) \equiv T \int_0^{1/T} d\lambda \langle A(t)^\dagger B(i\lambda) \rangle, \tag{12.11}$$

where the time evolution is governed by the Liouville operator $L = [H, \cdot]$, with H the Hamiltonian. The Laplace transform of $\tilde{C}_{AB}$ is directly related to the retarded propagators G^R appearing in the Kubo formula by [255]

$$\tilde{C}_{AB}(\omega) = (A|\frac{i}{\omega - L}|B) = \frac{T}{i\omega}\left[G^R_{AB}(\omega) - G^R_{AB}(i0)\right]. \tag{12.12}$$

The static susceptibilities thus equal

$$\chi_{AB} = \lim_{\omega \to 0} \int \frac{d\lambda}{\pi} \frac{\operatorname{Im} G^R(\lambda)}{\omega - \lambda} = \frac{1}{T} \tilde{C}_{AB}(0). \tag{12.13}$$

Box 12.1 (Continued)

Accordingly, the DC conductivity is given in terms of the $\tilde{C}_{AB}$ as

$$\sigma = \frac{1}{T} \lim_{\omega \to 0} \tilde{C}_{\vec{j}\vec{j}}(\omega). \tag{12.14}$$

In the computation of the dynamical susceptibilities the effect of the fast operators is still present in the evolution operator L. We now define an effective "slow" evolution in terms of the slow operators only by projecting onto the space of slow operators with

$$P = |A_i)\chi_{ij}^{-1}(A_j|. \tag{12.15}$$

Defining the complement $Q = 1 - P$, one has

$$\tilde{C}_{AB} = (A|\frac{i}{\omega - LP - LQ}|B). \tag{12.16}$$

Using the operator identity

$$\frac{1}{X+Y} = \frac{1}{X} - \frac{1}{X}Y\frac{1}{X+Y}, \tag{12.17}$$

this equals

$$\tilde{C}_{AB} = (A|\frac{1}{\omega - LQ}|B) - (A|\frac{L}{\omega - LQ}|D)\chi_{DC}^{-1}\tilde{C}_{CB}. \tag{12.18}$$

On solving this for $\tilde{C}_{AB}$, one finds

$$\tilde{C}_{CB} = \chi_{CD}\left(\chi_{AD} + (A|\frac{L}{\omega - LQ}|D)\right)^{-1}(A|\frac{i}{\omega - LQ}|B). \tag{12.19}$$

Finally, by formally expanding out the middle term one finds that

$$\tilde{C}_{CB} = T\chi\left(\hat{M} - i\omega\chi\right)^{-1}\chi, \tag{12.20}$$

with the memory matrix defined by

$$M_{AB}(\omega) = \frac{1}{T}(\dot{A}|Q\frac{i}{\omega - QLQ}Q|\dot{B}) \tag{12.21}$$

with $|\dot{A}) \equiv L|A)$. (The linear term in L is assumed to vanish by virtue of time-reversal invariance.)

We can now use this to compute the DC conductivity efficiently. Consider the charge current $\vec{J}$ and momentum current $\vec{P}$ as the slow operators. Then by writing out Eq. (12.14) explicitly for the $\vec{J}\vec{J}$ correlator in terms of the memory-matrix components in $\{\vec{J}, \vec{P}\}$ space, the electrical conductivity can be computed as

$$\sigma_{\vec{J}\vec{J}} = \begin{pmatrix} \chi_{\vec{J}\vec{J}} & \chi_{\vec{J}\vec{P}} \end{pmatrix} \begin{pmatrix} (X^{-1})_{\vec{J}\vec{J}} & (X^{-1})_{\vec{J}\vec{P}} \\ (X^{-1})_{\vec{P}\vec{J}} & (X^{-1})_{\vec{P}\vec{P}} \end{pmatrix} \begin{pmatrix} \chi_{\vec{J}\vec{J}} \\ \chi_{\vec{P}\vec{J}} \end{pmatrix}, \tag{12.22}$$

where we have used the shorthand $X_{AB} \equiv -i\omega\chi_{AB} + M_{AB}(\omega)$.

Let us now show why momentum conservation still determines the nature of the current, even if it is very weakly broken, i.e. $\dot{P} \sim \epsilon \ll 1$. Under this assumption, if we assume in addition that the frequency is parametrically smaller than ϵ^2, we will have $X_{\vec{P}\vec{P}} \sim M_{\vec{P}\vec{P}} \sim \epsilon^2$, $X_{\vec{J}\vec{J}} \sim 1$ and $X_{\vec{J}\vec{P}} \sim \epsilon^2$. The contributions to the conductivity scale as

$$\sigma_{\vec{J}\vec{J}} \sim \begin{pmatrix} \chi_{\vec{J}\vec{J}} & \chi_{\vec{J}\vec{P}} \end{pmatrix} \begin{pmatrix} 1 & 1 \\ 1 & 1/\epsilon^2 \end{pmatrix} \begin{pmatrix} \chi_{\vec{J}\vec{J}} \\ \chi_{\vec{P}\vec{J}} \end{pmatrix} \sim \chi_{\vec{J}\vec{P}} \frac{1}{\epsilon^2} \chi_{\vec{P}\vec{J}}. \tag{12.23}$$

The conductivity therefore is essentially given by

$$\sigma_{\vec{J}\vec{J}} = \frac{\chi_{\vec{J}\vec{P}}^2}{-i\omega\chi_{\vec{P}\vec{P}} + M_{\vec{P}\vec{P}}}. \tag{12.24}$$

We immediately see that, if momentum is conserved and $M_{\vec{P}\vec{P}}$ vanishes, the conductivity at low frequency correctly reduces to the perfect-metal form $\sigma(\omega) = \left(\chi_{\vec{J}\vec{P}}^2/\chi_{\vec{P}\vec{P}}\right)(i/\omega + \delta(\omega))$, with the prefactor $\chi_{\vec{J}\vec{P}}^2/\chi_{\vec{P}\vec{P}} = \rho^2/(\varepsilon + P)$, as we explained in the main text.

The final step is to compute the finite value of $M_{\vec{P}\vec{P}}$ for an infinitesimally weak lattice perturbation. By construction for the Hamiltonian deformed by the operator $\mathcal{O}(k_\mathrm{L})$, the equation of motion of the total momentum operator $\vec{P}$ becomes

$$\dot{\vec{P}} = i[H, \vec{P}] = g\vec{k}_\mathrm{L}\mathcal{O}(k_\mathrm{L}). \tag{12.25}$$

By definition the operator $\mathcal{O}(k_\mathrm{L})$ is a "fast" operator that resides in the subspace projected out by Q, and, by combining Eqs. (12.12), (12.21) and (12.25), we obtain

$$\begin{aligned} M_{\vec{P}\vec{P}} &= \frac{1}{T} \lim_{\omega \to 0} (\dot{\vec{P}} | \frac{i}{\omega - L} | \dot{\vec{P}}) \\ &= \frac{g^2 k_\mathrm{L}^2}{T} \lim_{\omega \to 0} \tilde{C}_{\mathcal{O}\mathcal{O}}(k_\mathrm{L})\Big|_{g=0} \\ &= g^2 k_\mathrm{L}^2 \lim_{\omega \to 0} \frac{\mathrm{Im}\, G_{\mathcal{O}\mathcal{O}}^\mathrm{R}(\omega, k_\mathrm{L})}{\omega}\Big|_{g=0}. \end{aligned} \tag{12.26}$$

Box 12.1   (Continued)

The universal expression due to weak momentum relaxation to leading order in g follows.

$$\sigma_{\vec{j}\vec{j}} = \lim_{\omega \to 0} \frac{\chi^2_{\vec{j}\vec{P}}}{M_{\vec{P}\vec{P}}(\omega)} = \frac{\chi^2_{\vec{j}\vec{P}}}{\chi_{\vec{P}\vec{P}}} \frac{1}{\Gamma_K}, \tag{12.27}$$

with the momentum relaxation rate defined as

$$\Gamma_K = \lim_{\omega \to 0} \frac{M_{\vec{P}\vec{P}}(\omega)}{\chi_{\vec{P}\vec{P}}}. \tag{12.28}$$

It should be clear to the reader that the memory-matrix approach is just an extension of phenomenological coarse-graining reasoning to the case in which hydrodynamical conservations are weakly violated. The alterations to the long-time behaviour of the fluid can still be captured in a perturbation theory departing from the "hydrodynamical" limit. The surprise is that for quantum matter, be it the Fermi liquid or the RN strange metal, one finds a behaviour of the momentum relaxation at low temperatures Eqs. (12.2), (12.9) that is markedly different from what one would find in truly classical liquids.

The next question is obvious. In chapter 7 we learned that hydrodynamics is a forte of holographic duality. Does it capture the memory-matrix results as effortlessly? It does. In fact, the careful revisiting of the memory-matrix theory in the above was motivated by the desire to interpret holographic results in this regard in terms of a universal boundary language [433, 434]. Though breaking translations in the boundary is challenging, the same perturbative control over linear response as in the memory matrix allows one to circumvent many of these technicalities in the bulk. In the simplest such holographic set-up, the operator which encodes the lattice momentum is chosen to be the background electrostatic potential itself (box 12.2). From the dictionary, it is immediately clear how to accomplish this in the bulk: the electrical flux emanating from the RN black-brane horizon should be subjected to a periodic spatial modulation, and subsequently one should seek solutions for the coupled gauge field and metric equations of motions of the bulk that are dual to the currents in the boundary. The full solutions of these coupled equations of metric and gauge fields for arbitrary frequency and potential strength require the solution of a system of PDEs in the bulk, instead of the ODEs that we have encountered up to this point. But in the weak perturbation limit, where linear response holds, we can focus on fluctuations around the homogeneous translationally invariant geometry instead. The dual image of the memory matrix will be that at finite momentum there exists a new mode coupling between the graviton and the "corrugated" electric field, which is tractable for small g. This construction reproduces exactly the memory-matrix expression Eq. (12.5) (box 12.2). The one piece of extra information holography yields is that it also reveals the explicit scaling of the relaxation rate, in terms of the scaling of the G^R_{OO} correlation function. In the RN metal we have the familiar AdS$_2$ form $G^R_{OO} \sim T^{2\nu_k - 1}$.

Box 12.2   Weak periodic potentials and holographic DC transport

In this box we will explain a semi-holographic model where the operator $\mathcal{O}(k_L)$ in Eq. (12.3) is associated with the charge-density operator. As an application of the memory-matrix method – which is a purely field-theoretical calculation – we directly use Eq. (12.5) to obtain the scaling behaviour of the relaxation rate that yields eventually the DC conductivity of the lattice distorted RN strange metal. In the second part, we will construct a fully holographic model where the operator $\mathcal{O}(k_L)$ in (12.3) is associated with an extra scalar operator. We compute the DC conductivity directly from holography, demonstrating that the final result does indeed adhere to the universal form of the memory matrix Eq. (12.5).

The RN strange metal in a weak lattice potential

For simplicity we focus on a lattice potential in the electrostatic sector, i.e. we modulate the chemical potential of the RN strange metal. In terms of the memory-matrix approach the operator $\mathcal{O}(k_L)$ we consider is therefore $\mathcal{O}(k_L) = J^t(k_L)$. All the dynamics of interest is therefore still captured by the Einstein–Maxwell action

$$S = \int d^4x \sqrt{-g} \left[\frac{1}{2\kappa^2} \left(R + \frac{6}{L^2} \right) - \frac{1}{4e^2} F_{\mu\nu} F^{\mu\nu} \right]. \tag{12.29}$$

As can be seen from the memory-matrix expression, we only need to obtain the low-frequency IR Green function of the operator J^t at a finite momentum in the RN background without the translational-symmetry-breaking term. From the matching method, we know that we can focus on the near-horizon geometry. At small $T \ll \mu$ this is a black hole in AdS$_2 \times \mathbb{R}^2$ with metric and electrostatic potential

$$ds^2 = \frac{L_2^2}{\zeta^2} \left(-f(\zeta)dt^2 + \frac{d\zeta^2}{f(\zeta)} \right) + \frac{r_*^2}{L^2} d\vec{x}^2, \quad A_t = \frac{eL}{\sqrt{6\kappa}} \left(\frac{1}{\zeta} - \frac{1}{\zeta_+} \right), \tag{12.30}$$

where

$$\zeta = \frac{L_2^2}{r - r_*}, \quad L_2 = \frac{L}{\sqrt{6}}, \quad f(\zeta) = 1 - \frac{\zeta^2}{\zeta_+^2}.$$

Box 12.2   (Continued)

From chapter 7 we know that the linear response is encoded in the fluctuations around this background. Choosing the lattice momentum in the x direction, we have again a coupled system formed by $(k_x = k, \omega)$ Fourier components of the metric- and gauge-field perturbations $\delta g_{xx}, \delta g_{yy}, \delta g_{tt}, \delta g_{xt}, \delta A_t, \delta A_x$. Their equations of motion can be written in terms of two decoupled second-order equations. The sector of relevance depends on the linear combinations

$$\Phi_\pm = \delta g_{yy} + \frac{\zeta^2}{\sqrt{6}\bar{k}^2}\left(1 \pm \sqrt{1 + 2\bar{k}^2}\right)\left(\delta A_t' - \sqrt{\frac{3}{2}}\frac{\delta g_{tt}}{f}\right), \tag{12.31}$$

where

$$\bar{k} = \frac{k}{\mu} = \frac{ke}{\sqrt{6}\kappa}. \tag{12.32}$$

These fluctuations satisfy

$$\Phi_\pm'' + \frac{f'}{f}\Phi_\pm' + \left(\frac{\omega^2}{f^2} - \frac{1 + \bar{k}^2 \pm \sqrt{1 + 2\bar{k}^2}}{\zeta^2 f}\right)\Phi_\pm = 0 \tag{12.33}$$

and can be solved in terms of hypergeometric functions after imposing in-falling boundary conditions at the horizon:

$$\Phi_\pm = (2\zeta_+)^{\nu_\pm}\Gamma(a_\pm)\Gamma(1 + \nu_\pm)f(\zeta)^{-i\zeta_+\omega/2}\zeta^{\frac{1}{2}-\nu_\pm}$$
$$\times\, _2F_1\left(\frac{a_\pm}{2}, \frac{a_\pm+1}{2}, 1 - \nu_\pm; \frac{\zeta^2}{\zeta_+^2}\right) - (\nu_\pm \leftrightarrow -\nu_\pm), \tag{12.34}$$

with $a_\pm = \frac{1}{2} - i\zeta_+\omega - \nu_\pm$, where the scaling exponents [325] are defined as

$$\nu_\pm = \frac{1}{2}\sqrt{5 + 4\bar{k}^2 \pm 4\sqrt{1 + 2\bar{k}^2}}. \tag{12.35}$$

Note that this IR scaling dimension is the same as Eq. (9.42) in AdS space-time with $m_{\rm eff}^2 L_2^2 = 1 + \bar{k}^2 \pm \sqrt{1 + 2\bar{k}^2}$.

To match the near-horizon solution Eq. (12.34) onto the solution away from the horizon, one expands the solution near the boundary $\zeta \sim 0$ of the AdS$_2 \times \mathbb{R}^2$ region,

$$\Phi_\pm \propto \zeta^{1/2}\left(\zeta^{-\nu_\pm} + \mathcal{G}_\pm(\omega)\zeta^{\nu_\pm}\right). \tag{12.36}$$

The locally quantum critical Green functions are

$$\mathcal{G}_\pm(\omega) = -(\pi T)^{2\nu_\pm}\frac{\Gamma(1 - \nu_\pm)\Gamma\left(\frac{1}{2} - i\omega/(2\pi T) + \nu_\pm\right)}{\Gamma(1 + \nu_\pm)\Gamma\left(\frac{1}{2} - i\omega/(2\pi T) - \nu_\pm\right)}$$
$$= \omega(\pi T)^{2\nu_\pm-1}\frac{\pi}{2}\frac{\Gamma(1 - \nu_\pm)\Gamma(\frac{1}{2} + \nu_\pm)}{\Gamma(1 + \nu_\pm)\Gamma(\frac{1}{2} - \nu_\pm)}\tan(\pi\nu_\pm) + \cdots, \tag{12.37}$$

with the temperature of the black hole $T = 1/(2\pi \zeta_+)$. Relying on the now familiar matching procedure introduced in section 9.3, one finds for the propagator at low frequencies [325]

$$G_\pm(\omega) = \frac{A + B\mathcal{G}_\pm(\omega)}{C + D\mathcal{G}_\pm(\omega)}, \tag{12.38}$$

where A, B, C, D are real constants that are generically non-zero and independent of ω and T. In particular, this implies that at the lowest frequencies

$$\operatorname{Im} G_\pm(\omega) \propto \operatorname{Im} \mathcal{G}_\pm(\omega). \tag{12.39}$$

Together with Eq. (12.39), this yields the result for the momentum relaxation rate,

$$\Gamma \sim \lim_{\omega \to 0} \frac{\operatorname{Im} G^R_{J^t J^t}(\omega)}{\omega} \sim T^{2\nu - 1}. \tag{12.40}$$

The memory matrix from holography

Blake, Tong and Vegh [436] showed how holography reproduces the exact memory-matrix calculation when the lattice deformation is encoded in a scalar operator $\mathcal{O}(k_L)$. The holographic lattice model is constructed by introducing a spatially modulated source for field ϕ dual to the scalar operator [436]:

$$S = \int d^4x \sqrt{-g} \left[\frac{1}{2\kappa^2} \left(R + \frac{6}{L^2} \right) - \frac{1}{4e^2} F_{\mu\nu} F^{\mu\nu} - \frac{1}{2} g^{\mu\nu} \, \partial_\mu \phi \, \partial_\nu \phi \right.$$
$$\left. - \frac{1}{2} m^2 \phi^2 \right]. \tag{12.41}$$

Near the AdS$_4$ boundary, the scalar behaves as

$$\phi(r, x, y) \sim \phi_s(x, y) r^{\Delta - d} + \phi_r(x, y) r^\Delta. \tag{12.42}$$

For simplicity, we consider the lattice in one direction. We impose the source $\phi_s = \epsilon \cos(k_L x)$ with ϵ a small number. Thus we can treat the lattice perturbatively. From the equation of motion, we can see that the metric field and gauge field will receive the corrections from the backreaction of scalar field only at the order $\mathcal{O}(\epsilon^2)$, i.e.

$$g_{\mu\nu} = g^{(0)}_{\mu\nu}(r) + \epsilon^2 [g^H_{\mu\nu}(r) + g^I_{\mu\nu}(r)\cos(2k_L x)],$$
$$A_\mu = A^{(0)}_\mu(r) + \epsilon^2 [A^H_\mu(r) + A^I_\mu(r)\cos(2k_L x)], \tag{12.43}$$
$$\phi = \epsilon \phi_0(r)\cos(k_L x).$$

To compute the conductivity, the three perturbations $\delta A_x, \delta g_{tx}$ and $\delta \phi$ are coupled to each other (in the gauge $g_{rx} = 0$). To leading order we can consider the homogeneous metric and Maxwell potential, with the modulation

Box 12.2   (Continued)

present only in the scalar field. The conductivity can then be computed in the standard way,

$$\sigma(\omega) = \frac{1}{i\omega} \frac{\delta A_x^{(1)}}{\delta A_x^{(0)}}\bigg|_{r\to\infty}, \tag{12.44}$$

with the final result

$$\sigma_{\text{DC}} = \frac{\rho^2}{\epsilon + P} \frac{1}{\Gamma}, \tag{12.45}$$

where

$$\Gamma = \frac{s}{4\pi} \frac{M^2(r_{\text{h}})}{\epsilon + P}, \quad M^2(r) = \frac{1}{2}\epsilon^2 k_L^2 \phi_0(r)^2 \tag{12.46}$$

and s, ϵ, P are quantities in Eqs. (8.10) and (8.11).

At $T = 0$, the near-horizon geometry of Eq. (12.43) is $\text{AdS}_2 \times \mathbb{R}^2$, i.e. Eq. (8.23). In the IR, the regular solution for ϕ falls off as $\phi_0 \sim \zeta^{\frac{1}{2} - \nu_{k_L}}$, where $\nu_{k_L} = \sqrt{\frac{1}{4} + m^2 L^2/6 + 2k_L^2/\mu^2}$. The conformal dimension of the dual operator $\mathcal{O}(k_L)$ in the momentum space is $\Delta_{k_L} = \nu_{k_L} - \frac{1}{2}$. Thus $\phi_0 \sim \zeta^{-\Delta_{k_L}}$.

At finite and low temperature $\zeta_+ \sim T^{-1}$, thus

$$\rho_{\text{DC}} \sim \epsilon^2 k_L^2 T^{2\Delta_{k_L}}. \tag{12.47}$$

This is precisely the result Eq. (12.5) obtained by Hartnoll and Hofman [433].

12.2 Periodic potentials in holographic superconductors and the optical conductivity

As we already emphasised, the bulk problem becomes seriously complicated when translational invariance in the boundary is explicitly broken. Only the DC conductivity in the weak-potential limit is easily tractable. To address the optical conductivity at finite potential strength requires the solution of the PDEs in the bulk. This can only be accomplished numerically. This tour de force was accomplished by Horowitz, Santos and Tong [438, 439, 440]. Inherently this makes the solutions difficult to interpret qualitatively, but compared with the Galilean continuum one undoubtedly adds realism to the comparison with the condensed matter systems. Remarkably, the first results were surprisingly similar to real optical-conductivity data in high-T_c superconductors, reproducing puzzling features that have never been fully understood [441]. The validity of these results still awaits confirmation.

One can also add holographic superconductivity to the lattice-deformed strange metal. The qualitative properties of the optical conductivity of this holographic model are strikingly similar to the observations in optimally doped cuprates [440]. This includes behaviours that did not find any explanation in terms of the conventional "Eliashberg paradigm", and one can defend the case that as mere phenomenology this holographic construction is far superior to anything else that has been proposed. To a degree this might be coincidental, but it has at least the rare benefit (in condensed matter theory) that these similarities with experiments were realised *after* the computations had been completed. Before we turn to the results of Horowitz, Santos and Tong, let us first discuss some of the specifics of the approach [438]. In their first incarnation, they add an extra neutral real scalar field Φ to the Einstein–Maxwell action, as in Eq. (12.41) in box 12.2,

$$S = \frac{1}{16\pi G} \int d^4x \sqrt{-g} \left[R + \frac{6}{L^2} - \frac{1}{4} F_{ab} F^{ab} - \frac{1}{2} \nabla_a \Phi \, \nabla^a \Phi - V(\Phi) \right], \quad (12.48)$$

with a simple "minimal" potential that gives the scalar field a mass,

$$V(\Phi) = -\frac{\Phi^2}{L^2}. \quad (12.49)$$

A later incarnation built on an earlier idea [437] considering pure Einstein–Maxwell gravity where the lattice is induced by a periodic gauge potential [439]. The first model with the additional neutral scalar field acting as the source of the lattice has the virtue that it is the same generic set-up as we considered in the perturbative memory-matrix approach, but now we can go beyond perturbation theory. The explicit translational symmetry breaking is imposed through the GKPW rule as the boundary condition on the asymptotic behaviour of the auxiliary scalar field,

$$\Phi = \frac{\phi_1}{r} + \frac{\phi_2}{r^2} + \cdots, \quad (12.50)$$

by choosing the source term ϕ_1 to be a periodically modulated function with UV strength A_0 and a periodicity $2\pi/k_0$ in the x direction: $\phi_1 = A_0 \cos(k_0 x)$. A unidirectional translational-symmetry-breaking source term in the x direction is chosen here just for convenience; it has no further interesting consequences in this context. The most general static, electrically charged black-hole solution that is not translationally invariant in the x direction has the form

$$ds^2 = r^2 \left[-\left(1 - \frac{1}{r}\right) P(r) Q_{tt} \, dt^2 + Q_{xx} \left(dx + \frac{Q_{xr}}{r^4} \, dr\right)^2 \right.$$

$$\left. + Q_{yy} \, dy^2 + \frac{Q_{rr} dr^2}{r^2 P(r)(1 - 1/r)} \right], \tag{12.51}$$

with

$$\Phi = \frac{\phi(x, r)}{r} \tag{12.52}$$

and

$$A = \left(1 - \frac{1}{r}\right) \psi(x, r) dt. \tag{12.53}$$

Here $Q_{ij}(x, r)$ for $i, j \in t, x, y, r$, are functions of x and r to be determined by solving the equations of motion, while $P(r)$ can be chosen to be

$$P(r) = 1 + \frac{1}{r} + \frac{1}{r^2} - \frac{\mu_1^2}{2r^3}, \tag{12.54}$$

corresponding to an "average" black-hole temperature $T_H = P(1)/(4\pi L)$. One can now determine numerically a fully backreacted solution and one obtains smooth and well-behaved modulated profiles both in the gauge fields and in the geometry. These modulations vanish on approaching the extremal horizon, encoding for the desired irrelevant nature of the periodic potential.

Within this fully non-perturbative holographic lattice background one can now try to assess the consequences of momentum loss for transport. In particular, one can now compute numerically the optical conductivity in this background. The typical results for the finite-temperature RN metal in $2 + 1$ dimensions [438] are shown in Fig. 12.1. Similar results (not shown) are obtained for a $(3 + 1)$-dimensional boundary [439]. In the low-energy regime one finds precisely the Drude form $\sigma(\omega) = K\tau_K/(1 - i\omega\tau_K)$. Moreover, the momentum relaxation rate behaves precisely according to the "memory-matrix law" for the RN metal of the previous section $1/\tau_K \sim T^{2\nu_-(k_0)-1}$. This is fully according to expectation, since the periodic potential is irrelevant in the IR, but constitutes a high-precision test both of the arguments presented in the first section and, even more so, of the correctness of the complicated numerics.

The surprise happens at the intermediate frequencies $\omega \simeq \mu$. Here Horowitz, Santos and Tong find that the absolute value of the optical conductivity behaves as (see Fig. 12.2),

$$|\sigma(\omega)| = \frac{B}{\omega^\gamma} + C, \tag{12.55}$$

where the exponent $\gamma = 2/3$ within numerical accuracy. Over roughly a decade $1 \lesssim \omega\tau_K \lesssim 10$, the conductivity exhibits at these high frequencies an approximate scaling behaviour, governed by a dimension $-2/3$ that "falls out of the air". It

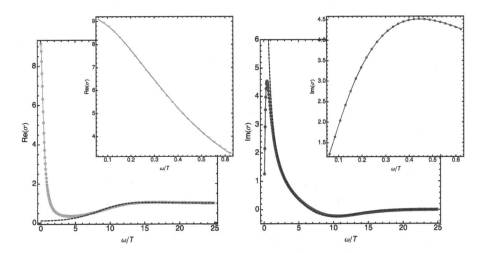

Figure 12.1 The real and imaginary parts of the optical conductivity in a holographic metal with (solid line) and without lattice (dashed line) [438]. When $\omega \to 0$, the imaginary part of the optical conductivity becomes finite in the presence of the lattice. The lack of momentum conservation broadens the delta-function peak for Re σ into the Drude form. The insets which focus on the situation at low frequencies show that the numerical data points are in excellent agreement with the Drude line shape $\sigma(\omega) = D\tau_K/(1 - i\omega\tau_K)$, while $\tau_K \sim T^{2\nu_-(k_0)-1}$, obeying the expectation from the memory-matrix consideration of section 12.1. Figure source [438]. (Reprinted with kind permission from Springer Science and Business Media, © 2012, SISSA, Trieste.)

is claimed that this scaling behaviour is remarkably insensitive to the parameters defining the problem. At the time of writing there is no explanation of the origin of this scaling; it also still awaits confirmation by an independent computation.

What is so remarkable about these holographic results is that, quite a number of years ago, high-quality optical-conductivity data obtained for the strange metals realised in optimally doped cuprate superconductors [149] revealed a behaviour that is strikingly similar. In the "hydrodynamical" regime ($\hbar\omega \leq k_BT$) the data again fit precisely to a Drude form characterised by a momentum relaxation time $\tau = 0.7\hbar/(k_BT)$ (associated with the linear-in-temperature resistivity, see section 12.5 for potential explanations). Most pointedly, at "intermediate" energies (up to 0.5 eV, which has a similar status in cuprates to the Planck scale in high-energy physics) a strikingly perfect scaling behaviour was found, both in the real and in the imaginary part of the conductivity, described by the power law $\sigma(\omega) \sim 1/(i\omega)^{2/3}$. This has in common with holography not only the "dichotomy" in terms of a Drude low-energy behaviour and an "intermediate-scale emerging conformality" but also the same value for the power-law scaling exponent. Immediately after the experimental paper [149] had been published, it was pointed out this "dichotomy"

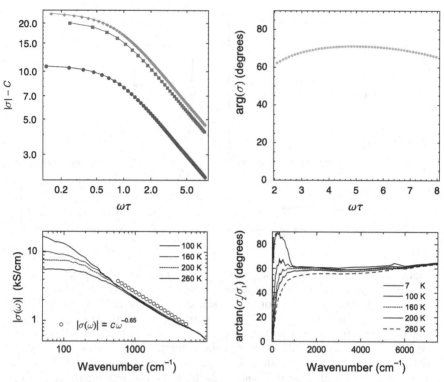

Figure 12.2 The power-law scaling of the optical conductivity of a holographic metal in a lattice observed by Horowitz, Santos and Tong. From the log–log plot on the top left one can extract that $|\sigma(\omega)| = B/\omega^{2/3} + C$ for an intermediate range $2 < \omega\tau < 8$, with τ the Drude relaxation time (the regime for $\omega\tau > 5$ is not shown). On the top right the nearly constant phase angle of the conductivity $\arg(\sigma)$ is shown, which is consistent with the $-2/3$ power. This closely resembles actual experimental results in cuprate semiconductors. The bottom left and right show the real part and the phase angle of the measured optical conductivity. They also show an intermediate regime with nearly identical scaling behaviour $\sigma = B/\omega^{2/3}$. Figure sources [438] (reprinted with kind permission from Springer Science and Business Media, © 2012, SISSA, Trieste) and [149].

is very hard to explain on the basis of standard scaling theory [442], and these results were shelved as significant but plainly puzzling.

We require a better understanding, one that is at this time still missing, to assess whether this similarity between holography and experiment is a coincidence or whether it is rooted in a yet-to-be-discovered general principle. There is at least one crucial difference. The scaling regime $1 \lesssim \omega\tau_K \lesssim 10$ in the holographic computation is bounded by the relaxation time τ_K. This sensitively depends on the temperature as we have seen. No such temperature dependence is seen in the cuprate data, however. Further detailed and independent computations of transport in holographic lattice models are being computed as we speak in order to shed more light on these results.

Though they are perhaps less striking, there are quite a number of additional features that are suggestive regarding the applicability of holography to the physics of the real cuprates in this context. This becomes more evident when holographic superconductivity is also taken into account. Horowitz and Santos computed holographically the evolution of the optical conductivity as a function of temperature, from the normal state to deep inside the superconductor [440]. The results are eerily realistic, compared with the experimental results in optimally doped cuprates. The intermediate-energy $\omega^{-2/3}$ scaling regime is not at all affected by the onset of superconductivity. However, as anticipated from the discussion in section 10.2, a gap opens up with a BCS temperature dependence in the low-energy regime, which is now much more visible because it eats away the Drude peak. At low temperature this saturates at the "correct" value of $2\Delta/(k_B T_c) \simeq 8$. The Ferrel–Grover–Tinkham sum rule (relating the superfluid density to the decrease in $\sigma_1(\omega)$) is satisfied, although there are signs of spectral weight transfer coming from high energy $\omega \sim \mu$. A strong signal indicating that non-Fermi-liquid physics is at work is present in the form of the rapid drop of the momentum relaxation rate upon entering the superconducting state. This signals the development of coherence in the superconducting state from a normal state that is completely incoherent. This rather spectacular effect was observed already in the early 1990s [443], and its explanation has been a challenge for conventional theory. Perhaps the most striking anomaly is the prediction of a finite normal-fluid fraction in this holographic s-wave superconductor *even at zero temperature*. Although this is not completely understood [440], it is probably related to the "Lifshitz quantum critical phase" in the holographic superconductor, which we discussed at length in section 10.5. There are some indications in the experimental literature for the existence of such a zero-temperature normal fluid that coexists in the IR [444], and this might be a prediction that should be further studied by dedicated experiments.

12.3 Lattice potentials and the fermion spectral functions of the Reissner–Nordström metal

We highlighted the holographic fermion spectral functions in chapter 9 as a fundamental source of information on holographic quantum matter, complementary to the collective responses. It is natural to ask what happens to these fermions in the presence of a periodic potential: is there a notion of "band structure" in the strange metals? The bulk description is, even in the probe limit, technically very challenging. One faces a "band-structure" problem in the bulk, where the bulk fermions are diffracting against the spatial potential, in combination with the space-time curvature and the special boundary conditions near the horizon and boundary. At present this problem has been solved only in the fixed background of a RN black hole, in the leading order of the perturbation associated with the weak-periodic-potential limit [359].

Given that the focus is on the influence of a weak periodic potential in the deep infrared, one can further simplify matters by eliminating the auxiliary neutral scalar of the previous section. Instead, the spatial modulation is directly encoded in the chemical potential in the boundary,

$$ds^2 = -r^2 f(r)dt^2 + \frac{dr^2}{r^2 f(r)} + r^2(dx^2 + dy^2),$$

$$f(r) = 1 - \frac{1+Q^2}{r^3} + \frac{Q^2}{r^4},$$

$$A_t = 2Q\left(1 - \frac{1}{r}\right) + A_t^1 \tag{12.56}$$

$$A_t^{(1)} = \mu_1(x)\left(1 - \frac{1}{r}\right),$$

$$\mu_0 = 2Q, \quad \mu_1 = 2\epsilon\cos(k_L x),$$

where the single-harmonic potential is unidirectional as in the previous section. It is characterised by an Umklapp wave vector $k_L = 2\pi/a$ (a is the "lattice constant") and a strength $\epsilon/\mu_0 \to 0$. Although this looks quite simple, the computation of even the leading-order perturbative correction to the fermion spectral functions in terms of the small parameter ϵ is a lengthy affair. Differently from the computation of the DC conductivity in the full perturbative holographic part in the box 12.2, in order to get the fermionic spectral function the interactions between different Brillouin zones have to be accounted for, which makes the computation much more complicated. We summarise the results here and we refer the reader to the literature for the details [359].

The aspect which is most straightforward to understand is how the "domain-wall" fermions respond to the potential. Despite the serious technical difficulties, at the end of the day one recovers just the textbook Bloch wave theory in the boundary spectral functions insofar as this "confining sector" is concerned. We have already emphasised in chapters 9 and 11 that the boundary Fermi gas is in direct correspondence to the gas formed at the domain wall in the bulk. Unsurprisingly, these fermions do indeed react to the presence of the periodic potential as free fermions. For a unidirectional potential along the x direction we find that a band gap opens in the "background" dispersion right at the scattering wave vector (k_L, k_y). For small ϵ the size of the gap behaves as

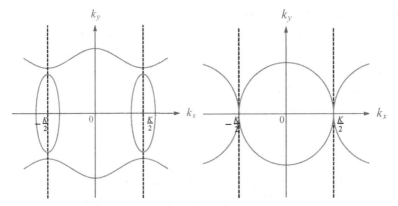

Figure 12.3 A cartoon of the band structure for different Fermi momenta k_F. The system under consideration has a lattice structure only in the x direction. The undulating curve is the Fermi surface ($\omega = 0$) and the dashed line is the first Brillouin-zone boundary. In the figure K is the lattice momentum $K = k_L$. The left plot is for $k_F > k_L/2$ and the right plot is for $k_F = k_L/2$. We have a band gap at the first Brillouin-zone boundary $k_x = \pm k_L/2$ for generic $k_F > k_L/2$ (i.e. $k_y \neq 0$), which will close when $k_F = k_L/2$ (i.e. $k_y = 0$). Note that this picture is a result in the extended-Brillouin-zone scheme. Figure source [359].

$$\Delta(k_L, k_y) \simeq \epsilon \sqrt{1 - \frac{1}{\sqrt{1 + (2k_y/k_L)^2}}}. \tag{12.57}$$

The band gap vanishes for this unidirectional potential when the transverse momentum $k_y = 0$. This might look unfamiliar, but holographic fermions have a chiral property that causes them to react to potentials in the same way as do the helical surface states of three-dimensional topological insulators: at $k_y = 0$ the gap disappears, since such fermions do not scatter in a backward direction. The new quasi-Fermi surface in the presence of the weak potential can now be constructed as usual (Fig. 12.3). The Umklapp surfaces are straight lines in the k_y direction, centred at the Umklapp momenta. When these Umklapp surfaces intersect the Fermi surface, they reconstruct into Fermi-surface "pockets" due to the opening of a gap centred at the Umklapp momenta.

Insofar as the domain-wall/gauge-singlet fermions are concerned, holography is just a slightly awkward detour to recover familiar wisdom. However, the situation is quite different regarding the influence of the potential on the quantum critical AdS$_2$ sector in the deep infrared. In the "domain-wall" fermion regime where the quasiparticles form, they provide the self-energy as explained in chapter 9; in the *absence* of domain-wall fermions, the deep-IR Green functions directly control the boundary fermion propagators. In the presence of a weak potential the bulk computation reveals that the fermions in the AdS_2 bulk do form standard Bloch waves at Bloch momenta $k + nk_L$. Even though the momentum dependence in the local quantum critical sector appears in the exponent characterising the propagator, at first order in the lattice perturbation ϵ the local quantum critical propagators of the fermions in the boundary acquire the form

$$\mathcal{G}(\omega, \vec{k}) = \alpha_{\vec{k}} \mathcal{G}_0(\omega, \vec{k}) + \mathcal{G}_1(\omega, \vec{k})$$
$$= \alpha_{\vec{k}} \, \omega^{2\nu_{\vec{k}}} + \beta_{\vec{k}}^{(-)} \omega^{2\nu_{\vec{k}-\vec{k}_L}} + \beta_{\vec{k}}^{(+)} \omega^{2\nu_{\vec{k}+\vec{k}_L}} + \cdots . \qquad (12.58)$$

One recognises the zeroth-order term with coefficient $\alpha_{\vec{k}}$ as the zeroth-order result, which has already been presented in chapter 9. The amplitudes $\alpha_{\vec{k}}$, $\beta_{\vec{k}}^{(\pm)}$ are such that $\beta \ll \alpha$ for $\epsilon \ll \mu_0$ [359].

The meaning of this "Bloch sum" in the dual field theory is quite surprising. The AdS/CFT dictionary associates the free problem in the bulk with the physics of strongly interacting conformal fields in the boundary. In the $AdS_2 \times \mathbb{R}^2$ deep interior in the absence of a lattice, one associates a new *independent* CFT_1 with every point in momentum space. We remind the reader that these CFT_1s are not completely independent: as discussed in chapters 8 and 9, the quantum criticality is "quasi-local" since the propagators reveal that they are spatially correlated over a length scale $\xi_{\text{space}} = \sqrt{2}/\mu_{\nu_{k=0}}$, as determined by the chemical potential [159]. Given this spatial correlation, the effect of breaking the continuous translational invariance is that these CFT_1s start to interact, but, despite the exponential dependence on the momenta, the effect is a regular sum over Bloch copies.

This "Bloch wave of CFT_1s" Eq. (12.58) has a remarkable consequence for the boundary physics. Consider the spectral functions in the extended-zone scheme, at some momentum k large compared with the Umklapp momentum k_L so that one resides in some higher zone. Starting at high energy, the small periodic potential cannot exert any influence, and accordingly one finds the standard "algebraic pseudogap" spectral function $\sim \omega^{2\nu_{\vec{k}}}$. However, when the potential becomes noticeable, the CFT_1s associated with the "lattice copies" at $k \pm k_L$ start to contribute. From Eq. (9.40) one infers that half-way along the Brillouin zone for $k > k_L/2$ one of the lattice copies as characterised by the prefactor $\beta_{\vec{k}}^{(\cdots)}$ becomes "less irrelevant" $\sim \omega^{2\nu_{\vec{k}-\vec{k}}} > \omega^{2\nu_{\vec{k}}}$, as illustrated in Fig. 12.4. The effect is that of a crossover from the high-energy scaling characterised by the dimension ν_k to a scaling associated with the more relevant "Umklapped" scaling dimension ν_{k-k_L}, depending where one looks in a particular higher Brillouin zone (Fig. 12.5). This Bloch wave of CFT_1 implies that at low energy the CFT_1 associated with the Brillouin zone with an index lower by one k_L in the extended-zone scheme takes over. In this way, the periodic potential gets completely encoded in the energy scaling behaviour of the fermion spectral function.

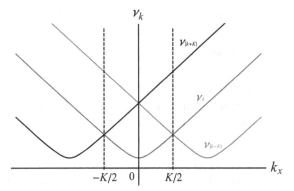

Figure 12.4 The behaviour of the different powers of $\omega^{2\nu_{k-K}}$, $\omega^{2\nu_k}$ and $\omega^{2\nu_{k+K}}$ in the lattice AdS$_2$ metal spectral function as a function of momentum k. The IR of the Green function is controlled by the lowest branch: $\omega^{2\nu_{k+K}}$ in the $\ell = -1$ Brillouin zone, $\omega^{2\nu_k}$ in the $\ell = 0$ Brillouin zone and $\omega^{2\nu_{k-K}}$ in the $\ell = 1$ Brillouin zone. Figure source [359].

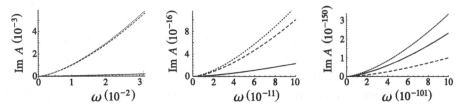

Figure 12.5 This sequence of AdS$_2$ metal spectral functions for a fixed generic k shows how the Umklapp contribution takes over at low frequencies. From left to right we zoom in on lower and lower frequencies on the $\ell = 0$ spectral function, inside the first Brillouin zone. We show the full corrected spectral function $A_{\text{full}}(\omega, \vec{k}) \sim \text{Im}\, G$ (dotted), the original "bare" holographic spectral function $A_{\text{pure AdSRN}}(\omega, \vec{k}) \sim \text{Im}\, G_0$ (dashed), and the Umklapp correction due to the periodic chemical potential modulation $\delta A_{\text{lattice}}(\omega, \vec{k}) \sim \text{Im}\, \delta G$ (solid). Figure source [359].

Though only the leading-order effect in ϵ has been computed explicitly, it is almost unavoidable that at every subsequent order in perturbation theory one can Umklapp to one further Brillouin zone:

$$\mathcal{G}_{\text{full}}(\omega, \vec{k}) \sim \omega^{2\nu_{\vec{k}}} + \epsilon^2 (\omega^{2\nu_{\vec{k}-\vec{K}}} + \omega^{2\nu_{\vec{k}+\vec{K}}}) + \epsilon^4 (\omega^{2\nu_{\vec{k}-2\vec{K}}} + 2\omega^{\nu_{\vec{k}+2\vec{K}}})$$
$$+ \cdots + \epsilon^{2n} (\omega^{2\nu_{\vec{k}-n\vec{K}}} + \omega^{2\nu_{\vec{k}+n\vec{K}}}) + \cdots . \tag{12.59}$$

At low enough energies, one expects some "descendance" to the first zone due to the Umklapp upon lowering the energy.

This result is surprising but actually rather sensible: the periodic potential simply sums the exponential momentum dependence of the local quantum critical excitations over the individual Brillouin zones in the extended zone scheme. Could this perhaps shed light on the "nodal–antinodal" dichotomy observed by electron spectroscopy in the cuprate normal state as discussed in chapter 3? The principle revealed by this holographic calculation, namely that the weak periodic potential reorganises the scaling behaviour of the spectral function in momentum space, is suggestive. We stressed in chapter 3 that the sudden changes seem to occur at the surface in the Brillouin zone associated with a doubling of the periodicity in real space, where the puzzle is that there is no sign of this required translational symmetry breaking among the hidden orders. Could this be related to the $k_L/2$ effect illustrated in Fig. 12.4?

At this stage this is no more than suggestive: to assess whether holography can shed light on the nodal–antinodal dichotomy, Fermi arcs and so forth, one has to extend these computations to strong potentials at zero temperature. This is a formidable technical challenge that one hopes may be overcome in the near future. However, there is a very useful take-home message for experimentalists. In the ARPES community it is taken for granted that the photoemission spectral functions should be precisely periodic in the extended Brillouin zone, since the crystal is periodic in real space. For nearly free fermions in conventional band-structure theory this is indeed a truism: they are quantum-mechanical waves governed by the Fourier transformation. However, as illustrated by the un-particle RN metal in the above, for anything that is not a Fermi liquid there is no a-priori reason for the response to show a precise periodicity in the extended zone as a function of the single-particle momentum. Hence, this can be used in principle to extract the non-Fermi-liquid features directly from the experimental data.

12.4 Unidirectional potentials becoming strong: Bianchi VII geometry and the quantum smectic

The explanation of the difference between metals and insulators is an early highlight of the quantum-mechanical theory of electrons in solids. The story is familiar. The nearly free fermions of the Fermi liquid diffract like quantum-mechanical waves against the periodic potential of the ion lattice and band gaps open up at the Umklapp surfaces. Filling up the bands using the Pauli principle, band insulators form when the chemical potential lies inside the band gap. For an even number of electrons per unit cell the valence band can be completely filled and, when the band gap is large enough, the system turns into an insulator. Alternatively, more recently the phenomenon of Anderson localisation was discovered: the fact that wave functions can completely localise in a random potential. Finally, there are the Mott insulators, which arise at any integer filling when the electrons repel each other sufficiently strongly. These are the rules we know to apply when constructing insulators in "particle" physics. But what are the principles governing "un-particle" physics in this regard? Is it possible to form insulators by combining the strange metals of holography with periodic lattices? If so, what is the nature of such "strange insulators"?

In fact, from the previous section on fermion spectral functions one already obtains a hint regarding what one should look for. The strange metals are quantum critical phases characterised by scaling properties, and we found that a periodic potential has the capacity to reorganise the scaling properties. Returning our attention to the collective responses and the conductivity in particular, could it be that a sufficiently strong periodic potential could remove all charged excitations in the IR? By definition an insulator would form. When all charged excitations are gapped, the conductivity must vanish. Holographically, this means that the translational symmetry breaking must somehow change the geometry drastically such that it caps off at finite radial distance, and no flux should remain at the cap.

In the previous sections of this chapter we encountered the great technical difficulties one faces when one imposes translational symmetry breaking by hand. Exploring the regime where potentials become strong is particularly hard. However, there are special classes of bulk geometries that describe a boundary where the translational symmetry is broken, for which the bulk equations of motion have only a radial evolution and all of its other spatial dependence is predetermined. One such family follows from the so-called Bianchi classification of homogeneous but anisotropic spaces that were developed in GR for cosmological purposes [445]. These arise quite naturally in the holographic context [446]. Some other clever constructions that wire in translational symmetry breaking in the bulk are briefly touched on in box 12.3.

We focus here on the so-called Bianchi VII geometry. What allows the equations to have only non-trivial evolution in the radial direction is an extra symmetry in the problem. As realised by Donos and Hartnoll [447], this is a particularly attractive way to model a periodic potential in the boundary. The specific highly symmetric way in which translation invariance is broken here is through the presence of a helical vector field propagating in the $(3 + 1)$-dimensional boundary space-time along the x_1 spatial direction, while the vectors precess in the x_2–x_3 plane with a pitch p. This is associated with the existence of three invariant one-forms,

$$\omega_1 = dx_1, \quad \omega_2 + i\omega_3 = e^{ipx_1}(dx_2 + i \, dx_3), \tag{12.60}$$

and illustrated in Fig. 12.6. Notice that this helix breaks translations in the x_1 direction only.

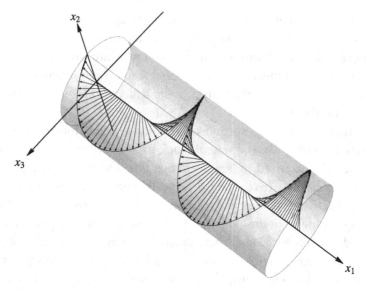

Figure 12.6 A cartoon depiction of the helix structure of the three-dimensional homogeneous space of Bianchi VII. The helix occurs at each point in the x_2-x_3 plane, and is thus translationally invariant along those directions. The helix breaks outright translations in the x_1 direction, though it preserves a particular combination of translations in the x_1 direction with rotations in the x_2-x_3 plane.

This now has to be implemented in the holographic bulk in a similar way to what was done in section 12.2. However, instead of a neutral scalar that is sourced at the boundary by a simple translational-symmetry-breaking field Eq. (12.50), one now considers a vector field B_μ. This exists in addition to the usual Maxwell field A_μ, which takes care of the finite density. The $(4 + 1)$-dimensional bulk theory is thus described by the Einstein–Maxwell–Maxwell action,

$$S = \int d^5x \sqrt{-g} \left[R + 12 - \frac{1}{4} F_{\mu\nu} F^{\mu\nu} - \frac{1}{4} W_{\mu\nu} W^{\mu\nu} \right] - \frac{\kappa}{2} \int B \wedge F \wedge W, \quad (12.61)$$

where $W = dB$ and $F = dA$, and we have added one of the possible Chern–Simons terms. As usual the boundary condition $A^{(0)} = \mu\, dt$ is used to impose the finite density. Introducing a source for the massive vector field through the boundary condition $B^{(0)} = \lambda\omega_2$ the translational symmetry is explicitly broken, in the form of a unidirectional helix in the space of the boundary.

As we further elaborate in box 12.3, the bulk problem can be solved with an appropriate Ansatz for the metric of the Bianchi VII geometry. The fully backreacted bulk solutions are obtained from equations of motion that are straightforward ordinary differential equations in the radial direction. This reveals that there are three classes of near-horizon solutions [447], depending on the strength of the potential λ. The first one corresponds to a translationally invariant $AdS_2 \times \mathbb{R}^3$ near horizon, with irrelevant deformations encoding for the irrelevance of the potential in the deep infrared of the field theory. This is in the same class as the weak-lattice-potential theories discussed in section 12.1, where one finds a "memory-matrix" momentum relaxation rate, and a Drude behaviour governing the transport. However, for increasing strength of the potential the scaling dimension of the periodic potential becomes at some point *relevant* in the infrared. Two things can happen: the metallic solution becomes unstable because an IR Breitenlohner–Freedman bound is violated. This is the generic solution for $\kappa \gg 1$. For small values of κ the $AdS_2 \times \mathbb{R}^3$ phase can remain stable, but by tuning the pitch p of the helix (or other parameters) one can reach a new quantum critical point, beyond which the potential becomes relevant again. This $\kappa \ll 1$ situation is analogous to the deformation behaviour in the dilaton electron star in Fig. 11.7.

When the potential becomes relevant, the Bianchi VII geometry imprints all the way down to the deep interior and one has to determine the fully backreacted solution. This becomes very anisotropic with regard to the x_1 direction versus the $x_2 - x_3$ spatial plane. In the latter directions, the RN extremal horizon stays intact but the geometry and the fields completely reconstruct in the x_1 direction. This shows that the metric is indeed capped off, but in the x_1 direction only. Moreover, the B field is now relevant in the deep interior and, via the Chern–Simons term, it can act as a source for the electric fields in the bulk [447, 448]. This has the capacity to absorb all the flux so that the geometry is horizon-less. This is similar to the role played by the scalar field in the holographic superconductor, but we will see in a moment that this encodes for very different physics in the boundary.

One can now compute the optical conductivity in these backgrounds, and this reveals the punchline. In Fig. 12.7 the conductivity in the x_1 direction is shown, comparing the regimes where the potential is irrelevant (the metallic phase) and relevant (the insulating phase). As expected, in the irrelevant regime it just behaves according to the expectations for the strange metal exposed to a weak periodic potential: one finds the Drude peak with a finite width at low energy. However, in the regime where the potential is relevant the conductivity decreases upon lowering the energy, to vanish completely at $\omega = 0$, $T = 0$. The system has turned into an insulator in the x_1 direction, but it does so in a special way. For the insulating phase a semi-analytical calculation shows that the real part of the optical conductivity behaves as

$$\text{Re}\,\sigma \sim \omega^{4/3} \tag{12.62}$$

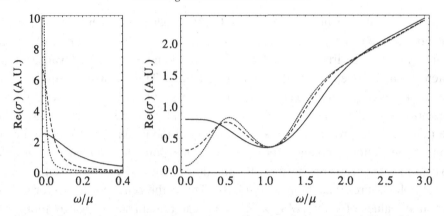

Figure 12.7 Optical conductivity in the x_1 direction for metallic (left) and insulating (right) phases of the field theory dual to the holographic system with helical Bianchi VII symmetry. The responses for three different temperatures are given, with the solid line for the highest T/μ and the dotted line for the lowest T/μ. In the left plot, for the metallic phase, the conductivity exhibits a Drude peak. In the right plot one sees the insulating gap appear at lowest T/μ. Figure source [447]. (Reprinted by permission from Macmillan Publishers Ltd © (2013).)

at zero temperature and small frequencies $\omega/\mu \to 0$. There is no true gap, but the conductivity vanishes as a power law; most notably the delta-function characteristic of translation invariance is now gone. The exponent 4/3 appears to be independent of parameters for reasons that are not yet well understood. This shows that the relevant flow of the potential goes hand in hand with an *irrelevant* continuous renormalisation flow of the currents. The conductivity of the system as a whole now reveals a literal "algebraic pseudogap behaviour", where the "gap" now refers to a transport gap.

This exercise revealed a real surprise from the viewpoint of condensed matter physics [148]. It is related to the anisotropic nature of the transport. Donos and Hartnoll chose a unidirectional potential just for mathematical convenience, motivated by the Bianchi VII choice for the metric. Although the potential is trivially encoded in the bulk geometry, they found that it can act strongly on the x_1 direction, turning it in that direction into an "algebraic insulator", while the perpendicular spatial x_2-x_3 plane is not affected at all. The finite-density system remains a *perfect metal* in that perpendicular plane while it insulates in the direction of the potential. "Particle-physics insulators" cannot possibly accomplish this phenomenon. With the quantum mechanics of the free particles, it can be proven that (de)localisation is, in the scaling limit, always isotropic: the system is either a conductor or an insulator in all directions [86]. This is also again clearly illustrated holographically. All of the "capped" geometries which have been constructed so far are isotropic, but this one is not (box 12.3).

There is one precedent for such a behaviour, in the only context where quantum critical phases can be dealt with in terms of established condensed matter theory [449]. Luttinger liquids in $1+1$ dimensions are of this kind, and Emery *et al.* [450] considered the problem of a system of Luttinger liquids forming a "striped array" in $2+1$ dimensions, to subsequently switch on couplings between the individual Luttinger liquids. Conventional wisdom would insist that a dimensional crossover would immediately happen, with the system locking into a $(2+1)$-dimensional anisotropic metal that is conducting in both spatial directions. However, it turns out that there is a small corner in parameter space, associated with very strong intra-Luttinger liquid couplings, and forward scattering dominating the inter-Luttinger couplings, where the system remains insulating perpendicular to the "stripes" while in the direction of the Luttinger liquids a perfect metal is realised. Strikingly, along the insulating direction one finds the same irrelevance of the currents which gives rise to an algebraic pseudogap as found in the holographic set-up. Such states were named "quantum smectics", in analogy with the classical "half-way solid" liquid-crystal phase. The holographic narrative highlighted in this section reveals that quantum smectics in higher dimensions appear to be easy to realise for the strange metals of holography.

Box 12.3   Constructing a helical lattice using Bianchi VII geometry

In this box we will show how to construct the solution of the action Eq. (12.61) and specify the procedure to compute the conductivity. Consider the following Ansatz for the bulk metric:

$$ds^2 = -U(r)dt^2 + \frac{dr^2}{U(r)} + e^{2v_1(r)}\omega_1^2 + e^{2v_2(r)}\omega_2^2 + e^{2v_3(r)}\omega_3^2, \qquad (12.63)$$

and for the gauge fields

$$A = a(r)dt, \quad B = b(r)\omega_2, \qquad (12.64)$$

where the definitions of the one-forms ω_i have been given in Eq. (12.60). There are six functions $a(r), b(r), U(r), v_i(r)$ for $i = 1, 2, 3$, which need to be solved from the equations of motion.

Box 12.3 **(Continued)**

At $r \to \infty$, the boundary behaviour including the source terms should be of the form

$$U = r^2 + \cdots, \quad a = \mu + \frac{v}{r^2} + \cdots, \quad b = \lambda + \cdots, \quad v_i = \ln r + \cdots. \quad (12.65)$$

This means that there is an explicit translational-symmetry breaking since the vector field B_2 explicitly depends on the spatial coordinates through ω_2.

Several solutions are possible in the deep interior $r \to 0$. The metallic state corresponds to a deep-IR geometry that is translationally invariant and has the form of an $AdS_2 \times \mathbb{R}^3$ space-time. In the same manner as in the construction of the electron star, box 11.2, this geometry has deformations characterised by a scaling exponent δ:

$$U = 12r^2(1 + u_1 r^\delta), \quad v_i = v_0(1 + v_{i1} r^\delta),$$
$$a = 2\sqrt{6}r(1 + a_1 r^\delta), \quad b = b_1 r^\delta. \quad (12.66)$$

If the metallic state is stable, all deformations are *irrelevant* (i.e. $\delta > 0$), and one can integrate up to the RG flow to the asymptotically AdS_5 UV conformal field theory.

The scaling dimension δ will in general depend on the helix amplitude λ and pitch p, of the field B and the Chern–Simons coupling κ. Upon tuning to a regime where δ becomes relevant, a new IR must form. This deep interior geometry corresponds to an insulating state of the boundary field theory. At leading order it is given by

$$U = u_0 r^2 + \cdots, \quad e^{v_1} = e^{v_{10}} r^{-1/3} + \cdots, \quad e^{v_2} = e^{v_{20}} r^{2/3} + \cdots, \quad (12.67)$$
$$e^{v_3} = e^{v_{30}} r^{1/3} + \cdots, \quad a = a_0 r^{5/3} + \cdots, \quad b = b_0 + b_1 r^{4/3} + \cdots.$$

Note that the translational-symmetry-breaking "source" B field does not vanish in the deep IR for this case.

For a fixed κ the transition from relevance to irrelevance of the scaling dimension δ in the metallic state can either occur universally or as a function of the deformation parameters λ, p. In the latter case there is a critical value λ_c, p_c in the RG flow mediating the metal–insulator transition. This implies that there is a third critical solution for the IR geometry, characterised by the scaling behaviour

$$U = u_0 r^2, \quad v_1 = v_{10}, \quad e^{v_2} = e^{v_{20}} r^\alpha, \quad e^{v_3} = e^{v_{30}} r^\alpha, \quad a = a_0 r, \quad b = b_0 r^\alpha, \quad (12.68)$$

where the constants α, v_{i0}, a_0, u_0 and b_0 can be determined numerically from the equations of motion. The deformations away from this critical point can also be found by perturbing this geometry at order r^δ. The inherent critical nature means that the IR fixed point will always have relevant deformations, but, depending on the parameters, these can have either complex or real scaling dimensions. This is exactly analogous to the unstable fixed point of the dilaton electron star in Fig. 11.7. For real scaling dimensions the metal–insulator transition is of second order; complex scaling signals that the metal–insulator transition is of first order.

Let us finally comment on some special features that arise when one computes the optical conductivity. As we learned in section 8.3, the influence of the momentum conservation on the currents in the boundary translates into a coupling between the equations of motion of the gauge fields and the metric perturbation. The field B_μ is of course a dynamical field in the bulk. When it is switched on by sourcing it at the boundary, its detrimental effect on momentum conservation dualises in the bulk into a coupling of its fluctuations both to the gauge field and to the graviton. One therefore has to perturb the background by

$$\delta A = e^{-i\omega t} A(r)\omega_1, \quad \delta B = e^{-i\omega t} B(r)\omega_3,$$
$$\delta ds^2 = e^{-i\omega t} \left(C(r)dt \otimes \omega_1 + D(r)\omega_2 \otimes \omega_3 \right). \tag{12.69}$$

On substituting these perturbations into the equations of motion of Eq. (12.61) one obtains the ODE for the fluctuating fields. Note that we are considering five-dimensional bulk gravity here. A notable feature is that one finds that the near-boundary asymptote $r \to \infty$ of the gauge field A_μ is altered by an extra logarithm term [451],

$$\delta A_{x_1} = A^{(0)} + \frac{A^{(2)} + \frac{1}{2} A^{(0)} \omega^2 \ln r}{r^2} + \cdots . \tag{12.70}$$

The logarithmic behaviour originates from a scaling anomaly in the four-dimensional boundary theory. The retarded Green function can be obtained by following exactly the same procedure as in section 5.1. After adding a counter-term to cancel out the $\ln r$ term [452], we can compute the holographic conductivity which is defined as the retarded two-point Green function of the dual current J_{x_1},

$$\sigma = \frac{2}{i\omega} \frac{A^{(2)}}{A^{(0)}} + \frac{i\omega}{2}, \tag{12.71}$$

with an in-falling near-horizon boundary condition for δA. This computation has to be done numerically. One finds either the finite width of the Drude peak or the "algebraic gap" in the insulating regime through the coupling between the graviton and the B-field fluctuations, where the latter "carry away" the conserved momentum; see Fig. 12.7.

Box 12.3 (Continued)

Homogeneous spaces that break translations: beyond Bianchi VII

The Bianchi type VII geometry discussed in the main text is introduced on pragmatic grounds to avoid the computation of partial differential equations, as required in the inhomogeneous spaces associated with the more straightforward lattices discussed in the first three sections. The price one has to pay is in the form of a quite particular symmetry breaking (the unidirectional helix), which is less obviously related to the kind of symmetry breaking associated with the crystals of condensed matter physics. In addition, not everything can be computed using ODEs in this set-up. For instance, to determine the conductivity in the "metallic" $(x_2 - x_3)$ direction PDEs have to be solved in the bulk. Although one can read off from the geometry that this has to behave like a perfect metal, the full solution of the "metallic"-direction conductivity is at the time of writing not yet known.

One can construct other homogeneous bulk geometries that "break translations" in the sense that the momentum in the boundary is no longer conserved, while the background geometry can be determined in terms of purely radial ODEs in the bulk. Let us illustrate these ideas with two examples. One such realisation is the so-called Q-lattice construction [453, 454], where a neutral complex scalar at a given momentum is used to break translational symmetry. The Ansatz of the scalar profile in the bulk is $\phi = e^{ikx} \psi(r)$, and one avoids here the constraints on the homogeneous spaces associated with the Bianchi classification. Since the scalar is neutral, it is easy to check that the contribution of the scalar field to the equations of motion do not depend on the x coordinate as can be seen from the expression of the energy–momentum tensor for the scalar field,

$$T_{\mu\nu}(\phi) = \partial_{(\mu}\phi \, \partial_{\nu)}\phi^* - g_{\mu\nu}\big(|\partial\phi|^2 + V(|\phi|)\big). \tag{12.72}$$

In this way, only ODEs need to be solved in order to account for the effects of the backreaction of this scalar field on the background. Both metallic and unidirectional solutions are found, where the metals are like the ones found from Bianchi VII with an $AdS_2 \times \mathbb{R}^2$ (for the $(2 + 1)$-dimensional boundary), plus irrelevant deformations associated with the periodic neutral scalar.

These systems also exhibit both metallic and insulating solutions [453]. The metallic state corresponds to a solution with the near-horizon metric equal to $AdS_2 \times \mathbb{R}^2$ at zero temperature. However, in this simple set-up the insulating phase does not exist at zero temperature and starts to show up only above a certain temperature. It was subsequently shown that introducing a dilaton coupling into the action allows there to be a zero-temperature insulating phase [454].

To compute the conductivity at zero momentum one finds that the graviton and gauge field now couple to the fluctuations of the neutral scalar $\delta\phi$, taking the role of the B-field of the Bianchi VII solution,

$$\delta g_{tx} = \delta h_{tx}(r)e^{-i\omega t},$$
$$\delta A_x = \delta a_x(r)e^{-i\omega t}, \qquad (12.73)$$
$$\delta\phi = ie^{ikx}e^{-i\omega t}\,\delta\phi(r).$$

One finds similar results in the low-frequency regime as for the Bianchi VII helix solution. Although in this construction momentum is no longer conserved, it is far from clear how to interpret the nature of the translational symmetry breaking in the boundary. Differently from the "axion" and "massive-gravity" constructions that will come next, there is still a sense of periodicity since the scalar field is a periodic function in space. For a real lattice this would mean that, upon considering finite momentum, it has to be the case that modes separated by an Umklapp momentum should couple, as we highlighted in section 12.3 for the fermion case. However, here the equations of motion for $\delta A_x(k_a)$ at a finite momentum k_a couple only to $\delta A_x(-k_a)$, and the periodicity of the Q-lattice remains hidden.

Perhaps the simplest and most flexible way to encode the breaking of the translations in the bulk is the "axion" construction found by Andrade and Withers [455]. This amounts to a maximally "featureless" way to impose momentum conservation, and it turns out to be an explicit dynamical realisation of the massive-gravity construction of the next section. One introduces a massless neutral scalar – the phase of the Q-lattice – with an "axion" profile,

$$\phi \propto \alpha_i x^i, \qquad (12.74)$$

where the α_i are constants. By taking all α_i finite, translations are broken in all directions, even though the space stays effectively homogeneous. To obtain such a spatially isotropic solution the scalar field is just taken to be massless,

$$\mathcal{L}(\phi) = -\frac{1}{2}\sum_{I=1}^{d}(\partial\phi_I)^2, \qquad (12.75)$$

as can be implemented in an arbitrary number of spatial dimensions d. Since the backreaction of the scalar contributes through $\partial_\mu\phi_I$, the background can be obtained merely by solving ODEs, and even analytical solutions can be obtained. It is just an explicit, dynamical realisation of the massive gravity of the next section. There can also exist insulators when a dilaton field is included [456]. One can study the fluctuation associated with the conductivity of the system around the fixed background to find exactly the same equations for all frequencies and momenta as those obtained in massive gravity. Accordingly, one finds an identical expression for the

---

Box 12.3   (Continued)

DC resistivity as follows from massive gravity – Eq. (12.88) below – after redefinition of the various quantities. In terms of the "potential strength" (or "graviton mass") $\alpha^2 = (1/d) \sum_{a,I=1}^{d} \alpha_{Ia} \alpha_{Ia}$ and the black hole horizon r_0,

$$\sigma_{DC} = r_0^{d-2} \left(1 + (d-1)^2 \frac{\mu^2}{\alpha^2} \right). \tag{12.76}$$

The generalisation to the Lifshitz case with or without hyperscaling-violating geometry can be found in Ref. [456].

12.5 The dual of translational-symmetry breaking as gravity with a mass

We have witnessed in the previous sections a flow from a "brute-force" real lattice imposed by the boundary on the bulk, via the more subtle Bianchi VII helices all the way to the more abstract (from the boundary perspective) Q-lattice and axion constructions discussed in box 12.3. In fact, by considering the breaking of translational invariance from its most fundamental symmetry perspective, one can deduce the geometrical principle that governs the general nature of the corresponding bulk.

Symmetry principles play a central role in the construction of the correspondence. In chapter 5 we used this to deduce that global space-time symmetries generated by the energy–momentum tensor in the boundary are dual to the gauged coordinate-invariance symmetries of the space-time metric perturbations in the bulk. When momentum is conserved in the boundary due to global translation symmetry of the boundary space, the bulk should be generally coordinate-invariant, and Einsteinian gravity is the corresponding minimal gauge theory associated with these space-time diffeomorphisms. The implied gauge invariance demands that the gravitons are massless and this masslessness of the bulk excitations dualises in the conservation of the momentum in the boundary. However, breaking the global symmetry in the boundary should correspond to Higgsing the gauge symmetry in the bulk as exemplified by the holographic superconductor. In the context of internal symmetries this means, among other things, that the gauge field becomes massive. But this wisdom extends also to the gauged space-time symmetries that are encoded in the dynamics of gravity. Thus, if we desire to break the global translations explicitly in the boundary, the implication has to be that in one way or another the gravity in the bulk will become "massive".

Despite their flagrant tension with diffeomorphism symmetry, massive theories of gravity have a long and convoluted history in the relativity community. Besides the mere theoretical interest, one finds also some motivation in cosmology. For instance, one way to deal with dark matter and dark energy is to modify Einsteinian gravity, and it has been argued that massive gravity could be an explanation – when the graviton mass is very small, its consequences would become manifest only on cosmological scales [457, 458]. However, the conflict with the breaking of general covariance has always prevented a consistent formulation of such a theory. Next to deep problems with the causality structure, the existence of extra longitudinal polarisations creates obstacles in trying to consistently quantise the theory. In massive gravity, this is particularly painful: at the linear level this longitudinal polarisation doesn't even decouple in the limit where the mass is taken back to zero. At the non-linear level, one even encounters extra unphysical Boulware–Deser ghost fields that cannot be gauged away [457]. Only very recently did de Rham, Gabadadze and Tolley [459] construct a fomalism (known as "dRGT") that is claimed to be ghost-free. We will use this theory below. However, its consistency will never be a sore point, since we will see later that massive gravity is an excellent effective description of translational-symmetry breaking through other means in theories with massless fully diffeomorphism-invariant gravity.

We should put "breaking the diffeomorphism invariance" in quotation marks because this is the same careless language as has often been used when discussing the Higgs phase of a Yang–Mills theory. Gauge invariance can of course not be broken, but instead the Higgs condensate renders invariant combinations of the field strength massive. In the same way, massive gravity means that the geometrical curvature costs energy of the order of the graviton mass. Amusingly, this theme has quite a prominent history in the rather modern "field theory of elasticity". This has long roots in the field of the mathematics of "metallurgy", while it was formulated in a general theoretical-physics language by Kleinert in the 1980s [460, 461]. The essence is that in a crystalline solid the translational symmetry and rotational symmetry of space are clearly broken. Nevertheless, one can reformulate the theory of elasticity in a geometrical, GR-like language. In a crystal one manifestly has a rigid coordinate system as defined by the positions of the atoms, and this acts as a fixed frame. The diffeomorphisms associated with coordinate changes from this frame are the smooth elastic deformations that accordingly cost energy and are therefore not gauged. The curvature in this "crystalline space" is now completely contained in the disclinations. These are the topological defects associated with the breaking of the isotropy of this space. These react to curvature in much the same way as fluxoids in a type II superconductor react to the gauge curvature (magnetic field). In two spatial dimensions this analogy becomes literal and, there is a substantial literature in the physics of soft condensed matter dealing with this theme [462]. One can now demonstrate very precisely that the disclinations (and thereby curvature) are *confined*. It takes an infinite amount of energy to create an isolated disclination in the crystal in a flat background space [463]. Upon quantum melting the crystal it can turn into a *quantum-nematic* crystal: this is the zero-temperature version of the nematic liquid crystals familiar from flat-screen technology and physically characterised by the breaking of the isotropy of an otherwise homogeneous (translationally invariant) space. One can now demonstrate that in a Lorentz-invariant setting such a state carries massless gravitons with curvature sources that are entirely in the form of the quantised disclinations [464]. On melting this further such that the curvature sources multiply and collectively form an isotropic fluid, one ends up with the "dust-like" pressure-less fluid matter: the familiar textbook source on the right-hand side of the Einstein equation in introductory general relativity.

Though massive gravity is a very recent development, the underlying symmetry understanding makes clear that this is a fruitful and insightful way to study translational-symmetry breaking in holography. To illustrate how these matters work, let us focus on the seminal contribution due to Vegh [465] which started this development. This is a minimal dRGT implementation of the massive-gravity idea, where all knowledge of the origin of translational-symmetry breaking is suppressed. Subsequently we will show how more detailed models exactly confirm this approach.

The claimed-to-be-ghost-free dRGT massive gravity corresponds to an action for a $(3+1)$-dimensional bulk dual to a $(2+1)$-dimensional boundary theory [459],

$$S = \int d^4x \sqrt{-g} \left[\frac{1}{16\pi G} \left(R + \frac{6}{L^2} \right) - \frac{1}{4e^2} F^2 + m^2 \sum_{i=1}^{4} c_i U_i(g, f) \right], \quad (12.77)$$

where U_i are symmetric polynomials of the eigenvalues of the 4×4 matrix $K_{\nu}^{\mu} = \sqrt{g^{\mu\alpha} f_{\alpha\nu}}$:

$$U_1 = \text{Tr}\, K,$$
$$U_2 = (\text{Tr}\, K)^2 - \text{Tr}\, K^2,$$
$$U_3 = (\text{Tr}\, K)^3 - 3\, \text{Tr}\, K\, \text{Tr}\, K^2 + 2\, \text{Tr}\, K^3, \quad (12.78)$$
$$U_4 = (\text{Tr}\, K)^4 - 6\, \text{Tr}\, K^2 (\text{Tr}\, K)^2 + 8\, \text{Tr}\, K^3\, \text{Tr}\, K + 3(\text{Tr}\, K^2)^2 - 6\, \text{Tr}\, K^4.$$

This is simply Einstein–Maxwell theory except for the last term, which couples the Einstein metric g to a fixed background metric $f_{\mu\nu}$. The latter is put in by hand, and explicitly wires in a frame that "breaks the diffeomorphism invariance". In this "massive-gravity" term, the c_i are constants, and m is the mass of the graviton.

Clearly the Einstein–Maxwell action will be recovered for vanishing mass of the graviton $m \to 0$.

Many of the hazards of massive gravity and ghost-like excitations in particular are associated with the time direction. Breaking general covariance in spatial directions only is less hazardous. A distinct advantage of the dRGT construction is that one is free to choose in which direction one wants to "fix the frame". Spatial translational-symmetry breaking is encoded in the particular choice of the background metric,

$$f_{xx} = 1, \ f_{yy} = 1 \tag{12.79}$$

and all other entries zero. The diffeomorphism now reduces to the isometry group which keeps $f_{\mu\nu}$ unchanged. The general covariance in the t–r plane is now fully preserved, while it is "broken" in the x–y plane. The beauty is that one keeps general coordinate invariance in the t–r plane where it plays an essential role for the consistency of the dictionary.

For this choice of $f_{\mu\nu}$ there are only two independent mass terms in the action. It acquires the simple form

$$S = \int d^4x \sqrt{-g} \left[\frac{1}{16\pi G} \left(R + \frac{6}{L^2} \right) - \frac{1}{4e^2} F^2 + \alpha \, \mathrm{Tr}\, K + \beta \left((\mathrm{Tr}\, K)^2 - \mathrm{Tr}\, K^2 \right) \right],$$

$$\tag{12.80}$$

where α and β have the dimension of mass. Given that the radial and time directions are not affected, the generic solutions of this system are very similar to standard Einsteinian theory. Importantly, there is a charged black-brane solution in this system:

$$ds^2 = L^2 \left(\frac{dr^2}{r^2 f(r)} + r^2 \left(-f(r)dt^2 + dx^2 + dy^2 \right) \right), \tag{12.81}$$

with

$$A(r) = \mu \left(1 - \frac{r_0}{r} \right) dt, \tag{12.82}$$

which is very similar to the usual RN solution. The mass terms affect only the emblackening factor,

$$f(r) = 1 + \frac{\alpha L}{2r} + \frac{\beta}{r^2} - \frac{M}{r^3} + \frac{Q^2}{r^4}, \quad Q = \frac{2\sqrt{\pi G}\mu r_0}{eL}, \tag{12.83}$$

where μ corresponds as usual to the dual chemical potential. For $\alpha = 0$, $\beta = 0$ this solution will reduce to the standard AdS RN solution, while the near-horizon geometry at zero temperature is still $AdS_2 \times \mathbb{R}^2$, unaffected by the mass terms. The position of the horizon r_0 is determined by the largest root of $f(r) = 0$, and the mass M can be expressed in terms of r_0 as

$$M = r_0^3 \left(1 + \frac{\alpha L}{2r_0} + \frac{\beta}{r_0^2} + \frac{4\pi G \mu^2}{e^2 L^2 r_0^2} \right). \qquad (12.84)$$

The temperature of this black brane is

$$T = \frac{r_0}{4\pi} \left(3 + \frac{\alpha L}{r_0} + \frac{\beta}{r_0^2} - \frac{4\pi G \mu^2}{e^2 L^2 r_0^2} \right). \qquad (12.85)$$

In particular, we see that there is again a natural zero-temperature solution. Since $T \sim f'(r_0)$, this solution is the one where $f(r)$ has a double zero, as usual. As a consequence, the near-horizon geometry contains an AdS_2 factor. The free energy equals

$$\Omega = S_{\text{bulk}} + S_{\text{boundary}} = -\frac{VL^2}{2\kappa^2} \left(r_0^3 - \beta r_0 + \frac{\mu^2 4\pi G r_0}{e^2 L^2} \right) + \epsilon_0(\alpha, \beta), \qquad (12.86)$$

where $\epsilon(\alpha, \beta)$ is an undetermined constant piece that is independent of T or μ, and therefore has no consequences for thermodynamic quantities. Except for the modification of the horizon due to the mass terms, in any other regard this massive-gravity RN black brane behaves identically to the usual one.

However, in contrast to the thermodynamics, the transport changes in qualitative ways due to the presence of the graviton mass – obviously so, because the breaking of momentum conservation was the whole purpose of this exercise. The optical conductivity is computed as in chapter 7, by considering the coupled system of graviton and gauge-field fluctuations. The novelty is that now the metric fluctuations g_{tx} and g_{rx} acquire a mass. In the charged-black-hole background this mass depends on the radial coordinate and the mass parameters α, β as,

$$m_{\text{graviton}}^2(r) = -2\beta - \alpha L r. \qquad (12.87)$$

To secure the stability of both the bulk and the boundary, we insist that $m_{\text{graviton}}^2(r)$ is positive for all r. The mass term proportional to α has a peculiar local dependence on r. In fact, the relationship between α and translational-symmetry breaking is obscure, whereas β takes care of this in a natural manner. We shall retain α just for the sake of completeness. For the optical conductivity at finite frequencies only numerical solutions are available. Highlighting once again the consistency among the various holographic systems, it turns out that this massive-gravity system also shows the "intermediate-energy scaling regime" which we briefly discussed in section 12.2. In the regime $T < \omega < \mu$ the numerical results again appear to display something akin to a scaling form $|\sigma(\omega)| \sim A/\omega^\alpha + B$, with the difference that now the scaling exponent α is no longer fixed. It is instead a function of the graviton mass [465]. This finding adds credibility to the idea that there is indeed a general reason behind this phenomenon, although the finding that the value of the exponent is undetermined in this setting robs it of some of its predictive power. As before, the cause of this intermediate scaling is still obscure.

On the other hand, elegant analytic results can be obtained for the DC conductivity in the massive-gravity system [466]. One finds explicitly

$$\sigma_{DC} = \frac{1}{e^2}\left(1 + \frac{16\pi G e^2 \rho^2}{L^2 m^2(r_0)r_0^2}\right), \tag{12.88}$$

where $\rho = \mu/(e^2 r_0)$ is the charge density and $m^2(r_0)$ is the graviton mass evaluated at the horizon. For small mass, this is of the Drude form

$$\sigma_{DC} = \frac{\rho^2}{\epsilon + P}\tau_K + \cdots , \tag{12.89}$$

with a momentum relaxation time τ_K [467],

$$\tau_K^{-1} = \frac{r_0^2 L^2}{16\pi G e^2}\frac{m^2(r_0)}{\epsilon + P}. \tag{12.90}$$

Note that the latter combination scales as r_0^2. It is therefore proportional to the entropy, and a more appropriate way to express the relaxation rate is

$$\tau_K^{-1} = \frac{s}{4\pi}\frac{m^2(r_0)}{\epsilon + P}, \tag{12.91}$$

in terms of the entropy density (s), energy density (ϵ), pressure (P) and graviton mass.

The DC resistivity as computed from massive gravity, Eq. (12.91), has some remarkable physical implications. It reveals a very general and simple message regarding the transport properties of quantum critical metals. The momentum relaxation controlling the resistivity is governed by entirely different physical principles from those controlling the transport in particle-like systems, including the Fermi liquid. One can in fact understand this DC resistivity without the help of holography [467]. The key assumption is that the system behaves hydrodynamically even in the presence of the Umklapp scattering. This is never the case in Fermi liquids, as we argued in chapter 2, since momentum relaxes already at the microscopic scale, long before local equilibrium will be realised. However, in an un-particle system, due to the very fast relaxation and the absence of large-momentum quasiparticles communicating with the lattice, it is expected that the system first collectivises, while its hydrodynamical behaviour is destroyed only at later times by the momentum relaxation.

With this insight, the deep message contained in Eq. (12.91) can now be understood in terms of elementary hydrodynamics. Firstly, as Stokes himself explained, in a system characterised by a length l where the breaking of translations becomes manifest to the fluid, the momentum relaxation rate that will show up in the Drude form of the resistivity is given by $1/\tau_K = D/l^2$, where D is the diffusivity associated with the transverse sound channel. This can be directly related to the behaviour of the sound in the clean limit, as we discussed for the RN metal in section 8.3. We learned that its dispersion is governed by a relaxational pole $\omega_k = i\Gamma_k = iDk^2$, where Γ_k is the attenuation rate. In the presence of featureless translational-symmetry breaking characterised by a length l, this dispersion relation turns into $\omega_k = iD\left((1/l^2) + k^2\right)$ and one reads off that the momentum relaxation rate at $k = 0$ of relevance to DC transport equals $1/\tau_K = D/l^2$.

For a relativistic fluid the diffusivity can be expressed in terms of the transport coefficients using the relation $D = \eta/(\varepsilon + P)$, where η is the shear viscosity, ε the energy and P the pressure. This reduces to $D = \eta/(\rho m)$ in the non-relativistic limit, with ρm the mass density. At this point we add that we are dealing with a strongly coupled quantum critical fluid: we know that for this special case the viscosity is "minimal", $\eta/s = A\hbar/k_B$, where $A \geq 1/(4\pi)$. On substituting this into the Drude formula, one finds for the DC resistivity

$$\rho_{DC} \sim 1/\tau_K \sim D/l^2 \sim \eta/l^2 \sim s/l^2. \tag{12.92}$$

The featureless nature of the result suggests that one should consider this as the result of quenched disorder.

The remarkable nature of this result is that, provided that l is temperature-independent, the resistivity is proportional to the entropy. This appears absurd at first sight, but it is a simple consequence of the "perfect-fluid" behaviour of the finite-temperature quantum critical metal. Even stranger is that, when the entropy obeys a Sommerfeld law $s \sim T$, one finds automatically a linear resistivity: the famous enigma exhibited by the cuprate superconductors.

This linear-resistivity wisdom can be confirmed in an explicit holographic example that is elucidated in box 12.4. It should be emphasised, however, that in the present context holography plays only the modest role of illustrating a general principle. The combination of the basic principles of hydrodynamics and the "perfect-metal" principle in the thermal regime of quantum critical systems implies that the resistivity *has* to be proportional to the entropy.

Turning to the numbers in the real cuprates, one finds a quantitative expression for the resistivity in terms of quantities of the cuprate strange metals that can be measured independently [468],

$$\rho_{DC}(T) = \frac{A\hbar}{\omega_p^2 m_e l^2} \frac{S_e(T)}{k_B}, \tag{12.93}$$

where $A \geq 1/(4\pi)$ is the constant associated with the viscosity–entropy-density ratio, ω_p is the plasma frequency $\simeq 1$ eV, m_e is the electron mass and $S_e(T)$ is the *measured* electronic entropy of the cuprate metal, which is Sommerfeld: $S_e(T)/k_B \simeq k_B T/E_c$, where the "Fermi energy" $E_c \simeq 1$ eV. The momentum relaxation rate is, according to experiment, itself "Planckian", $1/\tau_K \simeq \hbar/(k_B T)$ [441] – within the framework of this interpretation its "Planckian" magnitude is to some degree coincidental (see also [469]). On combining these numbers, one finds out that l is of the order of a few nanometres, a quite reasonable value given the expectation associated with the known, rather strong chemical disorder intrinsic to the cuprate planes. In fact, perhaps the best reason to take this explanation very seriously is that it resolves a very old puzzle. Despite the undisputed "dirty" nature of the cuprate chemistry, it was noted in the late 1980s that upon extrapolating the linear resistivity to zero temperature it precisely vanishes at zero temperature in many of the optimally doped superconductors. This used to be a conundrum, since quasiparticles will always be subject to elastic scattering, resulting in a residual resistivity. This mystery is resolved in this hydrodynamical interpretation since at zero temperature the system turns into a truly perfect fluid with a vanishing viscosity as the entropy vanishes.

Box 12.4   Linear resistivity: the conformal-to-AdS$_2$ metal and massive gravity

The observation that in holographic massive-gravity theories, the resistivity is proportional to the entropy density, naturally asks for a comparison of the local quantum critical model with a Sommerfeld specific heat $s \sim T$ with the cuprates. Here we will present some of the details of this holographic computation, showing how massive gravity acts on the "conformal-to-AdS$_2$" metal [468] which we introduced in section 8.4. For related work on the influence of spatial disorder using the "axion" model of box 12.3 on various scaling geometries, see [456].

We can add the massive-gravity term of Eq. (12.80) to the Einstein–Maxwell-dilaton action Eq. (8.46). Moreover, we can choose the dilaton potentials such that the near-horizon geometry will be conformal to AdS$_2$:

$$S = \frac{1}{2\kappa_4^2} \int d^4x \sqrt{-g} \left[R - \frac{1}{4} e^\phi F_{\mu\nu} F^{\mu\nu} - \frac{3}{2} \partial_\mu \phi \, \partial^\mu \phi \right.$$
$$\left. + \frac{6}{L^2} \cosh \phi - \frac{1}{2} m^2 \left(\mathrm{Tr}(\mathcal{K})^2 - \mathrm{Tr}\left(\mathcal{K}^2\right) \right) \right]. \quad (12.94)$$

Here $\mathcal{K}_\alpha^\mu \mathcal{K}_\nu^\alpha = g^{\mu\alpha} f_{\alpha\nu}$, and the non-zero elements of the fixed reference metric $f_{\mu\nu}$ are again $f_{xx} = f_{yy} = 1$.

Box 12.4   (Continued)

As for the Einstein–Maxwell system with massive-gravity terms, "fixing the spatial frame" does not change the gross features of the bulk solutions, and accordingly one finds a charged black-hole solution of the same form as in the absence of the graviton-mass terms,

$$ds^2 = \frac{r^2 g(r)}{L^2}\left(-h(r)dt^2 + dx^2 + dy^2\right) + \frac{L^2}{r^2 g(r)h(r)}\,dr^2,$$

$$A_t(r) = \sqrt{\frac{3Q(Q+r_0)}{L^2}\left(1 - \frac{m^2 L^4}{2(Q+r_0)^2}\right)}\left(1 - \frac{Q+r_0}{Q+r}\right),$$

$$h(r) = 1 - \frac{m^2 L^4}{2(Q+r)^2} - \frac{(Q+r_0)^3}{(Q+r)^3}\left(1 - \frac{m^2 L^4}{2(Q+r_0)^2}\right),$$

$$\phi(r) = \frac{1}{3}\ln(g(r)), \qquad g(r) = \left(1 + \frac{Q}{r}\right)^{3/2}.$$

$$(12.95)$$

The temperature T and chemical potential μ of the dual field theory are

$$T = \frac{r_0\left(6(1+Q/r_0)^2 - m^2 L^4/r_0^2\right)}{8\pi L^2 \left(1 + Q/r_0\right)^{3/2}} \sim \sqrt{r_0 Q} + \cdots,$$

$$\mu = \frac{\sqrt{3Q(Q+r_0)\left(1 - m^2 L^4/(2(Q+r_0)^2)\right)}}{L^2} \sim Q + \cdots.$$

$$(12.96)$$

In the near-horizon limit $r - r_0 \ll r_0$ at zero temperature the geometry is conformal to AdS$_2$ with $ds^2 = 1/r^{3/2}\,ds^2_{\text{AdS}_2}$ and $-\theta/z = 1$. It is straightforward to see that at low temperatures $T/\mu \sim \sqrt{r_0/Q}$ (consider Q large).

For a generic EMD theory there is a correction to the universal formula for the DC conductivity [470], but in this case it happens to remain unchanged and we can use Eq. (12.88). This needs only the entropy and charge density as input. The entropy follows from the horizon area and the charge density from Gauss's law. We find

$$s/\mu^2 = \frac{2\pi L^2}{3\kappa_4^2}\sqrt{r_0/Q}\sqrt{1 + r_0/Q}\left(1 + \frac{3\bar{m}^2}{2(1+r_0/Q)}\right)$$

$$\sim T/\mu,$$

$$(12.97)$$

and the charge density σ_q behaves at low T as

$$\sigma_q/\mu^2 = \frac{L^2}{2\sqrt{3}\kappa_4^2}\sqrt{1 + r_0/Q}\sqrt{1 + \frac{3\bar{m}^2}{2(1+r_0/Q)}}$$

$$\sim (T/\mu)^0.$$

$$(12.98)$$

The universal result of Blake and Tong [466] then ensures a linear resistivity ρ_{DC} at low temperatures,

$$\rho_{DC} = \frac{s}{4\pi\sigma^2}m^2 = \frac{2\kappa_4^2}{L^2}\frac{1}{\sqrt{1+Q/r_0}}\frac{m^2}{\mu^2} \sim T/\mu \quad \text{at low } T, \tag{12.99}$$

corresponding to the desired "linear-in-temperature" resistivity.

The solution in Eq. (12.95) also exists in the axion/Q-lattice systems, with the graviton-mass term in Eq. (12.94) replaced by the action of the scalar field. Thus this linear resistivity behaviour for ρ_{DC} also exists in those models.

There is an important subtlety. In order for the temperature dependence of the resistivity to be entirely due to the entropy, it has to be the case that the "mean free path" l is temperature-independent. Even for a quantum critical system this will be quite generally not the case. The strength of the disorder potential itself will be subject to renormalisation – a famous example is the Harris criterion, which demonstrates that in low dimensions disorder is always relevant in a critical state [25]. However, as the holographic example in box 12.4 illustrates, there are special circumstances under which this is apparently not the case, since in this example l is temperature-independent. The reason is that this particular example departs from the conformal-to-AdS$_2$ metal, which is a local quantum critical phase. The renormalisation of the disorder strength is a spatial affair, and, since length scales do not renormalise in the $z \to \infty$ AdS$_2$ critical state, the disorder strength does not run.

This can be independently verified using the memory matrix. Assuming that the translational-symmetry-breaking operator $\mathcal{O}(k)$ is sourced by random impurities, one can apply the memory matrix by averaging over momenta, Eq. (12.6),

$$\rho_{DC} \sim \int d^2k\, k^2 \lim_{\omega\to 0} \frac{\text{Im } G^R_{\mathcal{O}\mathcal{O}}(\omega, k)}{\omega}. \tag{12.100}$$

We can now again use the unusual feature of the local quantum critical metal that the spectral function is known to scale as $\sim T^{2\nu_k-1}$. The essence of massive gravity is that all low-energy dynamics is governed by the gravitational energy–momentum tensor. Indeed, from a hydrodynamical perspective the most sensible operator to communicate momentum relaxation is the energy–momentum tensor itself: for the operator $\mathcal{O}(k)$ we should choose $T_{\mu\nu}$. This yields a powerful insight into the memory-matrix response, since the low-frequency two-point correlation function of the stress tensor is itself fully determined by hydrodynamics [471]. On substituting, one finds at small k

$$\rho_{DC} \sim \int d^2k \, k(\eta k^2 + \cdots).$$ (12.101)

This reproduces a viscous contribution that is consistent with taking the mean free path l to be a constant.

There is, however, a subtlety in the application of the memory matrix. This is in the right integration limits of Eq. (12.100). They are temperature-independent only for the strictly locally quantum critical state with $z \to \infty$. Thinking of the random impurities as a random superposition of different lattices, we can encode them in a set of weakly coupled operators $\int dk \, \mathcal{O}(k)$ as in the second half of box 12.2. To retain control over the UV theory, these must be relevant, or at most marginal, as is the case with $\mathcal{O} = T_{\mu\nu}$ above. Generically, as we run down to the IR each mode of $\mathcal{O}(k)$ becomes appreciable at the scale $L \sim 1/k$. For the DC conductivity at finite T we wish to know the effect at the horizon T; in a Lifshitz geometry this corresponds to a length scale $L \sim T^{-1/z}$. Following the memory-matrix approach, one therefore finds generically

$$\rho_{DC} \sim \int^{T^{1/z}} d^d k \, k^2 \lim_{\omega \to 0} \text{Im}\left(\frac{G_{\mathcal{O}\mathcal{O}}^R}{\omega}\right).$$ (12.102)

Using the fact that the emergent quantum critical scaling uniquely determines the temperature dependence of the Green function [472, 473],

$$\lim_{\omega \to 0} \text{Im}\left(\frac{G_{\mathcal{O}\mathcal{O}}^R}{\omega}\right) \sim T^{(2\Delta - 2z - d)/z},$$ (12.103)

one finds that

$$\rho_{DC} \sim T^{2(1+\Delta-z)/z},$$ (12.104)

where Δ is the scaling dimension of $\mathcal{O}$ in the Lifshitz IR. This insight can be confirmed with an explicit holographic computation in an EMD Lifshitz model [470]. The holographic computation gives a further crucial insight, however. It cleanly shows where the perturbative approach underlying the general expression Eq. (12.102) breaks down and that at that moment ρ_{DC} has the universal value

$$\rho \sim T^{2/z} s.$$ (12.105)

At that point the IR dimension of the operator is always marginal, and this connects it to the result Eq. (12.101) where $\mathcal{O} = T_{\mu\nu}$.

In conclusion, the massive-gravity approach is a typical example where holography has been used to discover a very general, simple principle: for a truly locally quantum critical fluid at finite temperature in a weak random potential its resistivity is proportional to its entropy. This should be inspirational for the experimentalists because it is quite suggestive of where to look for new types of experiments that can shed light on these matters. The gross difference is of course the claim that the "strongly interacting" hydrodynamics of the quark–gluon-plasma kind is at work instead of the kinetic gas physics of the Fermi liquid. This gross difference is actually quite hidden in available experimental data, and the experimentalists are challenged to find out ways to obtain direct access to this hidden information in the laboratory. In fact, at the time of writing we are facing an embarrassment of riches on the theoretical side, in the form of an equally credible alternative explanation of the linear resistivity rooted in a logic that is similar to but physically very different from the arguments above. This departs from the "critical Fermi-surface state" arising in the Hertz–Millis context as discussed in section 2.5. One can now make the case that in the presence of disorder the quantum critical order parameter takes over the momentum relaxation, giving rise to a linear resistivity at higher temperatures [474], which can in turn be confirmed by an explicit holographic computation [470].

12.6 Holographic crystallisation: the spontaneous breaking of translational symmetry

Up to this point in this chapter we have studied *explicit* symmetry breaking: how static background lattices influence the properties of the holographic fluids. It is natural to ask whether translational symmetry might be broken *spontaneously*, leading to the most familiar form of a symmetry-broken state: the conventional solid. Once again making the connection to condensed matter firmer, holography turns out to possess a quite natural mechanism to produce such phases. On the gravity side, however, the technical hurdles one must overcome in order to construct quantitative solutions are in this context even more demanding than for explicitly broken translational symmetry. It is therefore still relatively unexplored territory. We conclude this chapter with a brief overview of the present state, reviewing some first qualitative results.

Taking the holographic superconductor as a guiding example of spontaneous symmetry breaking, one notices immediately that in this case scalars with small momenta are also unstable, in addition to the zero-momentum mode. To a first approximation the momentum has the simple effect of increasing the effective mass slightly. Of course, in a plain vanilla superconductor the zero-momentum condensate will be energetically favoured. We must therefore seek a mechanism that changes the preferred free-energy ground state into a condensate forming at finite momentum. This will break the translations spontaneously.

Perhaps surprisingly for the condensed matter reader, a first category of extensions of the bulk theories that can give rise to dominating finite-momentum instabilities involves *topological* terms: Chern–Simons in odd space-time dimensions

[475, 476] and theta terms (non-dynamical axions) in even dimensions [477, 478]. In the ordered state, the boundary duals in this category will therefore also break parity and time reversal, next to translations. These correspond physically to periodic patterns of *spontaneous currents*, possibly accompanied by a density modulation. These are therefore similar to the spontaneous-current phases proposed in condensed matter physics, in the form of d-density waves and flux phases [479, 480], as well as Varma's current-loop phase [481] (for a general classification, see [482]). A second category of instabilities was discovered within the confines of Einstein–Maxwell-dilaton theory [483], where translational symmetry is broken while parity (P) and time reversal (T) are preserved. These correspond to conventional crystals, involving only periodic modulations of the densities.

Let us first focus on the Chern–Simons (CS) construction in an AdS$_5$ holographic model discovered by Nakamura, Ooguri and Park [475, 476]. We have already seen in the discussion on anomalies in chapter 7 that CS terms arise quite naturally in the $(4 + 1)$-dimensional-bulk theory, dual to the $(3 + 1)$-dimensional boundary. In a finite-density background the CS term accomplishes precisely what is necessary to find a preferred instability to a finite-momentum condensate. This is easily seen in the simpler example of flat $(4 + 1)$-dimensional Minkowski space-time. Take the action

$$S = \int d^5x \left(-\frac{1}{4e^2} F_{IJ} F^{IJ} + \frac{\alpha}{3!} \epsilon^{IJKLM} A_I F_{JK} F_{LM} \right), \tag{12.106}$$

where α is the CS coupling. Unlike many previous examples, the dimensionality here will be important: in $4 + 1$ dimensions CS has the same number of derivatives as the Maxwell term and, counting dimensions, it is in the first instance a marginal deformation.

Now take a background with a constant electric field along the x_4 direction, i.e. $\bar{F}_{04} = -\bar{F}_{40} = E$, while the other components of F_{IJ} are zero. The linear fluctuation of the gauge fields around this background can be obtained by substituting $F_{IJ} = \bar{F}_{IJ} + f_{IJ}$ into the equation of motion,

$$\partial_J(\sqrt{-g} F^{JI}) + \frac{\alpha}{2} \epsilon^{IJKLM} F_{JK} F_{LM} = 0. \tag{12.107}$$

One finds

$$(\partial^\mu \partial_\mu + \partial^k \partial_k) f_i - 4\alpha E \epsilon_{ijk} \partial_j f_k = 0, \tag{12.108}$$

with $\mu = t, x_4$, $\{i, j, k\} = x_1, x_2, x_3$ and $f_i = \frac{1}{2} \epsilon_{ijk} f_{jk}$. We have in frequency–momentum space $f_i \propto e^{-i\omega t + i p_4 x_4 + i k_i x^i}$, which implies a modified dispersion relation for the fluctuations,

$$\omega^2 - p_4^2 = (k \pm 2\alpha E)^2 - 4\alpha^2 E^2. \tag{12.109}$$

This indicates a finite-momentum tachyonic instability within a range of momenta $0 < k < 4|\alpha E|$, with the dominant instability at $k = \pm 2\alpha E$. This reveals the essence of the mechanism responsible for the translational-symmetry breaking in this holographic setting.

Qualitatively the same happens when one adds the CS term to the AdS–Einstein–Maxwell action. For the background we consider the zero-temperature extremal AdS–RN black hole, with a near-horizon metric and electric field,

$$ds^2 = \frac{-dt^2 + dr^2}{12r^2} + d\vec{x}^2, \quad F_{tr} = \frac{E}{12r^2}, \quad E = \pm 2\sqrt{6}, \tag{12.110}$$

where the curvature radius of AdS_2 is equal to $1/\sqrt{12}$. Of course, the difference from the flat geometry is that in AdS a negative mass squared does not necessarily imply an instability in the AdS–RN space-time. Instead, the gauge-field fluctuation becomes tachyonic when it starts to violate the BF bound in the near-horizon AdS_2, as we discussed in chapter 10 for the holographic superconductor. Taking into account also the fact that the fluctuations of the gauge field couple to the metric fluctuations, the minimum effective mass squared becomes

$$m_{\min}^2 = \frac{E^2(-64\alpha^6 - 24\alpha^4 + 6\alpha^2 - (16\alpha^4 + 4\alpha^2 + 1)^{3/2} + 1)}{2(4\alpha^2 + 1)^2}. \tag{12.111}$$

This $m_{\min}^2$ violates the BF bound of AdS_2 when $\alpha > \alpha_{\text{crit}} = 0.2896\ldots$.

The "hair" that forms when $\alpha > \alpha_{\text{crit}}$ is now a condensate of the transverse gauge field that translates into a state in the boundary characterised by a spontaneous, periodic pattern of *currents* of the form,

$$\langle \vec{J}(\vec{x}) \rangle = \text{Re} \left(\vec{u} e^{i\vec{k}_c \cdot \vec{x}} \right) \tag{12.112}$$

where $\vec{k}_c$ is the preferred non-zero ordering momentum, while the constant vector $\vec{u}$ is circularly polarised according to

$$\vec{k}_c \times \vec{u} = \pm i |k_c| \vec{u}. \tag{12.113}$$

A subtlety in this model is that the dominant instability is temperature-dependent. In the ordered phase, the preferred k increases when the temperature decreases. To find out at which momentum the ordering sets, one has to determine at which temperature the system becomes unstable towards a symmetry breaking with a particular $|k|$. As illustrated in Fig. 12.8, the ordering will occur at the wave vector associated with the highest T_c. The fully backreacted solution for the ordered state was found in [484], confirming that this helical current phase that breaks P and T is thermodynamically preferred, while the phase transition is of second order. In addition, in the zero-temperature limit the backreacted geometry approaches a zero-entropy ground state [484].

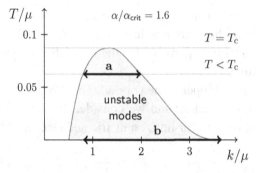

Figure 12.8 The unstable momentum modes in a $(4+1)$-dimensional holographic Chern–Simons metal as a function of the temperature for a particular value of the Chern–Simons coupling $\alpha = 1.6\alpha_{\text{crit}}$. For a temperature $T < T_c$ there is a range of unstable modes (denoted by **a**). What should be noted is that the true range of unstable modes at $T = 0$ is larger than a near-horizon analysis would reveal. The latter would give one only the range denoted by **b**. Figure adapted from [475]. (Reprinted with permission from the American Physical Society, © (2010).)

The CS action can be defined only in odd space-time dimensions. The topological terms allowed in even dimensions are the theta/axion terms of the form $S_\theta = \int \theta(\phi)\varepsilon^{\mu\nu\rho\sigma} F_{\mu\nu} F_{\rho\sigma}$, where ϕ is a pseudo-scalar [477, 478]. Such terms can be identified in top-down constructions that are based on $(10+1)$-dimensional supergravity [485]. In $3+1$ dimensions this term has two derivatives, and such axion theories have very similar physics to their $(4+1)$-dimensional CS siblings. In particular, one can show that the system again has an instability at a finite momentum in a constant-electric-field background. Consider at first again a flat space-time action,

$$S = \int d^4x \left(-\frac{\tau(\phi)}{4} F_{\mu\nu} F^{\mu\nu} + \frac{1}{2}(\partial_\mu\phi)^2 - V(\phi) + \frac{\theta(\phi)}{2}\varepsilon^{\mu\nu\rho\sigma} F_{\mu\nu} F_{\rho\sigma} \right),$$

(12.114)

where

$$\tau(\phi) = 1, \quad V(\phi) = \frac{1}{2}m^2\phi^2, \quad \theta(\phi) = \alpha\phi.$$

(12.115)

The equation of motion for this case is

$$(\partial^2 + m^2)\phi - \frac{1}{2}\varepsilon^{\mu\nu\rho\sigma} F_{\mu\nu} F_{\rho\sigma} = 0,$$

$$\partial_\mu F^{\mu\nu} - 2\alpha\varepsilon^{\mu\nu\rho\sigma} \partial_\mu(\phi F_{\rho\sigma}) = 0.$$

(12.116)

In a constant-radial-electric-field background $F_{tr} = -F_{rt} = E$ and $\phi = 0$, the linear fluctuations φ, $f_{\mu\nu}$ of the system are

$$(\partial_a \partial^a + \partial_i \partial^i + m^2)\varphi - 4\alpha E f_{12} = 0,$$
$$(\partial_a \partial^a + \partial_i \partial^i) f_{12} - 4\alpha E \, \partial_i \partial^i \varphi = 0,$$

(12.117)

where $a = t, r$ and $i = x_1, x_2$. In momentum space the fluctuations are of the form $e^{-i\omega t + ip_r r + ik_i x^i}$, and the dispersion relation becomes

$$\omega^2 - p_r^2 = k^2 + \frac{1}{2}m^2 \pm \frac{1}{2}\sqrt{m^4 + 64\alpha^2 E^2 k^2},$$

(12.118)

indicating that in the range $0 < k < \sqrt{16\alpha^2 E^2 - m^2}$ the modes become tachyonic.

The above instability arguments work again in the same way in an asymptotic AdS space. Assuming a unidirectional ("stripe") modulation in the x direction with wave vector k_c, a periodic pattern is found in the boundary dual, which is of the form [477]

$$\langle J_t \rangle - J_0 \sim \cos(2k_c x), \quad \langle J_x \rangle = 0, \quad \langle J_y \rangle \sim \sin(k_c x),$$

(12.119)

where J_0 is the background charge density. This corresponds to a stripe pattern, with periodic currents flowing in the positive- and negative-y direction. Every time the current density goes through zero, the charge density becomes maximal. This would look quite like the stripes seen in cuprates, after substituting a staggered (antiferromagnetic) magnetisation for the current direction [87]. The fully back-reacted P- and T-breaking solutions for this ordered phase were found in Refs. [486, 487, 488, 489], while the phase transitions turn out to be again of second order. In addition, the preferred period of the stripes is temperature-dependent, and monotonically increases as the temperature is lowered [488].

Finally, a holographic construction for spontaneous translational-symmetry breaking has been found, where the ordered finite-momentum state has the characteristics of a real crystal. This state preserves both P and T while only the charge densities are modulated. This construction is based on an Einstein–Maxwell-dilaton model [483] with an action

$$S = \int d^4x \sqrt{-g}\left[R - \frac{1}{2}(\partial\phi)^2 - V(\phi) - \frac{1}{4}\tau(\phi)F^2 \right].$$

(12.120)

The potentials are specifically chosen to be of the form

$$V = v_0\left(1 - v_1(\phi - \phi_0) - \frac{v_2}{2}(\phi - \phi_0)^2 + \cdots\right),$$
$$\tau = \tau_0\left(1 - \tau_1(\phi - \phi_0) - \frac{\tau_2}{2}(\phi - \phi_0)^2 + \cdots\right),$$

(12.121)

where $v_0 < 0$, $\tau_0 > 0$, such that the usual $AdS_2 \times \mathbb{R}^2$ geometry,

$$ds^2 = -r^2 \, dt^2 + \frac{dr^2}{r^2} + (dx^2 + dy^2), \quad A_t = Er, \quad \phi = \phi_0, \tag{12.122}$$

is a solution of Eq. (12.120). This is then identified with the IR limit of the $T = 0$ uniform black-hole phase. One now studies the behaviour of the linearised perturbations which are spatially modulated in the x direction in this background. There are in total seven fluctuations $\{\delta g_{tt}(k), \delta g_{xx}(k), \delta g_{yy}(k), \delta g_{tx}(k), \delta A_t(k), \delta A_x(k), \delta\phi(k)\}$, which are coupled, and each of which is proportional to $\cos(kx)$ or $\sin(kx)$ depending on the parity under $x \to -x$. The equations of motion for these fluctuations can be written into three gauge invariant degrees of freedom $\mathbb{V}$ of the form

$$\left(r^2 \, \partial_r^2 + 2r \, \partial_r \frac{\omega^2}{r^2} - \mathbb{M}^2 \right) \mathbb{V} = 0, \tag{12.123}$$

with the 3×3 mass matrix $\mathbb{M}$ given by

$$\mathbb{M}^2 = \begin{pmatrix} 2 + 2\tau_1^2 + k^2 & -2k^2 & 2\tau_1(2 - k^2 - \tau_2 - v_2) \\ -1 & k^2 & -2\tau_1 \\ -\tau_1 & 0 & k^2 + v_2 + \tau_2 \end{pmatrix}. \tag{12.124}$$

To find out whether instabilities occur, one has to inspect the eigenvalues of the mass matrix $\mathbb{M}^2$: the AdS_2 BF bound is violated if any of the three $m_i^2 < -1/4$. This can easily be achieved by choosing τ_1 and $\tau_2 + v_2$ appropriately. As an example, one can choose $V = v_0 e^{-\gamma\phi}$, $\tau = e^{\gamma\phi}$, with γ constant, such that the eigenvalues of $\mathbb{M}$ are $k^2, 1 + k^2 \pm \sqrt{1 + 2(1 + \gamma^2)k^2}$. For $\gamma > 1$, there is an unstable mode for $k \neq 0$. On computing the VEVs in the ordered phase, one finds

$$\langle J^t \rangle \propto \cos(k_c x_1), \quad \langle O_\phi \rangle \propto \cos(k_c x_1). \tag{12.125}$$

Therefore this corresponds to a pure density modulation in space while P and T are preserved. This is of course precisely coincident with the broken symmetry underlying the solid phase of matter which we encounter in daily life.

In summary, it appears that all the basic ingredients required to address the physics of spontaneous translational-symmetry breaking using holography have been identified. There is still much left to explore. For instance, the phonons associated with crystalline order are the most mundane examples of Goldstone bosons, but still have to be computed explicitly in the holographic framework. We anticipate an elegant description of the bulk dual, given that on general symmetry grounds this has to be related to an appropriate form of massive gravity in the bulk. In holographic crystals, there is a wealth of other condensed matter phenomena that can be studied in principle. For instance, one can contemplate constructing combinations of crystalline and strange-metal subsystems that are coupled to study electron–phonon coupling using holography.

13

AdS/CMT from the top down

In the previous chapters, we focussed mainly on the phenomenological *bottom-up* models. Here the gravitational bulk theory in four or five dimensions is phenomenologically put together in a similar way to Ginzburg–Landau theory. The actual Lagrangian of the boundary theory thereby remains completely unknown. In addition, it is not clear either whether the bulk theory is a well-defined and self-consistent quantum gravity theory. The advantage of these bottom-up models is that the gravity theory is relatively simple and one is free to add new ingredients in order to realise different behaviours in the boundary theory. However, to make sure that the phenomena found are self-consistent and/or to understand the dynamics in terms of the dual field theory more fully, it is necessary to find an explicit system in string theory where both the field theory and the exact dual gravity theory are known. Instead of the bottom-up approach, this calls for a *top-down* approach that starts directly from string/M-theory. The canonical example is the seminal construction by Maldacena, and the generalisations which were subsequently discovered share the property that the action of the dual field theory can be directly identified, including its weakly coupled limit. Since string theory is thought to be a fully consistent quantum theory, this guarantees that any phenomenon described by a top-down theory is physical.

The disadvantage of the top-down approach is that it is technically much more involved. There are far more fields in the gravity theory, often including whole infinite Kaluza–Klein towers that represent the additional dimensions of string theory. In practice one therefore resorts to a consistent truncation of this full top-down theory. This reduces the number of fields, but in such a way that the solution is still guaranteed to be a solution of the full theory. There is an important caveat, deserving special emphasis: a stable solution in a consistent truncation may turn out to be unstable in the full theory where all the truncated field fluctuations are reinstated. Although it will be ignored henceforth, one should be aware of this potential source of trouble.

These consistent truncations of top-down constructions closely resemble the bottom-up models, but one often finds additional features. Top-down truncations will generically be characterised by the following extra features:

(1) a few more fields,
(2) a non-miminal coupling of the fields and
(3) an inherent restriction on the allowed range of parameters.

All the physical phenomena as described in the previous chapters can be robustly verified in this way, with mostly some small and innocuous changes. However, in certain cases top-down models can include interesting interaction terms that it is easy to overlook in the phenomenological approach. The resulting physics is then very different from what one would have thought from the bottom-up approach.

Amongst the top-down constructions, one can identify a special class that is quite particular to the string-theoretical origin of the correspondence. These are the so-called "probe-brane constructions" of explicit-gauge theory/gravity duals. These probe-brane constructions, which are also called intersecting D-brane models, can account for "flavour" degrees of freedom: objects that transform in the vector representation of the large-N symmetry [490]. The consistent truncations of these models describe the physics of a dynamical, higher-dimensional defect that is embedded in the curved AdS geometry.

In the following we will present an illustrated tour highlighting both the regular top-down and the probe-brane models, and we will show how they validate the holographic description of the condensed matter phenomena discussed in the previous chapters. Section 13.1 describes how to consistently truncate explicit string constructions and how this generically leads to a very particular AdS action with essentially no free parameters. The detailed nature is important for consistency. We shall show that these consistency requirements leave plenty of room for the occurrence of holographic superconductivity, the holographic non-Fermi liquids, and the Einstein–Maxwell-dilaton theories which include the conformal-to-AdS$_2$ metal. These results fully validate the holographic condensed matter physics discovered in the previous chapters. Section 13.2 describes the special probe-brane top-down models that derive from Dp/Dq-brane intersections. These models are relatively easy to construct and thus serve as an ideal validation arena for bottom-up models. We will focus on one such model, showing how the explicit knowledge of the field theory is used to directly identify the superconducting order parameter with a pair operator (section 13.2.2). In addition, the rather straightforward computation of transport coefficients of flavour currents will be explained. Probe-brane models can also serve as natural configurations for defect or impurity models. We will illustrate this in the last section, section 13.2.3, where it is explained how the essence of the classic Kondo problem can be holographically described in terms of a probe-brane construction.

Once again, we warn the reader that most of this chapter is intended as a first introduction. The technical aspects necessary to construct explicit top-down models require a significant background in string theory, and deserve a whole book to themselves [491]. We will therefore only sketch the origins of the models, highlighting the qualitative features. The main purpose of this chapter is to show that the phenomenology of bottom-up models is fully justified by explicit top-down constructions, with the important message that top-downs always provide more fields, more features, more restrictions and more insight into the physics from the perspective of the boundary field theory.

13.1 Top-down AdS/CMT models from supergravity

The most straightforward top-down models – those without probe branes – are invariably derived in a two-step approach. The starting point is always a set of N Dp-branes with p spatial dimensions embedded in a ten-dimensional space-time, as in chapter 4. The collective dynamics of these soliton "surfaces where open strings can end" is the open-string/CFT side of the correspondence. On the closed-string side, one can describe the same configuration in the low-energy/long-distance limit as black holes in ten-dimensional supergravity theory. The simplest such supergravity theories are those defined in the background formed by ten-dimensional Minkowski space. The canonical Maldacena construction is a typical example. However, one could just as well consider the supergravity in a more complicated background. Conventionally, one chooses a conventional flat Minkowski space along the $p + 1$ space-time directions of the brane and a more complicated $(9 - p)$-dimensional manifold $\mathcal{M}_{9-p}$. For technical control it is useful to preserve some but not all of the supersymmetry of the underlying ten-dimensional theory. If $p = 3$, this is equivalent to a specific condition on the $\mathcal{M}_6$ dimensional manifold, which was first formulated by Calabi and later firmly established by Yau [170]. For this reason Calabi–Yau manifolds play a prominent role in such explicit constructions. For other choices of p there are other manifolds that play an equivalent role, see e.g. [492], but if $\mathcal{M}$ is even-dimensional, one in practice often takes variations on Calabi–Yau manifolds: for instance, an eight-dimensional Calabi–Yau manifold for eleven-dimensional supergravity to obtain the closed-string theory dual to the strong-coupling limit of D2-branes.

The first step is then to take the near-horizon limit for these black holes, in exactly the same way as explained in chapter 4 in the context of Maldacena's example. In flat Minkowski space this is most conveniently done in radial coordinates, centred around the location $r = 0$ of the Dp-branes, with the $(8 - p)$-dimensional sphere at infinity containing the remnant information of the space-time outside the near-horizon region. For the more complicated Calabi–Yau models there exists a similar radial coordinate parametrisation. Writing the radial coordinates for the Calabi–Yau $\mathcal{M}_{9-p}$ metric (with $p = 2n + 1$, see above) as

$$ds^2 = dr^2 + r^2 \, ds^2_{\mathrm{SE}^{8-p}} \,, \tag{13.1}$$

one can think of these radial coordinates as a cone with the space SE^{8-p} as the base; in, for instance, flat space, SE^{8-p} is the $(8 - p)$-dimensional sphere, though for flat space the tip of the cone at $r = 0$ is not singular. For a Calabi–Yau metric $\mathcal{M}_{9-p}$ the space $ds^2_{\mathrm{SE}^{8-p}}$ is known as a Sasaki–Einstein space. Taking the near-horizon limit in these radial cone coordinates now combines the flat space-time along the branes with the radial direction of the $\mathcal{M}_{9-p}$ space to an anti-de Sitter space AdS$_{p+2}$. The remaining part of the metric is the base metric $r^2 \, ds^2_{\mathrm{SE}}|_{r=r_{\mathrm{h}}}$ evaluated at the horizon and the resulting total space-time is AdS$_{p+2} \times$ SE$_{8-p}$. Somewhat more complicated set-ups, starting either from $(10 + 1)$-dimensional supergravity or already a partial compactification to $d < 9 + 1$ dimensions, can give rise to AdS$_{p+2} \times$ SE$_q$ metrics where $q \neq 8 - p$.

13.1.1 Kaluza–Klein reduction and consistent truncations

In this language the original Maldacena proposal consists of N D3-branes at the tip of a cone with S^5 as the base. We never contemplate in any of the bottom-up gravity models what happens to the dependence of the fields on the location of the sphere. There is a simple energetic reason for the intuition that this can be ignored: any momentum along the sphere would cost extra energy. Because the sphere, or alternatively the base of any cone, is a compact space, the momentum along these spaces is quantised, with the implication that the excitation of finite-momentum modes in the compact directions involves an energy gap. This justification for ignoring the dynamics along compact directions is known as a Kaluza–Klein (KK) reduction. It harks back to an idea first proposed by Theodor Kaluza in 1921 that was aimed at unifying the four-dimensional gravity force with the electromagnetic force, by assuming that the gravity theory lives in a five-dimensional space-time with one extra spatial dimension in addition to the observed world. Several years later, in 1926 Oskar Klein provided the crucial ingredient that one can ignore the dynamics in this extra direction if it is a compact circle with a small radius r_{c}: quantum mechanics then discretises the momenta, and provides an energy gap of size $E_{\mathrm{gap}} \sim \hbar c / r_{\mathrm{c}}$.

Let us examine this KK reduction in more detail by considering an arbitrary space-time dimension. We choose the gravity theory to live in $(d + 2) + n$ dimensions, with the aim of being able to compactify it to $d + 2$ dimensions. We can choose n to be an arbitrary number appropriate to a particular purpose. The simplest compact topologies of the extra dimensions are either tori, T^n – obtained by periodically identifying flat space and formally equivalent to direct products of circles S^1 – or spheres S^n. The next step is to decompose the fields into a complete set of functions on the internal space; for spheres these are generalised spherical harmonics, while for tori they correspond to simple Fourier series. For example, for a compactification $T^1 \simeq S^1$ with the extra dimension denoted as $y \simeq y + 2\pi R$, a scalar field is decomposed as

$$\phi(x, y) = \sum_{m=-\infty}^{\infty} \phi(x, m)e^{im\,y/R}, \tag{13.2}$$

where m is an integer for periodic boundary conditions, or a half-integer for anti-periodic ones. A vector field A_μ, $\mu = 0, \ldots, d + 2$ decomposes into two parts: a lower-dimensional vector A_i, $i = 0, \ldots, d + 1$ with

$$A_i(x, y) = \sum_{m=-\infty}^{\infty} A_i(x, m)e^{im\,y/R} \tag{13.3}$$

and a lower-dimensional scalar when the index of the vector field points along the internal direction,

$$\tilde{\phi}(x, y) \equiv A_y(x, y) = \sum_{m=-\infty}^{\infty} A_y(x, m)e^{im\,y/R}, \tag{13.4}$$

and similarly for any higher-order tensor field with multiple indices. For example, the metric $g_{\mu\nu}$ reduces to a lower-dimensional metric g_{ij}, a lower-dimensional vector $B_i = g_{iy}$ and a lower-dimensional scalar $\Phi = g_{yy}$.

If and only if the full $d + 2 + n$ space-time is a product between the internal compact space and the remaining space-time, the Laplacian which governs the fluctuations of minimally coupled fields decomposes into a sum of the space-time Laplacian and the Laplacian of the internal space-time:

$$\Box\phi = \left(\Box_{\text{space-time}} + \Box_{\text{internal}}\right)\phi = 0. \tag{13.5}$$

This is easily seen to be the case for a scalar field. For a product space-time the metric can be chosen to be of the form

$$ds^2(x, y) = g^{\text{st}}_{\mu\nu}(x)dx^\mu\,dx^\nu + g^{\text{int}}_{ab}(y)dy^a\,dy^b, \tag{13.6}$$

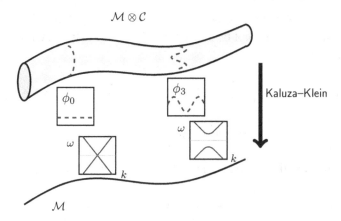

Figure 13.1 A cartoon for Kaluza–Klein reduction of a product space-time $\mathcal{M} \otimes \mathcal{C}$. Modes that carry gradient energy (momentum) in the internal directions of the compact space $\mathcal{C}$ appear as massive modes in the effective lower-dimensional theory on the space-time $\mathcal{M}$. At energies below this mass scale we may therefore truncate the theory to the modes which are constant on the internal space.

and thus

$$\Box \phi = \frac{1}{\sqrt{-g^{\mathrm{st}}}} \partial_\mu \left(\sqrt{-g^{\mathrm{st}}} g_{\mathrm{st}}^{\mu\nu} \partial_\nu \phi \right) + \frac{1}{\sqrt{-g^{\mathrm{int}}}} \partial_a \left(\sqrt{-g^{\mathrm{int}}} g_{\mathrm{int}}^{ab} \partial_b \phi \right) = 0 . \quad (13.7)$$

Combined with the decomposition of $\phi(x, y)$ into a complete set of functions in the internal space, this means that one can evaluate the internal Laplacian to obtain

$$\Box_{\mathrm{space\text{-}time}} \phi(x, m) - f(m)\phi(x, m) = 0, \quad (13.8)$$

where $f(m)$ is a positive semidefinite function of the eigenfunction labels m. From this consideration it is immediately obvious that the eigenvalues of the internal Laplacian gap out the higher-order functions on the internal space, with a magnitude set by the scale of the compact dimensions. This means that at low energies we can truncate the full infinite set of fields $\phi(x, m)$ to only the finite set of those which are constant in the internal directions $\phi(x, 0)$. (See Fig. 13.1.)

This, however, is a statement that pertains only to the fluctuations. The subtlety with Kaluza–Klein compactifications is whether this factorisation property can be extended from the infinitesimal fluctuations to the full action. In fact, for direct product space-times this truncation to constant fields remains fully consistent even for the full non-linear equations of motion. The meaning of a *consistent trunca-tion* is that a solution to the truncated set of (non-linear) equations of motion is automatically a solution to the full set of equations of motion of the original higher-dimensional theory. In the general situation where the background space-time is not a direct product, or when the background has other fields with VEVs, this trunca-tion to constant fields need not be consistent. To illustrate how this works, consider an action of the type

$$S = \int d^{d+2+n}x \sqrt{-g} \left[-\frac{1}{2}\partial_\mu \phi_0 \partial^\mu \phi_0 - \frac{1}{2}\partial_\mu \phi_1 \partial^\mu \phi_1 - V(\phi_0) - \phi_1 V'(\phi_0) \right.$$
$$\left. -\frac{1}{2}\phi_1^2 V''(\phi_0) + \cdots \right].$$
(13.9)

For this action a solution to the truncated set of equations of motion,

$$\Box \phi_0 - \frac{\partial}{\partial \phi_0} V(\phi_0) = 0,$$
(13.10)

is also a solution to the full set of equations of motion with $\phi_1 = 0$ if *and only if* $V'(\phi_0) = 0$. This follows by direct inspection of the full equations of motion

$$\Box \phi_0 - \frac{\partial}{\partial \phi_0} V(\phi_0) - \phi_1 \frac{\partial}{\partial \phi_0} V'(\phi_0) - \frac{1}{2}\phi_1^2 \frac{\partial}{\partial \phi_0} V''(\phi_0) = 0,$$
$$\Box \phi_1 - V'(\phi_0) - \phi_1 \frac{\partial}{\partial \phi_0} - \phi_1 V''(\phi_0) = 0.$$
(13.11)

For $\phi_1 = 0$ the second equation reduces to $V'(\phi_0) = 0$. The solution to the reduced theory is therefore a solution of the full theory only if this holds. Note, however, that a consistent truncation does not guarantee that this solution is stable in the full theory. If $V'(\phi_0)|_{\phi_0 = \phi_0^{\text{sol}}} = 0$ but $V''(\phi_0)|_{\phi_0 = \phi_0^{\text{sol}}} < 0$, the truncation is consistent but the solution is unstable in the "hidden" direction ϕ_1. Although this subtle fact is generally acknowledged, it is usually silently assumed that this potential source of trouble can be ignored.

This truncation to an effective lower-dimensional theory with a finite number of fields *in a consistent fashion* is the critical second step in top-down construc-tions of holographic duals. With a silent nod to the caveat above, it guarantees that they can be "uplifted" to fully consistent solutions of a string theory. From the dual field-theory perspective, this means that the solution is part of a fully quantum-consistent theory at all orders in N and the 't Hooft coupling λ. The dual of this fully quantum-consistent theory is a complete string theory in an $\text{AdS}_p \times \mathcal{M}_q$ background. In practice one immediately considers the large-N limit at large 't Hooft coupling to approximate the string theory by a $(9+1)$-dimensional or $(10+1)$-dimensional supergravity. This supergravity theory is compactified on $\mathcal{M}_q$ to a four- or five-dimensional AdS low-energy effective-gravity theory with an infinite tower of different matter fields and gauge fields, which are subsequently truncated consistently. There is quite some freedom of choice in the last step; in practice one always wishes to keep only the minimal number of fields.

It is immediately obvious that a near-infinite number of consistently truncated low-energy supergravity theories must exist, descending from the many ways to choose the compact manifold $\mathcal{M}_q$. Each has in turn its own set of solutions, and these are by construction part of the solution space of the original ten- or eleven-dimensional theory. This corresponds to the "landscape" of string theory. From the perspective of applications to CMT, it means that the space of effective consistent holographic theories is large: in fact, it is large enough to allow the description of the phenomena described in this book. But a consistent truncation of a top-down construction almost always adds something more to the simplest bottom-up constructions. This is the focus of this chapter. It is interesting to ask to what extent such "remnants" of the demands from quantum consistency are unavoidable.

13.1.2 A top-down holographic superconductor

An illustrative example is the top-down construction of a holographic super-conductor from $(10 + 1)$-dimensional supergravity theory. Building on earlier linearised studies of Ref. [493], Gauntlett, Sonner and Wiseman constructed solutions of Kaluza–Klein reduced $(3 + 1)$-dimensional AdS gravity times an arbitrary seven-dimensional Sasaki–Einstein manifold, which can be consistently uplifted to eleven-dimensional supergravity solutions [494]. The $(3 + 1)$-dimensional low-energy effective action obtained by Gauntlett *et al.* is [494, 495]

$$
S = \frac{1}{16\pi G} \int d^3x \, dt \sqrt{-g} \left[R - \frac{(1 - b^2)^{3/2}}{(1 + 3b^2)} F_{\mu\nu} F^{\mu\nu} - \frac{3}{2(1 - \frac{3}{4}|\Phi|^2)^2} |D\Phi|^2 \right.
$$
$$
- \frac{3}{2(1 - b^2)^2} (\nabla b)^2 - \frac{24(-1 + b^2 + |\Phi|^2)}{(1 - \frac{3}{4}|\Phi|^2)^2 (1 - b^2)^{3/2}}
$$
$$
\left. + \frac{2b(3 + b^2)}{4(1 + 3b^2)} \epsilon^{\mu\nu\rho\sigma} F_{\mu\nu} F_{\rho\sigma} \right]. \tag{13.12}
$$

Here $D_\mu \Phi = \partial_\mu \Phi - 4i A_\mu \Phi$ and b is an additional real scalar field. Through holography, these $AdS_4 \times SE_7$ compactifications correspond to superconductors in three dimensions. There are three distinct aspects that are immediately apparent. In the first place, in this top-down construction the effective mass and charge in the lower-dimensional theory are no longer free parameters: the ratio between Newton's constant and the $U(1)$ charge is unity and in these units the field Φ dual to the order parameter has charge $q = 4$ and mass $m_\Phi^2 L^2 = -2$, where the AdS radius L^2 follows from the constant term in the potential $V(b = \Phi = 0) \equiv -6/L^2$ with $L = 1/2$. In chapter 10 we showed that it is the relative strength of these parameters which determines whether a scalar VEV arises or not. It immediately follows that a phenomenon in a bottom-up theory need not necessarily exist in *any*

top-down theory. In this case it works out fine, of course: holographic superconductivity is in this sense a physical phenomenon. A second finding is that the scalar
field is no longer minimally coupled – the kinetic term is that of a non-linear sigma
model – and the potential for the scalar field is also fixed in the top-down construction. Moreover, this potential actually depends on the third unconventional aspect:
an extra real field b is present, which alters this potential dynamically. In addition,
notice that the ranges of the scalar fields are limited to $|\Phi| < 2/\sqrt{3}$ and $|b| < 1$,
respectively. These ranges correspond precisely to the existence of a Kaluza–Klein
gap that allows one to ignore all the higher-order harmonics in the fields. The gap
closes when either field approaches the boundary of the range, with the ramification
that the low-energy approximation Eq. (13.12) is no longer valid.

The simplest solution of this action describing a holographic superconductor is
characterised by $b = 0$. This interpolates between AdS_4 in the UV and a black hole
with scalar hair in the IR (Fig. 13.2). This solution by itself already affirms that
the holographic superconductor is a real physical phenomenon. Despite the more
involved form of the action, the phenomenology is nearly identical to the minimal
bottom-up models for holographic superconductivity in chapter 10. The extra field
b does not play a critical role in the phenomenology, but it does have an effect.
As we saw in chapter 10, due to the fact that the AdS_2 radius of the charged black
hole differs from the AdS_4 radius at infinity, even a neutral scalar field can condense in the presence of a chemical potential. The thermodynamically preferred
solution is therefore a variant of the holographic superconductor, which *also* has a
non-vanishing VEV for the operator dual to the new field b. This VEV can be interpreted as the breaking of an additional $\mathbb{Z}_2$ symmetry in the boundary, and in the
preferred ground state both the usual $U(1)$ symmetry and the extra $\mathbb{Z}_2$ symmetry
are therefore broken.

Since a minimal superconductor is characterised by a sole $U(1)$ symmetry, one
can try to achieve this in this model by explicitly deforming the theory to break the
$\mathbb{Z}_2$. The holographic description of the breaking of this additional $\mathbb{Z}_2$ is as usual.
Near the AdS boundary the solution of the extra field b will have the universal
behaviour

$$b(r) = \frac{J_b}{r} + \frac{\langle \mathcal{O} \rangle}{r^2} + \cdots . \tag{13.13}$$

Following the standard dictionary, one can deform the theory by turning on a source
J_b for the operator $\mathcal{O}_b$ dual to b with strength equal to the coefficient of the leading
term in the expansion near the boundary. Physically this means that we explicitly
break the $\mathbb{Z}_2$ symmetry; mathematically it means that we allow arbitrary profiles of
the field b that satisfy the equations of motion. The effect of this deformation is to
suppress the tendency to exhibit superconductivity. This cannot easily be deduced

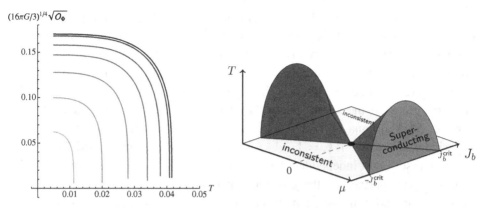

Figure 13.2 The top-down holographic superconductor. Left: the VEV of the order parameter $\mathcal{O}_\Phi$ as a function of the temperature T/μ. The different curves are for different values of a deformation by a source J_b/μ for the extra neutral operator $\mathcal{O}_b$. This suppresses the ordered state. Right: the phase diagram of the top-down holographic superconductor as a function of T/μ and the source J_b/μ. Figures adapted from [495]. (Reprinted with kind permission from Springer Science and Business Media, © 2010, SISSA, Trieste.)

by inspection of the action: the effect of the geometry plays a large role in the dynamics of the neutral field b. An interesting aspect is that the deformation of b is able to suppress the thermal phase transition to superconductivity all the way to a quantum phase transition at $T = 0$. However, this happens precisely when b reaches the edge of its domain of validity at the horizon of the black hole. The quantum phase transition belongs therefore to a fully different regime, which lies outside the validity of the consistent truncation which leads to this action. In summary, this consistently truncated top-down theory has a phase diagram as drawn in Fig. 13.2. There is a superconducting dome with a maximum at the $\mathbb{Z}_2$ symmetric point. As we deform away from this point the critical temperature decreases to zero, which is simultaneously the edge of the validity of the consistently truncated action.

13.1.3 A top-down holographic Fermi liquid

As our next example, let us consider the fate of fermions in a holographic top-down model. The most straightforward $(3 + 1)$-dimensional top-down model already substantiates the non-Fermi-liquid fermion spectral functions found in the probe approximation to the RN metal in chapter 9 [362, 363]. From our insight that these spectral functions arise only if the RN metal is already unstable towards the electron-star solutions of chapter 11, this indirectly corroborates their existence as well, although the effects of backreaction have not yet been studied in full detail in the top-down model.

This most straightforward $(3 + 1)$-dimensional model is the unique $(10 + 1)$-dimensional supergravity theory compactified on a seven-dimensional sphere. Owing to the high symmetry of the seven-sphere S^7, the resulting Kaluza–Klein compactified theory is the unique $\mathcal{N} = 8$ AdS$_4$ supergravity. This theory has an $SO(8)$ gauge symmetry, which is the $(3 + 1)$-dimensional manifestation of the isometry group of the seven-sphere. What is important for us is that the theory contains 56 Majorana fermions χ_{ijk} transforming in the **56** of $SO(8)$; here the triple index ijk with $i, j, k, = 1, \ldots, 8$ is completely anti-symmetrised. The theory also possesses eight independent "vector-spinors" $\psi_{\mu i}$, – these "gravitini" are the spin-3/2 super-partners of the graviton which are the gauge fields of local super-symmetry. In addition, there are 70 scalars $\phi_{ijkl} = (1/4!)\epsilon_{ijklmnpq}\phi_{mnpq}$, but these will not play a role. Setting all scalars and fermions to vanish, the action is just the Einstein–Yang–Mills action,

$$\mathcal{L} = \frac{1}{16\pi G} \left(R + \frac{6}{L^2} - \frac{1}{2} \operatorname{Tr} F^2 \right), \tag{13.14}$$

with $F^{ij}_{\mu\nu}$ the $SO(8)$ field strength. With respect to the $SO(2) \sim U(1)$ generated by A^{12}_μ the theory has a straightforward AdS Reissner–Nordström solution. To study the fluctuations around this background, fortunately one needs only the quadratic part of the full action. The latter is complicated, but to quadratic order the action for the Majorana fermions is given by

$$\mathcal{L}_{\text{spin-1/2}} = -\frac{1}{12}\bar{\chi}^{ijk}\left(\gamma^\mu \overrightarrow{D}_\mu - \overleftarrow{D}_\mu \gamma^\mu\right)\chi_{ijk} - \frac{1}{2}\left(F^+_{\mu\nu ij} S^{ij,kl} O^{+\mu\nu kl} + \text{h.c.}\right). \tag{13.15}$$

Here $D_\mu \chi_{ijk} = \nabla_\mu \chi_{ijk} + (3/\sqrt{2}L)A^{m}_{\mu[i}\chi_{jk]m}$, $S^{ij,kl} = \delta^i_j\delta^k_l - \delta^k_j\delta^i_l$ and the operator $O^{+\mu\nu kl}$ is a Pauli coupling between the spin-1/2 fermions themselves and gravitini,

$$O^{+\mu\nu kl} = -\frac{\sqrt{2}}{144}\epsilon^{ijklmnpq}\bar{\chi}_{klm}\sigma^{\mu\nu}\chi_{npq} - \frac{1}{2}\bar{\psi}_{\rho k}\sigma^{\mu\nu}\gamma^\rho \chi^{ijk} + \text{``}\psi^2_{\rho k}\text{''} \text{ terms.} \tag{13.16}$$

Here $\sigma^{\mu\nu} = \frac{1}{4}[\gamma_\mu, \gamma_\nu]$. The complication is that the Pauli coupling is non-zero in the RN background. For this reason the second term in Eq. (13.16) can mix the fermions with gravitini. Fortunately, this is not the case [362, 363]. In this specific charged RN background 20 of the fermions χ_{ijk} with $i, j, k = 3, \ldots, 8$ are uncharged under rotations generated by A^{12}_μ; so are the 6 fermions χ_{12k}. The remaining 30 fermions which contain only one of the indices 1 and 2 are charged under the gauge field. Inspecting now the Pauli coupling, one directly observes that this only affects the 6 neutral fermions χ_{12k}. We are, however, interested in the 30 charged fermions to probe holographically for the existence of Fermi surfaces in the dual theory. We therefore do not have to worry about the mixing, and proceed with just these 30 fermions. Written in complex form, they form 15 complex spinors $\chi_{1jk} + i\chi_{2jk}$, $j, k = 3, \ldots, 8$, that satisfy the conventional Dirac equation

$$\Gamma^\mu (\nabla_\mu - iqa_\mu)\chi = 0 \tag{13.17}$$

for a massless fermion with charge $q = 1/(\sqrt{2}L)$. Again we see that in the top-down construction the previously free parameters are fixed. Now we can immediately refer to the results of chapter 9 to check for the existence of a Fermi surface at low T/μ. For this value of the charge and this mass one finds a normalisable mode of the Dirac equation, corresponding to the existence of a Fermi surface in the dual field theory at $k_{\rm F} \sim 0.9185$. Upon inspecting the dispersion and width, one finds that this Fermi liquid supports non-Fermi liquid excitations with $v_{k_{\rm F}} \simeq 0.2393$ [362, 363].

Box 13.1   The conformal-to-AdS$_2$ metal

In chapter 8 we showed how the conformal-to-AdS$_2$ metal is almost certainly a more reliable model of local quantum criticality than the plain extremal RN metal with its extensive ground-state entropy. Also this particular solution can be validated by a top-down construction [321]. The starting point is $(10 + 1)$-dimensional supergravity compactified on an S^7 sphere, as before. Its Kaluza–Klein reduction is the unique $\mathcal{N} = 8$ $SO(8)$ gauged supergravity theory [54]. The Cartan subgroup of $SO(8)$ is $U(1)^4$, and in principle this theory can have black-hole solutions with different charges under each of the $U(1)$s [496, 497]. The technical issue is that the Yang–Mills action contains a non-minimal coupling to the 70 scalar fields that schematically behaves as

$$\mathcal{L} = \cdots - \frac{1}{4}F_{\mu\nu ij}\left(2S^{ij,kl}(\phi) - \delta_j^{[i}\delta_l^{k]}\right)F^{\mu\nu}{}_{kl}, \tag{13.18}$$

with $S^{ij,kl}(\phi) = \delta_j^{[i}\delta_l^{k]} + $ "ϕ_{ijkl}" terms. To build a consistent truncation we can always set the charged fields to vanish. There are, however, three neutral scalar fields. Choosing the four $U(1)$s to be generated by A_μ^{12}, A_μ^{34}, A_μ^{56} and A_μ^{78}, these are [496]

$$\lambda_1 = \phi_{1234} + \phi_{5678}, \ \lambda_2 = \phi_{1256} + \phi_{3478}, \ \lambda_3 = \phi_{1278} + \phi_{3456}; \tag{13.19}$$

recall from earlier that only the combination $\phi_{ijkl} + (1/4!)\epsilon_{ijklmnpq}\phi_{mnpq}$ with $ijkl$ completely anti-symmetrised is a physical degree of freedom.

Box 13.1   (Continued)

Einsteinian gravity plus the four $U(1)$s plus these three scalars can form a consistent truncation with the action

$$\mathcal{L} = R - \frac{1}{2}(\partial \lambda_i)^2 - \frac{2}{L^2} \sum_i \cosh \lambda_i - 2 \sum_{A=1}^{4} e^{\alpha_A^i \lambda_i} (F_{\mu\nu}^{(A)})^2, \qquad (13.20)$$

where

$$\alpha_A^i = \begin{pmatrix} 1 & 1 & -1 & -1 \\ 1 & -1 & 1 & -1 \\ 1 & -1 & -1 & 1 \end{pmatrix}. \qquad (13.21)$$

The single charge simple Reissner–Nordström black hole Eq. (13.14) is actually the black hole with the same charge under all four $U(1)$ symmetries. In this case the equations of motion for λ_i,

$$\Box \lambda_i - \frac{2}{L^2} \sinh \lambda_i - 2 \sum_A \alpha_A^i e^{\alpha_A^i \lambda_i} (F_{\mu\nu}^{(A)})^2 = 0, \qquad (13.22)$$

reduce to

$$\Box \lambda_i - \frac{2}{L^2} \sinh \lambda_i - 2 \big(F_{\mu\nu}^{(\mathrm{diag})}\big)^2 \sum_{A=1}^{4} \alpha_A^i e^{\alpha_A^i \lambda_i} = 0.$$

Since $\sum_{A=1}^{4} \alpha_A^i = 0$, this has $\lambda_i = 0$ as a solution to its equation of motion, and therefore the scalar fields can be consistently set to zero in backgrounds with four equal $U(1)$ charges.

The conformal-to-AdS$_2$ metal follows from a similar consideration. Consider now a black hole with no charge under the first $U(1)$, $F_{\mu\nu}^{(1)} = 0$, but three equal charges under the remaining $U(1)$s, $F_{\mu\nu}^{(2)} = F_{\mu\nu}^{(3)} = F_{\mu\nu}^{(4)} \equiv F_{\mu\nu}^{(\mathrm{three\text{-}ch})}$. The equations of motion for λ_i become

$$\Box \lambda_i - \frac{2}{L^2} \sinh \lambda_i - 2(F_{\mu\nu}^{(\mathrm{three\text{-}ch})})^2 \sum_{A=2}^{4} \alpha_A^i e^{\alpha_A^i \lambda_i} = 0. \qquad (13.23)$$

By diagonalising the 3×3 matrix α_A^i, where $A = 2, \ldots, 4$, it is easy to show that $\lambda_1 = \lambda_2 = \lambda_3$ solves two out of the three equations, with the remaining equation being

$$\Box \lambda - \frac{6}{L^2} \sinh \left(\frac{\lambda}{3}\right) + 12 \left(F_{\mu\nu}^{(\mathrm{three\text{-}ch})}\right)^2 e^{-\lambda/3} = 0. \qquad (13.24)$$

By substituting the constraint that the first $U(1)$ vanishes and that the last three have equal charge plus the equality of the three scalar fields directly into the action one obtains

$$\mathcal{L} = R - \frac{1}{6}(\partial\lambda)^2 + \frac{6}{L^2}\cosh\left(\frac{\lambda}{3}\right) - 2e^{-\lambda/3}(F_{\mu\nu})^2. \tag{13.25}$$

This equals the action for the conformal-to-AdS$_2$ metal Eq. (8.58) after changing variables to $\phi = \lambda/\sqrt{3}$, $\tilde{A}_\mu = 2\sqrt{2}A_\mu$ [321].

13.2 Probe-brane holography from Dp/Dq-brane intersections

In the straight-up top-down constructions discussed in the previous section the starting configuration in string theory always singles out a single type of D-brane. The low-energy description of its collective modes is then dual to the near-horizon limit of the closed-string description of the same set-up. However, in string theory one has the freedom to concoct more intricate D-brane set-ups to start with. In particular, one can have different types of D-branes at the same time, and they may even intersect. Starting from such brane configurations and following the same steps as above leads then to a special new class of AdS/CFT dual pairs known as "probe-brane holography". To understand these probe-brane top-down constructions, we shall need an additional piece of information about D-branes. We have already seen in chapter 4 that one should think of D-branes as solitons, where the collective modes of the soliton correspond to the open strings ending on the brane. In a flat space-time one of these collective modes is always the Goldstone zero mode of translations. The action of this Goldstone mode is uniquely fixed by invariance under coordinate transformations both of the space-time and of the local coordinates of the Dp-brane surface. Parametrise the surface spanned by a Dp-brane in $9 + 1$ dimensions by the embedding of local coordinates ξ^i with $i = 0, \ldots, p$ into the space-time coordinates $X^\mu(\xi^i)$ with $\mu = 0, \ldots, d + 2$, similarly to the way the worldline of a particle is defined by a one-parameter curve $x^\mu(\tau)$ in space-time, and a string is parametrised by $X^\mu(\tau, \sigma)$ with τ, σ the local coordinates on the worldsheet (see Eq. (4.35)). The unique action for the Goldstone mode then becomes

$$S_{\mathrm{D}p} = -T_p \int d\xi^{p+1} \sqrt{-\det\left(g_{\mu\nu}(X(\xi))\frac{\partial X^\mu(\xi)}{\partial\xi^i}\frac{\partial X^\nu(\xi)}{\partial\xi^j}\right)} + \cdots, \tag{13.26}$$

where T_p is the tension (energy density) of the Dp-brane. One of the key aspects of the open string that parametrises the Dp-brane collective modes is that it also contains a gauge-field excitation along the surface of the brane. Realising that its field strength F_{ij} transforms in the same way as the induced metric $g_{ij} = g_{\mu\nu}(X)\partial_i X^\mu \partial_j X^\nu$ under Dp-brane local coordinate changes, there is a very natural way to add the gauge-field dynamics to the action:

$$S_{\mathrm{D}p} = -T_p \int d\xi^{p+1}\sqrt{-\det\left(g_{\mu\nu}\, \partial_i X^\mu\, \partial_j X^\nu + 2\pi\alpha' F_{ij}\right)} + \cdots . \qquad (13.27)$$

We have included a term $2\pi\alpha'$, where α' has dimensions of length squared, to ensure that F_{ij} has standard scaling dimensions. This action is in fact not fixed by coordinate invariance, but is dictated instead by one of the T-duality symmetries in string theory that we mentioned in chapter 4 [498]. In essence T-duality states that one should think about the gauge dynamics A_i, which is the end point of the open string moving along the Dp-brane, in the same way as one does about $\partial_i X^\mu$, which encodes how the end point of the open string tries to move the Dp-brane.

Equation (13.27) is a well-known non-linear extension of electrodynamics known as the Dirac–Born–Infeld (DBI) action, which was inspired by the special-relativistic generalisation of the kinetic energy. It was originally proposed by Born and Infeld in the 1930s [499, 500] as an attempt to eliminate the infinite self-energy of a point charge in classical electrodynamics. On choosing the trivial embedding $\partial_i X^\mu = \delta_i^\mu$ and expanding for $\alpha' F \ll 1$, the standard Maxwell action is recovered,

$$\begin{aligned}
S_{\mathrm{DBI}} &= -T_p \int d^{p+1}x \sqrt{-\det(\eta_{ij} + 2\pi\alpha' F_{ij})} \\
&= -T_p \int d^{p+1}x \left(1 + \frac{(2\pi\alpha')^2}{4} F_{\mu\nu}F^{\mu\nu} + \cdots \right),
\end{aligned} \qquad (13.28)$$

with coupling constant $g^2 = 1/((2\pi\alpha')^2 T_p)$. One of the two key aspects of the probe-brane constructions is this non-linearity of the effective action. Let us first show how this non-linearity can be exploited holographically to compute non-linear responses in the dual field theory. Subsequently we will discuss the other important feature of intersecting D-brane models: it is straightforward to deduce the explicit form of the boundary field theory from the Dp/Dq-brane intersections. This will make it possible to unambiguously link the order parameter of a "flavour" holographic superconductor to the pair operator formed from UV fermions.

Box 13.2   The Dirac–Born–Infeld action in string theory and its non-Abelian extension

Fundamentally, the low-energy worldvolume action for D-branes is directly computed from the string scattering amplitudes [501, 502]; see for instance the textbooks [171, 172]. A powerful aspect of this action is that it shows how the open-string fields couple to the closed-string fields. This is directly illustrated by the fact that the open string moves in a space-time and the excitations of space-time itself in terms of the gravitons are the closed-string excitations. In addition to the graviton, a perturbative $(9 + 1)$-dimensional nearly flat-space string theory has massless bosonic excitations corresponding to an anti-symmetric tensor "Kalb–Ramond" field $B_{\mu\nu}$, a scalar Φ, known as the dilaton, and a number of anti-symmetric $(0, q)$ tensor fields $C^{(q)}_{\mu_1\mu_2...\mu_q}$ known as Ramond–Ramond fields.

In the limit where the derivatives of all these fields are small, while the fields themselves need not be, the full non-linear DBI action for the massless excitations of a Dp-brane including all the couplings to the massless bosonic closed-string fields is

$$S_p = - T_p \int d^{p+1}\xi \, \mathrm{Tr}\left(e^{-\phi}\sqrt{-\det(g_{ij} + B_{ij} + 2\pi\alpha' F_{ij})}\right)$$

$$- i\mu_p \int e^{B+2\pi\alpha' F} \sum_{q=0}^{p} C^{(q)} . \tag{13.29}$$

Here we have used the shorthand notation $g_{ij} = g_{\mu\nu}\,\partial_i X^\mu\,\partial_j X^\nu$ and $B_{ij} = B_{\mu\nu}\,\partial_i X^\mu\,\partial_j X^\nu$, and the same for the Ramond–Ramond fields $C^{(q)}$. The last term is a Chern–Simons-like term that should be understood as follows. Each field is an anti-symmetric tensor field $B = B_{ij}\,d\xi^i\,d\xi^j$, $C^{(q)} = C^{(q)}_{i_1i_2...i_q}\,d\xi^{i_1}\,d\xi^{i_2}\ldots d\xi^{i_q}$, and the exponential $e^{B+2\pi\alpha F}$ should be expanded up to precisely $p - q$ terms $d\xi^i$, since together with $C^{(q)}$ these form an invariant measure that can be integrated. Finally, $T_p = 2\pi/(4\pi^2\alpha')^{(p+1)/2}$ denotes the tension of the D-branes, while $\mu_p = 2\pi/(4\pi^2\alpha')^{(p+1)/2}$ is the charge of the Dp-brane under the $C^{(p)}$ Ramond–Ramond field: notice that these expressions are equal. This is a consequence of supersymmetry. A Dp-brane is a Bogomol'nyi–Prasad–Sommerfield (BPS) object: its energy saturates the BPS bound $E \geq Q$. Because of this the gravitational attraction between two Dp-branes precisely balances their Ramond–Ramond charge repulsion.

In chapter 4 we discussed a most interesting characteristic of the D-brane effective action in string theory. When two or more D-branes of the same type are placed on top of each other, the strings stretching between them become massless and some of these massless degrees of freedom enhance the gauge symmetry from $U(1)$ to $U(N)$, with N equal to the number of D-branes. The curious fact is that this matrix-enhancement applies also to the collective translational mode. The N independent translational modes $X^{\mu\perp\xi^i}_a$, $a = 1, \ldots, N$ become an $N \times N$ matrix-valued field $X^{\mu\perp\xi^i}_{ab}$. Given its

---

Box 13.2   (Continued)

origin, it is clear that a $U(N)$ transformation is a relabelling of the original D-branes and thus this field is now charged under the $U(N)$ gauge symmetry. However, it is no longer possible to immediately write down the action. Since matrices do not commute, there is an ordering ambiguity in the non-Abelian DBI action [503]. The precise ordering has not been resolved, but in most cases a completely symmetrised ordering can be employed. This then yields the non-Abelian DBI action in flat space

$$S_p = -T_p \, \mathrm{STr} \int d^{p+1}\xi \, \left(e^{-\phi}\sqrt{-\det(\eta_{\mu\nu}D_i X^\mu D_j X^\nu + B_{ij} + 2\pi\alpha' F_{ij})}\right)$$

$$- i\mu_p \, \mathrm{STr} \int e^{B+2\pi\alpha' F} \sum_{q=0}^{p} C^{(q)}, \tag{13.30}$$

where $D_i X^{\mu \perp \xi} = \partial_i X^\mu - i[A_i, X^\mu]$, and the symbol "STr" means "take the symmetrised trace after expanding the square root".

When two D-branes of different types are placed on top of each other, there are additional massless states, if the number of "unshared directions" is a multiple of four. An "unshared direction" refers to a dimension into which only one of the D-branes extends. These massless states do not give rise to enhanced gauge symmetries, but instead give rise to two sets of regularly charged fields. The first set transforms in the vector (i.e. fundamental) representation of the gauge group of one D-brane and in the conjugate vector representation of the gauge group of the other D-brane. The other set transforms the other way around. Specifically, for a Dp/Dq-brane set-up with p strictly less than q and where the Dq-brane is parallel to and on top of the Dp-brane, but extends in $q - p = 4n$ more dimensions, the field content on the common $p + 1$ dimensions is a $U(N_p)$ gauge field, a $U(N_q)$ gauge field, $q - p$ scalars in the adjoint representation of $U(N_p)$, $q - p$ scalars in the adjoint of $U(N_q)$ and one complex field in the representation $(N_p, \bar{N}_q)$ with its conjugate in the representation $(\bar{N}_p, N_q)$.

In the most basic holographic set-up the low-energy dynamics of the linearised DBI action becomes the field theory on the boundary, e.g. Maldacena's canonical example. The trick to keep the non-linear part of the action in holography is to let it be the effective action of an additional $N_f = 1$ or $N_f = 2$ "flavour" Dq-brane in the original string-theory set-up, in addition to the large number N_c of Dp-branes to which the open–closed-string duality can be applied in the low-energy/near-horizon limit [490]. In the limit $N_c \gg N_f$ the backreaction of the N_f probe branes on the geometry can be completely ignored. Therefore, in the closed-string low-energy approximation for the N_c Dp-branes, we still have the extremal charged black-brane-like metric as in Eq. (4.39). In this background we can now put the additional N_f Dq-brane as a probe. Since the non-linear DBI action encodes the coupling to the full closed-string fields for small gradients, but not limited to small perturbations, the extremal N_c black-brane background can be substituted into the DBI action to obtain the full non-linear action for the low-energy degrees of freedom of the N_f Dq-branes.

The resulting dynamics then depends on how the original Dp- and Dq-branes are oriented relative to each other. A straightforward set-up is to choose them to be parallel in $p + 1 < q + 1$ dimensions, where the Dq-brane extends in a further $q - p$ directions. This situation is usually summarised in a table, which looks as follows for the case of D3/D7-branes:

	x^0	x^1	x^2	x^3	x^4	x^5	x^6	x^7	x^8	x^9
N_c D3	x	x	x	x						
N_f D7	x	x	x	x	x	x	x	x		

On taking the near-horizon limit of the D3-branes, the four directions x^0, x^1, x^2, x^3 together with the radial direction of the dimensions $x^4, x^5, x^6, x^7, x^8, x^9$ form an AdS$_5$ space. Since the polar coordinates for the $x^4, x^5, x^6, x^7, x^8, x^9$ directions contain also the polar coordinates for x^4, x^5, x^6, x^7, this radial direction contains the radial direction along the remaining four dimensions of the D7-brane. The D7-brane will therefore extend in some form into the AdS radial direction, in addition to sharing the $3 + 1$ dimensions of the boundary; we will see shortly how this radial embedding is determined. Its other three extra dimensions will span a three-sphere inside the five-sphere formed from $x^4, x^5, x^6, x^7, x^8, x^9$. There are two qualitatively different ways for the D7-brane to accomplish this. One way is to locate both the D3-branes and the D7-branes in the original set-up at the origin of the directions x^8, x^9. It is then clear that as one moves out radially along the D7-brane nothing special happens. This *black-brane embedding* thus corresponds to a configuration where the D7-brane does not move along the S^5 (see Fig. 13.3). On the other hand, if in the original set-up the D7-brane was located away from the D3-brane in the $x^8 - x^9$ plane, then as one moves outward from the D3-branes one has a region where no brane is present in the S^5. Only at some further distance away are D7-branes that wrap an S^3 portion of the S^5 at each point encountered. This does not mean that the surface of the D7-brane suddenly appears or that it is discontinuous. The beauty is that this can be done smoothly by letting the point at which the D7-brane starts to wrap the S^3 be coincident with the north or south pole of the S^5, in precisely the same way as one can slip a rubber band on a globe (see Fig. 13.3). This is called the *Minkowski embedding*; the details of these two embeddings are provided in box 13.3.

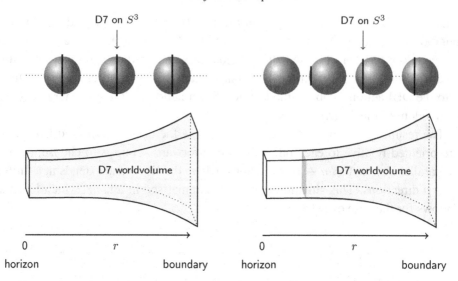

Figure 13.3 The two qualitatively different D7-brane embeddings. On the left we see the AdS configuration that arises when the N_f D7-branes are located on top of the large number N_c of D3-branes. In this *black-brane embedding* the location of the S^3 sphere which the D7-branes span in the S^5 around the D3-branes does not change as a function of the radial direction. The D7-brane worldvolume fills the whole of AdS in this case. If, on the other hand, the D7-branes are located some distance away from the D3-branes, then they can appear on an S^5 sphere only once one has moved out far enough from the interior location of the D3-branes. This *Minkowski embedding* is depicted on the right. In that case the D7-brane worldvolume fills only part of AdS. It can nevertheless be done smoothly by letting the radius of the compact S^3 which the D7-branes span start from zero: at the edge in AdS they are located at the north or south pole of the S^3.

   The explicit form of the field theory which is dual to these D3/D7-brane con-figurations follows directly from the open-string excitations on the branes. A first conclusion is that one has a large-N_c gauge theory at strong coupling, associated with the D3-branes. The intrinsic D7-brane dynamics is, on the other hand, *not* present in the dual field theory. This remains only explicitly encoded in the DBI action on the gravity side, but this is fully consistent with the fact that the D7-branes are considered as probes. These sectors do interact, however, through the open strings that stretch between the D3- and the D7-branes. Since these open strings have one end point on the D3-branes, they carry one index $1, \ldots, N_c$: they transform in the vector representation of the large-N_c gauge group. The other end point ends on one of the N_f D7-branes, and there are therefore N_f different *flavours* of matter in the vector N_c representation. Through the standard GKPW relation between the boundary values of AdS fields and the operators in the dual field the-ory, the dynamics in AdS of the D7-brane degrees of freedom can be interpreted as the dual description of the N_f flavours of vector-charged matter content. As illustrated by the geometric cartoon in Fig. 13.4, it encodes the dynamics of the

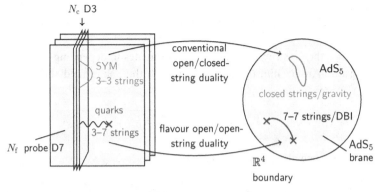

Figure 13.4 A cartoon of probe-brane holography. The starting point (left) is an intersecting N_c Dp/N_f Dq-brane configuration. In the probe limit where $N_f \ll N_c$ the closed-string geometry that arises after open/closed-string duality is the one controlled by the N_c Dp-branes. Here one can take Maldacena's near-horizon limit that is dual to the low-energy limit on the intersecting branes to generate an AdS/CFT duality. Since the geometry is controlled solely by the N_c Dp-branes the AdS side (right) is the conventional AdS$_{p+2}$ space-time. The flavour physics of both the p–q strings that stretch between the different branes and the q–q strings that stretch between the Dq-branes is captured by the Dirac–Born–Infeld action in the AdS space-time background. Figure adapted from [506].

strings stretched between the D3- and the D7-branes. In particular, the gauge field appearing in the bulk DBI action is the dual of the global $U(N_f)$ flavour symmetry in the boundary.

Box 13.3 Black-brane and Minkowski embeddings of probe branes

Here we will show explicitly how the two possible embeddings for the D3/D7 system arise. We ignore the dynamics of the Maxwell field for the moment, i.e. we are considering the brane embedding at finite temperature with zero density.

The convenient way to construct the probe D7-brane action is to choose a physical embedding where seven of the local coordinates on the D7-brane worldvolume are identified with the space-time coordinates in AdS$_5 \times S^5$. We write the metric as

$$ds^2 = \frac{r^2}{L^2}\left(-dt^2 + dx_1^2 + dx_2^2 + dx_3^2\right) + \frac{L^2}{r^2}\,dr^2 + L^2\,d\Omega_5^2$$

with $d\Omega_5^2$ is the metric on the S^5. Parametrise this metric as

$$d\Omega_5^2 = d\theta^2 + \sin^2\theta\,d\Omega_3^2 + \cos^2\theta\,d\phi^2.$$

The D7-brane will wrap the three-sphere $d\Omega_3^2 = d\chi_1^2 + \sin^2\chi_1\left(d\chi_2^2 + \sin^2\chi_2\,d\chi_3^2\right)$. Its position is therefore a point in the remaining two directions (θ, ϕ). The physical embedding choice we make is

$$(\xi^0, \xi^1, \xi^2, \xi^3) = (x^0, x^1, x^2, x^3), \quad \xi^4 = r, \quad (\xi^5, \xi^6, \xi^7) = (\chi_1, \chi_2, \chi_3).$$

Box 13.3   (Continued)

The symmetries then restrict the remaining coordinates to depending solely on r. On substituting this Ansatz into the DBI action with the gauge field set to vanish we obtain

$$S_{\text{DBI}} = -T_7 \, \text{Vol}_{S^3} \, \text{Vol}_{\mathbb{R}^{3,1}} \int dr \, r^3 \sin^3\theta \sqrt{1 + r^2(\partial_r\theta)^2 + r^2\cos^2\theta(\partial_r\phi)^2}. \quad (13.31)$$

By symmetry we can also set $\partial_r\phi = 0$, and we have

$$S_{\text{DBI}} = -T_7 \, \text{Vol}_{S^3} \, \text{Vol}_{\mathbb{R}^{3,1}} \int dr \, r^3 \sin^3\theta \sqrt{1 + r^2(\partial_r\theta)^2}. \quad (13.32)$$

On defining $\chi = \cos\theta$ as the new dynamical variable, the action becomes

$$S_{\text{DBI}} = -T_7 \, \text{Vol}_{S^3} \, \text{Vol}_{\mathbb{R}^{3,1}} \int dr \, r^3 (1-\chi^2)\theta\sqrt{(1-\chi^2) + r^2(\partial_r\chi)^2}. \quad (13.33)$$

It is straightforward to show that near the boundary

$$\chi = \frac{m}{r} + \frac{c}{r^3} + \cdots, \quad (13.34)$$

where m and c are the source (mass) and VEV of the flavour operators, respectively. The numerical solutions for (13.33) can be found in [504], but we can understand the dynamics of the system qualitatively. The solutions to $\theta(r)$ are precisely curves in the $\theta-r$ plane that start and end at some fixed boundary value $\theta_c = \pm\arccos\chi$. This is exactly analogous to the Wilson-line embeddings we studied in chapter 6 and in box 6.4. There are two qualitatively different solutions. The first one is characterised by the two ends hanging straight down into the horizon from θ_c and $-\theta_c$; in the other, these ends form a closed curve that descends to some minimum value in the radial direction r_{min}. It is also readily understood which solution is thermodynamically preferred. The boundary value of $\chi \equiv \cos\theta = m/r + c/r^3 + \cdots$ corresponds to the mass of the lightest open string. If $T \ll m$ the solution should not depend on the temperature. This will be the case for the closed curve. If r_{min} is considerably larger than the horizon size, its dynamics will be essentially unaffected. On the other hand, for $T \gg m$ the mass deformation should be irrelevant. One is essentially back at the massless solution for which we know that the black-brane solution with $\theta = $ constant is the right configuration.

As in the Wilson-loop example, this change in configuration is associated with a (first-order) deconfinement phase transition. For $T \ll m$ we have a confined system of "mesons" formed by the open strings stretching back and forth between the probe brane and the large-N_c-branes. For $T \gg m$ these mesons can deconfine. They melt, and one has independent strings for each of the constituents.

There is yet a quite different kind of boundary physics that can be encoded using the probe branes. We can also consider a "defect" in the boundary theory, such as a $(0 + 1)$-dimensional impurity, or a $(2 + 1)$-dimensional domain wall embedded in a $(3 + 1)$-dimensional space-time. Such a defect theory can be modelled by a system where the D-branes do not lie parallel but instead intersect. After constructing the holographic dual pair from the low-energy/near-horizon limit, the intersections themselves will turn out to be dual to the defects in the boundary. In section 13.2.3 this will be illustrated in more detail for a holographic realisation of Kondo-like interacting-impurity physics. Let us directly specialise to the simple case of a point impurity – higher-dimensional defects can be constructed using the same basic logic. For the point defect, consider N_f D5-branes intersecting with the N_c D3-branes *at a point* in the sense that they share only the time direction. The intersection table then looks as follows:

	x^0	x^1	x^2	x^3	x^4	x^5	x^6	x^7	x^8	x^9
N_c D3	x	x	x	x						
N_f D5	x				x	x	x	x	x	

We have not quite addressed yet the question of what determines the choice of the dimensionality of the N_f Dq-branes. This depends on the precise low-energy dynamics in the field theory one wishes to model, together with the constraints from string theory. The most basic constraint is that any string-theory set-up is characterised by either odd-dimensional branes (type IIB string theory) or even-dimensional branes (type IIA string theory). Another common characteristic of the various string theories is that the number of "unshared" dimensions is a multiple of four. Notice that the number of unshared directions is not just $|q - p|$: for instance, in the D3/D7 set-up these correspond to the four directions x^4, x^5, x^6, x^7, while in the "Kondo" D3/D5 set-up one has the eight directions $x^1, x^2, x^3, x^4, x^5, x^6, x^7, x^8$. This requirement that the number of dimensions should differ by $4n$ originates from the fact that only when this condition is satisfied do the strings stretching between the two different branes carry massless degrees of freedom [171, 172]. Such configurations are often also supersymmetric, but as usual this can be ignored in practice, and one can focus on either the bosonic or the fermionic degrees of freedom. When these conditions are satisfied, one can go ahead with using the remaining degrees of freedom to "engineer" a particular probe-brane holographic duality with the aim of being able to study one's favourite boundary field theory. However, this involves a deeper insight into the workings of the string-theoretical constructions that is beyond the scope of this chapter, and we refer the reader to the literature.

13.2.1 Holographic flavour transport

In the previous chapter, chapter 12, we saw that in a finite-density system with translational invariance there cannot be any dissipative current response at zero frequency. However, this rule can be circumvented in principle with probe-brane set-ups. The reason is that the momentum of the probe degrees of freedom which are charged under the N_f flavour symmetry can be transferred to the large bath of uncharged degrees of freedom associated with the N_c matrix-valued fields. Since we are not accounting for any backreaction of the probe branes, this momentum can be "lost", and this is a computationally efficient way to realise "heat-bath" dissipation [505]. The physics underlying this particular dissipation mechanism is obviously very different from the type of momentum relaxation we discussed in the previous chapter, and this is reflected in the behaviour of transport properties.

Let us consider the D3/D7 system described above as an explicit example. In order to compute the physical properties in the boundary, we first need to fill in some further details. Since the N_f D7-branes are probes, the background AdS geometry is fully determined by the dual of the field theory of the large number N_c of D3-branes. At finite temperature this becomes the standard Schwarzschild black hole in AdS$_5 \times S^5$,

$$ds^2 = \frac{r^2}{L^2}\left(-f(r)dt^2 + dx_1^2 + dx_2^2 + dx_3^2\right) + \frac{L^2}{r^2 f(r)}dr^2 + L^2 d\Omega_5^2, \quad (13.35)$$

with the emblackening factor

$$f(r) = 1 - \frac{r_0^4}{r^4}. \quad (13.36)$$

Here $d\Omega_5^2$ is the metric on the S^5. We can parametrise this metric as follows:

$$d\Omega_5^2 = d\theta^2 + \sin^2\theta \, d\Omega_3^2 + \cos^2\theta \, d\phi^2. \quad (13.37)$$

As in box 13.3, the D7-brane will wrap the three-sphere $d\Omega_3^2 = dx_1^2 + \sin^2\chi_1(dx_2^2 + \sin^2\chi_2 \, dx_3^2)$, and it will be located at a point in the remaining two directions (θ, ϕ). The precise point depends on where one is along the radial direction, which is in turn determined by extremising the DBI action. On choosing the physical embedding where seven of the local coordinates on the D7-brane worldvolume are identified with the space-time coordinates,

$$(\xi^0, \xi^1, \xi^2, \xi^3) = (x^0, x^1, x^2, x^3), \quad \xi^4 = r, \quad (\xi^5, \xi^6, \xi^7) = (\chi_1, \chi_2, \chi_3), \quad (13.38)$$

the symmetries restrict the remaining coordinates to depending only on r. Moreover, we can always set $\partial_r\phi = 0$ by symmetry, and the sole dynamical field that determines the embedding is $\theta(r)$. From our qualitative description we conclude that this field should be dual to the lowest gauge-invariant operator of a string stretching (back and forth) between the probe brane and the D3-branes of the type $\text{Tr}(\phi_A^\dagger \phi^B)$, where the trace is over the gauged N_c indices and the flavour indices are in the range $A, B = 1, \ldots, N_f$, depending on which D7-brane the string ends on.

This coordinate choice plus the freedom to set $\partial_r \phi = 0$ can now be substituted into the DBI action to obtain the dynamics in the flavour sector. In the probe-brane approximation this is *all* the dynamics that we will solve. The AdS geometry stays constant throughout – this is the dual version of the fact that the neutral bath in the boundary is not affected by the dynamics of the charged "microscopic" sub-sector. Accordingly, one does not have to solve Einstein's equations, and this is a significant technical simplification of the probe-brane set-ups. There is no free lunch, however, since the price one must pay is that one must now deal with the rather complicated dynamics associated with the non-linear DBI action. The good news is that this simplifies in many situations of practical interest.

We are primarily interested in solutions characterised by finite flavour density. Following the GKPW description, this corresponds to solutions to the DBI action that asymptotically behave as

$$A_0 = \mu - \frac{\rho}{r^3} + \cdots . \tag{13.39}$$

By symmetry and gauge invariance we are therefore allowed only to consider the dynamics of A_0. On substituting this into the DBI action together with the coordinate choice Eq. (13.38), we obtain [507]

$$S_{\text{DBI}} = -N_f T_7 \int d^8\xi \sqrt{-\det\left(g_{\mu\nu}^{\text{D7}} + 2\pi\alpha' F_{\mu\nu}\right)}$$

$$= -\mathcal{N} V_{\mathbb{R}^{1,3}} \int dr \, r^3 \sin^3\theta \sqrt{1 + r^2 f(r)(\partial_r\theta)^2 - (\partial_r \mathcal{A}_0)^2}, \tag{13.40}$$

with $\mathcal{N} = N_f T_7 \, \text{Vol}_{S^3}$, and we have redefined the electrostatic potential as $\mathcal{A}_0 = (2\pi\alpha') A_0$. Since the action depends only on $\partial_r \mathcal{A}_0$, there is a constant of the motion

$$\Pi = \mathcal{N} \frac{r^3 \sin^3\theta \, \partial_r \mathcal{A}_0}{\sqrt{1 + r^2 f(r)(\partial_r\theta)^2 - (\partial_r \mathcal{A}_0)^2}}, \tag{13.41}$$

in terms of which we can solve for $\partial_r \mathcal{A}_0$:

$$\partial_r \mathcal{A}_0 = \Pi \sqrt{\frac{1 + r^2 f(r)(\partial_r\theta)^2}{\mathcal{N}^2 r^6 \sin^6\theta + \Pi^2}} . \tag{13.42}$$

In the special case in which the D7-brane wraps the equator at $\theta = \pi/2$, Eq. (13.42) is a solution of the full equations of motion by symmetry and the embedding $\theta(r)$ does not depend on the radial direction. In this case we immediately see that

$$\partial_r \mathcal{A}_0 = \frac{\Pi}{\sqrt{\mathcal{N}^2 r^6 + \Pi^2}}, \tag{13.43}$$

and thus asymptotically

$$\frac{\mathcal{A}_0}{2\pi\alpha'} = \mu - \frac{\Pi}{2(2\pi\alpha')\mathcal{N}}\frac{1}{r^2} + \cdots . \tag{13.44}$$

Including the normalisation, this allows us to identify the charge density as $\langle\rho\rangle = \Pi/(2\pi\alpha'\mathcal{N})$. The exact solution for $\mathcal{A}_0$ in this black-brane embedding with $\theta = \pi/2$ where the lowest excitation is massless can be found in terms of elliptic functions. Other black-brane solutions with $\partial_r\theta \neq 0$ and Minkowski embeddings in the presence of a source for $\theta(r)$ need to be numerically determined; we briefly discuss them in box 13.3.

Now we can study transport. Remarkably, we do not need to understand the precise solutions to extract the response of the probe system subjected to a constant driving force. To interrogate the transport properties of finite-density holographic matter, the approach so far has been to use the standard linear-response formalism associated with infinitesimal fluctuations around a fixed background. In the probe-brane construction it is, however, easy to directly address the way a finite current dissipates. In order to compute the DC conductivity, the full solution in a constant background electric field can be determined by considering the deformation $A_x = -Et + a_x(r)$. Upon adding A_x to the set of dynamical fields, the full DBI action becomes

$$S_{\text{DBI}} = -\mathcal{N}V_{\mathbb{R}^{1,3}}\int dr\, r^3 \sin^3\theta$$
$$\times \left[1 + \left(r^2 f(r) - \frac{E^2}{r^2}\right)(\partial_r\theta)^2\right.$$
$$\left. - \left((\partial_r\mathcal{A}_0)^2 - f(r)(\partial_r\mathcal{A}_x)^2 + \frac{E^2}{r^4 f(r)}\right)\right]^{1/2}, \tag{13.45}$$

where we set $L = 1$ and define $\mathcal{A}_x = 2\pi\alpha' a_x$. The constant of the motion Π for $\partial_r\mathcal{A}_0$ is appropriately modified and one now also has a constant of the motion Σ associated with $\partial_r\mathcal{A}_x$,

$$\Pi = r^3 \sin^3\theta\, \partial_r\mathcal{A}_0 \left[r^2\left[1 + \left(r^2 f(r) - \frac{E^2}{r^2}\right)(\partial_r\theta)^2\right]\right.$$
$$\left. - \left(r^2(\partial_r\mathcal{A}_0)^2 - r^2 f(r)(\partial_r\mathcal{A}_x)^2 + \frac{E^2}{r^2 f(r)}\right)\right]^{-1/2}, \tag{13.46}$$
$$\Sigma = r^3 \sin^3\theta f(r)\partial_r\mathcal{A}_x \left[r^2\left[1 + \left(r^2 f(r) - \frac{E^2}{r^2}\right)(\partial_r\theta)^2\right]\right.$$
$$\left. - \left(r^2(\partial_r\mathcal{A}_0)^2 - r^2 f(r)(\partial_r\mathcal{A}_x)^2 + \frac{E^2}{r^2 f(r)}\right)\right]^{-1/2}.$$

The coupled system of $\partial_r A_0$ and $\partial_r A_x$ can be solved in terms of Π and Σ. One has

$$r^2(\partial_r A_0)^2 = \Pi^2 \frac{f(r)r^4\left(1 + (r^2 f(r) - E^2/r^2)(\partial_r \theta)^2\right) - E^2}{f(r)r^6 \sin^6\theta + f(r)r^2\Pi^2 - r^2\Sigma^2},$$

$$f(r)r^2(\partial_r A_x)^2 = \frac{\Sigma^2}{f(r)} \frac{f(r)r^4\left(1 + (r^2 f(r) - E^2/r^2)(\partial_r \theta)^2\right) - E^2}{f(r)r^6 \sin^6\theta + f(r)r^2\Pi^2 - r^2\Sigma^2}.$$

(13.47)

This immediately shows that, near the boundary (for configuration $\theta = \pi/2$),

$$\frac{A_x}{2\pi\alpha'} = -Et - \frac{\Sigma}{2r^2} + \cdots.$$

(13.48)

It follows that the conductivity can be directly computed simply by dividing the second constant of motion by the electric field: $\sigma_{DC} \sim -\Sigma/E$. What remains is to determine Σ in terms of the applied electric field E and the charge density $\rho \sim \Pi$. One can obtain this from solving the equations of motion, but there is actually a shortcut. Notice that the left-hand side of Eq. (13.47) is positive or zero. The numerator on the right-hand side clearly is not. It switches sign at some particular radius $r_* > r_0$. The system can therefore be consistent only if the denominator switches sign at exactly the same radius r_* [505]. On substituting the solution for r_* from the numerator into the denominator, we find that

$$\sigma_{DC} = \sqrt{\frac{N_f^2 N_c^2 T^2}{16\pi^2} \sqrt{e^2 + 1} \cos^6\theta(r_*) + \frac{d^2}{e^2 + 1}},$$

(13.49)

where $e = 2E/(\pi\alpha' T^2)$, $d = \Pi/(\pi\alpha' T^2)$ and $r_*^2 = (\sqrt{e^2 + 1} - e)^{-1} r_0^2$. It is instructive to rewrite this as

$$\sigma_{DC} = \left[\left(\frac{N_f^2 N_c^2 T^2}{16\pi^2} \cos^6\theta(r_*) + d^2\right) \right.$$

$$\left. + \left[\frac{N_f^2 N_c^2 T^2}{16\pi^2}(\sqrt{e^2 + 1} - 1)\cos^6\theta(r_*) - \frac{d^2 e^2}{e^2 + 1}\right]\right]^{1/2}.$$

(13.50)

One now sees that the total conductivity is the root mean square of an intrinsic electric-field-independent contribution that corresponds to the linear response and a non-linear enhancement [508].

Differently from the DC conductivity, it turns out that the full non-linear dependence of the *optical* conductivity cannot be readily obtained from the probe-brane system. One has again to resort to the linear-response limit. This follows exactly the same steps as in chapters 7 and 8, with the simplification that in the probe approximation there is no longer mixing with a graviton mode. This is of course related to the fact that the flavour currents of the subsystem are no longer governed by momentum conservation, given that this relaxes away in the bath. This is confirmed by numerical solutions indicating that the optical conductivity shows a Drude behaviour with a finite momentum relaxation time [508].

After turning on a background magnetic field, one can study the Hall conductivity, magneto-transport and magnetic phase transition for the probe-brane systems [509, 510]. Moreover, in [508] the above method is used to study the conductivity for probe branes in a Lifshitz background, and a similar behaviour to that for the strange metal was obtained.

13.2.2 Holographic superconductivity with flavour

Let us now illustrate the second advantage of probe-brane constructions: the explicit form of the dual field theory is easy to construct and we can therefore try to understand the physics which is deduced holographically in terms of the manifest degrees of freedom of the field theory. In section 10.2.3 we briefly discussed a bottom-up construction for the holographic p-wave superconductor, where an $SU(2)$ vector field that is charged under its $U(1)$ subgroup was found to condense. The probe-brane system that realises literally this kind of p-wave superconductor is the one we just discussed: the N_c-D3/N_f-D7 system with the D7-branes parallel to the D3-branes. We now choose $N_f = 2$ [370, 511], such that the flavour symmetry is the desired $U(2)$. Let us now demonstrate that from this construction we can directly infer that the standard pairing mechanism is at work in this holographic superconductor.

We have already learned how to construct the gravity side of this probe-brane duality. On the field-theory side, we know that we have a $U(N_c)$ gauge group from the D3-branes and that there are N_f objects in the vector representation of N_c. In fact, we know from chapter 4 that the field theory corresponding to the D3-branes is $\mathcal{N} = 4$ supersymmetric Yang–Mills: the field content is a vector A_μ, four Majorana fermions ψ_A and six scalars $\Phi_{AB} = -\Phi_{BA}$, all in the adjoint representation of $U(N_c)$. We now have to add the sector coming from the strings stretching between the D3- and the D7-branes. From the string-theory origin one can deduce that in this configuration the D7-branes break half of the supersymmetries. The field content is therefore arranged in multiplets of $\mathcal{N} = 2$ $(3 + 1)$-dimensional supersymmetry. There are two possible multiplets. The one we need here does *not* contain a vector, since there are no additional gauge degrees of freedom. This is called the hypermultiplet, and it consists of a single Dirac fermion χ_i and two complex scalars ζ_{1i}, ζ_{2i}. All of the fields in each hypermultiplet are charged under the $U(N_c)$ gauge group, and this uniquely determines all the interactions in the theory. Furthermore, the two hypermultiplets rotate into each other under the global flavour symmetry $U(2) = U(1)_B \otimes SU(2)_I$. The gauge field on the probe brane in the bulk describes the currents associated with this symmetry. We shall focus solely on the $SU(2)$ part. Their microscopic expression follows immediately from the field theory: $J^a_\mu = i\bar\chi\sigma^a\gamma_\mu\chi + i\bar\zeta\sigma^a \overset{\leftrightarrow}{\partial}_\mu\zeta$, with σ^a the Pauli matrices.

Let us now focus on the dual probe action to find out whether it describes the superconductivity in the boundary. This should happen at temperatures that are low compared with the chemical potential in the $U(1)$ subgroup of $SU(2)$. It is straightforward to see that the probe-brane DBI analysis will be almost identical to our discussion of the holographic superconductor in chapter 10. For small gradients, the DBI action reduces to the bottom-up p-wave holographic superconductor action Eq. (10.23) plus the dynamics associated with the embedding function $\theta(\rho)$. This field does couple non-minimally to the gauge field A_μ. However, the constant choice $\theta = \pi/2$ remains a solution, and in this case the lowest-order dynamics is identical to that of the bottom-up holographic superconductor [370, 511]. Thus we see that at finite chemical potential μ, which induces a charge density J^3_0, the probe model yields a second-order holographic mean-field phase transition at low $T/\mu \ll 1$ to a spontaneously symmetry broken state with an expectation value for J^1_z.

Having identified the microscopic nature of $J^1_z = i\bar\psi\sigma^1\gamma_z\psi + \cdots$, it becomes clear that the superconducting transition is at its heart a standard BCS pair condensation. The order parameter J^1_z is the supersymmetric version of a fermion-pair operator. The holographic system is of course richer than a weakly coupled BCS superconductor insofar as the microscopic field theory has many more degrees of freedom and is very strongly coupled. This richness is directly reflected in the phase diagram (Fig. 13.5). For low masses of the stretched strings, the system has a BCS-like phase transition from a "deconfined" phase where the gauge charge is fractionalised directly into the superconductor. At high mass for the lowest flavour excitation, on the other hand, the fermions first pair up without condensing: the system first experiences a confinement transition to a state with gauge-singlet pairs. Only later at lower temperatures do the paired fermions condense into an ordered state.

13.2.3 Holographic impurity models

Let us end this chapter with two examples of the use of brane intersections to describe the physics of "defects". Using the brane intersections introduced at the beginning of this section one can "engineer" situations in the boundary where the matter in the fundamental representation is confined to live on a point, a line, a plane, etc., embedded in a higher-dimensional space that supports the gauge fields in the adjoint representation. The simplest situation one can imagine is to confine the matter in the fundamental representation in a spatial point, which interacts

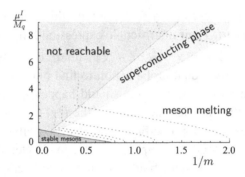

Figure 13.5 The phase diagram for a probe-brane p-wave superconductor. Here M_q labels the temperature rescaled by the coupling of the D3-brane gauge theory $M_q = \frac{1}{2}\sqrt{4\pi g_{YM}^2 N_c T}$. One can reach the superconducting phase from a deconfined phase of "melting mesons". This is similar to conventional BCS theory. Or one can reach the superconducting phase from a "stable-meson" phase where the microscopic degrees of freedom first pair up, and only condense at a lower temperature. The region labelled "not reachable" is the region where the probe approximation breaks down. Figure source [511]. (© IOP Publishing. Reproduced by permission of IOP Publishing. All rights reserved.)

with a CFT living in $3 + 1$ dimensions. This has some similarity to the strongly interacting impurity problems we discussed in chapter 3: the Anderson and Kondo impurity problems. These are icons of 1970s condensed matter physics; from a modern perspective this is the simplest context where the non-trivial effects of "Mottness" are at work. The starting point is a non-interacting Fermi sea, where one inserts a single site where the Hubbard U is exerting its influence (the Anderson impurity). When U is large, while the occupancy of the interacting site is close to unity, the strongly interacting local fermion turns into a spin, subject to an antiferromagnetic ("Kondo") exchange interaction with the spins of the fermions in the Fermi gas. One of the highlights of the 1970s was the discovery that this simple problem exhibits a very similar renormalisation flow to that in QCD, from an "asymptotically free" impurity spin in the UV to a strongly coupled "confined" Kondo singlet in the deep IR.

The classic problem of an $SU(2)$ $s = 1/2$ spin in a Fermi gas can be bosonised, and it was exactly solved a long time ago. However, one can seek to generalise this impurity physics to systems where the "host" is not a simple Fermi gas but instead a strongly interacting field-theoretical medium. In addition, one can change the symmetries controlling the impurity itself. A traditional alteration in condensed matter physics is to enlarge the symmetry controlling the spin from $SU(2)$ to $SU(N)$, while keeping the total spin S small. This of course refers to a *vector* theory that becomes free in the large-N limit. Another natural modification is to introduce multiple channels or "flavours" that are in the literal condensed matter context associated with atomic angular-momentum degeneracy. One arrives in this way at a model of the form [512, 513]

$$H_K = \sum_{\vec{k},i,\alpha} \epsilon(\vec{k}) \psi^\dagger_{\vec{k}i\alpha} \psi_{\vec{k}i\alpha} + \lambda_K \sum_{k,k',i,\alpha,\beta} S_{\alpha\beta} \psi^\dagger_{\vec{k}i\alpha} \psi_{\vec{k}'i\beta}, \tag{13.51}$$

where $\psi^\dagger_{\vec{k}i\alpha}$, $\psi_{\vec{k}i\alpha}$ are the creation and annihilation operator for an electron with momentum $\vec{k}$, i labels the channel $i = 1, \ldots, K$ and α is the spin of $SU(N)$. As usual, $\epsilon(k) = k^2/(2m) - \epsilon_F$ is the dispersion relation of a finite-density Fermi gas. $S_{\alpha\beta}$ is the impurity spin of the defect located at the origin; it transforms in a representation R of $SU(N)$ [514, 515]. As we will see below, the multichannel symmetry and the $SU(N_f)$ have a natural incarnation in the holographic set-up as global flavour symmetries. However, in order to identify a classical bulk one requires in addition a matrix large-N limit, and in this regard the holographic impurity models are very different from the condensed matter versions. Two such holographic impurity models that are based on brane intersections have been constructed: a direct supersymmetric top-down version [512, 516] and a single-flavour version focussing on an effective reduction of the theory to the radial dynamics around the defect [513].

The starting point of the supersymmetric construction [512, 516] is to couple the usual $(3 + 1)$-dimensional $\mathcal{N} = 4$ supersymmetric $SU(N_c)$ gauge theory to N_f $(0 + 1)$-dimensional fermions χ^I_α, where $I = 1, \ldots, N_f$ and $\alpha = 1, \ldots, N$, with total fermion number k (the "defect" density) such that $\sum_{\alpha=1}^N \bar{\chi}^I_\alpha \chi^\alpha_I = k$. This gives the action

$$S = S_{\mathcal{N}=4} + \int dt\, i \left[\bar{\chi}^I_\alpha \partial_t \chi^\alpha_I + \bar{\chi}^I (T_A) \left(A^A_0(t, 0) + n^a \phi^A_a(t, 0) \right) \chi_I \right]$$
$$+ \int dt\, B^J_I \left(\bar{\chi}^I_\alpha \chi^\alpha_J - k\delta^I_J \right), \tag{13.52}$$

where T_A are the generators in the adjoint of $SU(N)$ and n_a is a unit vector in R^6. B^J_I is a Lagrange multiplier that sets the fermion number. The defect part will break half of the $\mathcal{N} = 4$ supersymmetry. In this holographic set-up the electrons are the matrix-valued fermions of the D3-branes. The impurity spin S_{ij} arises from the VEV of the defect fermion bilinear $(\bar{\chi}^I_i \chi_{Ij} - k\delta_{ij})$. This shows that the symmetries hard-wired into this construction can be related to the symmetries of the K-multichannel $SU(N)$ Kondo model by taking the channel degeneracy K equal to N, where representation R of the impurity spin is the anti-symmetric representation of $SU(N)$ with k indices. Compared with the original model, this impurity spin now has an additional microscopic N_f degeneracy.

This field theory turns out to arise exactly from a bulk dual describing N D3-branes intersecting N_f D5-branes at a point. The defect in the kth anti-symmetric representation can be modelled by the introduction of a Wilson loop [512, 517]; we saw in chapter 6 that the dual description of the latter is an explicit fundamental string. The brane configuration is therefore the following:

	x^0	x^1	x^2	x^3	x^4	x^5	x^6	x^7	x^8	x^9
N_c D3	x	x	x	x						
N_f D5	x				x	x	x	x	x	
k F1	x								x	

with F1 denoting the fundamental strings. From the open-string side, the k fundamental strings between the D3- and D5-branes lead to an additional k fermions localised on the D3/D5 intersection, and the effective action is precisely coincident with the impurity part in Eq. (13.52).

On the gravity side, we now need to understand the embedding of the probe D5-branes and how to account quantitatively for the fundamental strings. The D3-branes are again considered from the closed-string perspective and these give rise to an $AdS_5 \times S^5$. The D5-branes will have the time direction in common with the D3-branes plus potentially the radial AdS direction. These will also span an S^4 in the S^5, similarly to the D7-branes in the D3/D7-brane construction before. To make this manifest, the $AdS_5 \times S^5$ metric can be written as

$$ds^2 = L^2\left(du^2 + \cosh^2 u \, ds^2_{AdS_2} + \sinh^2 u \, d\Omega_2^2\right)$$
$$+ L^2\left(d\theta^2 + \sin^2\theta \, d\Omega_4^2\right). \tag{13.53}$$

The D5 probe worldvolume spans the $AdS_2 \times S^4$. The contribution from the k fundamental strings can be accounted for by insisting that the gauge field on the brane carries k units of flux. In particular, this means that the Chern–Simons term Eq. (13.29) now contributes and can no longer be ignored. Its effect is that the constant embedding condition is shifted and the D5-branes are located at

$$u = 0, \quad \theta = \theta_k \tag{13.54}$$

with $k = (N_c/\pi)\left(\theta_k - \frac{1}{2}\sin(2\theta_k)\right)$.

The further analysis follows the usual playbook. For instance, an interesting quantity when one is dealing with impurity problems is the impurity entropy or "g-function", which is defined as

$$\ln g = S_{imp} = \lim_{T\to 0}\lim_{V\to\infty}\left[S(T) - S_{host}(T)\right]. \tag{13.55}$$

S_{host} is the entropy associated with the higher-dimensional fields which form the "host" for the impurity, and this can be easily identified since it scales with the volume. The impurity entropy can be computed on the gravity side, with the result

$$S_{\text{imp}} = \sqrt{\lambda} \frac{\sin^3 \theta_k}{3\pi} N_f N_c. \tag{13.56}$$

Very differently from the condensed matter impurity models, the impurity entropy now scales with the 't Hooft coupling λ, highlighting the fundamental difference coming from the fact that the "host" is not a free-fermion system but instead a strongly interacting CFT.

A different holographic approach to the Kondo model was pursued in [513]. This departs from a simplification that is very specific to a free-Fermi-gas host system. The interacting impurity is assumed to communicate with its host solely in the s-wave channel, with the consequence that the higher-dimensional Fermi gas can be replaced by a $(1+1)$-dimensional system on the half line. This was first realised by Wilson, and was the key to his Nobel-prize-winning numerical real-space renormalisation-group solution of the Kondo problem. Since $(1 + 1)$-dimensional systems can be bosonised, this is also the key ingredient in all later developments, including the CFT treatment [515]. This "dimensional-reduction" procedure is unique to the Fermi gas, and it surely fails when one is dealing with strongly interacting host systems. The holographic construction that now follows should therefore be interpreted in a literal fashion, as describing a zero-space-dimensional impurity in a strongly interacting one-dimensional host.

The task is to construct a brane-intersection set-up that both describes the $(1 + 1)$-dimensional host as a defect and describes the impurity as a $(0 + 1)$-dimensional "defect on the defect". This can be accomplished by the following set-up [513]:

	x^0	x^1	x^2	x^3	x^4	x^5	x^6	x^7	x^8	x^9
N_c D3	x	x	x	x						
N_5 D7	x	x			x	x	x	x	x	x
N_7 D5	x				x	x	x	x	x	

The degrees of freedom from the large number N of D3-branes dualise into the usual $(3+1)$-dimensional strongly interacting CFT which represents a background mediating the interactions between the defect degrees of freedom. The strings stretching between the D3- and the D7-brane sectors intersect in a joint $(1 + 1)$-dimensional subspace that represents the host system in the boundary: the massless degrees of freedom on this intersection form the chiral fermions of the $(1 + 1)$-dimensional CFT. The lowest degrees of freedom of the D3/D5 strings in turn describe the impurity, just as in the supersymmetric model discussed above. The "Kondo" interaction is directly mediated by the D5/D7 strings. For the impurity with $k = 1$, the coupling between the impurity fermions χ and the chiral fermions ψ of the $(1 + 1)$-dimensional CFT can be rewritten in the (matrix) large-N_c limit as

$$S_{\text{int}} = -\lambda_K \int dt \, \bar{\psi}^\alpha \chi_\alpha \bar{\chi}^\beta \psi_\beta = -\lambda_K \int dt \, \text{Tr}(\bar{\psi}\chi)\text{Tr}(\bar{\chi}\psi). \qquad (13.57)$$

The bulk field dual to the operator $\mathcal{O} = \text{Tr}(\bar{\psi}\chi)$ turns out to be a complex scalar associated with the D5/D7 strings. The incarnation of the "Kondo singlet" in the present context is the pairing of ψ with χ, and we observe that this now corresponds to the condensation of the operator $\mathcal{O}$. The "Kondo" coupling term Eq. (13.57) just corresponds to a double-trace deformation as discussed in chapter 5, while we learned in chapter 10 on holographic superconductivity that such double-trace transformations can drive the system through a thermal transition where $\mathcal{O}$ acquires a VEV.

The bottom line is that the "Kondo singlet" will be formed at a mean-field transition. This can be viewed as the *matrix* large-N_c generalisation of the standard "slave-boson" mean-field treatments of condensed matter physics which were briefly discussed in chapter 3. These rely on a similar condensation occurring in the *vector* large-N limit of the impurity problem. As in the latter case, one foresees that already the leading-order $1/N_c$ corrections will change the phase transition into a smooth crossover, respecting the fact that phase transitions are not possible in $0 + 1$ dimensions.

13.2.4 Further top-down models

There are many more examples of top-down models in holography. Without prejudice we mention here models of the quantum Hall effect [518, 519, 520]; here the magnetic field is encoded in the flux through a probe-brane surface. Systems with a lattice of defects and dimerisation are studied in [516, 521], and can be considered as a holographic avatar of the Hubbard model. We have already seen that the zero-sound characteristic of a Fermi liquid has a natural extension in holographic models, and top-down constructions that study this specifically are [522, 523, 524]. Mixed probe-brane/bottom-up models have been able to exhibit the formation of Abrikosov vortex lattices in magnetic fields [525]. A particularly peculiar property of probe-brane models is that they can escape the large-N mean-field transitions. In

[526] it was shown that a specific D3/D5 top-down model exhibits instead a quantum BKT transition as a function of density/unit magnetic flux. Finally we should mention that the classification of ground states in intersecting D-brane models shares many mathematical aspects with the classification of topological insulators [527, 528]. This has been used in particular to establish families of strongly coupled fractional topological insulators holographically [529].

14

Outlook: holography and quantum matter

We are nearing the end of this journey through the landscape of holographic matter. It is a colourful place where there is much to be seen. Since the first direct application to condensed matter in 2007 [24], the pace of exploration of this landscape has been remarkable. The rich sceneries we have described in this book have all been discovered in the last seven years. But what does it all mean? In truth, this is still a mystery at present.

One aspect is crystal clear. A whole new side of Einstein gravity has been discovered, by focussing on what the correspondence has to tell us about matter. Inspired by common-sense condensed matter questions on the boundary, the holographists had to look for unusual solutions of the Einstein equations in the bulk. In this search, they discovered whole new categories of unexpected gravitational universes. The most prominent of these is the "hairy black hole" describing holographic superconductivity: it flagrantly violates the no-hair theorem. Much more followed, up to the very recent realisation that translational-symmetry breaking in the boundary goes hand in hand with massive gravity in the bulk. These discoveries may only be the tip of the iceberg. They nearly all concern equilibrium matter dual to stationary solutions of the gravity in the bulk. Currently, specifically in the context of AdS holography, there has been astonishing progress in numerical time-dependent solutions in gravity. This also brings the non-equilibrium physics in the boundary into reach. The spectacular results for the duals of classical and superfluid turbulence in section 10.2.1 which are deduced from black holes with *fractal* horizons demonstrate the unexpected richness that exploration in this direction may reveal. AdS/CMT has, at the very least, given us tremendous new insights into the physics of black holes. Clearly, deeply buried in the equations of general relativity, there is still an enormous potential wealth for holography to uncover.

14.1 The UHOs: the unidentified holographic objects

In spite of its motivational role, the mystery is the condensed matter side of AdS/CMT. Do the remarkable findings by holography actually have anything to say about the reality seen by experimentalists in condensed matter laboratories? The honest answer so far is "no", even though the similarities to strange metals and quark–gluon plasma are striking. With tongue in cheek, holographic matter is sometimes referred to as "avatars": incarnations of an idea. The expression "avatar" comes from Hinduism, where it refers to the bodily manifestation of a deity on earth. Does holography signal such a divine message regarding the nature of matter; and what idea in condensed matter physics does it represent?

14.1.1 Capping off geometry: confinement and the return of normalcy

What gives one confidence that holography does have something meaningful to say about physical reality, is the capacity of AdS/CFT to describe all familiar forms of matter in a natural gravitational dual. We have highlighted how thermo-dynamical principles are holographically encoded in the universality of black-hole physics. This extends to small departures from equilibrium. The correspondence between Navier–Stokes fluids in the boundary and AdS gravity is a stunning result. The *structure* of the hydrodynamical theory is hard-wired into the bulk gravita-tional dynamics; the UV information of the specific model enters only through the free parameters of the Navier–Stokes theory. As we stressed, even these num-bers are UV-sensitive only to a degree. Given the conformal invariance of the zero-temperature quantum theory, Planckian dissipation is hard-wired and the hydrodynamics has to be nearly "perfect", with very small transport coefficients. For the super-conformal large-N Yang–Mills theory or other theories with an AdS gravity dual, one finds a ratio of the viscosity to the entropy density of $1/(4\pi)$. This will be parametrically different for the simple global $U(1)$ theory describ-ing the superfluid Bose Mott-insulator quantum critical state. But this difference between numbers of order one should certainly not invalidate the holographic point of view.

The confidence-building continues with the "cohesive" zero-temperature states. If one is faced with any ambiguity in holography, the first instruction is to "cap off" the deep interior either with a hard or soft wall, or with an AdS soliton. At finite charge density, one then finds solutions that describe impeccably the familiar states of matter as discussed in chapters 2 and 3 on condensed matter: the "Higgs scalar hair" encoding for the normal superconductor, or the "bulk Fermi gas" for the Fermi liquid. Although the programme has not quite been completed yet, the other symmetry-breaking product vacua (crystals, magnets) will in all likelihood be described in this way. All these amount to a solid sanity check on the proper operation of the holographic duality. Capped-off geometries correspond to the confining regime of the Yang–Mills theory in the boundary. In the large-N limit confined gauge singlets constitute a weakly interacting cohesive system at finite density. In this regime AdS/CFT correctly maps the weakly coupled gravitational physics to the emergent weakly coupled collective dynamics of the conventional states of zero-temperature matter.

14.1.2 Holographic matter and the emergence of quantum critical phases

The mystery is that all of the other results of holography at zero temperature and finite density presented in the preceding chapters cannot be identified within the established framework of condensed matter physics and/or quantum field theory. This is of course also its irresistible attraction. Can we, however, be sure that we are not being misled? How can it be that we not have seen any of these states before?

There could be a very good explanation in an overarching theme shared by all these surprising outcomes. This theme starts with the observation that many of these surprising outcomes have an emergent scaling symmetry in the IR. Scale invariance is an extremely powerful symmetry, especially in quantum physics. It is by itself already rather miraculous that within conventional bosonic field theory such scale-invariant quantum critical states are realised. In detail they are the familiar thermal critical states with real Minkowski time replacing periodic Euclidean time. Their existence therefore relies eventually on the profundity of Onsager's proof that a critical state is present at the second-order phase transition of the two-dimensional Ising model.

Now, though breaking a global symmetry is easy, generating one dynamically at long wavelength by emergence is very hard. It is also written in stone that such thermal critical states, and thereby the bosonic quantum critical states, require an infinitely fine tuning to be realised. They occur only at isolated points in coupling-constant space in space-time dimensions larger than two. These isolated unstable fixed points are ingrained in the intuition of condensed matter physicists as the "quantum critical points" at the origin of the quantum critical wedges in the coupling-constant–temperature diagrams discussed at length in section 2.1.3. There are just two exceptions to this iron rule. We have already mentioned $(1+1)$-dimensional systems where emergent scale invariance is ubiquitous for all theories controlled by continuous internal symmetry. These are the Luttinger liquids. The other exception is supersymmetric field theories in higher dimensions. Owing to non-renormalisation theorems implied by the exact cancellation between fermion and boson loops, it is straightforward to engineer super-conformal supersymmetric theories that exhibit "quantum criticality" regardless of the specific values of the UV coupling constant, see e.g. [530]. These special scale-invariant supersymmetric theories of course lie at the origin of the conformal nature of the top-down field theories described by AdS/CFT at zero density. It cannot be stressed enough, however, that scale invariance is a very unnatural circumstance without supersymmetry.

Even though AdS/CFT has precisely such an exceptional super-conformal theory at its origin, we need not heed this stringent warning for the purposes of AdS/CMT. Switching on a finite chemical potential manifestly destroys the primordial conformal invariance and breaks the supersymmetry badly as well. These rarefied properties no longer matter, if we take this as a starting point to generate new emergent theories in the IR. Instead of the Schrödinger equations for electrons and atoms in real matter as our departure point, we take this special theory. But this choice of starting point should not matter, due to the universality of the renormalisation group. Yet the staggeringly novel surprise is that the generic holographic IR is not a textbook ground state.

From the gravitational side it is easy to understand why. In the gravitational dual it is natural to anticipate that some form of space-time will survive in the deep interior. There is no obvious reason why a generic deformation ought to result in a capping off of the geometry in the interior. On the contrary, in many cases the geometry exhibits scaling. Not only does a large sector of the boundary field theory survive in the IR, but also it is some form of scale-invariant zero-temperature matter. The central finding of holography for finite-density matter is that *"strongly coupled" matter readily forms a quantum critical phase.* In total contradistinction to our field-theory knowledge, this form of quantum criticality does not appear to require strong fine tuning. It just appears.

This emergent scaling is the first pointer towards the shared theme. The systematic map of these deep-interior scaling geometries charted in chapter 8 reveals yet another great surprise. The emergent dynamical critical exponent z and hyperscaling-violation exponent θ can be a priori changed "at will". Depending on the UV and IR details, these exponents can take any value. This is distinctly unconventional physics. In a Wilsonian RNG setting the values these exponents can acquire are strongly constrained. As we discussed in the context of the Hertz–Millis theories in Chapter 2, a dynamical critical exponent $z = 1$ is natural, as expected for "isolated" order parameters generating emergent Lorentz invariance – Bose–Hubbard-type phase dynamics, antiferromagnetic order in insulators, etc. A diffusive exponent $z = 2$ is even more natural, since it describes the relaxational dynamics of non-conserved order parameters in the presence of a "heat bath". Similarly, $z = 3$ can occur when the order parameter is conserved as in a ferromagnet, but at least in continuum field theories, different values of z require very unnatural circumstances. According to holography, however, z can vary all over the place, even taking the "quasi-local quantum critical" value of $z \to \infty$.

A non-vanishing hyperscaling-violation exponent θ is even more alarming viewed from a Wilsonian standpoint. In a classical bosonic theory hyperscaling violation requires that the system is classically frustrated in one way or another, and classical frustration is a very fragile circumstance that needs very delicate fine tuning. Its intrinsically high entropy is easily destabilised to a more regular configuration.

There is a case where hyperscaling is not confounding. As we emphasised in our discussion in chapter 8, the only "cohesive", "quantum-orderly" state which naturally exhibits a hyperscaling violation $\theta = d - 1$ is the Fermi liquid. On purpose we included the non-standard discussion of the Fermi liquid in section 2.3, highlighting that this "un-classical" physics of the Fermi liquid is rooted in its long-range entanglement as a consequence of Pauli's anti-symmetrisation demand. This, together with the unnatural need for frustration to explain hyperscaling violation in classical matter, unveils the shared theme which might explain these whole new families of unfamiliar emergent states. These strange scaling properties might signal that holography is describing *long-range entangled compressible quantum matter.*

14.1.3 The RN metal: from fractionalisation in holography to entanglement

Let us analyse this in the simplest of all gravitational duals for finite-density matter. This is at the same time the most extreme example of an unidentified holographic object (UHO): the Reissner–Nordström "AdS_2" metal we discussed in chapter 8. Because of the peculiarities of the extremal black-hole near-horizon geometry with a finite horizon area at zero temperature, one finds that the emergent IR is a state with dynamical critical exponent "$z = \infty$" and hyperscaling violation $\theta = d$, equal to the dimensionality of space itself. The hyperscaling violation is directly correlated with the eye-opening extensive ground-state entropy of the state. By analogy with the classically frustrated systems, it is widely believed that this state is for this reason unphysical as a true zero-temperature state. The ground-state entropy implies that it is "infinitely unstable" with respect to anything. In the context of the "intermediate unstable fixed points" viewpoint of Anderson and of Liu, Iqbal and Mezei, this ground-state entropy is cherished as the secret of high-T_c superconductivity (section 3.6.2, chapters 10 and 11). As we discussed in chapter 3, in order to explain why the transition temperature is so high in cuprate superconductors, one has to explain why the normal metallic state is actually so unstable.

But does the RN black hole really describe a "frustrated metal"? The one inroad we have to understand this anomalous ground-state entropy is that it is proportional to N^2: somehow one is supposed to count the deconfined or fractionalised degrees of freedom. In the top-down models we know that this amounts to counting the gluon degrees of freedom of a non-Abelian Yang–Mills theory. It is, however, not entirely clear how one should think about these degrees of freedom in a condensed matter context. The terminology "fractionalised" was adapted to indicate this counting in the context of finite-density holography, in analogy to the phenomenon of fractionalisation in condensed matter physics which we sketched in section 2.4. This association is, however, more intuitive than precise in any sense. The deconfining "fractionalised" vacua of condensed matter physics are *by construction* in a regime where the "partons" are weakly coupled. The "holons" typically form a Bose condensate, while the spinons are thought to form a free Fermi gas, weakly coupled to the gauge bosons present in the deconfining regime. It is an inconvenience of holographic duality that one can study only the propagators of gauge singlets, and these reveal in principle only indirect information regarding the precise nature of the parton degrees of freedom. However, it is a clearly a fallacy to assume that the holographic deconfining degrees of freedom have anything to do with simple "particle" physics. Given the minimal viscosity, there is no doubt that there are no weakly coupled "partons". Rather, at finite temperature there are "un-particles" subject to the principle of the strongly interacting quantum critical state. For the zero-temperature Reissner–Nordström metal, as for many other UHOs, these are governed by the critical states dual to the AdS_2 or other scaling geometries.

But, the question remains, has this form of "fractionalisation" any relationship with anything ever discussed in condensed matter? There is the distinct worry that there might be a nasty form of "UV dependence" that cannot be circumvented. The large-N Yang–Mills theories suffer from an "embarrassment of riches" in the form of an incredibly large number of degrees of freedom. Perhaps this goes hand in hand with a very special "fractionalisation" phenomenon, which is rather alien to anything that can ever occur in electron systems. The strongest reason to believe that this is not so is a real experimental observation. Many of the characteristics of the deconfined "un-particle-like" phase are seen experimentally in the attempt to create a quark–gluon plasma through relativistic heavy-ion collisions. QCD is a non-Abelian Yang–Mills theory, but with $N = 3$, and even though QCD does behave in many ways like a large-N theory, as discussed at length in the companion book [12], it has been clear from the data that it is the strong coupling and (classical) scale invariance of QCD which are the best insights for a qualitative understanding of the data.

The main piece of experimental information which is clarifying in this regard is the hydrodynamical response. Holography clearly shows how the minimal viscosity associated with the Schwarzschild black hole is a striking example of a UV-insensitive quantity – modulo numbers of order unity, any quantum critical

UV is equally fine. This should put most of our worries to rest at finite temperature. Our aim here, on the other hand, is to understand the zero-temperature RN metal and its ground-state entropy. Can this be revealed by a "zero-temperature hydro-dynamics"? Even though the direct fluid–gravity correspondence in terms of a gradient expansion discussed in chapter 7 appears to fail for the extremal-Reissner–Nordström black hole, a linear response does turn out to work. The result for the viscosity is striking [332]. One finds the same $\eta = (1/(4\pi))s_0$, where now s_0 is the zero-temperature entropy. This is quite an enlightening result. Physically it means that the perturbation which is local in space – the sound mode – actually communicates with the whole of the degenerate ground-state manifold. This is very different from a classical frustrated system. Systems of the latter type are invariably highly non-ergodic. Degenerate classical ground states are locally "disconnected" in the sense that, to get from one to the other, very non-local changes are required, and a local perturbation can therefore "see" only a vanishingly small fraction of all ground states. A local disturbance in the Reissner–Nordström metal, however, appears to see all the ground states. It could be that the mean-field nature of the large-N limit somehow restores ergodicity. But a far more interesting possibility is that it is rooted in an inherent quantum property of this novel quantum matter. In principle it is imaginable that, analogously to the Fermi liquid, the degenerate state is *long-range entangled*, and in such a vacuum everything talks to everything via EPR "spooky action at a distance".

14.1.4 Holographic symmetry breaking is not short-range entangled product vacuum

From a theoretical perspective the Reisner–Nordström metal has a special status, plainly because it is the most extreme form of strange holographic matter. There are many more holographic strange metals characterised by a deep interior without a finite-area horizon but still showing an unconventional scaling geometry. Particularly "reasonable" ways of obtaining such strange metals are associated with the formation of cohesive states: in chapters 10 and 11 we described how, upon forming the "Higgs" or "electron" stars in the bulk, the deep-interior geometry reconstructs due to the backreaction with the bulk matter. As emphasised above, the surprise is that these geometries are generically not confining; they are again scaling geometries with non-trivial dynamical critical and hyperscaling-violation exponents.

This is not supposed to happen. As explained in chapters 2 and 3, "normal" spontaneous symmetry breaking goes hand in hand with a short-range entangled product-state vacuum, and such a vacuum only supports Goldstone bosons as massless excitations. A way to argue why holography does not follow these iron rules is by invoking the large number of degrees of freedom of the underlying matrix large-N super-Yang–Mills theory. Conceivably only some of them are charged under the holographic order parameter. These gap out, and the uncharged degrees of freedom then remain as a deep-infrared quantum critical phase. This separation into a "charged and an uncharged sector" at the scale of the order parameter is inconsistent, however. They cannot even be separated by an infinitesimal gap of order $1/N$, because upon removal of the symmetry breaking the emergent critical state collapses completely. Its criticality is protected by the existence of the order parameter.

In this regard, the holographic symmetry-breaking states are strikingly different from the standard Hartree–Fock variety forming the bread and butter of condensed matter physics. It is perhaps a useful metaphor to view them as similar to Hartree–Fock states formed in the Fermi gas, which only partially gap the Fermi surface. Unconventional superconductors can easily have point or line nodes, while density waves typically leave behind Fermi-surface pockets with gapless quasiparticles. As for the Fermi gas itself, this is a signature of its "baby entanglement". Certainly, we shall need a more dramatic form of long-range entanglement to understand the massless deep-infrared quantum critical phases which emerge due to the presence of the holographic order parameters.

14.2 Is holographic matter extreme quantum matter?

Though we cannot understand the predictions of holography in terms of existing condensed matter principles, it is beyond doubt that holography describes the formation of genuine matter of some kind. There is nothing deeply mysterious about holographic substance at zero density: we understand the rules there well. We run into trouble as soon as a finite density is turned on. Why can't we recognise the resulting states in well-known field-theory language? From what we know about quantum many-body systems, the difference has to be in the "sign structure": as soon as that charge conjugation is broken, there is no longer any reason why a quantum state with fermions should be described by a Boltzmannian statistical physics in disguise. As we emphasised in chapter 3, there is simply no formalism available that is able to handle such sign-full matter in full generality.

The physics of sign-full matter *is* long-range field-theoretical entanglement. This is the quantum matter theme introduced in chapter 2. "Fermion" signs are just a very effective means to go beyond short-range-entangled product states. The point in case is the Fermi liquid. It changes the rules radically compared with the bosonic states. We do not know any sign-full states beyond the Fermi liquid, but arguably different forms of long-range-entangled fermionic states should exist, for instance when the way to the Fermi gas is blocked by ingredients such as the Mottness of chapter 3.

We can now argue like a lawyer: since we understand all short-range-entangled product states, and since we do not understand anything about the UHOs, it necessarily has to be the case that the latter are long-range entangled. This is especially so since these are insistently forming quantum critical phases. Long-range entanglement does not directly imply quantum criticality, but we will put forward soon some very general arguments that a quantum critical phase does go hand in hand with long-range entanglement.

The following question immediately arises: what is the unique distinguishing feature of long-range entanglement. How do we classify its various forms? What plays the role of the conventional Ginzburg–Landau–Wilson order parameter for this "quantum order", and how does this relate to the physical properties of the system? In chapters 2 and 3 we discussed instances that can act as proofs of principle. The incompressible topological states of the fractional quantum Hall liquid as enumerated by Chern–Simons topological field theory are one example. A second is the spin liquids (realised in e.g. the BCS superconductor) controlled by the topology of the deconfining state of the emergent gauge fields. The great challenge is to extend this to the *compressible* states. As we argued, the only example here that is understood is the Fermi liquid, which is entirely controlled by "anti-symmetrisation entanglement". In all other regards, nothing is known. This is perhaps more than anything else the frontier in fundamental condensed matter physics.

14.2.1 The von Neumann and Rényi entropies

The first requirement is therefore to identify "observables" that directly measure the "quantum order" in the compressible systems. Ideally, the next step would be to identify the holographic dictionary entry for such observables, which in turn should yield unambiguous information revealing whether, and in what way, the UHOs relate to quantum matter.

Much of the recent activity aiming at "entanglement observables" has been the pursuit of the von Neumann and Rényi entanglement entropies. These are well known from the field-theory perspective, and Ryu and Takayanagi showed that the von Neumann entropy is encoded in a specific minimal-surface construction in the gravitational dual. This is the first instance where a property that is probing aspects of entanglement could be formulated in terms of a holographic dictionary entry. After much initial excitement, however, it is becoming increasingly clear that these quantum entropies have also severe limitations. They actually do not reveal much regarding the "classification" question raised in the previous paragraphs. The issue is that we do not in fact know the proper observables for many-body entanglement. The von Neumann entropy and the closely related Rényi entropy capture only bipartite entanglement, whereas true multi-partite entanglement is much richer. Nevertheless, they have been the primary source of inspiration to discern the distinguishing characteristics of quantum matter both from field theory and from holography. We have reserved the Ryu–Takayanagi construction as the very last piece of holographic technology to be discussed in this book, a status it deserves given its remarkable elegance. Before we delve into the specifics, let us first sketch some of the general background.

In a way the von Neumann bipartite entanglement entropy heralded the birth of quantum information. The reader should be aware that much of quantum information theory has evolved driven by the desire to build and use a quantum computer. This is, at the end of the day, a challenging engineering effort, and engineers have to rely on building blocks that are well understood. These building blocks are *quantum-mechanical*: single-bit "Schrödinger-cat" states and two-bit Bell pairs taking care of the entanglement. There is, however, a big lapse from few-body quantum mechanics to the infinite degrees of freedom in quantum field theory. There are well-known examples indicating that quantum-mechanical entanglement wisdoms do not reveal anything regarding the "field-theoretical" entanglement physics of quantum matter. We will see that the limitations of the entanglement entropies are rooted precisely in this difficulty.

In the case of two bits the von Neumann entropy does encode the entanglement with merciless precision. Starting from the full density matrix of the two-bit system,

$$\rho = \sum_{ij} a_{ij} |i\rangle_A |j\rangle_B \langle i|_A \langle j|_B, \tag{14.1}$$

where i, j take the values 0, 1, one takes the partial trace over bit B to obtain the reduced density matrix $\rho_A = \text{Tr}_B \rho$, to subsequently compute the entropy associated with the subsystem,

$$S_{vN} = -\text{Tr}_A \left[\rho_A \ln \rho_A \right]. \tag{14.2}$$

For a product state, such as $|0\rangle_A |1\rangle_B$, one finds that the von Neumann entropy vanishes, $S_{vN} = 0$, while for a maximally entangled Bell pair, such as $(|0\rangle_A |1\rangle_B + |1\rangle_A |0\rangle_B) / \sqrt{2}$, the entropy is maximally large, $S_{vN} = \ln 2$. This von Neumann entropy measures uniquely entanglement, as exemplified by its property that it is representation-independent. For instance, a state like $(|0\rangle_A |0\rangle_B + |0\rangle_A |1\rangle_B + |1\rangle_A |0\rangle_B + |1\rangle_A |1\rangle_B) / 2$ might appear to be entangled. However, it can also be written as a product $|+\rangle_A |+\rangle_B$ after a "single-bit" unitary transformation such that $|+\rangle = (|0\rangle + |1\rangle) / \sqrt{2}$. The von Neumann entropy does indeed vanish for this state.

Closely related are the nth Rényi entropies defined as

$$S_{\text{Rényi}}(n) = -\frac{1}{n-1} \ln \text{Tr}_A \rho_A^n. \tag{14.3}$$

Formally, for a normalised total density matrix $\text{Tr}\,\rho = 1$, the von Neumann entropy is obtained from the limit

$$S_{\text{vN}} = \lim_{n \to 1} S_{\text{Rényi}}(n). \tag{14.4}$$

For n bits, there are in principle n independent Rényi entropies. These therefore contain much more information than does the single von Neumann entropy. The maximal amount of bipartite information is of course contained in the reduced density matrix itself. In full generality it can be written as

$$\rho_A = e^{-\mathcal{H}_A} \tag{14.5}$$

as if it were a thermal ensemble at effective inverse temperature $\beta_A = 1$, described by an effective "entanglement Hamiltonian" $\mathcal{H}_A$ with eigenvalues describing the "entanglement spectrum" [531]. This spectrum is the maximal amount of data one can mine from the reduced density matrix. This is eventually the information one has to inspect in order to judge the significance of the entropies obtained by taking traces over this spectrum. It can be demonstrated that the entanglement spectrum can be reconstructed from the knowledge of the Rényi entropies for all n.

It is also known, however, that for $n > 2$ bits, the bipartite von Neumann and Rényi entropies fail to enumerate the entanglement of the original state. The clearest example of this is the Greenberger–Horne–Zeilinger state [532]

$$|\text{GHZ}\rangle = \frac{1}{\sqrt{2}}\left(|111\rangle + |000\rangle\right). \tag{14.6}$$

A trace over any of the three single-bit subsystems yields an unentangled mixed state, even though the state is in some sense maximally entangled as a three-bit state [533]. In quantum field theory the number of bits n is infinite, and arguably bipartite entropies are intrinsically incomplete. But, for lack of a better measure, this is what we are stuck with for the time being. The field-theoretical extension of quantum entropies was pioneered by Srednicki [534] and Holzhey, Larsen and Wilczek [535]. Their intuitive constructions are based on the bipartite division. EPR entanglement "paradoxes" are usually discussed in real space, and therefore it should somehow be appropriate to define the two subsystems required for the quantum-mechanical entropies as two complementary regions A and B obtained by splitting space into two. By tracing out the degrees of freedom of B and computing the entropies from the reduced density matrix defined on A, one acquires information regarding how the degrees of freedom in A and B are non-locally entangled.

Holzhey *et al.* [535] zoomed in on $(1+1)$-dimensional CFTs living in $\mathbb{R}^2$, where the bipartite von Neumann entropy can be explicitly computed due to the integrable nature of the CFT. By tracing out a spatial region $(-l/2, l/2)$, they obtained the famous result [535]

$$S_{\text{vN 1+1 CFT}} = \frac{c}{3} \ln \left(\frac{l}{\epsilon} \right), \tag{14.7}$$

where c is the central charge of the CFT and ϵ is the UV cut-off. The logarithmic dependence is suggestive of a long-range entanglement, with its magnitude universally determined by the central charge, which is an important property defining the CFT. A caveat is that the actual computation is fundamentally quite similar to the classic way of measuring the central charge, which was formulated a long time ago by Cardy and Bloete [536]: interpret the CFT as a thermal state in two spatial dimensions, then the central charge determines the correction to the free energy under finite-size scaling.

These field-theoretical space-bipartite entropies have proven remarkably powerful in enumerating the topological order of incompressible states, despite their fundamental shortcomings. Considering a gapped system, the leading-order contribution to the entanglement entropy will always scale with the area of the entanglement surface $\Sigma = l^{d-1}$ defined as the manifold defining the boundary between the areas $\mathcal{A}$ and $\mathcal{B}$, with linear dimension l. The reason is trivially that the degrees of freedom near this surface will always be short-range entangled. Computing up to the next order, in the specific contexts of $d = 2$-dimensional fractional quantum Hall states, it was found [537, 538] that

$$S_{\text{vN 2+1 FQHS}} = \alpha \frac{l}{\epsilon} - \gamma + \cdots. \tag{14.8}$$

Here ϵ is the UV cut-off and α is a non-universal constant. It is the second term which is of interest. This universal term is a UV-insensitive constant $\gamma = \ln \mathcal{D}$, which is directly correlated to the "quantum dimension" $\mathcal{D}$. The quantum dimension measures the number of quasiparticles associated with the particular quantum Hall state, and also sets the ground-state degeneracy as a function of the number of handles of the closed target manifold. Li and Haldane explained this remarkable connection between the entanglement entropy and the topological properties of the ground state [531] by showing that the universal low-energy end of the entanglement spectrum is precisely coincident with the spectrum at the physical edge! Given the bulk–edge correspondence of topological states discussed in section 2.2, the entanglement Hamiltonian precisely enumerates the information characterising the bulk. The elegant aspect is that the entanglement entropy allows one to extract this information from the bulk dynamics, without any need to introduce explicit boundaries. These two examples display a generic wisdom. The entanglement entropy in general follows an *area law*: the leading contribution to the entropy is UV-divergent and scales as the area of the bipartite entanglement region [534, 539]. The universal non-trivial information resides in the sub-leading terms, however. The claim [540, 541] is that, for a region $\mathcal{A}$ in the vacuum state, in all odd numbers of *spatial* dimensions $d = 2n + 1$ the entropy scales as

$$S_{\text{vN}} = \alpha \frac{l^{d-1}}{\epsilon^{d-1}} + \cdots + (-1)^{(d-1)/2} s_d \ln(l/\epsilon) + \cdots . \tag{14.9}$$

The logarithmic term has a similar status to that in $(1 + 1)$-dimensional CFTs. One can think of the UV-insensitive coefficient s_d as the generalisation of the conformal charge. In even spatial dimensions $d = 2n$, the entanglement entropy has the form

$$S_{\text{vN}} = \alpha \frac{l^{d-1}}{\epsilon^{d-1}} + \cdots + (-1)^{(d-1)/2} f_d + \cdots , \tag{14.10}$$

with now f_d the universal UV-insensitive component. It is thought that these general forms continue to be true for Rényi entropy and compressible systems. Using the Ryu–Takayanagi prescription for the entanglement entropy, which we will explain presently, this can indeed be verified in many examples in holography at *zero* density. In these scenarios, one moreover finds the very interesting result that in many cases the null-energy theorem in the bulk implies that both s_d and f_d are *monotonic* functions under RG flows (i.e. we have a generalised c-theorem) [542]. They therefore play the role of higher-dimensional generalisations of the central charge of $(1 + 1)$-dimensional CFTs. Such monotonic c-theorems provided very useful and strong checks on physical regimes where conventional methods do not apply [543]. We should note, however, that a universal exact c-theorem for the coefficients f_d and s_d does not yet exist, e.g. they do depend on the shape of the surface, while not depending on its size [541].

These wisdoms are blindly taken to hold in compressible systems at finite density as well. There is, however, one example in this class for which we can compute the entanglement entropy explicitly, and it does not obey these rules. This is the Fermi gas. Its leading-order contribution violates the generic area law and the Fermi gas thus reveals an "entanglement" that is longer-ranged than in any bosonic theory [544],

$$S_{\text{vN}} = c k_{\text{F}}^{d-1} A_\Sigma \ln(k_{\text{F}}^{d-1} A_\Sigma) + \cdots . \tag{14.11}$$

Here Σ is the boundary of the entanglement region $\mathcal{A}$, c is a constant and k_{F} is the Fermi momentum. We put quotation marks around the word entanglement on purpose. We know that the entanglement in this case directly results from Pauli exclusion – there is no relation of any kind to Einstein's spooky action at a distance dealing with two bits. In fact, as we will show directly below, analogous holographic calculations show that the scaling of the leading-order contribution is completely determined by the unfamiliar but very interesting scaling property of the quantum matter as a whole. It simply measures the hyperscaling-violation exponent θ. For the Fermi liquid $\theta = d - 1$ and one obtains the extra logarithm, but more exuberant behaviours are a priori possible.

The example of the Fermi liquid makes abundantly clear that in a compressible system the spatial bipartite von Neumann entropy does not necessarily have a relation with entanglement in the quantum-mechanical sense. As discussed in the above, it picks up bits and pieces of information that do relate to quantum matter properties in a direct (quantum dimension) or rather indirect fashion (e.g. hyperscaling violation). To call it "entanglement" entropy is therefore a bit of a misnomer. In fact, a much more severe critique is possible. In quantum critical systems, the entanglement entropy nevertheless contains a scale. In the context of holography, but also in a direct field-theoretic approach, one can demonstrate that the entanglement entropy has a distinct "phase transition" as a function of the size of the entanglement region [545]. This is clearly a serious flaw in the bipartite entanglement approach, which one should keep very much in mind.

14.2.2 *Holographic entanglement entropy*

Other than the Fermi liquid, the only compressible states at finite density which we know are holographic constructions. Thanks to a beautiful insight by Ryu and Takayanagi, we know how to compute the bipartite entanglement entropy in these cases [546]. Their argument is based on two observations. (1) In the special case that the entanglement region is the full space, the entanglement entropy should become the actual entropy. (2) If the system is in a thermal state, the actual entropy follows holographically from a quarter of the area A_{hor} of the black-hole horizon in units of Newton's constant

$$S_{\text{BH}} = \frac{A_{\text{hor}}}{4G_{\text{N}}}. \tag{14.12}$$

To capture the entanglement entropy holographically, the bulk needs to know about the entanglement regions $\mathcal{A}$ and $\mathcal{B}$. The natural way to arrange this is to consider a surface $\mathscr{A}$ in the bulk that ends precisely on the boundary $\partial\mathcal{A}$ of the entanglement region: the two boundaries $\partial\mathcal{A}$ and $\partial\mathscr{A}$ should coincide. This visually translates how the degrees of freedom at distances shorter than the "width" of $\mathcal{A}$ are encoded in the bulk. They are contained within the region between the surface $\mathscr{A}$ and the boundary (Fig. 14.1). Thus $\mathscr{A}$ is the boundary of some region $\Sigma_{\mathscr{A}}$ that can be thought of as a surface that delineates the physics of the bulk degrees of freedom within $\Sigma_{\mathscr{A}}$ from the remainder of AdS. Note that this is not a bulk entanglement

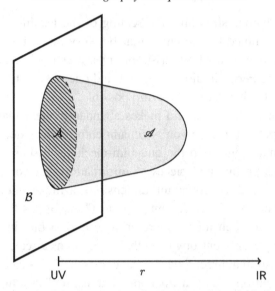

UV r IR

Figure 14.1 A sketch of the Ryu–Takayanagi proposal for the holographic entanglement entropy. Given the entanglement region $\mathcal{A}$ on the AdS boundary, one considers a surface $\mathscr{A}$ in the bulk that shares its boundary $\partial\mathscr{A}$ with the boundary $\partial\mathcal{A}$ of the entanglement surface. To agree with the thermal Bekenstein–Hawking entropy, it is natural to propose that the area of $\mathscr{A}$ is related to a boundary entropy $S_\mathcal{S} = A_\mathscr{A}/(4G_N)$. To determine the exact surface $\mathscr{A}$ Ryu and Takayanagi proposed that it should be the surface of minimal area.

surface – in the bulk everything is classical. The area of this dividing delineation surface $A_\mathscr{A}$ surface scales in the same way as the horizon encoding the black-hole entropy. Since, at finite temperature, the entanglement entropy in the limit at which $\mathcal{A}$ becomes the full space should become the thermal entropy, i.e. in the black-hole background the black-hole horizon should be the dividing surface, see Fig. 14.2, we are naturally led to the Ryu–Takayanagi conjecture that at leading order

$$S_\mathcal{A} = \frac{A_\mathscr{A}}{4G_N} + \cdots. \tag{14.13}$$

It remains to determine the precise shape of the surface $\mathscr{A}$. Since the full entropy also follows from the free energy as the extremum of the action, Ryu and Takayanagi naturally proposed that the correct surface $\mathscr{A}$ from which to extract the equilibrium entanglement entropy is an extremal surface obtained by minimising the embedding action

$$S = \int d^d\xi \sqrt{\det\left(g_{ij}\,\partial_{\xi^a}X^i(\xi)\partial_{\xi^b}X^j(\xi)\right)}. \tag{14.14}$$

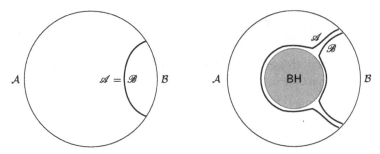

Figure 14.2 Left: for pure AdS space-time, the total system $\mathcal{A} \cup \mathcal{B}$ is in a pure state. In the bulk the minimal surface bounding $\mathcal{A}$ equals the minimal surface bounding $\mathcal{B}$, which means that we have $S_{\mathcal{A}} = S_{\mathcal{B}}$ for the dual system. Right: if the total system is in a mixed (thermal) state, the bounding surfaces of $\mathcal{A}$ and $\mathcal{B}$ need not be equal, hence $S_{\mathcal{A}} \neq S_{\mathcal{B}}$.

Here i, j run over spatial coordinates only. In equilibrium the spatial directions are clearly defined with respect to the rest frame of the matter. The value of the area is then the value of this action on the extremal solution.

This prescription exactly reproduces the Holzhey–Larsen–Wilczek entanglement entropy of $(1 + 1)$-dimensional CFTs, including the factor of 1/3 in front of the central charge (box 14.1). It passes many other non-trivial tests [547, 548]: the area law of the entanglement entropy follows straightforwardly from holography; the strong subadditivity of the entanglement entropy can be checked as area inequality in the bulk etc. Within theories with a two-derivative Einsteinian gravity dual the Ryu–Takayanagi prescription can in fact be proven to correctly enumerate the bipartite entanglement entropy [549, 550, 551]. Moreover, the sub-leading corrections can be computed [551]. The generalisation of the Ryu–Takayanagi formula to extensions of Einsteinian gravity beyond two derivatives was considered in [552, 553, 554, 555].

Box 14.1   Holographic entanglement entropy for CFTs at zero temperature and density

Here we show how the Ryu–Takayanagi proposal for the calculation of the holographic entanglement entropy correctly computes the known entanglement entropy in $(1 + 1)$-dimensional CFTs [535]

$$S = \frac{c}{3} \ln \left(\frac{l}{\epsilon} \right), \qquad (14.15)$$

Box 14.1 (Continued)

where c is the central charge for the CFT and ϵ is the UV cut-off. The gravity dual to a $(1+1)$-dimensional CFT is a $(2+1)$-dimensional AdS space. In Poincaré coordinates its metric equals,

$$ds^2 = L^2\left(\frac{dr^2}{r^2} + r^2(-dt^2 + dx^2)\right). \tag{14.16}$$

Since the UV cut-off explicitly appears in the answer, we should be careful and regulate the theory. Following the details of the GKPW prescription in chapter 5, this is done by considering the boundary to be at $r = \epsilon^{-1}$ and then carefully taking $\epsilon \to 0$ at the end.

Following the original Holzhey–Larsen–Wilczek set-up, we consider the entanglement entropy between region A defined by $-l/2 < x < l/2$ and the remaining space. The Ryu–Takayanagi prescription is then to consider a d-dimensional surface in the bulk – in this case where $d = 1$ a curve – which ends at $(r, x) = (1/\epsilon, -l/2)$ and $(r, x) = (1/\epsilon, l/2)$. The correct surface from which to extract the entanglement entropy is the extremal surface between these two points obtained by minimising its spatial area,

$$S = \int d\xi \sqrt{g_{ij}\, \partial_\xi X^i \, \partial_\xi X^j}. \tag{14.17}$$

In this $d = 1$-dimensional case the surface is simply a curve and the extremal curve is a geodesic.

Since we are interested in the equilibrium entanglement entropy, this geodesic does not depend on time and thus $\partial_t X^\mu = 0$. Furthermore, by symmetry the geodesic will extend from $(r, x) = (1/\epsilon, -l/2)$ into the bulk up to a minimal value $r_{\min}$ obtained at $x = 0$ and then return identically to $(r, x) = (1/\epsilon, l/2)$. We can therefore choose ξ to parametrise the distance in the r direction $\partial_\xi X^r = 1$. By then substituting the metric into the action one obtains

$$S = 2L \int_{r_c}^{1/\epsilon} d\xi \sqrt{\frac{1}{r^2} + r^2(x')^2}, \tag{14.18}$$

where $x' = \partial_r x$.

Such an action is very familiar from the Wilson loops we discussed in chapter 6. Since the dynamical variable appears only with a derivative, there is a constant of the motion,

$$\Pi = \frac{2Lr^2 x'}{\sqrt{1/r^2 + r^2(x')^2}}. \tag{14.19}$$

At the turning point r_c, $dr/dx = 0$ or $x' \to \infty$, and on taking this limit one immediately sees that $\Pi = 2Lr_c$. Therefore,

$$r^2 x' = r_c \sqrt{\frac{1}{r^2} + r^2 (x')^2} \qquad (14.20)$$

and hence

$$x' = \pm \frac{r_c}{r^2 \sqrt{r^2 - r_c^2}}. \qquad (14.21)$$

To fix r_c in terms of the entanglement region's size l, we use the fact that we also know that at the turning point the geodesic has traversed a distance $l/2$ from $x = -l/2$ to $x = 0$. In other words,

$$l/2 = \int_{-l/2}^{0} dx = \int_{1/\epsilon}^{r_c} dr \, x'. \qquad (14.22)$$

Using the relation Eq. (14.20), one thus obtains

$$l/2 = -\int_{r_c}^{\infty} dr \, \frac{r_c}{r^2 \sqrt{r^2 - r_c^2}} = \frac{1}{r_c}. \qquad (14.23)$$

The area of the geodesic now follows from substituting the solution Eq. (14.19) into the action. This gives

$$A = 2L \int_{r_c}^{\epsilon^{-1}} \frac{dr}{\sqrt{r^2 - r_c^2}} = 2L \ln\left(\frac{l}{\epsilon}\right). \qquad (14.24)$$

The holographic entanglement entropy then equals

$$S_{\text{EE}} = \frac{A}{4G_N} = \frac{L}{2G_N} \ln\left(\frac{l}{\epsilon}\right) = \frac{c}{3} \ln\left(\frac{l}{\epsilon}\right), \qquad (14.25)$$

where we have used the fact that the central charge for Einstein–Hilbert gravity in AdS$_3$ is $c = 3L/(2G)$ [556, 557]. The holographic entanglement entropy thus matches exactly the field-theory result Eq. (14.15). One can repeat the above procedure at finite temperature. The one subtlety is that in $(2 + 1)$-dimensional gravity there are no true black holes. Instead one has a space-time with a conical defect. This Banados–Teitelboim–Zanelli (BTZ) space-time with metric is

$$ds^2 = L^2 \left((r^2 - r_+^2) \, dt^2 + \frac{dr^2}{r^2 - r_+^2} + r^2 \, dx^2 \right). \qquad (14.26)$$

Box 14.1 (Continued)

It behaves in many respects just like a Schwarzschild black hole in higher dimensions. Computing the holographic entanglement entropy in this finite-temperature set-up again gives the same result as the $(1+1)$-dimensional CFT result [547]

$$S = \frac{c}{3} \ln \left(\frac{\beta}{\pi \epsilon} \sinh \left(\frac{\pi l}{\beta} \right) \right), \tag{14.27}$$

where $\beta = 1/T$ is the inverse of temperature.

These computations can easily be generalised to higher-dimensional $\text{AdS}_{d+2}/\text{CFT}_{d+1}$ cases. One finds that the entanglement entropy for a ball with radius l behaves as

$$S = \begin{cases} p_1 l^{d-1}/\epsilon^{d-1} + p_3 l^{d-3}/\epsilon^{d-3} + \cdots + p_{d-1}l/\epsilon + f_d + \mathcal{O}(\epsilon/l), \\ \qquad\qquad\qquad\qquad\qquad\qquad\qquad\qquad\qquad\qquad \text{even } d, \\ p_1 l^{d-1}/\epsilon^{d-1} + p_3 l^{d-3}/\epsilon^{d-3} + \cdots + p_{d-2}l^2/\epsilon^2 + s_d \ln(l/\epsilon) + \mathcal{O}(1), \\ \qquad\qquad\qquad\qquad\qquad\qquad\qquad\qquad\qquad\qquad\quad \text{odd } d, \end{cases} \tag{14.28}$$

where ϵ is the UV cut-off of the theory. The leading-order contribution is consistent with the area law. The most interesting terms are the sub-leading contributions f_d and s_d. Not only are they universal, but also they can be shown to obey a holographic c-theorem as long as the theory is unitary. This means that for any deformation of the UV CFT to a new fixed point in the IR $f_d^{UV} > f_d^{IR}$ and $s_d^{UV} > s_d^{IR}$. The coefficient s_d in d odd can, moreover, be identified with the coefficient of the Euler characteristic in the Weyl anomaly [558]. This is the failure of the theory to obey scaling at the quantum level. In even dimensions, no such scaling violation exists. Instead, it is thought that the object which s_d relates to in the field theory is the free energy of the $(d+1)$-dimensional CFT on a $(d+1)$-dimensional sphere [559, 560]. Note that this is not the free energy of the finite-temperature CFT; this would be the system on a d-dimensional sphere times Euclidean periodically identified time S^1.

14.2.3 Von Neumann entropy and strange metals

The Ryu–Takayanagi prescription for the holographic entanglement entropy now allows us to compute this for the strange metals which are the focus of our attention. Despite its shortcomings, can the von Neumann entanglement entropy nevertheless support our conjecture that these are forms of genuine, long-range-entangled quantum matter? The answer is perhaps predictable. We do not get what we expect, but instead we do learn something that is very useful: it probes the hyperscaling violation property, and is thereby an interesting diagnostic of the holographic strange metals.

On studying how the Ryu–Takayanagi minimal surface behaves, one quickly realises that for large entanglement regions the entanglement entropy is dominated by the contribution of the Ryu–Takayanagi surface near the horizon. Rather than meticulously tracing out the full computation, we can therefore simply zoom in on the near-horizon region. In the landscape of holographic scaling geometries described in section 8.4, the most general such region takes the form [334, 336, 344, 561]

$$ds^2 = r^{2\theta/d}\left(-\frac{dt^2}{r^{2z}} + \frac{dr^2 + dx_i^2}{r^2}\right).$$

(14.29)

As we discussed, under the transformation $t \to \lambda^z t$, $x_i \to \lambda x_i$, $r \to \lambda r$, the metric is not scale-invariant, but transforms as

$$ds \to \lambda^{\theta/d} ds,$$

(14.30)

with z and θ as the dynamical critical and hyperscaling-violation exponents, respectively. In particular, we learned that the thermal entropy density scales as

$$S \sim T^{(d-\theta)/z}.$$

(14.31)

As in the Fermi liquid, all these hyperscaling-violating geometries are supported by a charge density Q. What will be important for us is that we can think of this density as the UV cut-off for the near-horizon region. The IR hyperscaling-violating geometries of the holographic strange metals are thus the natural holographic generalisation of the Fermi liquid. In turn, the Fermi liquid is characterised from this perspective by free-fermion excitations that are gapless on a $(d-1)$-dimensional surface in momentum space (the Fermi surface), dispersing along the single dimension transverse to the surface with dynamic critical exponent $z = 1$. This implies a Sommerfeld entropy

$$S \sim T,$$

(14.32)

which is independent of d, since $\theta = d - 1$.

Now we can probe the entanglement entropy in these compressible finite-density systems. For general θ and z the entanglement entropy S_A of an arbitrary entanglement region A in theories dual to this class of bulk geometries scales as [336]

$$S_{\text{EE};A} \sim \begin{cases} Q^{(d-1)/d} A_A + \cdots, & \text{if } \theta < d-1, \\ Q^{(d-1)/d} A_A \ln(Q^{(d-1)/d} A_{\partial\Sigma}) + \cdots, & \text{if } \theta = d-1, \\ Q^{\theta/d} A_A^{\theta/(d-1)} + \cdots, & \text{if } \theta > d-1. \end{cases}$$

(14.33)

Through the holographic version of Luttinger's theorem, the charge density can be directly equated with a Fermi momentum $Q \sim k_F^d$. The hyperscaling-violating geometry with $\theta = d - 1$ thus reproduces the Fermi-liquid entanglement entropy exactly. The most interesting message obtained from studying the strange metals using the von Neumann entropy is that it reveals rather directly the hyperscaling-violation properties of their deep infrareds. Hyperscaling violation can be seen as the "distribution of masslessness" in momentum space [562]. As soon as this distribution becomes comparable to the area of the entanglement region, the entanglement entropy will violate the area law $S \sim A_A$. For ordinary bosonic field theories, there is only one massless point (at $k = 0$ when translational symmetry is not broken), and the area law is always obeyed. The only conventional state which naturally violates hyperscaling is the Fermi gas/Fermi liquid characterised by a $(d - 1)$-dimensional (Fermi) surface of masslessness. This is just comparable to the entanglement area, implying a minimal violation $S \sim A_A \ln A_A$. But in holography the hyperscaling-violating exponent can take any value up to $\theta = d$. For $\theta < d - 1$ the phase space of massless excitations is sufficiently small that the entanglement entropy behaves in the same way as for a simple bosonic CFT. The "Fermi-liquid-like" case $\theta = d - 1$ is just the "marginal" case with a logarithmic violation, while for $d - 1 < \theta < d$ a longer-ranged, algebraic behaviour is found.

Let us be very clear that the holographic state of matter with $\theta = d - 1$ shares the entanglement-entropy scaling with the Fermi liquid, but it is in no way the same. For **z** large enough, the state can be cohesive, and one can detect well-defined fermionic quasiparticles, but the thermodynamics will be controlled by the accompanying emergent finite-**z** strange metal which is stabilised by the cohesive fermions. One way to try to understand this strange metal is that it could be consistent with the formation of a Fermi liquid of deconfined partons. A crucial observation is that its conventional Fermi surface would be hidden from any measurement made by the (physical) UV gauge-singlet probes of holography. The reason is that the deconfined parton Fermi surface is not a gauge-invariant quantity. Therefore it can be detected only through thermodynamical measurements, or via measurements of the von Neumann entropy. The implicit assumption is, however, that the partons form a nearly free-fermionic "particle" system. Although this is the case in e.g. extremely high-density QCD, there is no a-priori reason for this to be the case in these strongly interacting, conformal states of holography, even when a more general notion of deconfinement applies. Surely, the deconfining states at zero density associated with the Schwarzschild black holes are not at all of this kind. Arguably, the Fermi-liquid-like hyperscaling violation can therefore have a very different "un-particle-physics" origin. This is quite evident from the observation that $\theta = d - 1$ is just a rather arbitrary point in a continuum of possible hyperscaling exponents. The novelty emphasised by Eq. (14.33) is that there are plenty of states with an even more drastic "momentum-space filling of masslessness" behaviour than that of the familiar Fermi gas.

The most extreme member of this family is $\theta = d$, where the von Neumann entropy becomes extensive and scales with the volume of the traced-out area. As we notably found out in chapter 8, this extensive entropy is familiar from the ground state of the Reissner–Nordström metal. The von Neumann entropy actually implies a fascinating consistency condition for this case, linking the semi-local quantum criticality to the ground-state entropy. First, the semi-local quantum criticality property implies that all propagators have a form $\sim (i\omega)^{2\nu_k - 1}$, while the exponents ν_k are finite for all k. Although the gauge singlets might not probe the real deep-IR excitations, total momentum conservation implies that there have to be massless excitations present *in all of momentum space*. This is in turn consistent with a "volume" hyperscaling violation $\theta = d$, and with the volume scaling of the von Neumann entropy. Correspondingly, when we let the traced-out volume $A_{\partial\Sigma}$ become as large as the system itself, the von Neumann entropy turns into the overall ground-state entropy. This is in fact how one deduces that the RN metal is the special case of a hyperscaling-violating geometry with $z \to \infty$ and $\theta = d$.

This is a fascinating but somewhat mysterious result. The ground-state entropy and the local quantum criticality appear to imply each other. Given that the semi-local quantum criticality appears to require some extreme form of quantum matter physics, it suggests that the ground-state degeneracy of the RN metal has a similar extreme quantum origin.

14.2.4 The quantum critical state and its entanglement

Having learned about the shortcomings of the von Neumann "entanglement" entropy with regard to measuring field-theoretical entanglement, let us return to the main storyline of this concluding chapter. As we emphasised at the beginning of this chapter, the insistent message of holography is that it predicts a ubiquity of states that do not belong to the established folklore of short-range-entangled product states. These states are quantum critical phases, albeit with scaling properties that are beyond Wilsonian RG. Could it be that compressible long-range-entangled quantum matter is automatically quantum critical?

There are quite simple but penetrating arguments that seem to leave no doubt that the converse is certainly true: even the simple bosonic quantum critical states should be irreducibly macroscopically entangled. Although part of the reasoning rests on the understanding from statistical physics, it appears that the essence of it applies equally well to the non-bosonic finite-density quantum critical phases. The argument runs as follows.

As we argued in chapter 2, stable states of bosons invariably form short-range-entangled product states, but now we should consider the system right at the quantum critical point. Heuristically, the defining property of the strongly interacting quantum critical state is that its quantum dynamics is governed by the powerful symmetry of scale invariance. At very small, microscopic scales near the UV cut-off (which is not far away from the lattice constant) one expects that right at the quantum phase transition the matter should be strongly entangled. Since the scale invariance rules absolutely, everything should be scale-invariant, including the entanglement. Henceforth, because of the scale invariance, the microscopic entanglement should be self-similar to the entanglement on the macroscopic scale, and therefore the macroscopic state has to be long-range entangled.

This is nearly a tautology and quite hand-waving. But one can argue it a bit more convincingly by undertaking a close inspection of the classical critical state realised in Euclidean signature. Although there are no closed solutions available for the partition sum of the classical strongly interacting critical state, the computer tells us what this looks like: such states involve large numbers of field configurations that all share the property of self-similarity. In summing over these configurations to obtain the exact partition function, it is well known that in order to avoid critical scaling down one has resort to fanciful non-local updates: the configurations spanning up the critical manifold are *globally* different. After time slicing and Wick rotation, this classical partition sum turns into a *coherent superposition* of all these configurations in real time. Given the non-local nature of its thermal incarnation, how can this state not be long-range entangled?

A last heuristic argument involves the nature of the excitations. Particle physics, in the sense of the Standard Model or referring to Goldstone bosons, is defined through sharp poles in spectral functions. This in turn implies that the vacuum has a short-range-entangled product structure, since particles imply an ensemble of harmonic oscillators and these form a product state. It is by now overly familiar that in the strongly interacting quantum critical state one finds instead the "branch-cuts" of "un-particle" physics. It appears to be impossible to construct a short-range-entangled product-state vacuum that can support the non-locality of the quantum number imposed by the scale invariance required for the branch-cut.

What does this entanglement motive mean for the physics of the quantum critical state? Could it be that the Planckian dissipation, the unique capacity of the state to convert work into heat at a record speed, is deeply tied to long-range entanglement? The difference with the incompressible entangled states is that there is now a dense manifold of low-energy states that are all long-range entangled. It is well understood that temperature is detrimental for entanglement, but could it be that it can also work in reverse gear, namely that, by injecting some energy into the zero-temperature state, the vast amount of entanglements could be responsible for turning it in the shortest possible time into heat?

This consideration mercilessly affirms the shortcomings of the bipartite von Neumann entropy. The above physics alludes to the strongly interacting quantum critical state. Above the upper critical dimension, on the other hand, the fixed point becomes Gaussian – dealing with the effective Lorentz-invariant quantum theory (e.g. the Bose-superfluid Mott insulator in $d + 1$ dimensions, where $d \geq 3$) the scaling dimension of the order-parameter operator hits the unitarity limit and the particle spectrum is recovered. This non-interacting critical state is short-range entangled, but its bipartite von Neumann entropy would still reveal an area law and a conformal charge, and would not discriminate it in any way from the strongly interacting entangled case.

In a way it appears obvious that the strongly interacting bosonic quantum critical state is subject to long-range entanglement. Nonetheless, the status of this claim is conjectural. It is at present impossible to arrive at more solid conclusions that are based on rigorous mathematical procedures. It does illustrate emphatically the central challenge in the pursuit of field-theoretical quantum information: there are as yet not general measures available to precisely enumerate the meaning of long-range entanglement in such seriously *quantum* field-theoretical systems.

14.2.5 Fermionic quantum criticality

There is one other aspect of holographic strange matter that we have repeatedly emphasised. The reason why holography is able to discover such new states of quantum matter is likely to be because it addresses strongly interacting fermions at a finite density. These cannot be addressed in conventional approaches, given the sign problem, as discussed extensively in chapter 2. However, with one's eyes focussed on quantum matter this nuisance changes to an asset. It is easy to understand that the "signs" are the weapons of choice to defeat classical matter. It might well be that the "sign problem" is to quite a large degree coincident with the present lack of understanding of the principles governing seriously long-range-entangled compressible matter.

Referring to the discussion in chapter 3 and the previous section, the attempts to understand quantum matter are biased towards bosonic matter, for the simple reason that fermionic matter is very poorly understood. The Fermi gas is the exception, but this is already the case in point: even the simple non-interacting limit is irreducibly long-range entangled because of the requirement of anti-symmetrisation. As suggested by the phenomenon of "Mottness" discussed in chapter 2, the structure of the entanglement associated with strongly interacting fermion systems is likely to be very different from that in the non-interacting case, but also under these circumstances the signs should form an insurmountable obstacle to the formation of classical matter.

There is yet another way to understand why signs are good for long-range entanglement. It is established wisdom that the ground states of bosonic theories are rather sparsely entangled, while the excited states are much more densely entangled. For instance, in critical theories one finds typically a volume scaling of the von Neumann entropy for excited states [563]. Although the ground state is characterised by positive-definite amplitudes, all excited states have signs in order to be orthogonal to the ground state and to each other, boosting the entanglement. For fermions, this "sign-driven" entanglement is just imposed on the vacuum state itself.

To take this line of questioning a step further, could it be that the remarkable "self-organised" quantum criticality of the holographic strange metals is a consequence of dense long-range entanglement, driven by the fermion signs? In fact, support for this assertion can be found in an explicit construction for a non-Fermi-liquid fermionic vacuum. Considering the Fermi liquid, its peculiarity is that it is characterised by a scale that is of an exclusively statistical origin: the Fermi energy. To obtain a scale-free state of fermion matter one must somehow tamper with Fermi statistics itself. It is obscure how to accomplish this in the canonical formulations. However, there is a little-known alternative formulation of the fermionic path integral where this becomes a more flexible affair [564]. This "constrained path integral" as discovered by Ceperley in the 1990s [565] is explained in some detail in box 14.2. The outcome is that the fermion path integral can be rewritten in a probabilistic (sign-free, bosonic) form as long as only world histories that do not change the sign of the full density matrix associated with these histories are admitted. This is in turn equivalent to viewing the zeros of the density matrix as corresponding to an infinitely strong steric potential. These zeros in turn form a "surface" in configuration space: the "nodal surface". In this way, the effects of fermion statistics are captured in a *geometrical* language.

For the present purposes, one can get away with the more familiar concept of the nodes of the wave function. Consider the first quantised wave function in real space $\Psi(\mathbf{R})$, where $\mathbf{R} = (\mathbf{r}_1, \ldots, \mathbf{r}_N)$ is the position in dN-dimensional configuration space (d is the spatial dimension and N is the number of particles). The nodal surface is defined by $\Psi(\mathbf{R}) = 0$, corresponding to a $(dN - 1)$-dimensional "surface" in configuration space. This can be visualised by taking cuts through configuration space, which can be obtained by fixing the positions of all particles except one randomly chosen particle and tracking the sign of the wave function when the chosen particle moves around. In the case of the simple Fermi gas this is simply computed from the Slater determinant, and one obtains a smooth surface as indicated in the upper left panel of Fig. 14.3. One immediately realises that this is a convenient way to think about the entanglement of the Fermi gas: determining

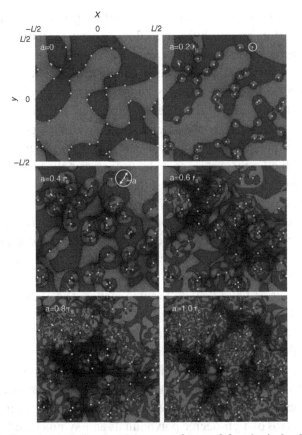

Figure 14.3 Cuts through the nodal hypersurfaces of fermionic backflow wave
functions in two dimensions for $N = 49$ particles, various values of the backflow
strength $\alpha = a/r_s$ and a small-distance cut-off $r_0/r_s = 0.1$. For $\alpha = 0$ we recover
the smooth nodal structure of free fermions. With increasing backflow strength,
additional clouds of nodal pockets start to develop. The linear dimension of these
clouds scales linearly with α. When the effective backflow range a becomes of
the order of the inter-particle spacing r_s, the nodal surface qualitatively changes
its geometry and seems to turn into a fractal. Figure source [564].

the zeros of the determinant corresponds to an extremely non-local problem. Upon
changing the position of one particle, the location of the nodal surface changes
even at an infinite distance away. In addition, the Fermi energy is directly encoded
in this nodal surface. The Fermi-gas nodal surface turns out to be characterised
by a geometrical scale (the "nodal pocket dimension"; see box 14.2) and, since
the nodal surface acts as a steric potential, this translates into a real dynamical
scale – the Fermi energy. As long as the nodal surface is a smooth manifold, this
scale (and thereby the Fermi energy) cannot be avoided. The remarkable conclusion is that the only way to diminish the Fermi energy in order to realise a fermionic

quantum critical state is to change the nodal surface geometry from a smooth to a fractal manifold. With an insightful "backflow" Ansatz for the wave function such a fractal nodal surface density matrix can be explicitly constructed (box 14.2): the way in which the smooth nodal surface gradually evolves into a fractal nodal surface is illustrated in Fig. 14.3. Little is known about the physics of such a state, but this does indeed show that the pole strength Z_k of the quasiparticles vanishes in the fractal state [564]. The characteristic quasiparticle signature of the Fermi scale is no longer present. This is actually the only non-holographic example of a Fermi liquid with a Fermi surface that shows increasingly strong quantum fluctuations, which eventually disappears at a quantum phase transition to a genuinely scale-invariant affair.

The nodal surface is an exact property of any fermionic state of matter – as a caveat, one needs particles in the UV to formulate the present first quantised version. Although a mathematical proof is still to be delivered, it seems unavoidable that a fractal geometry of the nodal surface is a prerequisite for any scale-invariant state of finite-density fermion matter. The nodal surface does enumerate the fermionic long-range entanglement, and it seems obvious that a fractal nodal surface is even more non-local than the smooth version, indicating that such critical fermionic states are more densely entangled than the Fermi liquid. The outlook is that it might perhaps become possible to classify completely the nodal structure of fermionic states (and thereby obtain a classification of the nature of their entanglement) on the basis of the nodal surface. Mathematically this is the question of the zeros of high-degree polynomials anti-symmetrised over many variables. This appears to have barely been explored as a mathematical subject, and it presents a considerable challenge. The only other examples we possess of controlled theory of such scale-invariant fermionic states of matter are the strange metals of holography. We address to the holographists the challenge of formulating the dictionary entry for the nodal surface, to check whether the strange metals do indeed exhibit fractal nodes.

Box 14.2   Fermionic quantum criticality and the fractal nodal surface

The claim in the main text is that the *nodal surface*, the zeros of the full density matrix, might be quite a useful measure of the many-particle entanglement in the case of matter formed from a finite density of fermions. This claim rests on the constrained path integral formulated by Ceperley, which delineates clearly how this nodal surface encapsulates the differences between bosonic and fermionic states of matter. So far this has been formulated only in terms of the first-quantised worldline representation for non-relativistic fermions.

Define the full real-space density matrix of a system of fermions to be $\rho_F(\mathbf{R}, \mathbf{R}'; \hbar\beta)$, where $\mathbf{R} = (\mathbf{r}_1, \ldots, \mathbf{r}_N)$ refers to the position in dN-dimensional configuration space (d is the spatial dimension and N is the number of particles) and $\beta = 1/(k_B T)$ is the inverse temperature. This can be written as a path integral over worldlines $\{\mathbf{R}_\tau\}$ in imaginary time τ ($0 \leq \tau \leq \hbar\beta$), weighted by an action $\mathcal{S}[\mathbf{R}_\tau]$,

$$\rho_F(\mathbf{R}, \mathbf{R}'; \hbar\beta) = \frac{1}{N!} \sum_{\mathcal{P}} (-1)^{\mathcal{P}} \int_{\mathbf{R} \to \mathcal{P}\mathbf{R}'} \mathcal{D}\mathbf{R}_\tau \, e^{-\mathcal{S}[\mathbf{R}_\tau]/\hbar}, \tag{14.34}$$

with

$$\mathcal{S}[\mathbf{R}_\tau] = \int_0^{\hbar\beta} d\tau \left\{ \frac{m}{2}\dot{\mathbf{R}}_\tau^2 + V(\mathbf{R}_\tau) \right\},$$

where the sum over all possible $N!$ particle permutations $\mathcal{P}$ accounts for the indistinguishability of the particles and the alternating sign imposes the Fermi–Dirac statistics. Here $p = \text{par}(\mathcal{P})$ denotes the parity of the permutation; even permutations enter with a positive sign, odd permutations with a negative sign. The term $V(\mathbf{R})$ is a shorthand notation both for external potentials and for particle interactions. The partition function is obtained as a trace over the diagonal elements of the density matrix,

$$\mathcal{Z}_N(\beta) = \int d\mathbf{R} \, \rho_F(\mathbf{R}, \mathbf{R}; \hbar\beta). \tag{14.35}$$

Ceperley proved [565] that the fermionic density matrix can be calculated as a path integral analogous to Eq. (14.34) but with the sum restricted to worldlines that do not cross the nodes of the density matrix itself. It follows directly from the antisymmetrisation requirement that the zeros of the full density matrix form for each given initial point $\mathbf{R}_0$ and inverse temperature β a $(dN-1)$-dimensional hypersurface in dN-dimensional configuration space: the "nodal surface",

$$\Omega_{\mathbf{R}_0,\beta} := \{\mathbf{R}|\rho_F(\mathbf{R}_0, \mathbf{R}; \hbar\beta) = 0\}. \tag{14.36}$$

Those hypersurfaces act as infinite potential barriers allowing only "node-avoiding" world histories $\mathbf{R}_\tau$ with $\rho_F(\mathbf{R}, \mathbf{R}_\tau; \tau) \neq 0$ for $0 \leq \tau \leq \hbar\beta$.

To calculate the partition function, we have to integrate over the diagonal density matrix $\rho_F(\mathbf{R}, \mathbf{R}; \hbar\beta)$, which is obtained as a path integral over all worldline configurations $\mathbf{R} \to \mathcal{P}\mathbf{R}$ that do not cross the nodal surface on any time slice and belong to the "reach",

$$\Gamma_\beta(\mathbf{R}) = \{\gamma : \mathbf{R} \to \mathbf{R}'|\rho_F(\mathbf{R}, \mathbf{R}(\tau); \tau) \neq 0\}. \tag{14.37}$$

Box 14.2 (Continued)

Because of the anti-symmetry of the fermionic density matrix under particle permutations $\mathcal{P}$,

$$\rho_F(\mathbf{R}, \mathcal{P}\mathbf{R}; \hbar\beta) = \rho_F(\mathcal{P}\mathbf{R}, \mathbf{R}; \hbar\beta)$$
$$= (-1)^P \rho_F(\mathbf{R}, \mathbf{R}; \hbar\beta), \tag{14.38}$$

all worldline configurations corresponding to odd permutations have to cross a node an odd number of times and are therefore completely removed from the partition function. They are exactly cancelled out by all node-crossing even permutations, and we are left with an ensemble of all node-avoiding worldline configurations corresponding to even permutations,

$$\rho_F(\mathbf{R}, \mathbf{R}; \hbar\beta) = \frac{1}{N!} \sum_{\mathcal{P},\text{even}} \int_{\gamma:\mathbf{R}\to\mathcal{P}\mathbf{R}}^{\gamma\in\Gamma_\beta(\mathbf{R})} \mathcal{D}\mathbf{R}_\tau \, e^{-\mathcal{S}[\mathbf{R}]/\hbar}. \tag{14.39}$$

Remarkably, this representation of an arbitrary-fermion problem does not suffer from the "negative probabilities" of the standard formulation. However, this does not solve the sign problem at all. The full density matrix represents the maximal knowledge of the system, and it follows from Ceperley's construction that the signs can be eliminated only when full knowledge of *all zeros* of the density matrix is available, which involves just marginally less information than the full solution. However, the advantage is that the sign problem is "geometrised": the nodal surface is a geometrical object that enters the dynamics as a simple steric constraint potential. One can now attempt to characterise the nature of the fermionic states in terms of the general geometrical properties of the nodal surface.

The non-interacting Fermi gas is informative in this regard. Formulating the Fermi gas in single-particle momentum space makes it very easy to show [64] that the Ceperley path integral just describes the "Mott insulator in a harmonic well" (Eq. (2.24)) as highlighted at the beginning of section 2.3. In this case the nodal surface just becomes the "Mott constraint" in momentum space. In real space it becomes more interesting. Focussing on zero temperature, the density matrix factorises into the product of first-quantised wave functions Ψ according to $\lim_{\beta\to\infty} \rho(\mathbf{R}, \mathbf{R}'; \beta) = \Psi(\mathbf{R})\Psi(\mathbf{R}')$. The nodal surface is therefore determined by the zeros of the wave function, which for the Fermi gas are set by the zeros of the Slater determinant $\Psi(\mathbf{r}_1, \ldots, \mathbf{r}_N) = \mathcal{N} \det\left(e^{i\mathbf{k}_i\cdot\mathbf{r}_j}\right)_{i,j=1,\ldots,N}$. The nodal surface can be visualised by taking d-dimensional "cuts" through configuration space: see the upper left panel of Fig. 14.3 for the case of a two-dimensional Fermi gas.

Given that the nodal surface encapsulates all of the fermionic information, how is the Fermi energy encoded in its geometry? The Pauli hypersurface is defined by the requirement that the wave function has to vanish when the positions of the fermions are coincident, and this is obviously a sub-manifold of the nodal surface. In the case that the nodal surface is a *smooth* manifold, it directly follows that the nodal surface is characterised by a scale [564]: the "nodal pocket" with a dimension of the order of the inter-particle spacing. Since the nodal surface acts like a steric potential, the nodal pocket acts like a confining potential and the Fermi energy is nothing other than the finite-size kinetic energy of a bosonic mode confined in the nodal pocket "box".

To get rid of the Fermi energy, one therefore has to get rid of the intrinsic scale in the nodal surface geometry. This can only be done by changing its smooth geometry to a fractal manifold: *a scale-invariant state of fermion matter is characterised by a fractal geometry of its nodal surface.*

A proof of principle for the existence of such fractal nodal surfaces is available in the form of a particular wave-function Ansatz [564]. This dates back to Feynman and corresponds to a "hydrodynamical backflow Ansatz for fermions". One writes a simple Fermi-gas determinant taking care of the anti-symmetrisation $\Psi(\mathbf{R}) = \det\left(e^{i\vec{k}_i \cdot \hat{r}_j}\right)$ but postulates that the coordinate is now a *collective* coordinate $\hat{\mathbf{r}}_i = \mathbf{r}_i + \sum_{j \neq i} \eta(|\mathbf{r}_i - \mathbf{r}_j|) (\mathbf{r}_i - \mathbf{r}_j)$, i.e. it is actually a function of the coordinates of all other particles as well. Upon taking for $\eta(r) = a^3/(r^3 + r_0^3)$ every single-particle state describes in principle the quantum-mechanical version of *hydrodynamical* backflow, as realised by Feynman: when particle i propagates forwards, the system of all other particles exhibits a dipolar flow in the opposite direction as in a classical hydrodynamical liquid in the incompressible regime. However, this depends on the free parameters a and r_0. The second one is a short-distance cut-off: at $r < r_0$ the effects of backflow just vanish and one is back in the Fermi gas. The important "IR" parameter is the "backflow" parameter a: when $a < r_s$ (the inter-particle distance) the backflow is not collective, whereas for $a > r_s$ the liquid is singularly different from a Fermi liquid. In terms of the bare particle coordinates one finds a greatly entangled state with a divergent number of single-particle Slater determinants.

In Fig. 14.3 we show how the nodal surface, as visualised by two-dimensional cuts through configuration space, changes when the backflow length a is varied, starting from the free Fermi gas $a = 0$ and passing into the regime $a > r_s$. For the Fermi gas one finds the smooth Fermi surface, but for increasing a it becomes increasingly rugged. A careful analysis shows that at distances shorter than a "correlation length" ξ_f the nodal surface is a fractal – geometrically scale-invariant – while at larger distances it is still smooth, like the Fermi gas. At a critical value a_c this correlation length diverges algebraically and the nodal surface has acquired a fractal geometry with a Hausdorff dimension $d_{\text{Hausdorff}} \simeq Nd - 0.5$ (the smooth Fermi-gas nodal surface has a dimension of $Nd - 1$).

14.2.6 Quantum criticality, long-range entanglement and anti-de Sitter space-time

In the above we have repeatedly emphasised the failure of the bipartite von Neumann entropy to fully encode the multi-party long-range entanglement at the heart of the quantum matter puzzle. This hampers holographic models as much as it does field-theory models. However, by asking a quite different kind of entanglement question, the correspondence reveals a remarkable insight. Up to this point we have focussed completely on what the bulk has to tell us about the boundary, but one can shift perspective and ask instead how the gravitational bulk can be reconstructed from the information available in the boundary. In a very recent development striking evidence has surfaced indicating that at least the classical geometry is encoded in quantum information in the boundary. This is a rapidly developing subject, but it provides further support to the connection between quantum entanglement and critical systems, and our story would not be complete without giving a first impression of how this works.

This development started with a seminal contribution by Van Raamsdonk [566, 567]. Consider two copies of a CFT that do not mutually interact. A state in the combined system is an *exact* product state in each of the independent Hilbert spaces,

$$|\Psi\rangle_{\mathrm{CFT}_1 \times \mathrm{CFT}_2} = |\Psi_1\rangle_{\mathrm{CFT}_1} \otimes |\Psi_2\rangle_{\mathrm{CFT}_2}. \tag{14.40}$$

Since the two theories do not interact, the gravity dual of this system is straightforward: for each copy of the CFT there is an independent AdS. The next step is to entangle the state in a very specific way

$$|\Psi\rangle_\beta = \sum e^{-\beta E_i/2} |\Psi_i\rangle_{\mathrm{CFT}_1} \otimes |\Psi_i\rangle_{\mathrm{CFT}_2}. \tag{14.41}$$

We have turned on some interaction between the CFTs, and by clever manipulations the state is arranged to be of this form. This state is, however, very well known: it is a thermal state, rewritten as a pure state in the Schwinger–Keldysh double form. Tracing its density matrix over the second CFT_2 yields a thermal density matrix,

$$\mathrm{Tr}_{\mathrm{CFT}_2}\left(|\Psi\rangle_{\beta\beta}\langle\Psi|\right) = \sum e^{-\beta E_i} |\Psi_i\rangle_{\mathrm{CFT}_1} \otimes \langle\Psi_i|_{\mathrm{CFT}_1}. \tag{14.42}$$

Van Raamsdonk observed that the gravity dual of this pure state $|\Psi\rangle_\beta$ is now a *single* space-time. As we saw in chapter 6, the Schwinger–Keldysh formulation

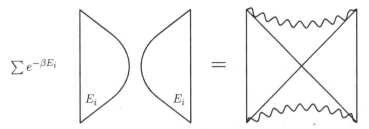

$$\sum e^{-\beta E_i}$$

Figure 14.4 The geometry description of two copies of a CFT entangled with a thermal density matrix is a single space-time given by the Kruskal extension of a black hole. Figure source [567].

is described by the Kruskal extension of the AdS Schwarzschild black hole (Fig. 14.4). The conclusion for the gravitational physics implied by this insight is remarkable. Before the entanglement, we had two independent, complete copies of AdS space-time. In terms of the standard Poincaré coordinates, this corresponds to two copies of the metric,

$$ds^2 = \frac{r^2}{L^2}\left(-dt^2 + dx^2\right) + L^2\frac{dr^2}{r^2}, \tag{14.43}$$

with r ranging from the Poincaré horizon $r = 0$ to the boundary $r = \infty$. *After* the entanglement one has a single space-time, where, moreover, new parts of space-time behind the horizon have been added to the overall geometry. It appears that *space-time emerges from the entanglement.*

Although this example concerns a single very special state, it does reveal the principle. In many other situations the same emergence of space-time from entanglement has now been identified (e.g. see box 14.3). In the next seminal development, Faulkner, Guica, Hartman, Myers, and Van Raamsdonk showed that this is not just a qualitative statement, but also one can quantitatively relate the emergent dual space-time to the entanglement information measured by the bipartite von Neumann entropy [568, 569]. This relies on two ingredients. Firstly, given the entanglement Hamiltonian H_{EE} and entanglement entropy S_{EE}, it follows from the definition of the former that under a small variation δ of the state of the system a "first law" is obeyed,

$$\delta S_{\mathrm{EE}} = \delta\langle H_{\mathrm{EE}}\rangle. \tag{14.44}$$

Secondly, given the ground state of a CFT, one can compute the entanglement Hamiltonian of ball-shaped regions explicitly. For a ball B of radius R centred around the origin in a $(d + 1)$-dimensional CFT in flat space, it equals

$$H_{\text{EE}, B_R} = 2\pi \int_{|x| < R} d^d x \frac{R^2 - x^2}{2R} T_{tt} , \qquad (14.45)$$

where T_{tt} is the tt component of the stress tensor. One now makes use of a special property of the CFT. By a conformal transformation one can map the causal development of a ball-shaped region to a non-compact hyperbolic space. The entanglement between the ball and its complement turns into the thermal entropy of this state of the CFT on hyperbolic space [560, 570]. This is completely analogous to the relation between Minkowski space and the Rindler wedge. The entanglement entropy between the Rindler regions in Minkowski space is a thermal entropy from the viewpoint of the Rindler observer.

These steps allow one to identify a special Rindler time T. In just the same way as time translations in the Minkowski CFT are related to an isometry in the dual AdS, so do time translations along T correspond to a special isometry of AdS. In Poincaré coordinates for AdS where the boundary is at $r = \infty$,

$$ds^2 = \frac{r^2}{L^2}(\eta_{\mu\nu} dx^\mu dx^\nu) + \frac{L^2}{r^2} dr^2, \qquad (14.46)$$

these are generated by the vector

$$\xi_T = -\frac{2\pi}{L} t \left(-r \, \partial_r + x^i \, \partial_i \right) + \frac{2\pi}{2L} \left(L^2 - \frac{L^4}{r^2} - t^2 - x^2 \right) \partial_t. \qquad (14.47)$$

At $t = 0$ and $r = \infty$ one recognises the entanglement Hamiltonian. This vector becomes null – its length squared vanishes – precisely on the Ryu–Takayanagi minimal surface, whose area encodes the entanglement entropy. The region Σ of AdS that therefore encodes the Rindler ball is the region between the boundary and the Ryu–Takayanagi surface (see Fig. 14.5). The volume of Σ can be encoded in its normal vector $\epsilon_{(\Sigma)}^\mu$. The manifest connection between the entanglement entropy and gravity in the bulk now follows from the existence of an anti-symmetric tensor field $\chi_{\mu\nu}$, with the property that its divergence is given by

$$\nabla^\mu \chi_{\mu\nu} = -2\xi^\rho \, \delta G_{\rho\nu} . \qquad (14.48)$$

Here $\delta G_{\mu\nu}$ are the linearised Einstein equations in AdS without an explicit source for matter. This anti-symmetric field has the property that its "flux" through the ball B in the boundary CFT equals

$$\int_B dB^{\mu\nu} \chi_{\mu\nu} = \int_B \chi^{rt} = \delta \langle H_{\text{EE}} \rangle. \qquad (14.49)$$

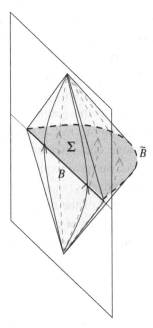

Figure 14.5 A sketch of the configuration used to prove that knowledge of the entanglement entropy of the ball B is equivalent to knowing the linearised gravitational dynamics of AdS. Here $\tilde{B}$ is the minimal surface whose boundary coincides with the boundary of B obtained from the Ryu–Takayanagi prescription. The region of AdS that therefore describes the dynamics within the ball B is the enclosed volume Σ. Time evolution in this region, generated by a Rindler time for B, is shown by the dashed curves. Figure source [569].

Here $dB^{\mu\nu}$ is double normal to the surface B in AdS. Integrating over the Ryu–Takayanagi surface, on the other hand, gives the "flux",

$$\int_{\tilde{B}} d\tilde{B}^{\mu\nu} \chi_{\mu\nu} = \delta S_{\text{EE}}. \tag{14.50}$$

Using the first law of entanglement, the difference between these two expressions has to vanish:

$$0 = \delta S_{\text{EE}} - \delta\langle H_{\text{EE}}\rangle = \int dB \cdot \chi - \int d\tilde{B} \cdot \chi. \tag{14.51}$$

The second expression is nothing other than the integral over the boundary of Σ. We can therefore use Stokes' theorem in reverse to conclude that

$$0 = \int d\Sigma^{\nu} \nabla^{\rho}\chi_{\rho\nu} = -2\int d\Sigma^{\nu}\xi^{\mu}\delta G_{\mu\nu}. \tag{14.52}$$

The first law of entanglement is thus *equivalent* to stating that the linearised Einstein equations have to be satisfied. To complete the proof one would need to extend this to the non-linear regime of gravity.

This result amounts to very strong support for the case we have been advocating. Certain quantum critical phases have a holographic description in terms of AdS geometry. The emergence of the geometry can be understood from the entanglement structure of the quantum critical theory, although apparently only the information on the short-range entanglement is required in order to "stitch together" the bulk space-time. As we have explained in this book, for such CFTs the geometry captures the relevant physics in an extremely efficient way. By extrapolation, finding out the precise nature of the entanglement is likely to be the most powerful way to characterise the physics of quantum critical systems. This should be a general statement, since neither entanglement nor quantum criticality *needs* holography.

Box 14.3   Rindler holography and entanglement

A situation that very clearly illuminates the emergence of space-time from entanglement is to consider the Rindler–Unruh entanglement of an accelerating observer [571, 572]. Consider a CFT living on a sphere S^d; as we know from the Hawking–Page transition in chapter 6, this has as its dual global AdS. The latter is topologically equivalent to a cylinder with the axial direction equal to time, and the surface equal to the space of the boundary field theory. The Poincaré coordinates describe only half of the surface of the cylinder. This half is isomorphic to Minkowski space. A uniformly accelerating observer in this Minkowski space, i.e. an accelerating observer probing only very local regions of the full CFT, sees only part of this space. The natural coordinates t, x, y_i for such an observer are

$$t = X \sinh(aT), \quad x = X \cosh(aT), \quad y_i = Y_i, \tag{14.53}$$

where the Minkowski space-time $ds^2 = -dt^2 + dx^2 + \sum_{i=2}^{d} dy_i^2$.

In these coordinates Minkowski space-time takes the form

$$ds^2 = -a^2 X^2 dT^2 + dX^2 + \sum_{i=2}^{d} dY_i^2. \tag{14.54}$$

This is Rindler space. It is straightforward to verify that an observer at rest in Rindler space corresponds to an observer moving along $x^2 - t^2 = X^2$ with velocity $v = dx/dt = t/x$ and constant acceleration given by

$$\frac{d}{dt}\left(-\frac{v}{\sqrt{1-v^2}}\right) = \frac{1}{X}$$

in the original Minkowski space. The two important aspects of this space are as follows.

- It covers only part of Minkowski space. Specifically, Rindler space covers only the wedges $x > 0$, $|t| < x$ (right wedge) and $x < 0$, $|t| < |x|$ (left wedge).
- Exactly as for the Hawking temperature of black holes, there is a Rindler temperature (this is known as the Unruh effect), $T_{\text{Rindler}} = 1/(2\pi X)$. The direct way to see this is to follow the prescription in box 6.1 and demand that the Euclideanised theory has no singularity. One finds that the near-horizon geometry of the Schwarzschild black hole is exactly the Rindler space-time.

Directly analogously to Hawking's black-hole calculation, a Rindler observer will see thermal radiation at a temperature T_{Rindler} from the "Rindler horizon" $X = 0$. In Minkowski space this horizon is the null surface $t = \pm x$ (the right plot in Fig. 14.6). From the Minkowski perspective this is a consequence of entanglement. The pure Minkowski ground state of a free field appears to the Rindler observer as a state entangled with the degrees of freedom in the *complementary* Rindler wedge (Fig. 14.6).

$$|0\rangle_{\text{Mink}} = \frac{1}{Z}\sum_i e^{-\beta_{\text{Rindler}}E_i/2}|E_i^{\text{Rindler}_{\text{L}}}\rangle|E_i^{\text{Rindler}_{\text{R}}}\rangle. \tag{14.55}$$

We now put this in a holographic context. Since the Rindler observer only sees part of the Minkowski space-time, the dual description of the CFT on Rindler space is only part of the AdS space-time. It is the part which is bounded by the light-like Rindler horizon extended along the radial direction. We can represent this pictorially by drawing global AdS as a filled-in cylinder (section 4.4). In the Minkowski space boundary of the Poincaré patch, which itself covers only half of the cylinder, we can now draw the Rindler wedges. The sensible states in the theory are the thermally entangled states between the left and right Rindler wedges. Even with its complement, the geometric extension of the two Rindler wedges to the interior fills only part of AdS (Fig. 14.6). We also know, however, that the entangled Rindler states are simply the normal states in the parent Minkowski CFT dual to the whole of AdS. "Disentangling" the Rindler state therefore corresponds to the "emergence" of the wedges of AdS which the Rindler patch does not cover.

Box 14.3   (Continued)

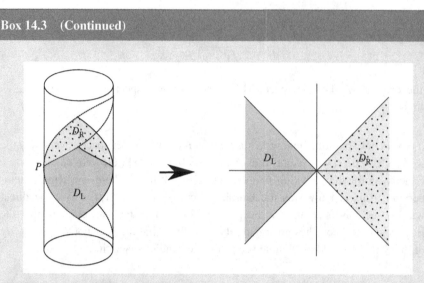

Figure 14.6 Left: thinking about how a Rindler observer in a CFT would be described holographically, one can consider the Rindler patch of the boundary Minkowski space. The holographic dual would be the geometry of the interior from this patch that is bounded by light rays. Even with the complementary Rindler wedge, this covers only part of the full AdS geometry. "Disentangling" the state is the same as the transformation back to the Minkowski basis of wave functions. Holographically this therefore means that the earlier missing pieces of the AdS space-time are now again covered by the dual geometry. "Disentangling" made them "emerge". Right: inside Minkowski space, Rindler coordinates cover only the two wedges bounded by light rays emanating from the origin. Rindler coordinates are the natural coordinates for a uniformly accelerating observer. This geometric picture shows clearly that the Rindler horizon is a coordinate singularity. In complete analogy with Hawking's computation of black holes, the only way this horizon can remain a coordinate singularity at the quantum level is if the state is thermal from the viewpoint of the Rindler observer. It is straightforward to show, from the wave functions of free particles in Rindler space, that this state is nothing but the original Minkowski vacuum, where the contribution from the complementary Rindler wedge is traced out. Figure source [572]. (© IOP Publishing. Reproduced by permission of IOP Publishing. All rights reserved.)

14.3 The final message for condensed matter

We have learned about worlds made from matter that come rolling out of the marvellous mathematical machines developed by the string theorists over a long period of time. These worlds are in some regards strikingly similar to the bread and butter of condensed matter physics. In other regards they are very different, if not plainly mysterious. As a theoretical pursuit, holography is perhaps the most interesting new development in fundamental theory altogether. Merging the conceptual progress coming from quantum information with the mathematical power of general relativity and aiming this machinery towards questions of condensed matter physics, we are forced to accept that broad new classes of quantum matter have a right to exist, at least in principle.

This is the proper viewpoint for the condensed matter physicist dealing with holography. Although definitive proof is missing, it would appear to be a very good bet to assume that holography describes states of quantum matter that at present cannot be described in any other way. It is established fact that at least superficially the unidentified holographic objects bear an eerie resemblance to the long-standing laboratory mysteries. The words "local quantum critical", "Planckian dissipation", "quantum smectics", "competing orders" and so forth were all invented to give names to experimental surprises. The results that rolled out of the holographic machinery turned out to fit these experimental names remarkably well. Is this a coincidence? At the very least, we have highlighted at several points in this book that there is no other theoretical approach that has produced answers that look so much like the mysterious aspects of the experimental observations.

At present there is no certainty: it might all be a coincidence, a delusion. However, when one is operating in the tradition of condensed matter physics, this is not something to worry about. Instead, theorists are supposed to figure out how to think differently, to demonstrate possibility with the sole aim of coming up with an unusual experiment that nobody had thought about. The usual course of affairs is that the experiment has a different outcome from what had been expected, which in turn forces the theorists to think even harder and better. This is the time-tested procedure which is at the root of the exquisite power of condensed matter physics.

The great strength of holography is precisely that it is an exceedingly fertile source for learning to think differently in this regard. It appears to reveal gross behaviours governed by new types of phenomenological principles, which should be accessible to experiment. To what extent are the cuprate strange metals precisely local quantum critical (section 8.2)? Is it the case that the DC transport is governed by "nearly perfect" hydrodynamics (section 12.5)? Is the anomalous anisotropy of the conductivity in the cuprates associated with the higher-dimensional quantum smectic (section 12.4)? Is the unconventional superconducting instability building up in a conformal fermionic metal (chapter 10)? Are the competing pseudogap orders in the cuprates perhaps of the non-Hartree–Fock kind, stabilising an emergent quantum critical IR (this chapter)?

Addressing these questions effectively in the laboratory is not easy, and will surely involve a departure from the present routine activity. Looking through the holographic eye-glass, one's attention is in the first place drawn to the strange-metal states, especially the one realised in optimally doped cuprates. All of the easy measurements were done a long time ago, and since the middle of the 1990s the strange metal has largely been ignored. The effect of this is that very little is known regarding properties that have come to the centre of our attention with holography. Is the strange metal truly conformal? Is z really infinite? How could one measure the hyperscaling-violation exponent directly? The standard machinery which is available in the laboratories can already bring us a long way. How do the incoherent backgrounds seen in ARPES at the antinodes of the cuprates actually behave, do they perhaps satisfy a scaling collapse governed by the local quantum criticality? How do the (faint) magnetic fluctuations that can be measured by neutron scattering precisely behave in the strange-metal phase, do they reveal the local quantum criticality? What happens with the residual resistivity, skin depth and Wiedemann–Franz law at very low temperature in optimally doped superconductors when T_c has been reduced in very large magnetic fields? Other experiments will need to be designed on purpose. The specific example of the Ferrel–Scalapino–Goldman probe measuring the pair susceptibility (section 10.3) might inspire the real experts – the experimentalists – to come up with something much better.

All of this is serious work. But we urge the condensed matter physicists to heed the call. The most resounding message of holography is that for the study of the fundamentals of physics the small items of laboratory machinery used to study high-T_c superconductors and related phenomena could well become the faithful twenty-first-century successors of the twentieth-century accelerators and space telescopes. A hairy black hole is more likely to be discovered in the lab next door than with the Hubble telescope or the LHC. Yet the virtual travails of astronauts near its horizon tell us more about its strange-metal incarnation than any microscope can see. To promote awareness of this opportunity has been one of our main motivations for writing this book. We hope that we have infected the reader with our fascination for this novel field in physics, to the degree that he or she will join the rebellion against Landau's iron fist.

References

[1] J. M. Maldacena, "The large N limit of super-conformal field theories and super-gravity," *Adv. Theor. Math. Phys.* **2**, 231 (1998) [republished *Int. J. Theor. Phys.* **38**, 1113 (1999)] [arXiv:9711200 [hep-th]].

[2] S. S. Gubser, I. R. Klebanov and A. M. Polyakov, "Gauge theory correlators from noncritical string theory," *Phys. Lett.* B **428**, 105 (1998) [arXiv:9802109 [hep-th]].

[3] E. Witten, "Anti-de Sitter space and holography," *Adv. Theor. Math. Phys.* **2**, 253 (1998) [arXiv:9802150 [hep-th]].

[4] Y. Nakayama, "A lecture note on scale invariance vs conformal invariance" [arXiv:1302.0884 [hep-th]].

[5] J. McGreevy, "Holographic duality with a view toward many-body physics," *Adv. High Energy Phys.* **2010**, 723105 (2010) [arXiv:0909.0518 [hep-th]].

[6] P. Breitenlohner and D. Z. Freedman, "Positive energy in anti-de Sitter backgrounds and gauged extended supergravity," *Phys. Lett.* B **115**, 197 (1982).

[7] O. Aharony, S. S. Gubser, J. M. Maldacena, H. Ooguri and Y. Oz, "Large N field theories, string theory and gravity," *Phys. Rep.* **323**, 183 (2000) [arXiv:9905111 [hep-th]].

[8] J. Erdmenger and H. Osborn, "Conserved currents and the energy momentum tensor in conformally invariant theories for general dimensions," *Nucl. Phys.* B **483**, 431 (1997) [arXiv:9605009 [hep-th]].

[9] M. Ammon and J. Erdmenger, *Gauge/Gravity Duality: Foundations and Applications*, Cambridge University Press, 2014.

[10] M. Natsuume, "AdS/CFT duality user guide" [arXiv:1409.3575 [hep-th]].

[11] J. Casalderrey-Solana, H. Liu, D. Mateos, K. Rajagopal and U. A. Wiedemann, "Gauge/string duality, hot QCD and heavy ion collisions" [arXiv:1101.0618 [hep-th]] (preliminary version of [12]).

[12] J. Casalderrey-Solana, H. Liu, D. Mateos, K. Rajagopal and U. A. Wiedemann, *Gauge/String Duality, Hot QCD and Heavy Ion Collisions*, Cambridge University Press, 2014.

[13] J. M. Maldacena, "TASI 2003 lectures on AdS/CFT," in *TASI 2003 Proceedings*, World Scientific, 2004 [arXiv:0309246 [hep-th]].

[14] G. T. Horowitz and J. Polchinski, "Gauge/gravity duality," in D. Oriti (ed.), *Approaches to Quantum Gravity: Toward a New Understanding of Space, Time and Matter*, Cambridge University Press, 2009, pp. 169–186 [arXiv:0602037 [gr-qc]].

[15] J. Polchinski, "Introduction to gauge/gravity duality," in *TASI 2010 Proceedings*, World Scientific, 2012 [arXiv:1010.6134 [hep-th]].

[16] J. Maldacena, "The gauge/gravity duality" [arXiv:1106.6073 [hep-th]].

[17] S. A. Hartnoll, "Lectures on holographic methods for condensed matter physics," *Class. Quant. Grav.* **26**, 224002 (2009) [arXiv:0903.3246 [hep-th]].

[18] C. P. Herzog, "Lectures on holographic superfluidity and superconductivity," *J. Phys.* A **42**, 343001 (2009) [arXiv:0904.1975 [hep-th]].

[19] G. T. Horowitz, *Introduction to Holographic Superconductors*, Springer, p. 313 (2011) [arXiv:1002.1722 [hep-th]].

[20] S. A. Hartnoll, "Horizons, holography and condensed matter," in *Black Holes in Higher Dimensions*, Cambridge University Press, 2011 [arXiv:1106.4324 [hep-th]].

[21] S. Sachdev, "What can gauge–gravity duality teach us about condensed matter physics?" *Ann. Rev. Condensed Matter Phys.* **3**, 9 (2012) [arXiv:1108.1197 [cond-mat.str-el]].

[22] N. Iqbal, H. Liu and M. Mezei, "Lectures on holographic non-Fermi liquids and quantum phase transitions," in *TASI 2010 Proceedings*, World Scientific, 2012 [arXiv:1110.3814 [hep-th]].

[23] P. M. Chesler and L. G. Yaffe, "Numerical solution of gravitational dynamics in asymptotically anti-de Sitter spacetimes," *JHEP* **1407**, 086 (2014) [arXiv:1309.1439 [hep-th]].

[24] C. P. Herzog, P. Kovtun, S. Sachdev and D. T. Son, "Quantum critical transport, duality, and M-theory," *Phys. Rev.* D **75**, 085020 (2007) [arXiv:0701036 [hep-th]].

[25] S. Sachdev, *Quantum Phase Transitions*, 2nd edn, Cambridge University Press, 2011.

[26] I. Herbut, *A Modern Approach to Critical Phenomena*, Cambridge University Press, 2007.

[27] M. Endres, T. Fukuhara, D. Pekker, M. Cheneau, P. Schauss, C. Gross, E. Demler, S. Kuhr and I. Bloch, "The 'Higgs' amplitude mode at the two-dimensional superfluid–Mott insulator transition," *Nature* **487**, 454 (2012) [arXiv:1204.5183 [cond-mat.quant-gas]].

[28] I. F. Herbut, V. Juričić and O. Vafek, "Relativistic Mott criticality in graphene," *Phys. Rev.* B **80**, 075432 (2009) [arXiv:0904.1019 [cond-mat.str-el]].

[29] P. H. Ginsparg, "Applied conformal field theory" [ar Xiv:9108028 [hep-th]].

[30] P. Di Francesco, P. Mathieu and D. Senechal, *Conformal Field Theory*, Springer, 1997.

[31] M. R. Gaberdiel, "An Introduction to conformal field theory," *Rep. Prog. Phys.* **63**, 607 (2000) [arXiv:9910156 [hep-th]].

[32] A. B. Zamolodchikov, "'Irreversibility' of the flux of the renormalization group in a 2-D field theory," *JETP Lett.* **43**, 730 (1986).

[33] D. L. Jafferis, I. R. Klebanov, S. S. Pufu and B. R. Safdi, "Towards the F-theorem: $N = 2$ field theories on the three-sphere," *JHEP* **1106**, 102 (2011) [arXiv:1103.1181 [hep-th]].

[34] Z. Komargodski and A. Schwimmer, "On renormalization group flows in four dimensions," *JHEP* **1112**, 099 (2011) [arXiv:1107.3987 [hep-th]].

[35] H. Elvang, D. Z. Freedman, L.-Y. Hung, M. Kiermaier, R. C. Myers and S. Theisen, "On renormalization group flows and the *a*-theorem in 6d," *JHEP* **1210**, 011 (2012) [arXiv:1205.3994 [hep-th]].

[36] A. J. Beekman, D. Sadri and J. Zaanen, "Condensing Nielsen–Olesen strings and the vortex–boson duality in $3 + 1$ and higher dimensions," *New J. Phys.* **13**, 033004 (2011) [arXiv:1006.2267 [cond-mat.str-el]].

[37] A. J. Beekman and J. Zaanen, "Electrodynamics of Abrikosov vortices: the field theoretical formulation," *Frontiers Phys.* **6**, 357 (2011) [arXiv:1106.3946 [cond-mat.supr-con]].

[38] M. Edalati, R. G. Leigh and P. W. Phillips, "Dynamically generated Mott gap from holography," *Phys. Rev. Lett.* **106**, 091602 (2011) [arXiv:1010.3238 [hep-th]].

[39] S. Chakravarty, B. I. Halperin and D. R. Nelson, "Two-dimensional quantum Heisenberg antiferromagnet at low temperatures," *Phys. Rev.* B **39**, 2344 (1989).

[40] E. Fradkin, *Field Theories of Condensed Matter Physics*, Cambridge University Press, 2013.

[41] T. Senthil, A. Vishwanat, L. Balents, S. Sachdev and M. P. A. Fisher, "Deconfined quantum critical points," *Science* **303**, 1490 (2004) [arXiv:cond-mat/0311326 [cond-mat.str-el]].

[42] L. Zhu, M. Garst, A. Rosch and Q. Si, "Universally diverging Gruneisen parameter and the magnetocaloric effect close to quantum critical points," *Phys. Rev. Lett.* **91**, 066404 (2003) [arXiv:0212335 [cond-mat.str-el]].

[43] J. Zaanen and B. Hosseinkhani, "Thermodynamics and quantum criticality in cuprate superconductors," *Phys. Rev.* B **70**, 060509 (2004) [arXiv:0403345 [cond-mat.supr-con]].

[44] R. Küchler, N. Oeschler, P. Gegenwart, T. Cichorek, K. Neumaier, O. Tegus, C. Geibel, J. A. Mydosh, F. Steglich, L. Zhu and Q. Si, "Divergence of the Gruneisen ratio at quantum critical points in heavy fermion metals," *Phys. Rev. Lett.* **91**, 066405 (2003).

[45] J. Zaanen, "Superconductivity: why the temperature is high," *Nature* **430**, 512 (2004).

[46] G. Policastro, D. T. Son and A. O. Starinets, "The shear viscosity of strongly coupled N = 4 supersymmetric Yang–Mills plasma," *Phys. Rev. Lett.* **87**, 081601 (2001) [arXiv:0104066 [hep.th]].

[47] M. A. Nielsen and I. L. Chuang, *Quantum Computation and Quantum Information*, Cambridge University Press, 2000.

[48] X. G. Wen, *Quantum Field Theory of Many Body Systems: From the Origin of Sound to an Origin of Light and Electrons*, Oxford University Press, 2004.

[49] T. Chakraborty and P. Pietiläinen, *The Fractional Quantum Hall Effect: Properties of an Incompressible Quantum Fluid*, Springer, 2012.

[50] J. Nissinen and C. A. Lütken, "The quantum Hall curve," [arXiv:1207.4693 [cond-mat.str-el]].

[51] A. Achucarro and P. Townsend, "A Chern–Simons action for three-dimensional anti-de Sitter supergravity theories," *Phys. Lett.* B **180**, 89 (1986).

[52] E. Witten, "(2 + 1)-Dimensional gravity as an exactly soluble system," *Nucl. Phys.* B **311**, 46 (1988).

[53] J. de Boer and J. I. Jottar, "Entanglement entropy and higher spin holography in AdS$_3$," *JHEP* **1404**, 089 (2014) [arXiv:1306.4347 [hep-th]].

[54] B. de Wit and H. Nicolai, "$N = 8$ supergravity," *Nucl. Phys.* B **208**, 323 (1982).

[55] M. Ammon, A. Castro and N. Iqbal, "Wilson lines and entanglement entropy in higher spin gravity," *JHEP* **1310**, 110 (2013) [arXiv:1306.4338 [hep-th]].

[56] C. Nayak, S. H. Simon, A. Stern, M. Freedman and S. Das Sarma, "Non-Abelian anyons and topological quantum computation," *Rev. Mod. Phys.* **80**, 1083 (2008) [arXiv:0707.1889 [cond-mat.str-el]].

[57] M. Z. Hasan and C. L. Kane, "Colloquium: topological insulators," *Rev. Mod. Phys.* **82**, 3045 (2010).

[58] X.-L. Qi and S. C. Zhang, "Topological insulators and superconductors," *Rev. Mod. Phys.* **83**, 1057 (2011) [arXiv:1008.2026 [cond-mat.mes-hall]].

[59] R. J. Slager, A. Mesaros, V. Juričić and J. Zaanen, "The space group classification of topological band insulators," *Nature Physics* **9**, 98 (2013) [arXiv:1209.2610 [cond-mat.mes-hall]].

[60] X. L. Qi, T. L. Hughes and S. C. Zhang, "Topological field theory of time-reversal invariant insulators," *Phys. Rev.* B **78**, 195424 (2008) [arXiv:0802.3537 [cond-mat.mes-hall]].

[61] C. W. J. Beenakker, "Search for Majorana fermions in superconductors," *Ann. Rev. Condensed. Matter Phys.* **4**, 113 (2013) [arXiv:1112.1950 [cond-mat. mes-hall]].

[62] C. Wang, A. C. Potter and T. Senthil, "Classification of interacting electronic topological insulators in three dimensions," *Science* **343**, 6171 (2014) [arXiv:1306.3238 [cond-mat.str-el]].

[63] H. Kleinert, *Path Integrals in Quantum Mechanics, Statistics, Polymer Physics and Financial Markets*, World Scientific, 2009.

[64] J. Zaanen, F. Kruger, J.-H. She, D. Sadri and S. I. Mukhin, "Pacifying the Fermi-liquid: battling the devious fermion signs," *Iranian J. Phys.* **8**, 39 (2008) [arXiv:0802.2455 [cond-mat.other]].

[65] M. Troyer and U.-J. Wiese, "Computational complexity and fundamental limitations to fermionic quantum Monte Carlo simulations," *Phys. Rev. Lett.* **94**, 170201 (2005) [arXiv:0408370 [cond-mat]].

[66] R. Shankar, "Renormalization group approach to interacting fermions," *Rev. Mod. Phys.* **66**, 129 (1994).

[67] J. Polchinski, "Effective field theory and the Fermi surface," in *TASI 1992 Proceedings*, [arXiv:9210046 [hep-th]].

[68] G. Baym and C. Pethik, *Landau Fermi Liquid Theory, Concepts and Applications*, Wiley, 2004.

[69] W. R. Abel, A. C. Anderson and J. C. Wheatley, "Propagation of zero sound in liquid He^3 at low temperatures," *Phys. Rev. Lett.* **17**, 74 (1966).

[70] P. R. Roach and J. B. Ketterson, "Observation of transverse zero sound in normal ^{3}He," *Phys. Rev. Lett.* **36**, 736 (1976).

[71] J.-H. She and J. Zaanen, "BCS superconductivity in quantum critical metals," *Phys. Rev.* B **80**, 184518 (2009).

[72] J.-H. She, B. J. Overbosch, Y.-W. Sun, Y. Liu, K. Schalm, J. A. Mydosh and J. Zaanen, "Observing the origin of superconductivity in quantum critical metals," *Phys. Rev.* B **84**, 144527 (2011) [arXiv:1105.5377 [cond-mat.str-el]].

[73] A. A. Abrikosov, L. P. Gor'kov, and I. Ye. Dzyaloshinskii, *Quantum Field Theoretical Methods in Statistical Physics*, 2nd edn, Pergamon Press, 1965.

[74] J. A. Hertz, "Quantum critical phenomena," *Phys. Rev.* B **14**, 1165 (1976).

[75] T. Moriya and A. Kawabate, "Effect of spin fluctuations on itinerant electron ferromagnetism," *J. Phys. Soc. Japan* **34**, 639 (1973).

[76] A. J. Millis, "Effect of a nonzero temperature on quantum critical points in itinerant fermion systems," *Phys. Rev.* B **48**, 7183 (1993).

[77] H. von Löhneisen, A. Rosch, M. Vojta and P. Wölfle, "Fermi-liquid instabilities at magnetic quantum phase transitions," *Rev. Mod. Phys.* **79**, 1015 (2007).

[78] E.-G. Moon and A. V. Chubukov, "Quantum-critical pairing with varying exponents," *J. Low Temp. Phys.* **161**, 263 (2010) [arXiv:1005.0356 [cond-mat.supr-con]].

[79] S. A. Hartnoll, D. M. Hofman, M. A. Metlitski and S. Sachdev, "Quantum critical response at the onset of spin density wave order in two-dimensional metals," *Phys. Rev.* B **84**, 125115 (2011) [arXiv:1106.0001 [cond-mat.str-el]].

[80] S.-S. Lee, "Low energy effective theory of Fermi surface coupled with U(1) gauge field in $2+1$ dimensions," *Phys. Rev.* B **80**, 165102 (2009) [arXiv:0905.4532 [cond-mat.str-el]].

[81] M. A. Metlitski and S. Sachdev, "Quantum phase transitions of metals in two spatial dimensions: II. Spin density wave order," *Phys. Rev.* B **82**, 075128 (2010) [arXiv:1005.1288 [cond-mat.str-el]].

[82] D. Dalidovich and S.-S. Lee, "Perturbative non-Fermi liquids from dimensional regularization," *Phys. Rev.* B **88**, 245106 (2013) [arXiv:1307.3170 [cond-mat.str-el]].

[83] A. L. Fitzpatrick, S. Kachru, J. Kaplan and S. Raghu, "Non-Fermi liquid fixed point in a Wilsonian theory of quantum critical metals," *Phys. Rev.* B **88**, 125116 (2013) [arXiv:1307.0004 [cond-mat.str-el]].

[84] T. Senthil and M. P. A. Fisher, "Z_2 gauge theory of electron fractionalization in strongly correlated systems," *Phys. Rev.* B **62**, 7850 (2000) [arXiv:9910224 [cond-mat.str-el]].

[85] E. Berg, M. A. Metlitski and S. Sachdev, "Sign-problem-free quantum Monte Carlo of the onset of antiferromagnetism in metals," *Science* **338**, 1606 (2012) [arXiv:1206.0742 [cond-mat.str-el]].

[86] P. W. Anderson, *The Theory of High-T_c Superconductivity*, Princeton University Press, 1997.

[87] J. Zaanen, "A modern, but way too short history of the theory of superconductivity at a high temperature," in H. Rogalla and P. H. Kes (eds), *100 Years of Superconductivity*, CRC Press, 2012, pp. 92–114 [arXiv:1012.5461 [cond-mat.supr-con]].

[88] H. Liu, "From black holes to strange metals," *Physics Today* **65**, 68 (2012).

[89] S. V. Kravchenko and M. P. Sarachik, "Metal–insulator transition in two-dimensional electron systems," *Rep. Prog. Phys.* **67**, 1 (2004).

[90] O. Gunnarsson and K. Schönhammer, "Electron spectroscopies for Ce compounds in the impurity model," *Phys. Rev.* B **28**, 4315 (1983).

[91] J. W. Allen, S. J. Oh, O. Gunnarsson, K. Schönhammer, M. B. Maple, M. S. Torikachvili and I. Lindau, "Electronic structure of cerium and light rare-earth intermetallics," *Adv. Phys.* **35**, 275 (1986).

[92] J. Zaanen, G. A. Sawatzky and J. W. Allen, "Band gaps and electronic structure of transition-metal compounds," *Phys. Rev. Lett.* **55**, 418 (1985).

[93] V. I. Anisimov, J. Zaanen and O. K. Andersen, "Band theory and Mott insulators: Hubbard U instead of Stoner I," *Phys. Rev.* B **44**, 943 (1991).

[94] J. Zaanen and A. M. Oles, "Canonical perturbation theory and the two band model for high-T_c superconductors," *Phys. Rev.* B **37**, 9423 (1988).

[95] A. C. Hewson, *The Kondo Problem to Heavy Fermions*, Cambridge University Press, 1993.

[96] P. Phillips, "Mottness: identifying the propagating charge modes in doped Mott insulators," *Rev. Mod. Phys.* **82**, 1719 (2010) [arXiv:1001.5270 [cond-mat.str-el]].

[97] Z.-Y. Weng, "Mott physics, sign structure, ground state wavefunction, and high-T_c superconductivity," *Frontiers Phys.* **6**, 370 (2011) [arXiv:1110.0546 [cond-mat.supr-con]].

[98] J. Zaanen and B. J. Overbosch, "Mottness collapse and statistical quantum criticality" *Phil. Trans. R. Soc.* A **369**, 1599 (2011) [arXiv:0911.4070 [cond-mat.str-el]].

[99] Z. Zhu, H.-C. Jiang, Y. Qi, C.-S. Tian and Z.-Y. Weng, "Strong correlation induced charge localization in antiferromagnets," *Sci. Rep.* **3**, 2586 (2013) [arXiv:1212.6634 [cond-mat.str-el]].

[100] Z. Zhu, H.-C. Jiang, D.-N. Sheng and Z.-Y. Weng, "Hole binding in Mott antiferromagnets: a DMRG study" [arXiv:1312.6893 [cond-mat.str-el]].

[101] P. W. Anderson, "The resonating valence bond state in La_2CuO_4 and superconductivity," *Science* **235**, 1196 (1987).

[102] T. H. Hansson, V. Oganesyan and S. L. Sondhi, "Superconductors are topologically ordered," *Annals of Physics* **313**, 497 (2004) [arXiv:cond-mat/0404327 [cond-mat.supr-con]].

[103] J. B. Kogut, "An introduction to lattice gauge theory and spin systems," *Rev. Mod. Phys.* **51**, 659 (1979).

[104] M. Levin and X.-G. Wen, "String-net condensation: a physical mechanism for topological phases," *Phys. Rev. B* **71**, 045110 (2005) [arXiv:0404617 [cond-mat.str-el]].

[105] S. Sachdev, "The quantum phases of matter" [arXiv:1203.4565 [hep-th]].

[106] S. A. Kivelson, D. S. Rohksar and J. P. Sethna, "Topology of the resonating valence-bond state: solitons and high-T_c superconductivity," *Phys. Rev. B* **35**, 8865 (1987).

[107] R. Moessner and S. L. Sondhi, "Resonating valence bond phase in the triangular lattice quantum dimer model," *Phys. Rev. Lett.* **86**, 1881 (2001).

[108] X. G. Wen, "Mean-field theory of spin-liquid states with finite energy gap and topological orders," *Phys. Rev. B* **44**, 2664 (1991).

[109] N. Read and S. Sachdev, "Large-N expansion for frustrated quantum antiferromagnets," *Phys. Rev. Lett.* **66**, 1773 (1991).

[110] A. Kitaev, "Anyons in an exactly solved model and beyond," *Annals of Physics*, **321**, 2 (2006) [arXiv:0506438 [cond-mat.mes-hall]].

[111] L. Balents, "Spin liquids in frustrated magnets," *Nature* **464**, 199 (2010).

[112] P. A. Lee, N. Nagoasa and X.-G. Wen, "Doping a Mott insulator: physics of high temperature superconductivity," *Rev. Mod. Phys.* **78**, 17 (2006) [arXiv:cond-mat/0410445 [cond-mat.str-el]].

[113] X.-G. Wen, "Quantum orders and symmetric spin liquids," *Phys. Rev. B* **65**, 165113 (2002) [arXiv:0107071 [cond-mat.str-el]].

[114] P. Coleman, "Heavy fermions: electrons at the edge of magnetism," in H. Kronmuller and S. Parkin (eds.), *Handbook of Magnetism and Advanced Magnetic Materials. Volume 1: Fundamentals and Theory*, John Wiley and Sons, 2007, pp. 95–148 [arXiv:0612006 [cond-mat.str-el]].

[115] B. Keimer, S. A. Kivelson, M. R. Norman, S. Uchida and J. Zaanen, "From quantum matter to high temperature superconductivity in copper oxides," *Nature* **518**, 179 (2015).

[116] S. Raghu, S. A. Kivelson and D. J. Scalapino, "Superconductivity in the repulsive Hubbard model: an asymptotically exact weak-coupling solution," *Phys. Rev. B* **81**, 224505 (2010).

[117] D. J. Scalapino, "A common thread: the pairing interaction for the unconventional superconductors," *Rev. Mod. Phys.* **84**, 1383 (2012) [arXiv:1207.4093 [cond-mat.supr-con]].

[118] C. M. Varma, "Considerations on the mechanisms and transition temperatures of superconductors," *Rep. Prog. Phys.* **75**, 052501 (2012) [arXiv:1001.3618 [cond-mat.supr-con]].

[119] M. R. Norman, "The challenge of unconventional superconductivity," *Science* **332**, 196 (2011).

[120] N. F. Berk and J. R. Schrieffer, "Effect of ferromagnetic spin correlations on superconductivity," *Phys. Rev. Lett.* **17**, 433 (1966).

[121] C. N. A. van Duin and J. Zaanen, "Interplay of superconductivity and magnetism in strong coupling," *Phys. Rev. B* **61**, 3676 (2000).

[122] S. R. White, "Density matrix formulation for quantum renormalization groups," *Phys. Rev. Lett.* **69**, 2863 (1992).

[123] U. Schollwoeck, "The density-matrix renormalization group," *Rev. Mod. Phys.* **77**, 259 (2005) [arXiv:0409292 [cond-mat.str-el]].

[124] F. Verstraete, J. I. Cirac and V. Murg, "Matrix product states, projected entangled pair states, and variational renormalization group methods for quantum spin systems," *Adv. Phys.* **57**, 143 (2008).

[125] P. Corboz, R. Orus, B. Bauer and G. Vidal, "Simulation of strongly correlated fermions in two spatial dimensions with fermionic projected entangled-pair states," *Phys. Rev.* B **81**, 165104 (2010) [arXiv:0912.0646 [cond-mat.str-el]].

[126] P. Corboz, T. M. Rice and M. Troyer, "Competing states in the $t-J$ model: uniform d-wave state versus stripe state" [arXiv:1402.2859 [cond-mat.str-el]].

[127] S. R. White and D. J. Scalapino, "Density matrix renormalization group study of the striped phase in the 2D $t-J$ Model," *Phys. Rev. Lett.* **80**, 1272 (1998).

[128] J. Zaanen and O. Gunnarsson, "Charged magnetic domain lines and the magnetism of the High-T_c superconducting oxides," *Phys. Rev.* B **40**, 7391 (1989).

[129] K. Machida, "Magnetism in La_2CuO_4 based compounds," *Physica* C **158**, 192 (1989).

[130] A. J. Heeger, S. A. Kivelson, J. R. Schrieffer and W.-P. Su, "Solitons in conducting polymers," *Rev. Mod. Phys.* **60**, 781 (1988).

[131] J. M. Tranquada, B. J. Sternlieb, J. D. Axe, Y. Nakamura and S. Uchida, "Evidence for stripe correlations of spins and holes in copper oxide superconductors," *Nature* **375**, 561 (1995).

[132] M. Vojta, "Lattice symmetry breaking in cuprate superconductors: stripes, nematics, and superconductivity," *Adv. Phys.* **58**, 699 (2009) [arXiv:0901.3145 [cond-mat.supr-con]].

[133] W. Metzner and D.Vollhardt, "Correlated lattice fermions in $d = \infty$ dimensions," *Phys. Rev. Lett.* **62**, 324 (1989).

[134] A. Georges and G. Kotliar, "Hubbard model in infinite dimensions," *Phys. Rev.* B **45**, 6479 (1992).

[135] G. Kotliar and D. Vollhardt, "Strongly correlated materials: insights from dynamical mean-field theory," *Physics Today* **57**, 53 (2004).

[136] A. Georges, G. Kotliar, W. Krauth and M. J. Rozenberg, "Dynamical mean-field theory of strongly correlated fermion systems and the limit of infinite dimensions," *Rev. Mod. Phys.* **68**, 13 (1996).

[137] G. Kotliar, S. Savrasov, K. Haule, V. Oudovenko, O. Parcollet and C. Matianetti, "Electronic structure calculations with dynamical mean-field theory," *Rev. Mod. Phys.* **78**, 865 (2006).

[138] T. Maier, M. Jarrell, T. Pruschke and M. H. Hettler, "Quantum cluster theories," *Rev. Mod. Phys.* **77**, 1027 (2005).

[139] S.-X. Yang, H. Fotso, S.-Q. Su, D. Galanakis, E. Khatami, J.-H. She, J. Moreno, J. Zaanen and M. Jarrell, "Proximity of the superconducting dome and the quantum critical point in the two-dimensional Hubbard model," *Phys. Rev. Lett.* **106**, 047004 (2011).

[140] C. Pfleiderer, "Superconducting phases of f-electron compounds," *Rev. Mod. Phys.* **81**, 1551 (2009) [arXiv:0905.2625 [cond-mat.supr-con]].

[141] P. Gegenwart, Q. Si and F. Steglich, "Quantum criticality in heavy-fermion metals," *Nature Physics* **4**, 186 (2008).

[142] J. Zaanen, "Quantum critical electron systems: The uncharted sign worlds," *Science* **319**, 1205 (2008).

[143] A. R. Schmidt, M. H. Hamidian, P. Wahl, F. Meier, A. V. Balatsky, J. D. Garrett, T. J. Williams, G. M. Luke and J. C. Davis, "Imaging the Fano lattice to 'hidden order' transition in URu_2Si_2," *Nature* **465**, 570 (2010).

[144] P. Aynajian, E. H. da Silva Neto, A. Gyenis, R. E. Baumbach, J. D. Thompson, Z. Fisk, E. D. Bauer and A. Yazdani, "Visualizing heavy fermions emerging in a quantum critical Kondo lattice," *Nature* **486**, 201 (2012).

[145] A. Schröder, G. Aeppli, E. Bucher, R. Ramazashvili and P. Coleman, "Scaling of magnetic fluctuations near a quantum phase transition," *Phys. Rev. Lett.* **80**, 5623 (1998).

[146] A. Schröder, G. Aeppli, R. Coldea, M. Adams, O. Stockert, H. von Löhneysen, E. Bucher, R. Ramazashvili and P. Coleman, "Onset of antiferromagnetism in heavy-fermion metals," *Nature* **407**, 351 (2000).

[147] P. Coleman, A. J. Schofield and A. M. Tsvelik, "How should we interpret the two transport relaxation times in the cuprates?" *J. Phys.: Condensed Matter* **8**, 9985 (1996) [arXiv:9609009 [cond-mat]].

[148] J. Zaanen, "Holographic duality: stealing dimensions from metals," *Nature Physics* **9**, 609 (2013).

[149] D. van der Marel, H. J. A. Molegraaf, J. Zaanen, Z. Nussinov, F. Carbone, A. Damascelli, H. Eisaki, M. Greven, P. H. Kes and M. Li, "Power-law optical conductivity with a constant phase angle in high T_c superconductors," *Nature* **425**, 271 (2003) [arXiv:0309172 [cond-mat.mes-hall]].

[150] K. Fujita, C. K. Kim, I. Lee, J. Lee, M. H. Hamidian, I. A. Firmo, S. Mukhopadhyay, H. Eisaki, S. Uchida, M. J. Lawler, E.-A. Kim and J. C. Davis, "Simultaneous transition in cuprate momentum-space topology and electronic symmetry breaking," *Science* **344**, 613 (2014) [arXiv:1403.7788 [cond-mat.supr-con]].

[151] U. Chatterjee, D. Ai, J. Zhao, S. Rosenkranz, A. Kaminski, H. Raffy, Z. Li, K. Kadowaki, M. Randeria, M. R. Norman and J. C. Campuzano, "Electronic phase diagram of high-temperature copper oxide superconductors," *PNAS* **108**, 9346 (2011).

[152] C. M. Varma, P. B. Littlewood, S. Schmitt-Rink, E. Abrahams and A. E. Ruckenstein, "Phenomenology of the normal state of Cu–O high-temperature superconductors," *Phys. Rev. Lett.* **63**, 1996 (1989).

[153] R. A. Cooper, Y. Wang, B. Vignolle, O. J. Lipscombe, S. M. Hayden, Y. Tanabe, T. Adachi, Y. Koike, M. Nohara, H. Takagi, C. Proust and N. E. Hussey, "Anomalous criticality in the electrical resistivity of $La_{2-x}Sr_xCuO_4$," *Science* **323**, 603 (2009).

[154] K. Fujita, M. H. Hamidian, S. D. Edkins, C. K. Kim, Y. Kohsaka, M. Azuma, M. Takano, H. Takagi, H. Eisaki, S. Uchida, A. Allais, M. J. Lawler, E.-A. Kim, S. Sachdev and J. C. Seamus Davis, "Direct phase-sensitive identification of a d-form factor density wave in underdoped cuprates," *PNAS* **111**, E3026 (2014) [arXiv:1404.0362 [cond-mat.supr-con]].

[155] J. Zaanen, "High temperature superconductivity: the sound of the hidden order," *Nature* **498**, 41 (2013).

[156] I. M. Vishik, E. A. Nowadnick, W. S. Lee, Z. X. Shen, B. Moritz, T. P. Devereaux, K. Tanaka, T. Sasagawa and T. Fujii, "A momentum-dependent perspective on quasiparticle interference in $Bi_2Sr_2CaCu_2O_{8+\delta}$," *Nature Physics* **5**, 718 (2009) [arXiv:0909.0762 [cond-mat.supr-con]].

[157] Y. He, Y. Yin, M. Zech, A. Soumyanarayanan, M. M. Yee, T. Williams, M. C. Boyer, K. Chatterjee, W. D. Wise, I. Zeljkovic, T. Kondo, T. Takeuchi, H. Ikuta, P. Mistark, R. S. Markiewicz, A. Bansil, S. Sachdev, E. W. Hudson and J. E. Hoffman, "Fermi surface and pseudogap evolution in a cuprate superconductor," *Science* **344**, 608 (2014).

[158] R. Comin, A. Frano, M. M. Yee, Y. Yoshida, H. Eisaki, E. Schierle, E. Weschke, R. Sutarto, F. He, A. Soumyanarayanan, Yang He, M. Le Tacon, I. S. Elfimov, J. E. Hoffman, G. A. Sawatzky, B. Keimer and A. Damascelli, "Charge order driven by Fermi-arc instability in $Bi_2Sr_{2x}La_xCuO_{6+\delta}$," *Science* **343**, 390 (2014).

[159] N. Iqbal, H. Liu and M. Mezei, "Semi-local quantum liquids," *JHEP* **1204**, 086 (2012) [arXiv:1105.4621 [hep-th]].

[160] N. Iqbal, H. Liu and M. Mezei, "Quantum phase transitions in semi-local quantum liquids" [arXiv:1108.0425 [hep-th]].

[161] S. W. Hawking and G. F. R. Ellis, *The Large Scale Structure of Space-time*, Cambridge University Press, 1973.

[162] S. Hawking and R. Penrose, *The Nature of Space and Time*, Princeton University Press, 1996.

[163] C. W. Misner, K. S. Thorne and J. A. Wheeler, *Gravitation*, W. H. Freeman and Company, 1973.

[164] J. D. Bekenstein, "Black hole hair: 25 years after," in *Second International A. D. Sakharov Conference on Physics*, 1996, pp. 216–219 [arXiv:9605059 [gr-qc]].

[165] J. M. Bardeen, B. Carter and S. W. Hawking, "The four laws of black hole mechanics," *Commun. Math. Phys.* **31**, 161 (1973).

[166] J. D. Bekenstein, "Black holes and entropy," *Phys. Rev.* D **7**, 2333 (1973).

[167] G. 't Hooft, "Dimensional reduction in quantum gravity" [arXiv:9310026 [gr-qc]].

[168] L. Susskind, "The world as a hologram," *J. Math. Phys.* **36**, 6377 (1995) [arXiv:9409089 [hep-th]].

[169] A. Strominger and C. Vafa, "Microscopic origin of the Bekenstein–Hawking entropy," *Phys. Lett.* B **379**, 99 (1996) [arXiv:9601029 [hep-th]].

[170] M. B. Green, J. H. Schwarz and E. Witten, *Superstring Theory. Volume 1: Introduction* and *Superstring Theory. Volume 2: Loop Amplitudes, Anomalies and Phenomenology*, Cambridge University Press, 1987.

[171] J. Polchinski, *String Theory. Volume 1: An Introduction to the Bosonic String*, Cambridge University Press, 1998.

[172] J. Polchinski, *String Theory. Volume 2: Superstring Theory and Beyond*, Cambridge University Press, 1998.

[173] A. N. Schellekens, "Life at the interface of particle physics and string theory," *Rev. Mod. Phys.* **85**, 1491 (2013) [arXiv:1306.5083 [hep-ph]].

[174] N. Seiberg, "Emergent spacetime" [arXiv:0601234 [hep-th]].

[175] J. Polchinski, "Dirichlet branes and Ramond–Ramond charges," *Phys. Rev. Lett.* **75**, 4724 (1995) [arXiv:9510017 [hep-th]].

[176] N. Arkani-Hamed, S. Dimopoulos and G. R. Dvali, "The hierarchy problem and new dimensions at a millimeter," *Phys. Lett.* B **429**, 263 (1998) [arXiv:9803315 [hep-th]].

[177] G. Shiu and S. H. H. Tye, "TeV scale superstring and extra dimensions," *Phys. Rev.* D **58**, 106007 (1998) [arXiv:9805157 [hep-th]].

[178] R. Maartens and K. Koyama, "Brane-world gravity," *Living Rev. Rel.* **13**, 5 (2010) [arXiv:1004.3962 [hep-th]].

[179] N. Beisert and M. Staudacher, "The $N = 4$ SYM integrable super spin chain," *Nucl. Phys.* B **670**, 439 (2003) [arXiv:0307042 [hep-th]].

[180] A. Cappelli and I. D. Rodriguez, "Matrix effective theories of the fractional quantum Hall effect," *J. Phys.* A **42**, 304006 (2009) [arXiv:0902.0765 [hep-th]].

[181] S.-S. Lee, "Low-energy effective theory of Fermi surface coupled with $U(1)$ gauge field in $2 + 1$ dimensions," *Phys. Rev.* B **80**, 165102 (2009) [arXiv:0905.4532 [cond-mat.str-el]].

[182] A. Liam Fitzpatrick, S. Kachru, J. Kaplan and S. Raghu, "Non-Fermi liquid behavior of large NB quantum critical metals," *Phys. Rev.* B **89**, 165114 (2014) [arXiv:1312.3321 [cond-mat.str-el]].

[183] S. Coleman, *Aspects of Symmetry*, Cambridge University Press, 1985.

[184] J. Zinn-Justin, *Quantum Field Theory and Critical Phenomena*, 4th edn, Clarendon Press, 2002.

[185] M. Moshe and J. Zinn-Justin, "Quantum field theory in the large-N limit: a review," *Phys. Rep.* **385**, 69 (2003) [arXiv:0306133 [hep-th]].

[186] G. 't Hooft, "A planar diagram theory for strong interactions," *Nucl. Phys.* B **72**, 461 (1974).

[187] A. V. Manohar, "Large N QCD" [arXiv:9802419 [hep-ph]].

[188] A. V. Ramallo, "Introduction to the AdS/CFT correspondence," in C. Merino (ed.), *Lectures on Particle Physics, Astrophysics and Cosmology, Proceedings of the Third IDPASC School, Santiago de Compostela, 2013*, Springer, 2015, 411 [arXiv:1310.4319 [hep-th]].

[189] E. Witten, "The $1/N$ expansion in atomic and particle physics," in G. 't Hooft (ed.), *Recent Developments in Gauge Theories, 1979* Cargèse Lectures, Plenum, 1980, HUTP-79/A078.

[190] E. Brezin and S. R. Wadia, *The Large N Expansion in Quantum Field Theory and Statistical Physics: From Spin Systems to Two-Dimensional Gravity*, World Scientific, 1993.

[191] D. J. Gross and W. Taylor, "Two-dimensional QCD is a string theory," *Nucl. Phys.* B **400**, 181 (1993) [arXiv:9301068 [hep-th]].

[192] J. Polchinski, "Scale and conformal invariance in quantum field theory," *Nucl. Phys.* B **303**, 226 (1988).

[193] D. Dorigoni and V. S. Rychkov, "Scale invariance + unitarity => conformal invariance?" [arXiv:0910.1087 [hep-th]].

[194] M. A. Luty, J. Polchinski and R. Rattazzi, "The a-theorem and the asymptotics of 4D quantum field theory," *JHEP* **1301**, 152 (2013) [arXiv:1204.5221 [hep-th]].

[195] S. Weinberg, *Gravitation and Cosmology: Principles and Applications of the General Theory of Relativity*, John Wiley & Sons, 1972.

[196] R. M. Wald, *General Relativity*, Chicago University Press, 1984.

[197] S. M. Carroll, *Spacetime and geometry: An Introduction to General Relativity*, Addison-Wesley, 2004.

[198] V. Balasubramanian, P. Kraus and A. E. Lawrence, "Bulk versus boundary dynamics in anti-de Sitter space-time," *Phys. Rev.* D **59**, 046003 (1999) [arXiv:9805171 [hep-th]].

[199] J. L. Petersen, "Introduction to the Maldacena conjecture on AdS/CFT," *Int. J. Mod. Phys.* A **14**, 3597 (1999) [arXiv:9902131 [hep-th]].

[200] S. de Haro, S. N. Solodukhin and K. Skenderis, "Holographic reconstruction of space-time and renormalization in the AdS/CFT correspondence," *Commun. Math. Phys.* **217**, 595 (2001) [arXiv:0002230 [hep-th]].

[201] D. T. Son and A. O. Starinets, "Minkowski space correlators in AdS/CFT correspondence: recipe and applications," *JHEP* **0209**, 042 (2002) [arXiv:0205051 [hep-th]].

[202] C. P. Herzog and D. T. Son, "Schwinger–Keldysh propagators from AdS/CFT correspondence," *JHEP* **0303**, 046 (2003) [arXiv:0212072 [hep-th]].

[203] D. Z. Freedman, S. D. Mathur, A. Matusis and L. Rastelli, "Correlation functions in the CFT(d)/AdS($d + 1$) correspondence," *Nucl. Phys.* B **546**, 96 (1999) [arXiv:9804058 [hep-th]].

[204] K. Skenderis, "Lecture notes on holographic renormalization," *Class. Quant. Grav.* **19**, 5849 (2002) [arXiv:0209067 [hep-th]].

[205] E. D'Hoker and D. Z. Freedman, "Supersymmetric gauge theories and the AdS/CFT correspondence," [arXiv:0201253 [hep-th]].

[206] M. J. G. Veltman, "Unitarity and causality in a renormalizable field theory with unstable particles," *Physica* **29**, 186 (1963).

[207] S. Minwalla, "Restrictions imposed by super-conformal invariance on quantum field theories," *Adv. Theor. Math. Phys.* **2**, 781 (1998) [arXiv:9712074 [hep-th]].

[208] J. de Boer, E. P. Verlinde and H. L. Verlinde, "On the holographic renormalization group," *JHEP* **0008**, 003 (2000) [arXiv:9912012 [hep-th]].

[209] E. Witten, "Multitrace operators, boundary conditions, and AdS/CFT correspondence" [arXiv:0112258 [hep-th]].

[210] W. Mueck, "An improved correspondence formula for AdS/CFT with multitrace operators," *Phys. Lett.* B **531**, 301 (2002) [arXiv:0201100 [hep-th]].

[211] E. Witten, "Anti-de Sitter space, thermal phase transition, and confinement in gauge theories," *Adv. Theor. Math. Phys.* **2**, 505 (1998) [arXiv:9803131 [hep-th]].

[212] G. W. Gibbons and S. W. Hawking (eds), *Euclidean Quantum Gravity*. World Scientific, 1993.

[213] E. Berti, V. Cardoso and A. O. Starinets, "Quasinormal modes of black holes and black branes," *Class. Quant. Grav.* **26**, 163001 (2009) [arXiv:0905.2975 [gr-qc]].

[214] D. Birmingham, "Topological black holes in anti-de Sitter space," *Class. Quant. Grav.* **16**, 1197 (1999) [arXiv:9808032 [hep-th]].

[215] S. S. Gubser, I. R. Klebanov and A. A. Tseytlin, "Coupling constant dependence in the thermodynamics of $N = 4$ supersymmetric Yang–Mills theory," *Nucl. Phys.* B **534**, 202 (1998) [arXiv:9805156 [hep-th]].

[216] S. S. Gubser, I. R. Klebanov and A. W. Peet, "Entropy and temperature of black 3-branes," *Phys. Rev.* D **54**, 3915 (1996) [arXiv:9602135 [hep-th]].

[217] G. W. Gibbons and S. W. Hawking, "Action integrals and partition functions in quantum gravity," *Phys. Rev.* D **15**, 2752 (1977).

[218] J. W. York, "Role of conformal three-geometry in the dynamics of gravitation," *Phys. Rev. Lett.* **28**, 1082 (1972).

[219] V. Balasubramanian and P. Kraus, "A stress tensor for anti-de Sitter gravity," *Commun. Math. Phys.* **208**, 413 (1999) [arXiv:9902121 [hep-th]].

[220] I. Papadimitriou and K. Skenderis, "Thermodynamics of asymptotically locally AdS spacetimes," *JHEP* **0508**, 004 (2005) [arXiv:0505190 [hep-th]].

[221] S. W. Hawking and D. N. Page, "Thermodynamics of black holes in anti-de Sitter space," *Commun. Math. Phys.* **87**, 577 (1983).

[222] J. M. Maldacena, "Wilson loops in large-N field theories," *Phys. Rev. Lett.* **80**, 4859 (1998) [arXiv:9803002 [hep-th]].

[223] S. J. Rey and J. T. Yee, "Macroscopic strings as heavy quarks in large-N gauge theory and anti-de Sitter supergravity," *Eur. Phys. J.* C **22**, 379 (2001) [arXiv:9803001 [hep-th]].

[224] A. Brandhuber, N. Itzhaki, J. Sonnenschein and S. Yankielowicz, "Wilson loops, confinement, and phase transitions in large-N gauge theories from supergravity," *JHEP* **9806**, 001 (1998) [arXiv:9803263 [hep-th]].

[225] O. Jahn and O. Philipsen, "The Polyakov loop and its relation to static quark potentials and free energies," *Phys. Rev.* D **70**, 074504 (2004) [arXiv:0407042 [hep-th]].

[226] S.-J. Rey, S. Theisen and J.-T. Yee, "Wilson–Polyakov loop at finite temperature in large-N gauge theory and anti-de Sitter supergravity," *Nucl. Phys.* B **527**, 171 (1998) [arXiv:9803135 [hep-th]].

[227] A. Brandhuber, N. Itzhaki, J. Sonnenschein and S. Yankielowicz, "Wilson loops in the large-N limit at finite temperature," *Phys. Lett.* B **434**, 36 (1998) [arXiv:9803137 [hep-th]].

[228] J. Erlich, "Recent results in AdS/QCD," *PoS Confinement* **8**, 032 (2008) [arXiv:0812.4976 [hep-th]].

[229] J. Polchinski and M. J. Strassler, "The string dual of a confining four-dimensional gauge theory" [arXiv:0003136 [hep-th]].

[230] I. R. Klebanov and M. J. Strassler, "Supergravity and a confining gauge theory: duality cascades and chi SB resolution of naked singularities," *JHEP* **0008**, 052 (2000) [arXiv:0007191 [hep-th]].

[231] M. Kruczenski, D. Mateos, R. C. Myers and D. J. Winters, "Towards a holographic dual of large $N(c)$ QCD," *JHEP* **0405**, 041 (2004) [arXiv:0311270 [hep-th]].

[232] T. Sakai and S. Sugimoto, "Low energy hadron physics in holographic QCD," *Prog. Theor. Phys.* **113**, 843 (2005) [arXiv:0412141 [hep-th]].

[233] T. Sakai and S. Sugimoto, "More on a holographic dual of QCD," *Prog. Theor. Phys.* **114**, 1083 (2005) [arXiv:0507073 [hep-th]].

[234] J. Erlich, E. Katz, D. T. Son and M. A. Stephanov, "QCD and a holographic model of hadrons," *Phys. Rev. Lett.* **95**, 261602 (2005) [arXiv:0501128 [hep-th]].

[235] C. P. Herzog, "A holographic prediction of the deconfinement temperature," *Phys. Rev. Lett.* **98**, 091601 (2007) [arXiv:0608151 [hep-th]].

[236] L. Da Rold and A. Pomarol, "Chiral symmetry breaking from five dimensional spaces," *Nucl. Phys.* B **721**, 79 (2005) [arXiv:0501218 [hep-ph]].

[237] S. Caron-Huot, P. Kovtun, G. D. Moore, A. Starinets and L. G. Yaffe, "Photon and dilepton production in supersymmetric Yang–Mills plasma," *JHEP* **0612**, 015 (2006) [arXiv:0607237 [hep-th]].

[238] A. Karch, E. Katz, D. T. Son and M. A. Stephanov, "Linear confinement and AdS/QCD," *Phys. Rev.* D **74**, 015005 (2006) [arXiv:0602229 [hep-th]].

[239] G. T. Horowitz and R. C. Myers, "The AdS/CFT correspondence and a new positive energy conjecture for general relativity," *Phys. Rev.* D **59**, 026005 (1999) [arXiv:9808079 [hep-th]].

[240] H. Boschi-Filho and N. R. F. Braga, "QCD/string holographic mapping and glueball mass spectrum," *Eur. Phys. J.* C **32**, 529 (2004) [arXiv:0209080 [hep-th]].

[241] D. K. Hong, T. Inami and H.-U. Yee, "Baryons in AdS/QCD," *Phys. Lett.* B **646**, 165 (2007) [arXiv:0609270 [hep-th]].

[242] S. S. Gubser, S. S. Pufu and F. D. Rocha, "Bulk viscosity of strongly coupled plasmas with holographic duals," *JHEP* **0808**, 085 (2008) [arXiv:0806.0407 [hep-th]].

[243] N. Iqbal and H. Liu, "Real-time response in AdS/CFT with application to spinors," *Fortsch. Phys.* **57**, 367 (2009) [arXiv:0903.2596 [hep-th]].

[244] W. Witczak-Krempa, E. Sorensen and S. Sachdev, "The dynamics of quantum criticality via quantum Monte Carlo and holography," *Nature Physics* **10**, 361 (2014) [arXiv:1309.2941 [cond-mat.str-el]].

[245] K. Skenderis and B. C. van Rees, "Real-time gauge/gravity duality: prescription, renormalization and examples," *JHEP* **0905**, 085 (2009) [arXiv:0812.2909 [hep-th]].

[246] G. C. Giecold, "Fermionic Schwinger–Keldysh propagators from AdS/CFT," *JHEP* **0910**, 057 (2009) [arXiv:0904.4869 [hep-th]].

[247] K. Damle and S. Sachdev, "Non-zero temperature transport near quantum critical points," *Phys. Rev.* B **56**, 8714 (1997) [arXiv:9705206 [cond-mat.str-el]].

[248] G. Policastro, D. T. Son and A. O. Starinets, "From AdS/CFT correspondence to hydrodynamics," *JHEP* **0209**, 043 (2002) [arXiv:0205052 [hep-th]].

[249] E. Shuryak, "Why does the quark–gluon plasma at RHIC behave as a nearly ideal fluid?" *Prog. Part. Nucl. Phys.* **53**, 273 (2004) [arXiv:0312227 [hep-th]].

[250] D. Teaney, "The effects of viscosity on spectra, elliptic flow, and HBT radii," *Phys. Rev.* C **68**, 034913 (2003) [arXiv:0301099 [nucl-th]].

[251] E. V. Shuryak, "What RHIC experiments and theory tell us about properties of quark–gluon plasma?," *Nucl. Phys.* A **750**, 64 (2005) [arXiv:0405066 [hep-th]].

[252] P. Kovtun, D. T. Son and A. O. Starinets, "Viscosity in strongly interacting quantum field theories from black hole physics," *Phys. Rev. Lett.* **94**, 111601 (2005) [arXiv:0405231 [hep-th]].

[253] S. A. Hartnoll, P. K. Kovtun, M. Muller and S. Sachdev, "Theory of the Nernst effect near quantum phase transitions in condensed matter, and in dyonic black holes," *Phys. Rev.* B **76**, 144502 (2007) [arXiv:0706.3215 [cond-mat.str-el]].

[254] S. Bhattacharyya, V. Hubeny, S. Minwalla and M. Rangamani, "Nonlinear fluid dynamics from gravity," *JHEP* **0802**, 045 (2008) [arXiv:0712.2456 [hep-th]].

[255] D. Forster, *Hydrodynamic Fluctuations, Broken Symmetry, and Correlation Functions*, Benjamin, 1975.

[256] L. D. Landau and E. M. Lifshitz, *Fluid Mechanics*, 2nd edn. Pergamon Press, 1987.

[257] R. Baier, P. Romatschke, D. T. Son, A. O. Starinets and M. A. Stephanov, "Relativistic viscous hydrodynamics, conformal invariance, and holography," *JHEP* **0804**, 100 (2008) [arXiv:0712.2451 [hep-th]].

[258] L. P. Kadanoff and P. C. Martin, "Hydrodynamic equations and correlation functions," *Annals of Physics* **24**, 419 (1963).

[259] N. Iqbal and H. Liu, "Universality of the hydrodynamic limit in AdS/CFT and the membrane paradigm," *Phys. Rev.* D **79**, 025023 (2009) [arXiv:0809.3808 [hep-th]].

[260] P. Kovtun, D. T. Son and A. O. Starinets, "Holography and hydrodynamics: diffusion on stretched horizons," *JHEP* **0310**, 064 (2003) [arXiv:0309213 [hep-th]].

[261] A. Buchel and J. T. Liu, "Universality of the shear viscosity in supergravity," *Phys. Rev. Lett.* **93**, 090602 (2004) [arXiv:0311175 [hep-th]].

[262] J. Mas, "Shear viscosity from R-charged AdS black holes," *JHEP* **0603**, 016 (2006) [arXiv:0601144 [hep-th]].

[263] D. T. Son and A. O. Starinets, "Hydrodynamics of R-charged black holes," *JHEP* **0603**, 052 (2006) [arXiv:0601157 [hep-th]].

[264] A. Buchel, J. T. Liu and A. O. Starinets, "Coupling constant dependence of the shear viscosity in $N = 4$ supersymmetric Yang–Mills theory," *Nucl. Phys.* B **707**, 56 (2005) [arXiv:0406264 [hep-th]].

[265] M. Brigante, H. Liu, R. C. Myers, S. Shenker and S. Yaida, "Viscosity bound violation in higher derivative gravity," *Phys. Rev.* D **77**, 126006 (2008) [arXiv:0712.0805 [hep-th]].

[266] J. Erdmenger, P. Kerner and H. Zeller, "Non-universal shear viscosity from Einstein gravity," *Phys. Lett.* B **699**, 301 (2011) [arXiv:1011.5912 [hep-th]].

[267] A. Rebhan and D. Steineder, "Violation of the holographic viscosity bound in a strongly coupled anisotropic plasma," *Phys. Rev. Lett.* **108**, 021601 (2012) [arXiv:1110.6825 [hep-th]].

[268] J. Polchinski and E. Silverstein, "Large-density field theory, viscosity, and '$2k_F$' singularities from string duals," *Class. Quant. Grav.* **29**, 194008 (2012) [arXiv:1203.1015 [hep-th]].

[269] M. Brigante, H. Liu, R. C. Myers, S. Shenker and S. Yaida, "The viscosity bound and causality violation," *Phys. Rev. Lett.* **100**, 191601 (2008) [arXiv:0802.3318 [hep-th]].

[270] R. C. Myers, M. F. Paulos and A. Sinha, "Holographic studies of quasi-topological gravity," *JHEP* **1008**, 035 (2010) [arXiv:1004.2055 [hep-th]].

[271] S. Jeon and L. G. Yaffe, "From quantum field theory to hydrodynamics: transport coefficients and effective kinetic theory," *Phys. Rev.* D **53**, 5799 (1996) [arXiv:9512263 [hep-th]].

[272] S. C. Huot, S. Jeon and G. D. Moore, "Shear viscosity in weakly coupled $N = 4$ super Yang–Mills theory compared to QCD," *Phys. Rev. Lett.* **98**, 172303 (2007) [arXiv:0608062 [hep-th]].

[273] J. Erlich, "How well does AdS/QCD describe QCD?," *Int. J. Mod. Phys.* A **25**, 411 (2010) [arXiv:0908.0312 [hep-ph]].

[274] C. Cao, E. Elliott, J. Joseph, H. Wu, J. Petricka, T. Schafer and J. E. Thomas, "Universal quantum viscosity in a unitary Fermi gas," *Science* **331**, 58 (2010).

[275] T. Schaefer and D. Teaney, "Nearly perfect fluidity: from cold atomic gases to hot quark–gluon plasmas," *Rep. Prog. Phys.* **72**, 126001 (2009).

[276] P. K. Kovtun and A. O. Starinets, "Quasinormal modes and holography," *Phys. Rev.* D **72**, 086009 (2005) [arXiv:0506184 [hep-th]].

[277] D. T. Son and A. O. Starinets, "Viscosity, black holes, and quantum field theory," *Ann. Rev. Nucl. Part. Sci.* **57**, 95 (2007) [arXiv:0704.0240 [hep-th]].

[278] A. Nata Atmaja and K. Schalm, "Photon and dilepton production in soft wall AdS/QCD," *JHEP* **1008**, 124 (2010) [arXiv:0802.1460 [hep-th]].

[279] I. Bredberg, C. Keeler, V. Lysov and A. Strominger, "From Navier–Stokes to Einstein" [arXiv:1101.2451 [hep-th]].

[280] K. S. Thorne, R. H. Price and D. A. Macdonald, *Black Holes: The Membrane Paradigm*, Yale University Press, 1986.

[281] P. Kovtun, "Lectures on hydrodynamic fluctuations in relativistic theories," *J. Phys.* A **45**, 473001 (2012) [arXiv:1205.5040 [hep-th]].

[282] J. Bhattacharya, S. Bhattacharyya and S. Minwalla, "Dissipative superfluid dynamics from gravity," *JHEP* **1104**, 125 (2011) [arXiv:1101.3332 [hep-th]].

[283] C. P. Herzog, N. Lisker, P. Surowka and A. Yarom, "Transport in holographic superfluids," *JHEP* **1108**, 052 (2011) [arXiv:1101.3330 [hep-th]].

[284] J. Bhattacharya, S. Bhattacharyya, S. Minwalla and A. Yarom, "A theory of first order dissipative superfluid dynamics," *JHEP* **1405**, 147 (2014) [arXiv:1105.3733 [hep-th]].

[285] K. Jensen, M. Kaminski, P. Kovtun, R. Meyer, A. Ritz and A. Yarom, "Parity-violating hydrodynamics in 2 + 1 dimensions," *JHEP* **1205**, 102 (2012) [arXiv:1112.4498 [hep-th]].

[286] D. T. Son and P. Surowka, "Hydrodynamics with triangle anomalies," *Phys. Rev. Lett.* **103**, 191601 (2009) [arXiv:0906.5044 [hep-th]].

[287] P. M. Chesler and L. G. Yaffe, "Horizon formation and far-from-equilibrium isotropization in supersymmetric Yang–Mills plasma," *Phys. Rev. Lett.* **102**, 211601 (2009) [arXiv:0812.2053 [hep-th]].

[288] A. Adams, P. M. Chesler and H. Liu, "Holographic turbulence," *Phys. Rev. Lett.* **112**, 151602 (2014) [arXiv:1307.7267 [hep-th]].

[289] S. Bhattacharyya, S. Minwalla and S. R. Wadia, "The incompressible non-relativistic Navier–Stokes equation from gravity," *JHEP* **0908**, 059 (2009) [arXiv:0810.1545 [hep-th]].

[290] M. Rangamani, "Gravity and hydrodynamics: lectures on the fluid–gravity correspondence," *Class. Quant. Grav.* **26**, 224003 (2009) [arXiv:0905.4352 [hep-th]].

[291] V. E. Hubeny, S. Minwalla and M. Rangamani, "The fluid/gravity correspondence," [arXiv:1107.5780 [hep-th]].

[292] J. D. Brown and J. W. York, "Quasilocal energy and conserved charges derived from the gravitational action," *Phys. Rev.* D **47**, 1407 (1993).

[293] M. Henningson and K. Skenderis, "The holographic Weyl anomaly," *JHEP* **9807**, 023 (1998) [arXiv:9806087].

[294] V. Juričić, O. Vafek and I. F. Herbut, "Conductivity of interacting massless Dirac particles in graphene: collisionless regime," *Phys. Rev.* B **82**, 235402 (2010) [arXiv:1009.3269 [cond-mat.mes-hall]].

[295] R. C. Myers, S. Sachdev and A. Singh, "Holographic quantum critical transport without self-duality," *Phys. Rev.* D **83**, 066017 (2011) [arXiv:1010.0443 [hep-th]].

[296] K. Chen, L. Liu, Y. Deng, L. Pollet and N. Prokof'ev, "Universal conductivity in a two-dimensional superfluid-to-insulator quantum critical system," *Phys. Rev. Lett.* **112**, 030402 (2013) [arXiv:1309.5635 [cond-mat.str-el]].

[297] D. Chowdhury, S. Raju, S. Sachdev, A. Singh and P. Strack, "Multipoint correlators of conformal field theories: implications for quantum critical transport," *Phys. Rev.* B **87**, 085138 (2013) [arXiv:1210.5247 [cond-mat.str-el]].

[298] S. Weinberg, *The Quantum Theory of Fields. Volume II: Modern Applications*, Cambridge University Press, 2001.

[299] D. E. Kharzeev, L. D. McLerran and H. J. Warringa, "The effects of topological charge change in heavy ion collisions: 'event by event P and CP violation'," *Nucl. Phys.* A **803**, 227 (2008) [arXiv:0711.0950 [hep-ph]].

[300] K. Fukushima, D. E. Kharzeev and H. J. Warringa, "The chiral magnetic effect," *Phys. Rev.* D **78**, 074033 (2008) [arXiv:0808.3382 [hep-ph]].

[301] D. E. Kharzeev, "The chiral magnetic effect and anomaly-induced transport," *Prog. Part. Nucl. Phys.* **75**, 133 (2014) [arXiv:1312.3348 [hep-ph]].

[302] C.-X. Liu, P. Ye and X.-L. Qi, "Chiral gauge field and axial anomaly in a Weyl semi-metal," *Phys. Rev.* B **87**, 235306 (2013) [arXiv:1204.6551 [cond-mat.str-el]].

[303] D. T. Son and B. Z. Spivak, "Chiral anomaly and classical negative magnetoresistance of Weyl metals," *Phys. Rev.* B **88**, 104412 (2013), [arXiv:1206.1627 [cond-mat.mes-hall]].

[304] A. A. Zyuzin and A. A. Burkov, "Topological response in Weyl semimetals and the chiral anomaly," *Phys. Rev.* B **86**, 115133 (2012) [arXiv:1206.1868 [cond-mat.mes-hall]].

[305] K. Landsteiner, "Anomalous transport of Weyl fermions in Weyl semimetals," *Phys. Rev.* B **89**, 075124 (2014) [arXiv:1306.4932 [hep-th]].

[306] A. V. Sadofyev and M. V. Isachenkov, "The chiral magnetic effect in hydrodynamical approach," *Phys. Lett.* B **697**, 404 (2011) [arXiv:1010.1550 [hep-th]].

[307] Y. Neiman and Y. Oz, "Relativistic hydrodynamics with general anomalous charges," *JHEP* **1103**, 023 (2011) [arXiv:1011.5107 [hep-th]].

[308] V. I. Zakharov, "Chiral magnetic effect in hydrodynamic approximation," in D. Kharzeev, K. Landsteiner, A. Schmitt and H.-U. Yee (eds), *Strongly Interacting Matter in Magnetic Fields*, Springer, 2013, p. 295 [arXiv:1210.2186 [hep-ph]].

[309] J. Erdmenger, M. Haack, M. Kaminski and A. Yarom, "Fluid dynamics of R-charged black holes," *JHEP* **0901**, 055 (2009) [arXiv:0809.2488 [hep-th]].

[310] N. Banerjee, J. Bhattacharya, S. Bhattacharyya, S. Dutta, R. Loganayagam and P. Surowka, "Hydrodynamics from charged black branes," *JHEP* **1101**, 094 (2011) [arXiv:0809.2596 [hep-th]].

[311] D. E. Kharzeev, K. Landsteiner, A. Schmitt and H.-U. Yee, "Strongly interacting matter in magnetic fields: an overview," in D. Kharzeev, K. Landsteiner, A. Schmitt and H.-U. Yee (eds), *Strongly Interacting Matter in Magnetic Fields*, Springer, 2013, p. 1 [arXiv:1211.6245 [hep-ph]].

558   References

[312] O. Saremi and D. T. Son, "Hall viscosity from gauge/gravity duality," *JHEP* **1204**, 091 (2012) [arXiv:1103.4851 [hep-th]].

[313] D. T. Son and C. Wu, "Holographic spontaneous parity breaking and emergent Hall viscosity and angular momentum," *JHEP* **1407**, 076 (2014) [arXiv:1311.4882 [hep-th]].

[314] H. Liu, H. Ooguri, B. Stoica and N. Yunes, "Spontaneous generation of angular momentum in holographic theories," *Phys. Rev. Lett.* **110**, 211601 (2013) [arXiv:1212.3666 [hep-th]].

[315] H. Liu, H. Ooguri and B. Stoica, "Angular momentum generation by parity violation," *Phys. Rev.* D **89**, 106007 (2014) [arXiv:1311.5879 [hep-th]].

[316] A. Gynther, K. Landsteiner, F. Pena-Benitez and A. Rebhan, "Holographic anomalous conductivities and the chiral magnetic effect," *JHEP* **1102**, 110 (2011) [arXiv:1005.2587 [hep-th]].

[317] K. Landsteiner, E. Megias and F. Pena-Benitez, "Anomalous transport from Kubo formulae," in *Strongly Interacting Matter in Magnetic Fields*, Springer, 2013, p. 433 [arXiv:1207.5808 [hep-th]].

[318] K. Landsteiner, E. Megias and F. Pena-Benitez, "Gravitational anomaly and transport," *Phys. Rev. Lett.* **107**, 021601 (2011) [arXiv:1103.5006 [hep-ph]].

[319] M. Greiner, O. Mandel, T. Esslinger, T. W. Hänsch and I. Bloch, "Quantum phase transition from a superfluid to a Mott insulator in a gas of ultracold atoms," *Nature* **415**, 39 (2002).

[320] T. Faulkner, H. Liu, J. McGreevy and D. Vegh, "Emergent quantum criticality, Fermi surfaces, and AdS_2," *Phys. Rev.* D **83**, 125002 (2011) [arXiv:0907.2694 [hep-th]].

[321] S. S. Gubser and F. D. Rocha, "Peculiar properties of a charged dilatonic black hole in AdS_5," *Phys. Rev.* D **81**, 046001 (2010) [arXiv:0911.2898 [hep-th]].

[322] K. Goldstein, S. Kachru, S. Prakash and S. P. Trivedi, "Holography of charged dilaton black holes," *JHEP* **1008**, 078 (2010) [arXiv:0911.3586 [hep-th]].

[323] C. Charmousis, B. Gouteraux, B. S. Kim, E. Kiritsis and R. Meyer, "Effective holographic theories for low-temperature condensed matter systems," *JHEP* **1011**, 151 (2010) [arXiv:1005.4690 [hep-th]].

[324] B. Gouteraux and E. Kiritsis, "Generalized holographic quantum criticality at finite density," *JHEP* **1112**, 036 (2011) [arXiv:1107.2116 [hep-th]].

[325] M. Edalati, J. I. Jottar and R. G. Leigh, "Holography and the sound of criticality," *JHEP* **1010**, 058 (2010) [arXiv:1005.4075 [hep-th]].

[326] M. Edalati, J. I. Jottar and R. G. Leigh, "Shear modes, criticality and extremal black holes," *JHEP* **1004**, 075 (2010) [arXiv:1001.0779 [hep-th]].

[327] R. A. Davison and N. K. Kaplis, "Bosonic excitations of the AdS_4 Reissner–Nordström black hole," *JHEP* **1112**, 037 (2011) [arXiv:1111.0660 [hep-th]].

[328] R. A. Davison and A. Parnachev, "Hydrodynamics of cold holographic matter," *JHEP* **1306**, 100 (2013) [arXiv:1303.6334 [hep-th]].

[329] C. P. Herzog, "The hydrodynamics of M theory," *JHEP* **0212**, 026 (2002) [arXiv:0210126 [hep-th]].

[330] C. P. Herzog, "The sound of M theory," *Phys. Rev.* D **68**, 024013 (2003) [arXiv:0302086 [hep-th]].

[331] L. D. Landau, "Oscillations in a Fermi liquid," *Zh. Éksp. Teor. Fiz.* **32**, 59 (1957) [*Soviet Phys. – JETP* **5**, 101 (1959)].

[332] M. Edalati, J. I. Jottar and R. G. Leigh, "Transport coefficients at zero temperature from extremal black holes," *JHEP* **1001**, 018 (2010) [arXiv:0910.0645 [hep-th]].

[333] M. Kaminski, K. Landsteiner, J. Mas, J. P. Shock and J. Tarrio, "Holographic operator mixing and quasinormal modes on the brane," *JHEP* **1002**, 021 (2010) [arXiv:0911.3610 [hep-th]].

[334] X. Dong, S. Harrison, S. Kachru, G. Torroba and H. Wang, "Aspects of holography for theories with hyperscaling violation," *JHEP* **1206**, 041 (2012) [arXiv:1201.1905 [hep-th]].

[335] S. Kachru, X. Liu and M. Mulligan, "Gravity duals of Lifshitz-like fixed points," *Phys. Rev.* D **78**, 106005 (2008) [arXiv:0808.1725 [hep-th]].

[336] L. Huijse, S. Sachdev and B. Swingle, "Hidden Fermi surfaces in compressible states of gauge–gravity duality," *Phys. Rev.* B **85**, 035121 (2012) [arXiv:1112.0573 [cond-mat.str-el]].

[337] S. S. Gubser and J. Ren, "Analytic fermionic Green's functions from holography," *Phys. Rev.* D **86**, 046004 (2012) [arXiv:1204.6315 [hep-th]].

[338] M. Spradlin and A. Strominger, "Vacuum states for AdS$_2$ black holes," *JHEP* **9911**, 021 (1999) [arXiv:9904143 [hep-th]].

[339] A. Almheiri and J. Polchinski, "Models of AdS$_2$ backreaction and holography" [arXiv:1402.6334 [hep-th]].

[340] K. Copsey and R. Mann, "Pathologies in asymptotically Lifshitz spacetimes," *JHEP* **1103**, 039 (2011) [arXiv:1011.3502 [hep-th]].

[341] G. T. Horowitz and B. Way, "Lifshitz singularities," *Phys. Rev.* D **85**, 046008 (2012) [arXiv:1111.1243 [hep-th]].

[342] S. Harrison, S. Kachru and H. Wang, "Resolving Lifshitz horizons," *JHEP* **1402**, 085 (2014) [arXiv:1202.6635 [hep-th]].

[343] N. Bao, X. Dong, S. Harrison and E. Silverstein, "The benefits of stress: resolution of the Lifshitz singularity," *Phys. Rev.* D **86**, 106008 (2012) [arXiv:1207.0171 [hep-th]].

[344] E. Shaghoulian, "Holographic entanglement entropy and Fermi surfaces," *JHEP* **1205**, 065 (2012) [arXiv:1112.2702 [hep-th]].

[345] J. Bhattacharya, S. Cremonini and A. Sinkovics, "On the IR completion of geometries with hyperscaling violation," *JHEP* **1302**, 147 (2013) [arXiv:1208.1752 [hep-th]].

[346] S. A. Hartnoll and E. Shaghoulian, "Spectral weight in holographic scaling geometries," *JHEP* **1207**, 078 (2012) [arXiv:1203.4236 [hep-th]].

[347] S. S. Gubser, "Breaking an Abelian gauge symmetry near a black hole horizon," *Phys. Rev.* D **78**, 065034 (2008) [arXiv:0801.2977 [hep-th]].

[348] S. A. Hartnoll, C. P. Herzog and G. T. Horowitz, "Building a holographic superconductor," *Phys. Rev. Lett.* **101**, 031601 (2008) [arXiv:0803.3295 [hep-th]].

[349] H. Liu, J. McGreevy and D. Vegh, "Non-Fermi liquids from holography," *Phys. Rev.* D **83**, 065029 (2011) [arXiv:0903.2477 [hep-th]].

[350] M. Cubrovic, J. Zaanen and K. Schalm, "String theory, quantum phase transitions and the emergent Fermi-liquid," *Science* **325**, 439 (2009) [arXiv:0904.1993 [hep-th]].

[351] S.-S. Lee, "A non-Fermi liquid from a charged black hole: a critical Fermi ball," *Phys. Rev.* D **79**, 086006 (2009) [arXiv:0809.3402 [hep-th]].

[352] S.-J. Rey, "String theory on thin semiconductors: holographic realization of Fermi points and surfaces," *Prog. Theor. Phys. Suppl.* **177**, 128 (2009) [arXiv:0911.5295 [hep-th]].

[353] V. Juričić, I. F. Herbut and G. W. Semenoff, "Coulomb interaction at the metal–insulator critical point in graphene," *Phys. Rev.* B **80**, 081405 (2009) [arXiv:0906.3513 [cond-mat.str-el]].

[354] R. Contino and A. Pomarol, "Holography for fermions," *JHEP* **0411**, 058 (2004) [arXiv:0406257 [hep-th]].

[355] J. P. Gauntlett, J. Sonner and D. Waldram, "Universal fermionic spectral functions from string theory," *Phys. Rev. Lett.* **107**, 241601 (2011) [arXiv:1106.4694 [hep-th]].

[356] T. Faulkner, N. Iqbal, H. Liu, J. McGreevy and D. Vegh, "Holographic non-Fermi liquid fixed points," *Phil. Trans. Roy. Soc.* A **369**, 1640 (2011) [arXiv:1101.0597 [hep-th]].

[357] M. Cubrovic, Y. Liu, K. Schalm, Y.-W. Sun and J. Zaanen, "Spectral probes of the holographic Fermi groundstate: dialing between the electron star and AdS Dirac hair," *Phys. Rev.* D **84**, 086002 (2011) [arXiv:1106.1798 [hep-th]].

[358] B. Pioline and J. Troost, "Schwinger pair production in AdS$_2$," *JHEP* **0503**, 043 (2005) [arXiv:0501169 [hep-th]].

[359] Y. Liu, K. Schalm, Y. W. Sun and J. Zaanen, "Lattice potentials and fermions in holographic non Fermi-liquids: hybridizing local quantum criticality," *JHEP* **1210**, 036 (2012) [arXiv:1205.5227 [hep-th]].

[360] T. Hartman and S. A. Hartnoll, "Cooper pairing near charged black holes," *JHEP* **1006**, 005 (2010) [arXiv:1003.1918 [hep-th]].

[361] T. Faulkner, N. Iqbal, H. Liu, J. McGreevy and D. Vegh, "Strange metal transport realized by gauge/gravity duality," *Science* **329**, 1043 (2010).

[362] O. DeWolfe, S. S. Gubser and C. Rosen, "Fermi surfaces in maximal gauged supergravity," *Phys. Rev. Lett.* **108**, 251601 (2012) [arXiv:1112.3036 [hep-th]].

[363] O. DeWolfe, S. S. Gubser and C. Rosen, "Fermi surfaces in $N = 4$ super-Yang–Mills theory," *Phys. Rev.* D **86**, 106002 (2012) [arXiv:1207.3352 [hep-th]].

[364] T. Faulkner, N. Iqbal, H. Liu, J. McGreevy and D. Vegh, "From black holes to strange metals" [arXiv:1003.1728 [hep-th]].

[365] J. Polchinski, "Low energy dynamics of the spinon–gauge system," *Nucl. Phys.* B **422**, 617 (1994). [arXiv:9303037 [cond-mat]].

[366] T. Faulkner and J. Polchinski, "Semi-holographic Fermi liquids," *JHEP* **1106**, 012 (2011) [arXiv:1001.5049 [hep-th]].

[367] S. A. Hartnoll, C. P. Herzog and G. T. Horowitz, "Holographic superconductors," *JHEP* **0812**, 015 (2008) [arXiv:0810.1563 [hep-th]].

[368] R. Ruffini and J. A. Wheeler, "Introducing the black hole," *Physics Today* **24**, 30 (1971).

[369] D. Anninos, S. A. Hartnoll and N. Iqbal, "Holography and the Coleman–Mermin–Wagner theorem," *Phys. Rev.* D **82**, 066008 (2010) [arXiv:1005.1973 [hep-th]].

[370] M. Ammon, J. Erdmenger, M. Kaminski and P. Kerner, "Superconductivity from gauge/gravity duality with flavor," *Phys. Lett.* B **680**, 516 (2009) [arXiv:0810.2316 [hep-th]].

[371] T. Albash and C. V. Johnson, "Vortex and droplet engineering in holographic superconductors," *Phys. Rev.* D **80**, 126009 (2009) [arXiv:0906.1795 [hep-th]].

[372] M. Montull, A. Pomarol and P. J. Silva, "The holographic superconductor vortex," *Phys. Rev. Lett.* **103**, 091601 (2009) [arXiv:0906.2396 [hep-th]].

[373] K. Maeda, M. Natsuume and T. Okamura, "Vortex lattice for a holographic superconductor," *Phys. Rev.* D **81**, 026002 (2010) [arXiv:0910.4475 [hep-th]].

[374] V. Keranen, E. Keski-Vakkuri, S. Nowling and K. P. Yogendran, "Inhomogeneous structures in holographic superfluids: I. Dark solitons," *Phys. Rev.* D **81**, 126011 (2010) [arXiv:0911.1866 [hep-th]].

[375] A. Adams, P. M. Chesler and H. Liu, "Holographic vortex liquids and superfluid turbulence," *Science* **341**, 368 (2013) [arXiv:1212.0281 [hep-th]].

[376] L. D. Landau, "Theory of the superfluidity of helium II," *Phys. Rev.* **60**, 356 (1941).

[377] L. Tisza, "The theory of liquid helium," *Phys. Rev.* **72**, 838 (1947).

[378] W. Israel, "Covariant superfluid mechanics," *Phys. Lett.* A **86**, 79 (1981).

[379] I. M. Khalatnikov and V. V. Lebedev, "Second sound in liquid helium II," *Phys. Lett.* A **91**, 70 (1982).

[380] W. Israel, "Equivalence of two theories of relativistic superfluid mechanics," *Phys. Lett.* A **92**, 77 (1982).

[381] D. T. Son, "Hydrodynamics of relativistic systems with broken continuous symmetries," *Int. J. Mod. Phys.* A **16** (suppl. 01C), 1284 (2001) [arXiv:0011246 [hep-th]].

[382] J. Sonner and B. Withers, "A gravity derivation of the Tisza–Landau model in AdS/CFT," *Phys. Rev.* D **82**, 026001 (2010) [arXiv:1004.2707 [hep-th]].

[383] G. T. Horowitz, J. E. Santos and B. Way, "A holographic Josephson junction," *Phys. Rev. Lett.* **106**, 221601 (2011) [arXiv:1101.3326 [hep-th]].

[384] E. Kiritsis and V. Niarchos, "Josephson junctions and AdS/CFT networks," *JHEP* **1107**, 112 (2011) [*Erratum ibid.* **1110**, 095 (2011)] [arXiv:1105.6100 [hep-th]].

[385] T. Faulkner, G. T. Horowitz, J. McGreevy, M. M. Roberts and D. Vegh, "Photoemission 'experiments' on holographic superconductors," *JHEP* **1003**, 121 (2010) [arXiv:0911.3402 [hep-th]].

[386] J.-W. Chen, Y.-J. Kao and W.-Y. Wen, "Peak–dip–hump from holographic superconductivity," *Phys. Rev.* D **82**, 026007 (2010) [arXiv:0911.2821 [hep-th]].

[387] S. S. Gubser and S. S. Pufu, "The gravity dual of a p-wave superconductor," *JHEP* **0811**, 033 (2008) [arXiv:0805.2960 [hep-th]].

[388] M. M. Roberts and S. A. Hartnoll, "Pseudogap and time reversal breaking in a holographic superconductor," *JHEP* **0808**, 035 (2008) [arXiv:0805.3898 [hep-th]].

[389] F. Benini, C. P. Herzog and A. Yarom, "Holographic Fermi arcs and a d-wave gap," *Phys. Lett.* B **701**, 626 (2011) [arXiv:1006.0731 [hep-th]].

[390] F. Benini, C. P. Herzog, R. Rahman and A. Yarom, "Gauge gravity duality for d-wave superconductors: prospects and challenges," *JHEP* **1011**, 137 (2010) [arXiv:1007.1981 [hep-th]].

[391] K. Y. Kim and M. Taylor, "Holographic d-wave superconductors," *JHEP* **1308**, 112 (2013) [arXiv:1304.6729 [hep-th]].

[392] M. Ammon, J. Erdmenger, V. Grass, P. Kerner and A. O'Bannon, "On holographic p-wave superfluids with back-reaction," *Phys. Lett.* B **686**, 192 (2010) [arXiv:0912.3515 [hep-th]].

[393] S. S. Gubser, F. D. Rocha and A. Yarom, "Fermion correlators in non-Abelian holographic superconductors," *JHEP* **1011**, 085 (2010) [arXiv:1002.4416 [hep-th]].

[394] J. Erdmenger, D. Fernandez and H. Zeller, "New transport properties of anisotropic holographic superfluids," *JHEP* **1304**, 049 (2013) [arXiv:1212.4838 [hep-th]].

[395] R. A. Ferrell, "Fluctuations and the superconducting phase transition: II. Onset of Josephson tunneling and paraconductivity of a junction," *J. Low Temp. Phys.* **1**, 423 (1969).

[396] D. J. Scalapino, "Pair tunneling as a probe of fluctuations in superconductors," *Phys. Rev. Lett.* **24**, 1052 (1970).

[397] J. T. Anderson and A. M. Goldman, "Experimental determination of the pair susceptibility of a superconductor," *Phys. Rev. Lett.* **25**, 743 (1970).

[398] A. M. Goldman, "The order parameter susceptibility and collective modes of superconductors," *J. Supercond. Nov. Magn.* **19**, 317 (2006).

[399] A. V. Chubukov, D. Pines and J. Schmalian, "Spin fluctuation model for d-wave superconductivity," in K. H. Bennemann and J. B. Ketterson (eds), *The Physics of Superconductors*, Vol. 1, Springer, 2004, pp. 495–590.

[400] K. Jensen, "Semi-holographic quantum criticality," *Phys. Rev. Lett.* **107**, 231601 (2011) [arXiv:1108.0421 [hep-th]].

[401] D. B. Kaplan, J.-W. Lee, D. T. Son and M. A. Stephanov, "Conformality lost," *Phys. Rev. D* **80**, 125005 (2009) [arXiv:0905.4752 [hep-th]].

[402] T. Nishioka, S. Ryu and T. Takayanagi, "Holographic superconductor/insulator transition at zero temperature," *JHEP* **1003**, 131 (2010) [arXiv:0911.0962 [hep-th]].

[403] G. T. Horowitz and B. Way, "Complete phase diagrams for a holographic superconductor/insulator system," *JHEP* **1011**, 011 (2010) [arXiv:1007.3714 [hep-th]].

[404] S. S. Gubser and A. Nellore, "Ground states of holographic superconductors," *Phys. Rev. D* **80**, 105007 (2009) [arXiv:0908.1972 [hep-th]].

[405] G. T. Horowitz and M. M. Roberts, "Zero temperature limit of holographic superconductors," *JHEP* **0911**, 015 (2009) [arXiv:0908.3677 [hep-th]].

[406] N. Iqbal, H. Liu, M. Mezei and Q. Si, "Quantum phase transitions in holographic models of magnetism and superconductors," *Phys. Rev. D* **82**, 045002 (2010) [arXiv:1003.0010 [hep-th]].

[407] P. W. Anderson, "In praise of unstable fixed points: the way things actually work," *Physica B: Condensed Matter* **318**, 28 (2002) [arXiv:0201431 [cond-mat]].

[408] P. C. W. Davies, "Thermodynamics of black holes," *Rep. Prog. Phys.* **41**, 1313 (1978).

[409] M. V. Medvedyeva, E. Gubankova, M. Cubrovic, K. Schalm and J. Zaanen, "Quantum corrected phase diagram of holographic fermions," *JHEP* **1312**, 025 (2013) [arXiv:1302.5149 [hep-th]].

[410] S. A. Hartnoll and P. Petrov, "Electron star birth: a continuous phase transition at nonzero density," *Phys. Rev. Lett.* **106**, 121601 (2011) [arXiv:1011.6469 [hep-th]].

[411] V. G. M. Puletti, S. Nowling, L. Thorlacius and T. Zingg, "Holographic metals at finite temperature," *JHEP* **1101**, 117 (2011) [arXiv:1011.6261 [hep-th]].

[412] A. Allais, J. McGreevy and S. J. Suh, "A quantum electron star," *Phys. Rev. Lett.* **108**, 231602 (2012) [arXiv:1202.5308 [hep-th]].

[413] A. Allais and J. McGreevy, "How to construct a gravitating quantum electron star," *Phys. Rev. D* **88**, 066006 (2013) [arXiv:1306.6075 [hep-th]].

[414] S. Sachdev, "A model of a Fermi liquid using gauge–gravity duality," *Phys. Rev. D* **84**, 066009 (2011) [arXiv:1107.5321 [hep-th]].

[415] E. Witten, "Baryons in the $1/N$ expansion," *Nucl. Phys. B* **160**, 57 (1979).

[416] E. Witten, "Baryons and branes in anti-de Sitter space," *JHEP* **9807**, 006 (1998) [arXir:9805112 [hep-th]].

[417] C. P. Herzog and J. Ren, "The spin of holographic electrons at nonzero density and temperature," *JHEP* **1206**, 078 (2012) [arXiv:1204.0518 [hep-th]].

[418] S. A. Hartnoll and A. Tavanfar, "Electron stars for holographic metallic criticality," *Phys. Rev. D* **83**, 046003 (2011) [arXiv:1008.2828 [hep-th]].

[419] N. Iizuka, N. Kundu, P. Narayan and S. P. Trivedi, "Holographic Fermi and non-Fermi liquids with transitions in dilaton gravity," *JHEP* **1201**, 094 (2012) [arXiv:1105.1162 [hep-th]].

[420] S. A. Hartnoll, D. M. Hofman and D. Vegh, "Stellar spectroscopy: fermions and holographic Lifshitz criticality," *JHEP* **1108**, 096 (2011) [arXiv:1105.3197 [hep-th]].

[421] N. Iqbal and H. Liu, "Luttinger's theorem, superfluid vortices, and holography," *Class. Quant. Grav.* **29**, 194004 (2012) [arXiv:1112.3671 [hep-th]].

[422] S. A. Hartnoll and L. Huijse, "Fractionalization of holographic Fermi surfaces," *Class. Quant. Grav.* **29**, 194001 (2012) [arXiv:1111.2606 [hep-th]].

[423] M. Cubrovic, K. Schalm and J. Zaanen, "The quantum phase transition from an AdS Reissner–Nordström black hole to an AdS electron star" (to be published).

[424] D. J. Gross and E. Witten, "Possible third order phase transition in the large-N lattice gauge theory," *Phys. Rev.* D **21**, 446 (1980).

[425] A. Bagrov, B. Meszena and K. Schalm, "Pairing induced superconductivity in holography," *JHEP* **1409**, 106 (2014) [arXiv:1403.3699 [hep-th]].

[426] Y. Liu, K. Schalm, Y. W. Sun and J. Zaanen, "BCS instabilities of electron stars to holographic superconductors," *JHEP* **1405**, 122 (2014) [arXiv:1404.0571 [hep-th]].

[427] J. de Boer, K. Papadodimas and E. Verlinde, "Holographic neutron stars," *JHEP* **1010**, 020 (2010) [arXiv:0907.2695 [hep-th]].

[428] X. Arsiwalla, J. de Boer, K. Papadodimas and E. Verlinde, "Degenerate stars and gravitational collapse in AdS/CFT," *JHEP* **1101**, 144 (2011) [arXiv:1010.5784 [hep-th]].

[429] J. M. Ziman, *Electrons and Phonons: The Theory of Transport Phenomena in Solids*, Oxford University Press, 1960.

[430] W. E. Lawrence and J. W. Wilkins, "Electron–electron scattering in the transport coefficients of simple metals," *Phys. Rev.* B **7**, 2317 (1973).

[431] W. Götze and P. Wölfle, "Homogeneous dynamical conductivity of simple metals," *Phys. Rev.* B **6**, 1226 (1972).

[432] A. Rosch and N. Andrei, "Conductivity of a clean one-dimensional wire," *Phys. Rev. Lett.* **85**, 1092 (2000).

[433] S. A. Hartnoll and D. M. Hofman, "Locally critical resistivities from umklapp scattering," *Phys. Rev. Lett.* **108**, 241601 (2012) [arXiv:1201.3917 [hep-th]].

[434] R. Mahajan, M. Barkeshli and S. A. Hartnoll, "Non-Fermi liquids and the Wiedemann–Franz law," *Phys. Rev.* B **88**, 125107 (2013) [arXiv:1304.4249 [cond-mat.str-el]].

[435] A. V. Andreev, S. A. Kivelson and B. Spivak, "Hydrodynamic description of transport in strongly correlated electron systems," *Phys. Rev. Lett.* **106**, 256804 (2011).

[436] M. Blake, D. Tong and D. Vegh, "Holographic lattices give the graviton a mass," *Phys. Rev. Lett.* **112**, 071602 (2014) [arXiv:1310.3832 [hep-th]].

[437] R. Flauger, E. Pajer and S. Papanikolaou, "A striped holographic superconductor," *Phys. Rev.* D **83**, 064009 (2011) [arXiv:1010.1775 [hep-th]].

[438] G. T. Horowitz, J. E. Santos and D. Tong, "Optical conductivity with holographic lattices," *JHEP* **1207**, 168 (2012) [arXiv:1204.0519 [hep-th]].

[439] G. T. Horowitz, J. E. Santos and D. Tong, "Further evidence for lattice-induced scaling," *JHEP* **1211**, 102 (2012) [arXiv:1209.1098 [hep-th]].

[440] G. T. Horowitz and J. E. Santos, "General relativity and the cuprates," *JHEP* **1306**, 087 (2013) [arXiv:1302.6586 [hep-th]].

[441] D. van der Marel, H. J. A. Molegraaf, J. Zaanen, Z. Nussinov, F. Carbone, A. Damascelli, H. Eisaki, M. Greven, P. H. Kes and M. Li, "Quantum critical behaviour in a high-T_c superconductor," *Nature* **425**, 271 (2003) [arXiv:0309172 [cond-mat.str-el]].

[442] D. Dalidovich and P. Phillips, "Nonlinear transport near a quantum phase transition in two dimensions," *Phys. Rev. Lett.* **93**, 27004 (2004) [arXiv:0310129 [cond-mat.supr-con]].

[443] D. A. Bonn, R. Liang, T. M. Riseman, D. J. Baar, D. C. Morgan, K. Zhang, P. Dosanjh, T. L. Duty, A. MacFarlane, G. D. Morris, J. H. Brewer, W. N. Hardy, C.

Kallin and A. J. Berlinsky, "Microwave determination of the quasiparticle scattering time in $YBa_2Cu_3O_{6.95}$," *Phys. Rev.* B **47**, 11314 (1993).

[444] J. Orenstein, "Optical conductivity and spatial inhomogeneity in cuprate superconductors," in *Handbook of High-Temperature Superconductivity. Theory and Experiment*, Springer, 2007.

[445] M. P. Ryan and L. C. Shepley, *Homogeneous Relativistic Cosmologies*, Princeton University Press, 1975.

[446] N. Iizuka, S. Kachru, N. Kundu, P. Narayan, N. Sircar and S. P. Trivedi, "Bianchi attractors: a classification of extremal black brane geometries," *JHEP* **1207**, 193 (2012) [arXiv:1201.4861 [hep-th]].

[447] A. Donos and S. A. Hartnoll, "Interaction-driven localization in holography," *Nature Physics* **9**, 649 (2013) [arXiv:1212.2998].

[448] E. D'Hoker and P. Kraus, "Charge expulsion from black brane horizons, and holographic quantum criticality in the plane," *JHEP* **1209**, 105 (2012) [arXiv:1202.2085 [hep-th]].

[449] J. Zaanen, "High-temperature superconductivity: the secret of the hourglass," *Nature* **471**, 314 (2011).

[450] V. J. Emery, E. Fradkin, S. A. Kivelson and T. C. Lubensky, "Quantum theory of the smectic metal state in stripe phases," *Phys. Rev. Lett.* **85**, 2160 (2000) [arXiv:cond-mat/0001077 [cond-mat.str-el]].

[451] G. T. Horowitz and M. M. Roberts, "Holographic superconductors with various condensates," *Phys. Rev.* D **78**, 126008 (2008) [arXiv:0810.1077 [hep-th]].

[452] M. Taylor, "More on counterterms in the gravitational action and anomalies" [arXiv:0002125 [hep-th]].

[453] A. Donos and J. P. Gauntlett, "Holographic Q-lattices," *JHEP* **1404**, 040 (2014) [arXiv:1311.3292 [hep-th]].

[454] A. Donos and J. P. Gauntlett, "Novel metals and insulators from holography," *JHEP* **1406**, 007 (2014) [arXiv:1401.5077 [hep-th]].

[455] T. Andrade and B. Withers, "A simple holographic model of momentum relaxation," *JHEP* **1405**, 101 (2014) [arXiv:1311.5157 [hep-th]].

[456] B. Goutéraux, "Charge transport in holography with momentum dissipation," *JHEP* **1404**, 181 (2014) [arXiv:1401.5436 [hep-th]].

[457] K. Hinterbichler, "Theoretical aspects of massive gravity," *Rev. Mod. Phys.* **84**, 671 (2012) [arXiv:1105.3735 [hep-th]].

[458] C. de Rham, "Massive gravity," *Living Rev. Rel.* **17**, 7 (2014) [arXiv:1401.4173 [hep-th]].

[459] C. de Rham, G. Gabadadze and A. J. Tolley, "Resummation of massive gravity," *Phys. Rev. Lett.* **106**, 231101 (2011) [arXiv:1011.1232 [hep-th]].

[460] H. Kleinert, *Gauge Fields in Condensed Matter. Volume 2: Stresses and Defects. Differential Geometry, Crystal Melting*, World Scientific, 1989.

[461] H. Kleinert, *Multivalued Fields in Condensed Matter, Electromagnetism, and Gravitation*, World Scientific, 2008.

[462] L. Giomi and M. Bowick, "Two-dimensional matter: order, curvature and defects," *Adv. Phys.* **58**, 449 (2009) [arXiv:0812.3064 [cond-mat.soft]].

[463] A. J. Beekman, K. Wu, V. Cvetkovic and J. Zaanen, "Deconfining the rotational Goldstone mode: the superconducting nematic liquid crystal in $2 + 1$D," *Phys. Rev.* B **88**, 024121 (2013) [arXiv:1301.7329 [cond-mat.str-el]].

[464] J. Zaanen and A. J. Beekman, "The emergence of gauge invariance: the stay-at-home gauge versus local-global duality," *Annals of Physics* **327**, 1146 (2012) [arXiv:1108.2791 [cond-mat.str-el]].

[465] D. Vegh, "Holography without translational symmetry" [arXiv:1301.0537 [hep-th]].

[466] M. Blake and D. Tong, "Universal resistivity from holographic massive gravity," *Phys. Rev.* D **88**, 106004 (2013) [arXiv:1308.4970 [hep-th]].

[467] R. A. Davison, "Momentum relaxation in holographic massive gravity," *Phys. Rev.* D **88**, 086003 (2013) [arXiv:1306.5792 [hep-th]].

[468] R. A. Davison, K. Schalm and J. Zaanen, "Holographic duality and the resistivity of strange metals," *Phys. Rev.* B **89**, 245116 (2014) [arXiv:1311.2451 [hep-th]].

[469] J. A. N. Bruin, H. Sakai, R. S. Perry and A. P. Mackenzie, "Similarity of scattering rates in metals showing T-linear resistivity," *Science* **339**, 804 (2013).

[470] A. Lucas, S. Sachdev and K. Schalm, "Scale-invariant hyperscaling-violating holographic theories and the resistivity of strange metals with random-field disorder," *Phys. Rev.* D **89**, 066018 (2014) [arXiv:1401.7993 [hep-th]].

[471] B. Bradlyn, M. Goldstein and N. Read, "Kubo formulas for viscosity: Hall viscosity, Ward identities, and the relation with conductivity," *Phys. Rev.* B **86**, 245309 (2012) [arXiv:1207.7021 [cond-mat.stat-mech]].

[472] S. Sachdev and J. W. Ye, "Universal quantum critical dynamics of two-dimensional antiferromagnets," *Phys. Rev. Lett.* **69**, 2411 (1992) [arXiv:9204001 [cond-mat]].

[473] S. Sachdev, "Universal relaxational dynamics near two-dimensional quantum critical points," *Phys. Rev.* B **59**, 14054 (1999) [arXiv:9810399 [cond-mat.str-el]].

[474] S. A. Hartnoll, R. Mahajan, M. Punk and S. Sachdev, "Transport near the Ising-nematic quantum critical point of metals in two dimensions," *Phys. Rev.* B **89**, 155130 (2014) [arXiv:1401.7012 [cond-mat.str-el]].

[475] S. Nakamura, H. Ooguri and C.-S. Park, "Gravity dual of spatially modulated phase," *Phys. Rev.* D **81**, 044018 (2010) [arXiv:0911.0679 [hep-th]].

[476] H. Ooguri and C.-S. Park, "Holographic end-point of spatially modulated phase transition," *Phys. Rev.* D **82**, 126001 (2010) [arXiv:1007.3737 [hep-th]].

[477] A. Donos and J. P. Gauntlett, "Holographic striped phases," *JHEP* **1108**, 140 (2011) [arXiv:1106.2004 [hep-th]].

[478] O. Bergman, N. Jokela, G. Lifschytz and M. Lippert, "Striped instability of a holographic Fermi-like liquid," *JHEP* **1110**, 034 (2011) [arXiv:1106.3883 [hep-th]].

[479] S. Chakravarty, R. B. Laughlin, D. K. Morr and C. Nayak, "Hidden order in the cuprates," *Phys. Rev.* B **63**, 094503 (2001).

[480] P. A. Lee, N. Nagaosa and X.-G. Wen, "Doping a Mott insulator: physics of high-temperature superconductivity," *Rev. Mod. Phys.* **78**, 17 (2006).

[481] A. Shekhter and C. M. Varma, "Considerations on the symmetry of loop order in cuprates," *Phys. Rev.* B **80**, 214501 (2009) [arXiv:0905.1987 [cond-mat.supr-con]].

[482] A. Allais, J. Bauer and S. Sachdev, "Bond order instabilities in a correlated two-dimensional metal," *Phys. Rev.* B **90**, 155114 (2014) [arXiv:1402.4807 [cond-mat.str-el]].

[483] A. Donos and J. P. Gauntlett, "Holographic charge density waves," *Phys. Rev.* D **87**, 126008 (2013) [arXiv:1303.4398 [hep-th]].

[484] A. Donos and J. P. Gauntlett, "Black holes dual to helical current phases," *Phys. Rev.* D **86**, 064010 (2012) [arXiv:1204.1734 [hep-th]].

[485] J. P. Gauntlett, S. Kim, O. Varela and D. Waldram, "Consistent supersymmetric Kaluza–Klein truncations with massive modes," *JHEP* **0904**, 102 (2009) [arXiv:0901.0676 [hep-th]].

[486] M. Rozali, D. Smyth, E. Sorkin and J. B. Stang, "Holographic stripes," *Phys. Rev. Lett.* **110**, 201603 (2013) [arXiv:1211.5600 [hep-th]].

[487] A. Donos, "Striped phases from holography," *JHEP* **1305**, 059 (2013) [arXiv:1303.7211 [hep-th]].

[488] B. Withers, "Black branes dual to striped phases," *Class. Quant. Grav.* **30**, 155025 (2013) [arXiv:1304.0129 [hep-th]].

[489] M. Rozali, D. Smyth, E. Sorkin and J. B. Stang, "Striped order in AdS/CFT correspondence," *Phys. Rev.* D **87**, 126007 (2013) [arXiv:1304.3130 [hep-th]].

[490] A. Karch and E. Katz, "Adding flavor to AdS/CFT," *JHEP* **0206**, 043 (2002) [arXiv:0205236 [hep-th]].

[491] L. E. Ibanez and A. M. Uranga, *String Theory and Particle Physics: An Introduction to String Phenomenology*, Cambridge University Press, 2012.

[492] P. S. Aspinwall, "Compactification, Geometry and Duality: $N = 2$," TASI 1999 Lectures, report DUKE-CGTP-00-01 [arXiv:0001001 [hep-th]].

[493] F. Denef and S. A. Hartnoll, "Landscape of superconducting membranes," *Phys. Rev.* D **79**, 126008 (2009) [arXiv:0901.1160 [hep-th]].

[494] J. P. Gauntlett, J. Sonner and T. Wiseman, "Holographic superconductivity in M-theory," *Phys. Rev. Lett.* **103**, 151601 (2009) [arXiv:0907.3796 [hep-th]].

[495] J. P. Gauntlett, J. Sonner and T. Wiseman, "Quantum criticality and holographic superconductors in M-theory," *JHEP* **1002**, 060 (2010) [arXiv:0912.0512 [hep-th]].

[496] M. J. Duff and J. T. Liu, "Anti-de Sitter black holes in gauged $N = 8$ supergravity," *Nucl. Phys.* B **554**, 237 (1999) [arXiv:9901149 [hep-th]].

[497] S. S. Gubser and I. Mitra, "The evolution of unstable black holes in anti-de Sitter space," *JHEP* **0108**, 018 (2001) [arXiv:0011127 [hep-th]].

[498] R. C. Myers, "Dielectric branes," *JHEP* **9912**, 022 (1999) [arXiv:9910053 [hep-th]].

[499] M. Born and L. Infeld, "Foundations of the new field theory," *Proc. R. Soc. Lond.* A **144**, 425 (1934).

[500] P. A. M. Dirac, "A reformulation of the Born–Infeld electrodynamics," *Proc. R. Soc.* A **257**, 32 (1960).

[501] R. G. Leigh, "Dirac–Born–Infeld action from Dirichlet sigma model," *Mod. Phys. Lett.* A **4**, 2767 (1989).

[502] A. A. Tseytlin, "On non-Abelian generalization of Born–Infeld action in string theory," *Nucl. Phys.* B **501**, 41 (1997) [arXiv:9701125 [hep-th]].

[503] A. A. Tseytlin, "Born–Infeld action, supersymmetry and string theory," in M. A. Shifman (ed.), *The Many Faces of the Superworld*, World Scientific, 2000, pp. 417–452 [arXiv:9908105 [hep-th]].

[504] S. Kobayashi, D. Mateos, S. Matsuura, R. C. Myers and R. M. Thomson, "Holographic phase transitions at finite baryon density," *JHEP* **0702**, 016 (2007) [arXiv:0611099 [hep-th]].

[505] A. Karch and A. O'Bannon, "Metallic AdS/CFT," *JHEP* **0709**, 024 (2007) [arXiv:0705.3870 [hep-th]].

[506] J. Erdmenger, N. Evans, I. Kirsch and E. Threlfall, "Mesons in gauge/gravity duals – a review," *Eur. Phys. J.* A **35**, 81 (2008) [arXiv:0711.4467 [hep-th]].

[507] A. O'Bannon, "Holographic thermodynamics and transport of flavor fields" [arXiv:0808.1115 [hep-th]].

[508] S. A. Hartnoll, J. Polchinski, E. Silverstein and D. Tong, "Towards strange metallic holography," *JHEP* **1004**, 120 (2010) [arXiv:0912.1061 [hep-th]].

[509] A. O'Bannon, "Hall conductivity of flavor fields from AdS/CFT," *Phys. Rev.* D **76**, 086007 (2007) [arXiv:0708.1994 [hep-th]].

[510] O. Bergman, J. Erdmenger and G. Lifschytz, "A review of magnetic phenomena in probe-brane holographic matter," in D. Kharzeev, K. Landsteiner, A. Schmitt and H.-U. Yee (eds), *Strongly Interacting Matter in Magnetic Fields*, Springer, 2013, p. 591 [arXiv:1207.5953 [hep-th]].

[511] M. Ammon, J. Erdmenger, M. Kaminski and P. Kerner, "Flavor superconductivity from gauge/gravity duality," *JHEP* **0910**, 067 (2009) [arXiv:0903.1864 [hep-th]].

[512] S. Harrison, S. Kachru and G. Torroba, "A maximally supersymmetric Kondo model," *Class. Quant. Grav.* **29**, 194005 (2012) [arXiv:1110.5325 [hep-th]].

[513] J. Erdmenger, C. Hoyos, A. Obannon and J. Wu, "A holographic model of the Kondo effect," *JHEP* **1312**, 086 (2013) [arXiv:1310.3271 [hep-th]].

[514] A. W. W. Ludwig, "Field theory approach to critical quantum impurity problems and applications to the multichannel Kondo effect," *Int. J. Mod. Phys.* B **8**, 347 (1994).

[515] I. Affleck, "Conformal field theory approach to the Kondo effect," *Acta Phys. Polon.* B **26**, 1869 (1995) [arXiv:9512099 [cond-mat]].

[516] S. Kachru, A. Karch and S. Yaida, "Adventures in holographic dimer models," *New J. Phys.* **13**, 035004 (2011) [arXiv:1009.3268 [hep-th]].

[517] A. Kolezhuk, S. Sachdev, R. R. Biswas and P. Chen, "Theory of quantum impurities in spin liquids," *Phys. Rev.* B **74**, 165114 (2006) [arXiv:0606385 [cond-mat]].

[518] J. L. Davis, P. Kraus and A. Shah, "Gravity dual of a quantum Hall plateau transition," *JHEP* **0811**, 020 (2008) [arXiv:0809.1876 [hep-th]].

[519] M. Fujita, W. Li, S. Ryu and T. Takayanagi, "Fractional quantum Hall effect via holography: Chern–Simons, edge states, and hierarchy," *JHEP* **0906**, 066 (2009) [arXiv:0901.0924 [hep-th]].

[520] O. Bergman, N. Jokela, G. Lifschytz and M. Lippert, "Quantum Hall effect in a holographic model," *JHEP* **1010**, 063 (2010) [arXiv:1003.4965 [hep-th]].

[521] S. Kachru, A. Karch and S. Yaida, "Holographic lattices, dimers, and glasses," *Phys. Rev.* D **81**, 026007 (2010) [arXiv:0909.2639 [hep-th]].

[522] A. Karch, D. T. Son and A. O. Starinets, "Zero sound from holography," *Phys. Rev. Lett.* **102**, 051602 (2009) [arXiv:0806.3796 [hep-th]].

[523] M. Kulaxizi and A. Parnachev, "Comments on Fermi liquid from holography," *Phys. Rev.* D **78**, 086004 (2008) [arXiv:0808.3953 [hep-th]].

[524] R. A. Davison and A. O. Starinets, "Holographic zero sound at finite temperature," *Phys. Rev.* D **85**, 026004 (2012) [arXiv:1109.6343 [hep-th]].

[525] Y. Y. Bu, J. Erdmenger, J. P. Shock and M. Strydom, "Magnetic field induced lattice ground states from holography," *JHEP* **1303**, 165 (2013) [arXiv:1210.6669 [hep-th]].

[526] K. Jensen, A. Karch, D. T. Son and E. G. Thompson, "Holographic Berezinskii–Kosterlitz–Thouless transitions," *Phys. Rev. Lett.* **105**, 041601 (2010) [arXiv:1002.3159 [hep-th]].

[527] S. Ryu and T. Takayanagi, "Topological insulators and superconductors from D-branes," *Phys. Lett.* B **693**, 175 (2010) [arXiv:1001.0763 [hep-th]].

[528] S. Ryu and T. Takayanagi, "Topological insulators and superconductors from string theory," *Phys. Rev.* D **82**, 086014 (2010) [arXiv:1007.4234 [hep-th]].

[529] A. Karch, J. Maciejko and T. Takayanagi, "Holographic fractional topological insulators in $2 + 1$ and $1 + 1$ dimensions," *Phys. Rev.* D **82**, 126003 (2010) [arXiv:1009.2991 [hep-th]].

[530] S. Franco, A. Hanany, D. Martelli, J. Sparks, D. Vegh and B. Wecht, "Gauge theories from toric geometry and brane tilings," *JHEP* **0601**, 128 (2006) [arXiv:0505211 [hep-th]].

[531] H. Li and F. D. M. Haldane, "Entanglement spectrum as a generalization of entanglement entropy: identification of topological order in non-Abelian fractional quantum Hall effect states," *Phys. Rev. Lett.* **101**, 010504 (2008) [arXiv:0805.0332 [cond-mat.mes-hall]].

[532] D. M. Greenberger, M. A. Horne and A. Zeilinger, "Going beyond Bell's theorem," in *Bell's Theorem, Quantum Theory, and Conceptions of the Universe*, Kluwer, 1989.

[533] D. Bouwmeester, J. W. Pan, M. Daniell, H. Weinfurter and A. Zeilinger, "Observation of three-photon Greenberger–Horne–Zeilinger entanglement," *Phys. Rev. Lett.* **82**, 1345 (1999) [arXiv:9810035 [quant-ph]].

[534] M. Srednicki, "Entropy and area," *Phys. Rev. Lett.* **71**, 666 (1993) [arXiv:9303048 [hep-th]].

[535] C. Holzhey, F. Larsen and F. Wilczek, "Geometric and renormalized entropy in conformal field theory," *Nucl. Phys.* B **424**, 443 (1994) [arXiv:9403108 [hep-th]].

[536] H. W. J. Bloete, J. L. Cardy and M. P. Nightingale, "Conformal invariance, the central charge, and universal finite size amplitudes at criticality," *Phys. Rev. Lett.* **56**, 742 (1986).

[537] A. Kitaev and J. Preskill, "Topological entanglement entropy," *Phys. Rev. Lett.* **96**, 110404 (2006) [arXiv:0510092 [hep-th]].

[538] M. Levin and X.-G. Wen, "Detecting topological order in a ground state wave function," *Phys. Rev. Lett.* **96**, 110405 (2006) [arXiv:0510613 [cond-mat.str-el]].

[539] L. Bombelli, R. K. Koul, J. Lee and R. D. Sorkin, "A quantum source of entropy for black holes," *Phys. Rev.* D **34**, 373 (1986).

[540] S. Ryu and T. Takayanagi, "Aspects of holographic entanglement entropy," *JHEP* **0608**, 045 (2006) [arXiv:0605073 [hep-th]].

[541] H. Liu and M. Mezei, "A refinement of entanglement entropy and the number of degrees of freedom," *JHEP* **1304**, 162 (2013) [arXiv:1202.2070 [hep-th]].

[542] R. C. Myers and A. Singh, "Comments on holographic entanglement entropy and RG flows," *JHEP* **1204**, 122 (2012) [arXiv:1202.2068 [hep-th]].

[543] T. Grover, "Entanglement monotonicity and the stability of gauge theories in three spacetime dimensions," *Phys. Rev. Lett.* **112**, 151601 (2014) [arXiv:1211.1392 [hep-th]].

[544] M. M. Wolf, "Violation of the entropic area law for fermions," *Phys. Rev. Lett.* **96**, 010404 (2006) [arXiv:0503219 [quant-ph]].

[545] A. Chandran, V. Khemani and S. L. Sondhi, "How universal is the entanglement spectrum?" *Phys. Rev. Lett.* **113**, 060501 (2014) [arXiv:1311.2946 [cond-mat.str-el]].

[546] S. Ryu and T. Takayanagi, "Holographic derivation of entanglement entropy from AdS/CFT," *Phys. Rev. Lett.* **96**, 181602 (2006) [arXiv:0603001 [hep-th]].

[547] T. Nishioka, S. Ryu and T. Takayanagi, "Holographic entanglement entropy: an overview," *J. Phys.* A **42**, 504008 (2009) [arXiv:0905.0932 [hep-th]].

[548] T. Takayanagi, "Entanglement entropy from a holographic viewpoint," *Class. Quant. Grav.* **29**, 153001 (2012) [arXiv:1204.2450 [gr-qc]].

[549] T. Hartman, "Entanglement entropy at large central charge" [arXiv:1303.6955 [hep-th]].

[550] T. Faulkner, "The entanglement Rényi entropies of disjoint intervals in AdS/CFT" [arXiv:1303.7221 [hep-th]].

[551] A. Lewkowycz and J. Maldacena, "Generalized gravitational entropy," *JHEP* **1308**, 090 (2013) [arXiv:1304.4926 [hep-th]].

[552] J. de Boer, M. Kulaxizi and A. Parnachev, "Holographic entanglement entropy in Lovelock gravities," *JHEP* **1107**, 109 (2011) [arXiv:1101.5781 [hep-th]].

[553] L. Y. Hung, R. C. Myers and M. Smolkin, "On holographic entanglement entropy and higher curvature gravity," *JHEP* **1104**, 025 (2011) [arXiv:1101.5813 [hep-th]].

[554] X. Dong, "Holographic entanglement entropy for general higher derivative gravity," *JHEP* **1401**, 044 (2014) [arXiv:1310.5713 [hep-th]].

[555] J. Camps, "Generalized entropy and higher derivative gravity," *JHEP* **1403**, 070 (2014) [arXiv:1310.6659 [hep-th]].

[556] J. D. Brown and M. Henneaux, "Central charges in the canonical realization of asymptotic symmetries: an example from three dimensional gravity," *Comm. Math. Phys.* **104**, 207 (1986).

[557] A. Strominger, "Black hole entropy from near horizon microstates," *JHEP* **9802**, 009 (1998) [arXiv:9712251 [hep-th]].

[558] R. C. Myers and A. Sinha, "Holographic c-theorems in arbitrary dimensions," *JHEP* **1101**, 125 (2011) [arXiv:1011.5819 [hep-th]].

[559] D. L. Jafferis, I. R. Klebanov, S. S. Pufu and B. R. Safdi, "Towards the F-theorem: $N = 2$ field theories on the three-sphere," *JHEP* **1106**, 102 (2011) [arXiv:1103.1181 [hep-th]].

[560] H. Casini, M. Huerta and R. C. Myers, "Towards a derivation of holographic entanglement entropy," *JHEP* **1105**, 036 (2011) [arXiv:1102.0440 [hep-th]].

[561] N. Ogawa, T. Takayanagi and T. Ugajin, "Holographic Fermi surfaces and entanglement entropy," *JHEP* **1201**, 125 (2012) [arXiv:1111.1023 [hep-th]].

[562] B. Swingle, "Entanglement entropy and the Fermi surface," *Phys. Rev. Lett.* **105**, 050502 (2010) [arXiv:0908.1724 [cond-mat.str-el]].

[563] J. Bhattacharya, M. Nozaki, T. Takayanagi and T. Ugajin, "Thermodynamical property of entanglement entropy for excited states," *Phys. Rev. Lett.* **110**, 091602 (2013) [arXiv:1212.1164 [hep-th]].

[564] F. Kruger and J. Zaanen, "Fermionic quantum criticality and the fractal nodal surface," *Phys. Rev. B* **78**, 035104 (2008) [arXiv:0804.2161 [cond-mat.str-el]].

[565] D. M. Ceperley, "Fermion nodes," *J. Statist. Phys.* **63**, 1237 (1991).

[566] M. Van Raamsdonk, "Comments on quantum gravity and entanglement" [arXiv:0907.2939 [hep-th]].

[567] M. Van Raamsdonk, "Building up spacetime with quantum entanglement," *Gen. Rel. Grav.* **42**, 2323 (2010) [republished *Int. J. Mod. Phys. D* **19**, 2429 (2010)] [arXiv:1005.3035 [hep-th]].

[568] N. Lashkari, M. B. McDermott and M. Van Raamsdonk, "Gravitational dynamics from entanglement 'thermodynamics'," *JHEP* **1404**, 195 (2014) [arXiv:1308.3716 [hep-th]].

[569] T. Faulkner, M. Guica, T. Hartman, R. C. Myers and M. Van Raamsdonk, "Gravitation from entanglement in holographic CFTs," *JHEP* **1403**, 051 (2014) [arXiv:1312.7856 [hep-th]].

[570] D. D. Blanco, H. Casini, L. Y. Hung and R. C. Myers, "Relative entropy and holography," *JHEP* **1308**, 060 (2013) [arXiv:1305.3182 [hep-th]].

[571] B. Czech, J. L. Karczmarek, F. Nogueira and M. Van Raamsdonk, "The gravity dual of a density matrix," *Class. Quant. Grav.* **29**, 155009 (2012) [arXiv:1204.1330 [hep-th]].

[572] B. Czech, J. L. Karczmarek, F. Nogueira and M. Van Raamsdonk, "Rindler quantum gravity," *Class. Quant. Grav.* **29**, 235025 (2012) [arXiv:1206.1323 [hep-th]].

Index

Printed in the United States
By Bookmasters